Nanofluidics
An Introduction

Nanofluidics
An Introduction

Zhigang Li

CRC Press
Taylor & Francis Group
Boca Raton London New York

CRC Press is an imprint of the
Taylor & Francis Group, an **informa** business

CRC Press
Taylor & Francis Group
6000 Broken Sound Parkway NW, Suite 300
Boca Raton, FL 33487-2742

Printed on acid-free paper

International Standard Book Number-13: 978-1-138-03551-5 (Hardback)
International Standard Book Number-13: 978-1-315-26624-4 (eBook)

Library of Congress Cataloging-in-Publication Data

Names: Li, Zhigang R., 1958- author.
Title: Nanofluidics : an introduction / Zhigang Li.
Description: Boca Raton : Taylor & Francis, a CRC title, part of the Taylor & Francis imprint, a member of the Taylor & Francis Group, the academic division of T&F Informa, plc, [2019] | Includes bibliographical references.
Identifiers: LCCN 2018031274 (print) | LCCN 2018037670 (ebook) | ISBN 9781315266244 (Master eBook) | ISBN 9781351969574 (Adobe Pdf) | ISBN 9781351969567 (ePUB) | ISBN 9781351969550 (Mobipocket) | ISBN 9781138035515 (hardback)
Subjects: LCSH: Nanofluids.
Classification: LCC TJ853.4.M53 (ebook) | LCC TJ853.4.M53 L53 2019 (print) | DDC 620.1/06—dc23
LC record available at https://lccn.loc.gov/2018031274.

Visit the Taylor & Francis Web site at
http://www.taylorandfrancis.com

and the CRC Press Web site at
http://www.crcpress.com

To my daughter, Rossie, and wife, Xuefen,
for their love and support.

Contents

Preface

Nanofluidics research has achieved a significant growth over the past decade. Although new phenomena and applications of nanoscale fluid flows are being reported continuously, it is necessary to summarize the current understanding of nanofluidics. This is important to researchers, who are interested in and devoted to this fascinating area.

There are a number of books dealing with microfluidics and nanofluidics, but few of them are entirely focused on nanofluidics. Even for books devoted to nanofluidics, there are different ways to discuss the topic. In addition to the introduction of numerical and experimental means for studying nanofluidics, this book describes new flow phenomena that are unique to the nanoscale with an emphasis on the physics underlying the flow phenomena. This, I believe, is important for researchers, especially graduate students, who have chosen nanofluidics as their research area. This, perhaps, is something that makes this book different from previous nanofluidics books.

This book is comprised of eight chapters. Chapter 1 briefly reviews the fundamental knowledge of the classic fluid mechanics. Chapter 2 describes what is unique to nanofluidics. A popular numerical method, molecular dynamics simulation, and nanofabrication approaches are discussed in Chapters 3 and 4, respectively. These two chapters are important for the discussion in later chapters. Chapters 5–7 deal with specific flows cases, i.e., nanoscale capillarity, pressure-driven flows of simple fluids in nanochannels, and flows over nanoobjects, respectively. The last chapter, Chapter 8, is concerned with nanoscale electrokinetic phenomena and presents a review of recent progress in nanoscale ion transport.

The goal of this book is to equip readers with foundational knowledge and a set of baseline skills in the areas of diverse flow phenomena at the nanoscale. I hope this book can serve as a stepping stone for beginners in the field and can motivate young researchers to explore the fantastic flow world at the nanoscale.

Parts of Chapters 6 and 7 are from the work of my former PhD students and post-doctoral research associates: Dr. Chong Liu (Institute of Mechanics,

Chinese Academy of Sciences), Dr. Jun Wang (Beijing University of Technology), Dr. Long Li (Qian Xuesen Laboratory of Space Technology, China), and Dr. Jingwen Mo (Southeast University, China). Without them, this book would not have been possible.

Finally, I thank my PhD student and postdoctoral research associate, Dr Xiang Gao, for drafting Chapter 8. A word of special thanks also goes to my PhD students, Miss Shuang Luo and Mr. Yakang Jin, for helping me develop some graphics.

Zhigang Li
Professor
Department of Mechanical and Aerospace Engineering
The Hong Kong University of Science and Technology
Clear Water Bay
Hong Kong

Author

Zhigang Li, PhD, is currently a professor in the Department of Mechanical and Aerospace Engineering at the Hong Kong University of Science and Technology (HKUST). He earned his PhD from the University of Delaware in 2005. He is a recipient of the Chinese Government Award for Outstanding Overseas Student in 2005. His research areas include nanoscale transport phenomena (nanofluidics, nanoscale heat transfer, and transport phenomena), droplet–surface interaction, and biosensing.

Chapter 1 Fundamentals of classic fluid mechanics

1.1 Introduction

Nanofluidics is concerned with fluid flows in nanoscale confinements. It can be quite different from the classic fluid transport in macro-systems. The difference is mainly caused by molecular interactions, which cause changes in fluid properties and lead to new transport phenomena. Although molecular interactions play important roles at the nanoscale, many ideas at the macroscale are valid. Therefore, before a new journey to the nanoscale, it is necessary to review the classic fluid mechanics. Readers who are familiar with the classic fluid mechanics and electrokinetics can skip this chapter and move to Chapter 2 directly. Further detailed information about the classic theories can be found in standard textbooks on fluid mechanics and electrokinetics.

1.2 What is a fluid

In the classic fluid mechanics, a fluid is defined as a substance that deforms continuously under a shearing stress of any magnitude. This definition is clear and presents a practical base for telling whether a substance is a fluid or not. Another approach to distinguish fluids from solids is to look into the motion of molecules or atoms. At the molecular scale, for solids, all the atoms are vibrating around and cannot move far away from their equilibrium positions, as illustrated in Figure 1.1a. Fluid molecules, however, can move freely in available space due to the weak molecular interactions (Figure 1.1b). Such difference between fluids and solids can be reflected in the mean square displacement (MSD) of molecules, which is defined as

$$\left\langle \Delta r^2(t) \right\rangle = \sum_{i=1}^{N} \left[r_i(t) - r_i(0) \right]^2 \Big/ N \tag{1.1}$$

where r is the position variable, t is time, and N is the total number of molecules. Since MSD is a measure of relative positions of atoms, it is bounded for solids with

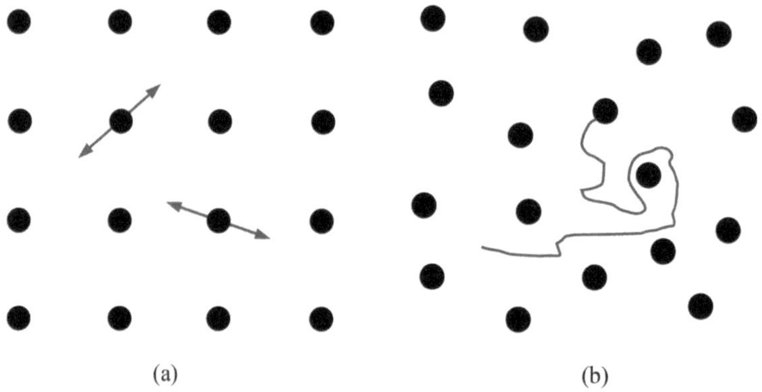

Figure 1.1 A molecular view of substances: (a) solids and (b) fluids (the solid line shows a possible trajectory of a molecule).

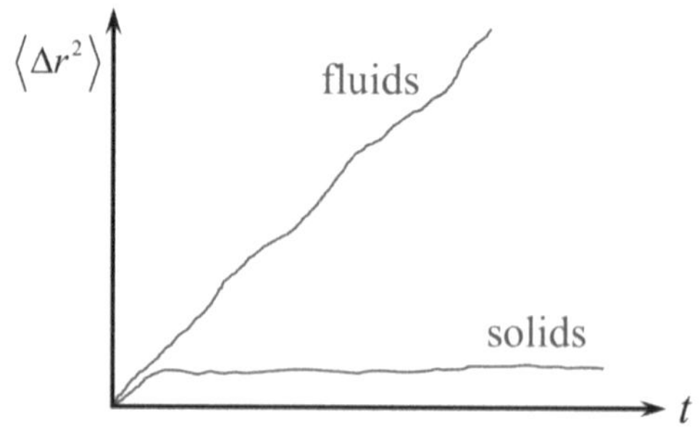

Figure 1.2 Mean square displacement for solids and fluids as a function of time.

respect to time. For fluids, however, it can be shown that the MSD is proportional to time (Reif, 1985). Therefore, fluids and solids can be distinguished based on the dependence of the MSD on time, as demonstrated in Figure 1.2. Practically, Equation 1.1 may not be a good way to characterize a substance because it is difficult to probe the motion of atoms. However, in nanofluidics, MSD is useful in studying the phase change of materials because it can be easily obtained in numerical simulations, as will be discussed later.

1.3 Physical properties of fluids

There are quite a few physical properties for fluids, including density, viscosity, surface tension, specific volume, specific weight, and bulk modulus. In this section, however, only density, viscosity, and surface tension are discussed because they are frequently used in nanofludics and most of the other properties are somewhat related to these three properties.

1.3.1 Density

The mass of a fluid per unit volume is defined as the density of the fluid. It is popularly denoted by the Greek symbol ρ. In the international system (SI) of units, the unit of density is kg/m^3. For bulk fluids, the value of density depends on pressure p and temperature T (for liquids, the pressure effect is minor). For a given pressure and temperature, ρ is a constant. For example, at standard conditions (15°C and 101.33 kPa), the density of water is 999 kg/m^3.

At the nanoscale, however, the density of a fluid may not be a constant. Due to the nanoconfinement, the density of a fluid near a solid surface can be greatly affected by the intermolecular interactions between the fluid and the solid surface (Liu and Li, 2009). This is also true for other properties, such as viscosity, as will be discussed later. The nonuniform physical properties of fluids make flow analysis at the nanoscale nontrivial.

1.3.2 Viscosity

Viscosity is a very important property for a fluid. Its definition is related to the shearing stress acting on a fluid. In the case of a fluid confined by two horizontal parallel plates, as shown in Figure 1.3a. If the bottom plate is stationary and the top plate moves under a horizontal force F, in the steady state, the top plate will move with a constant velocity, V_T, which is termed the terminal velocity. The fluid will also move under the effect of the moving top plate. For a simple fluid, the velocity of the fluid in the x-direction, u, is a linear function of the y-coordinate (Figure 1.3a). Furthermore, it can be shown that the shearing stress τ (Figure 1.3b), which is defined as the force per unit area, acting on the top plate by the fluid is proportional to the velocity gradient in the y-direction, given by

$$\tau = \mu \frac{du}{dy} \tag{1.2}$$

where μ is the dynamic viscosity (it is usually simply called viscosity). Therefore, mathematically, viscosity is just a proportional coefficient. However, the physical meaning of viscosity is unclear in Equation 1.2. Further analysis can show that viscosity actually is a measure of momentum transfer in a fluid. A fluid having strong momentum transfer ability has a large viscosity. The momentum transfer of a fluid depends on the intermolecular interactions among fluid molecules. Forces among liquid molecules are usually much larger than those among gas

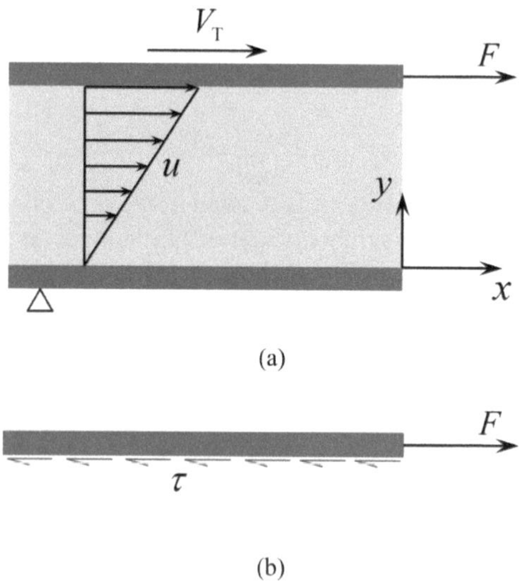

Figure 1.3 A fluid flow between two parallel plates caused by the moving top plate: (a) flow system and velocity distribution; (b) shearing stress acting on the top plate.

molecules. Hence, liquids exceed gases in momentum transfer and therefore have larger viscosities. Fluids that satisfy Equation 1.2 are called Newtonian fluids.

Viscosity is primarily a function of temperature. For liquids, the viscosity decreases with increasing temperature. For gases, however, the viscosity increases as the temperature is increased. This is because the density of a liquid is reduced when the temperature is increased, which increases the mean distance and reduces the molecular forces among liquid molecules. Consequently, the momentum transfer in the liquid is weakened. For gases, however, a high temperature can enhance the thermal motion of gas molecules and promote the collisions among molecules, which increase the momentum transfer ability of the gas.

Similar to density, at the nanoscale, the viscosity of a fluid may not be a constant under the effect of confinements (solid surfaces); the viscosity in the vicinity of a solid surface can fluctuate greatly due to the fluid–surface intermolecular interactions (Todd et al., 1995).

1.3.3 Surface tension

Surface tension is also caused by the molecular interactions among fluid molecules. Generally, the intermolecular force is mainly the van der Waals force, which consists of two parts: attractive and repulsive forces. These two parts are distance dependent. For two interacting molecules, there is an equilibrium distance r^*, where the force between them vanishes. When the separation between the two molecules is larger than r^*, the attractive force dominates. When the two molecules are very close and their distance is smaller than r^*, the force between them becomes repulsive. For a bulk fluid, all the fluid molecules, on average, experience equal forces in all the directions from neighboring molecules. At a liquid–air or liquid–liquid interface, however, the molecular forces are asymmetric. Liquid molecules in the interior region try to push the molecules on the surface outward. It is this repulsive force that creates an internal pressure. The repulsive force tends to stretch the surface and increase the mean distance among molecules on the surface. When the mean distance becomes sufficiently large, the forces among the molecules on the surface become attractive, which hold all the fluid molecules together to maintain the fluid in a certain shape. Therefore, in the surface, a tensile force exists in all the directions or along any line and the force per unit length along a line is called the surface tension, which is designated by γ. The SI unit of γ is N/m.

Surface tension is a strong function of temperature, as temperature can greatly affect the mean distance and consequently the molecular forces among fluid molecules. At 25°C, the surface tension of water is 71.9 mN/m. For nanoscale droplets, it has been shown that γ depends on the size of the droplets. It decreases with decreasing droplet size (Lu and Jiang, 2005), especially when the droplet diameter is below 2 nm. For droplets with diameters larger than 2 nm, the size effect becomes unimportant (Lu and Jiang, 2005).

1.4 Surface energy of solids and contact angle of droplets

Surface tension is a physical property of a fluid as it is determined by the molecular forces among the fluid molecules. However, when a fluid of finite volume is in contact with a solid surface, other than the forces among the fluid molecules, the interaction between the fluid and solid surface also plays an important role in affecting the shape of the fluid surface, especially for capillary phenomena and droplet dynamics on surfaces. The fluid–solid interaction is mainly determined by two factors: one is the force among fluid molecules, which governs the fluid–surface tension and the other is the force among solid atoms, which controls the so-called surface energy of solids. The surface energy of a solid is related to the energy needed to break the bonds among the solid atoms to create new surfaces. Since the intermolecular force is the most fundamental source for the surface energy, the surface tension of fluids, in this sense, is also related to the surface energy of fluids. Actually, surface tension is also defined as the surface energy density. Nonetheless, in this book, surface energy, unless otherwise stated, is used for solids.

The competition between the surface tension of a fluid and the surface energy of a solid determines the behavior of the fluid at the fluid–solid interface. For a droplet on a solid surface and capillary phenomena, another variable, called contact angle, is directly associated with the surface tension of the droplet and the surface energy of the solid. Let us consider a liquid droplet on a solid surface, as shown in Figure 1.4. Thus, Young's equation describes the following condition where the total force acting on the droplet is zero and the forces due to different surface tensions at the liquid–solid–vapor interface are in balance:

$$\gamma = \frac{\gamma_{sv} - \gamma_{ls}}{\cos\theta} \tag{1.3}$$

where γ_{sv} and γ_{ls} are the solid–vapor and liquid–solid surface tensions, and θ is the contact angle, which is the angle between the solid surface and the tangent line to the droplet surface (Figure 1.4). The contact angle is generally used to quantify the wettability of a solid surface. A solid surface is said to be hydrophilic if the contact angle of a water droplet on the surface is less than 90°. Such a surface usually has a relatively high surface energy. If $\theta > 90°$, the surface is considered as hydrophobic and the surface energy is typically low compared with that of hydrophilic surfaces. If $\theta > 150°$, the surface is called superhydrophobic. These three cases are illustrated in Figure 1.5.

Surface tension and contact angle can be used to describe many phenomena. A popular one is the rise of a liquid in a capillary tube. As demonstrated in Figure 1.6a, when a small open tube is inserted in water, a water column develops in the tube if the tube is hydrophilic. In this situation, the contact angle $\theta < 90°$ and the water meniscus in the tube is concave. The force due to the surface tension has a component in the upward direction, which picks up the water column. The height of the water column at the steady state can be calculated based on the free-body diagram in Figure 1.6b:

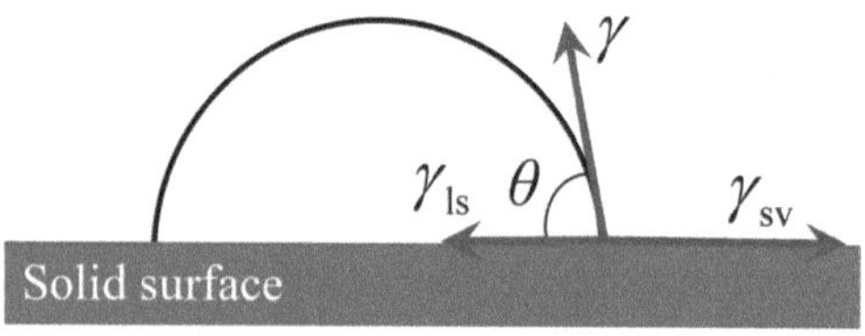

Figure 1.4 Contact angle θ of a droplet on a solid surface. γ_{ls} and γ_{sv} are the liquid–solid and solid–vapor surface tensions.

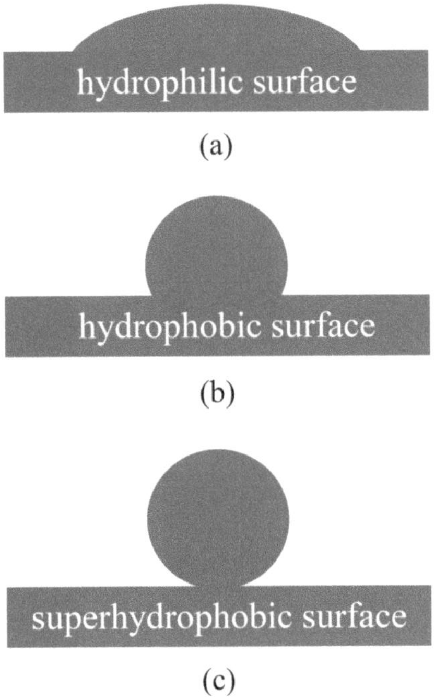

Figure 1.5 Droplets on surfaces of different wettabilities: (a) hydrophilic surface, (b) hydrophobic surface, and (c) superhydrophobic surface.

$$h = \frac{2\gamma \cos\theta}{\rho g R} \tag{1.4}$$

where g is the gravitational acceleration and R is the radius of the tube. Equation 1.4 is also valid for hydrophobic tubes. The difference is that the water level in hydrophobic tubes is lower than that outside the tube because the force caused by the surface tension is downward, as shown in Figure 1.6c.

1.5 Dimensionless numbers

In describing fluid flows, dimensionless numbers are widely employed. Dimensionless numbers are quantities that do not have a physical dimension or unit. They represent the ratio of two variables having the same unit and are a measure of the competition between the two variables. If a dimensionless number is much larger or smaller than unity, the role of one of the variables dominates over the other one. In most cases, the effects due to the "weaker" one can be neglected to simplify theoretical analyses. There are various dimensionless numbers for

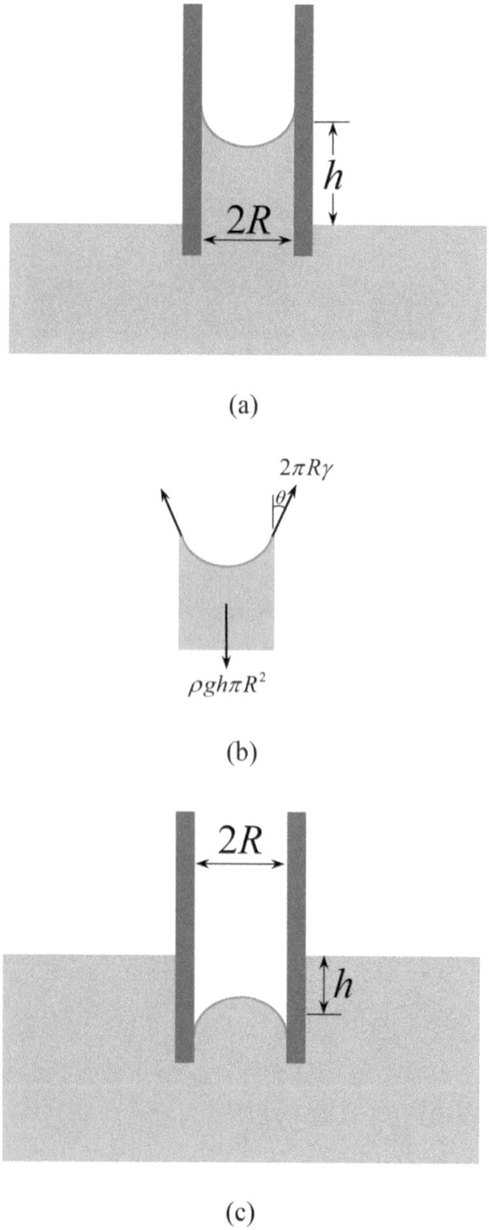

Figure 1.6 Capillarity in small tubes: (a) water rises in a hydrophilic tube, (b) free-body diagram of the water column in the tube, and (c) depression of water in a hydrophobic tube.

describing different flow phenomena. Herein, only some popular dimensionless numbers are discussed.

The Reynolds number, Re, is probably the most popular dimensionless number. It describes the competition between an inertial force and a viscous force, which is given by

$$\mathrm{Re} = \frac{\rho V L}{\mu} \tag{1.5}$$

where V is the flow velocity and L is a characteristic length. An important use of Re is to classify flow scenarios. For a pipe flow, the flow is laminar (no clear mixing between fluid layers) if Re<2100, while it becomes turbulent when Re>4000 (Munson et al., 2006).

The ratio between a viscous force and a force arising from surface tension is denoted by the capillary number, Ca, as

$$\mathrm{Ca} = \frac{\mu V}{\gamma} \tag{1.6}$$

At the micro- or nanoscale, Ca is usually quite small, indicating that the surface tension of a fluid plays an important role.

For the motion of particles in a flow, the competition between the diffusion of the particles and the convection of the fluid can be characterized by the Peclet number, Pe,

$$\mathrm{Pe} = \frac{VL}{D_\mathrm{p}} \tag{1.7}$$

where D_p is the diffusion coefficient of particles, which will be discussed in the next section. For large particles (small D_p) in a high Re flow, Pe is quite large, and the Brownian motion of the particles can be ignored. However, for small particles in a low Re flow, the diffusion of the particles plays a dominant role, as demonstrated by the trajectories of particles in Figure 1.7. Flows of proper Pe can be employed to take advantage of particle diffusion to achieve various processes, such as separation of particles (Li and Drazer, 2007).

The ratios between different characteristic times also form dimensionless numbers. The dynamics of polymers or long-chain molecules in a flow can be described by the Deborah number, De, which measures the competition between the relaxation time of polymers τ_p and the characteristic flow time τ_flow,

$$\mathrm{De} = \frac{\tau_\mathrm{p}}{\tau_\mathrm{flow}} \tag{1.8}$$

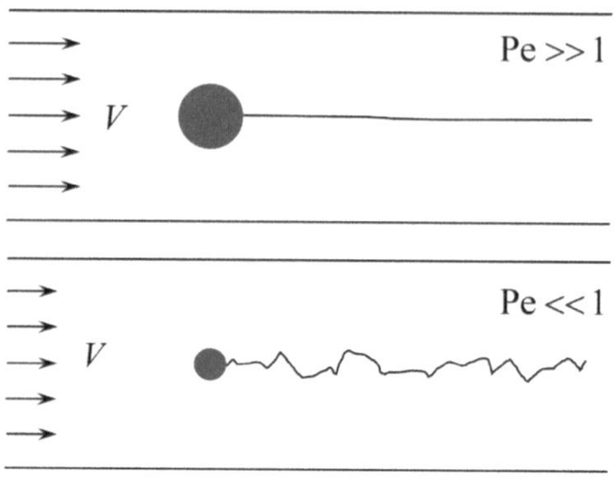

Figure 1.7 Trajectories of particles in flows of different Peclet numbers: high Pe flow (top) and low Pe flow (bottom).

There are many other dimensionless numbers for different flow phenomena (Squires and Quake, 2005). In these dimensionless numbers, all the competing parameters involved are macroscopic variables, such as viscosity, flow velocity, and characteristic length. At the nanoscale, some microscopic parameters may play an essential role, compete with macroscopic variables, and form new dimensionless numbers, which will be discussed in Chapter 6.

1.6 Transport properties

Transport phenomena, including mass, momentum, and energy transfer, are probably the most fundamental and important processes in nature. They are the basis of a variety of industrial and scientific problems and have tremendous applications in various areas. The topic of transport phenomena has been an important branch in engineering and science. Comprehensive discussion on transport phenomena can be found in standard textbooks (Bird et al., 2002). In this section, some basic transport properties of fluids and particles are reviewed.

1.6.1 Diffusion coefficient

Diffusion is an essential mechanism in mass transport. By diffusion a species moves from a high-concentration region to a low-concentration region. Fundamentally, diffusion is the consequence of intermolecular interactions. For a single-component fluid in equilibrium, i.e., temperature, density, and other properties are uniform throughout the fluid, fluid molecules move randomly and their "mobility" can be characterized by the so-called self-diffusion coefficient D_s, which can be evaluated by the MSD (Equation 1.1) of the fluid molecules,

$$D_s = \lim_{t \to \infty} \frac{\langle \Delta r^2(t) \rangle}{a \cdot t} \tag{1.9}$$

where a is a constant related to the dimension of the motion and $a = 2$, 4, and 6 for one-, two-, and three-dimensional diffusions, respectively.

For a nonequilibrium case, where the density or concentration of fluid molecules is nonuniform, the mean repulsive force among fluid molecules in the high-concentration area is higher than that in the low-concentration area. This causes the fluid molecules to diffuse from high concentration to low concentration (opposite to the direction of the concentration gradient), as depicted in Figure 1.8. The flux of the fluid, J, which is the number of molecules passing unit cross-sectional area per unit time, is calculated by Fick's first law of diffusion:

$$J = -D_t \frac{\partial n}{\partial x} \tag{1.10}$$

where n is the concentration of fluid molecules and D_t is referred to as the transport diffusion coefficient or Fickian diffusivity (or simply diffusivity). It is noted that the two diffusion coefficients in Equations 1.9 and 1.10 are different. D_s is for equilibrium systems. It is directly related to the random motion of molecules caused by molecular interactions. D_t is associated with the mass transport of a fluid caused by concentration variations. Both D_s and D_t depend on the concentration and temperature, while they become the same at the low-concentration limit (Sholl, 2006).

Since the diffusion of molecules is governed by intermolecular interactions, the diffusion coefficient of a species may be different in different mixtures. For

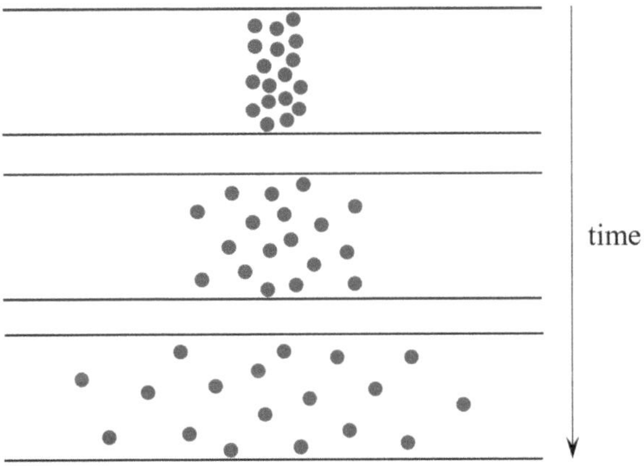

Figure 1.8 Schematic of the diffusion of fluid molecules from high-concentration to low-concentration regions.

example, at 20°C and infinite dilution, the diffusion coefficients of water in methanol and ethanol are 2.2×10^{-5} and $0.85 \times 10^{-5}\,cm^2/s$ (Lees and Sarram, 1971), respectively, which are different by a factor of about 3.

The concept of diffusion for fluid molecules also applies to ions and solid particles. For ions, other than the diffusivity, electric mobility is a very popular property in applications. For solid particles, the diffusivity can be evaluated by the drag force acting on the particles when they diffuse in a fluid. The diffusion coefficient, drag force, and electric mobility are correlated, as discussed in the following sections.

1.6.2 Drag force

When a particle is moving in a fluid, it experiences a resistant force, which is termed the drag force. Stokes (1851) determined the equation of motion for spherical particles in a fluid and obtained the drag force F_D on a sphere in low Re flows (or creeping flows),

$$F_D = -6\pi\mu RV \tag{1.11a}$$

where R is the radius of the sphere and V is the relatively velocity of the sphere to the fluid. The minus sign in Equation 1.11a indicates that the direction of the drag force is opposite to the relative velocity. Equation 1.11a is known as the Stokes law.

The Stokes law can also be applied to nonspherical particles, for which the hydrodynamic radius should be used to replace R in Equation 1.11a. Nevertheless, a more general expression for the drag force can be given as

$$F_D = -\alpha V \tag{1.11b}$$

where $\alpha = F_D/V$ is defined as the friction coefficient. For spherical particles, $\alpha = 6\pi\mu R$.

If the Brownian motion of a particle is considered such that the instantaneous force acting on the particle includes both the drag force and a rapid fluctuating force F' due to the molecular interactions between the particle and the surrounding fluid molecules, then the equation of motion for the particle can be written as

$$m\ddot{r} = -\alpha\dot{r} + F' \tag{1.12}$$

where m and r are the mass and position of the particle, $\dot{r} = dr/dt = V$ and $\ddot{r} = dV/dt$ are the velocity and acceleration of the particle. For low Re flows, the particle–fluid molecular interactions are approximately symmetric and the mean value of F' vanishes, i.e., $\langle F' \rangle = 0$.

Multiplying both sides of Equation 1.12 by position r yields

$$m\left[\frac{d}{dt}(r\dot{r})-\dot{r}^2\right]=-\alpha r\dot{r}+rF' \tag{1.13}$$

Taking average of both sides of Equation 1.13 and using the relationship $m\langle \dot{r}^2\rangle = 3k_\mathrm{B}T$,* where $k_\mathrm{B}=1.38\times10^{-23}$ J/K is the Boltzmann constant, it is obtained that

$$m\frac{d}{dt}\langle r\dot{r}\rangle = 3k_\mathrm{B}T-\alpha\langle r\dot{r}\rangle \tag{1.14}$$

where $\langle rF'\rangle=\langle r\rangle\langle F'\rangle=0$ is used. Equation 1.14 can be determined for $\langle r\dot{r}\rangle$, which is given by

$$\langle r\dot{r}\rangle=\frac{1}{2}\frac{d\langle r^2\rangle}{dt}=ce^{-\frac{\alpha}{m}t}+\frac{3k_\mathrm{B}T}{\alpha} \tag{1.15}$$

where c is a constant. If the particle is at the origin at $t = 0$, then r represents the displacement of the particle Δr and $c=-\dfrac{3k_\mathrm{B}T}{\alpha}$. Integrating Equation 1.15 gives

$$\langle \Delta r^2\rangle=\frac{6k_\mathrm{B}T}{\alpha}\left[t-\frac{m}{\alpha}\left(1-e^{-\frac{\alpha}{m}t}\right)\right] \tag{1.16}$$

For $t\to\infty$, Equation 1.16 is reduced to

$$\langle \Delta r^2\rangle=\frac{6k_\mathrm{B}T}{\alpha}t \tag{1.17}$$

As Equation 1.9 is also valid for calculating the diffusivity of particles, a comparison between Equations 1.9 and 1.17 gives the diffusion coefficient of particles

$$D_\mathrm{p}=\frac{k_\mathrm{B}T}{\alpha} \tag{1.18}$$

Equation 1.18 relates the particle diffusivity with the drag force. If $\alpha = 6\pi\mu R$, Equation 1.18 is called the Stokes–Einstein relation.

1.6.3 Electric mobility

For charged particles or ions, electric mobility is a widely used transport property. For a charged particle moving with a velocity V in a fluid under an electric field E, the electric mobility is defined as

* This relationship is from statistical thermodynamics, which gives that the mean kinetic energy of fluid molecules $m\langle \dot{r}^2\rangle/2 = 3k_\mathrm{B}T/2$.

$$Z = \frac{V}{E} \tag{1.19}$$

If the charge carried by the particle is q, at the steady state, the electrostatic force acting on the particle, qE, is equal to the drag force αV, which leads to the electric mobility in the limit of zero-field strength,

$$Z = \frac{q}{\alpha} \tag{1.20}$$

From Equations 1.11, 1.18, and 1.20, it is seen that the drag force, diffusion coefficient, and electric mobility are equivalent attributes of transport properties. If one of them is known, the other two can be readily obtained.

1.7 Navier–Stokes equation

The equations of motion for a fluid flow driven by a pressure gradient can be derived by considering all the body and surface forces on the fluid. For viscous and incompressible Newtonian fluids, in the Cartesian coordinates, the equations of motion are given by

$$x \text{ direction: } \rho\left(\frac{\partial u}{\partial t} + u\frac{\partial u}{\partial x} + v\frac{\partial u}{\partial y} + w\frac{\partial u}{\partial z}\right) = -\frac{\partial p}{\partial x} + \rho g_x + \mu\left(\frac{\partial^2 u}{\partial x^2} + \frac{\partial^2 u}{\partial y^2} + \frac{\partial^2 u}{\partial z^2}\right) \tag{1.21a}$$

$$y \text{ direction: } \rho\left(\frac{\partial v}{\partial t} + u\frac{\partial v}{\partial x} + v\frac{\partial v}{\partial y} + w\frac{\partial v}{\partial z}\right) = -\frac{\partial p}{\partial y} + \rho g_y + \mu\left(\frac{\partial^2 v}{\partial x^2} + \frac{\partial^2 v}{\partial y^2} + \frac{\partial^2 v}{\partial z^2}\right) \tag{1.21b}$$

$$z \text{ direction: } \rho\left(\frac{\partial w}{\partial t} + u\frac{\partial w}{\partial x} + v\frac{\partial w}{\partial y} + w\frac{\partial w}{\partial z}\right) = -\frac{\partial p}{\partial z} + \rho g_z + \mu\left(\frac{\partial^2 w}{\partial x^2} + \frac{\partial^2 w}{\partial y^2} + \frac{\partial^2 w}{\partial z^2}\right) \tag{1.21c}$$

where u, v, and w are the velocity components in the x-, y-, and z-directions, respectively, t is time, and p is pressure. Equations 1.21a–c are called the Navier–Stokes equation.

In cylindrical polar coordinates, the velocity components in the r-, θ-, and z-directions are denoted by V_r, V_θ, and V_z, and the Navier–Stokes equations can be simplified as follows:

$$r \text{ direction: } \rho\left(\frac{\partial V_r}{\partial t} + V_r\frac{\partial V_r}{\partial r} + \frac{V_\theta}{r}\frac{\partial V_r}{\partial \theta} - \frac{V_\theta^2}{r} + V_z\frac{\partial V_r}{\partial z}\right)$$

$$= -\frac{\partial p}{\partial r} + \rho g_r + \mu\left[\frac{1}{r}\frac{\partial}{\partial r}\left(r\frac{\partial V_r}{\partial r}\right) - \frac{V_r}{r^2} + \frac{1}{r^2}\frac{\partial^2 V_r}{\partial \theta^2} - \frac{2}{r^2}\frac{\partial V_\theta}{\partial \theta} + \frac{\partial^2 V_r}{\partial z^2}\right] \tag{1.22a}$$

$$\theta \text{ direction: } \rho\left(\frac{\partial V_\theta}{\partial t} + V_r\frac{\partial V_\theta}{\partial r} + \frac{V_\theta}{r}\frac{\partial V_\theta}{\partial \theta} + \frac{V_r V_\theta}{r} + V_z\frac{\partial V_\theta}{\partial z}\right)$$

$$= -\frac{1}{r}\frac{\partial p}{\partial \theta} + \rho g_\theta + \mu\left[\frac{1}{r}\frac{\partial}{\partial r}\left(r\frac{\partial V_\theta}{\partial r}\right) - \frac{V_\theta}{r^2} + \frac{1}{r^2}\frac{\partial^2 V_\theta}{\partial \theta^2} + \frac{2}{r^2}\frac{\partial V_r}{\partial \theta} + \frac{\partial^2 V_\theta}{\partial z^2}\right] \tag{1.22b}$$

$$z \text{ direction: } \rho\left(\frac{\partial V_z}{\partial t} + V_r\frac{\partial V_z}{\partial r} + \frac{V_\theta}{r}\frac{\partial V_z}{\partial \theta} + V_z\frac{\partial V_z}{\partial z}\right)$$

$$= -\frac{\partial p}{\partial z} + \rho g_z + \mu\left[\frac{1}{r}\frac{\partial}{\partial r}\left(r\frac{\partial V_z}{\partial r}\right) + \frac{1}{r^2}\frac{\partial^2 V_z}{\partial \theta^2} + \frac{\partial^2 V_z}{\partial z^2}\right] \tag{1.22c}$$

The above Navier–Stokes equations (see Equations 1.21 and 1.22) provide mathematical descriptions and are the governing equations for incompressible flows of Newtonian fluids. In many cases, they need to be combined with the mass conservation equation for incompressible flows, which is given by

$$\nabla \cdot \mathbf{V} = 0 \text{ or } \frac{\partial u}{\partial x} + \frac{\partial v}{\partial y} + \frac{\partial w}{\partial z} = 0 \tag{1.23}$$

If the Navier–Stokes equation and the mass equation can be determined, the velocity and pressure fields of a flow can be acquired analytically. Unfortunately, these partial differential equations contain nonlinear (viscous) terms and may not be determined for complex flows. Analytical solutions to the Navier–Stokes equation are obtainable only for some simple flows, and two of which will be discussed in the next two sections.

1.7.1 Pressure-driven flows between two parallel plates

Let us consider a steady, laminar flow between two fixed, horizontal, infinite parallel plates, as shown in Figure 1.9a. The distance between the two plates is $2h$.

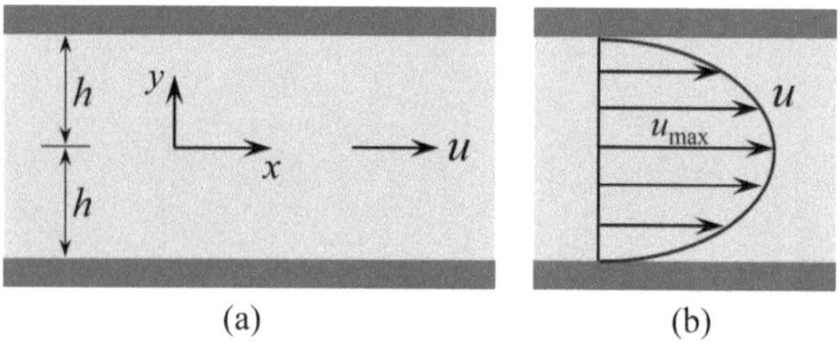

Figure 1.9 Pressure-driven flow between two parallel plates: (a) notation and coordinate system and (b) parabolic velocity profile.

The flow is driven by a pressure gradient in the x-direction, $\partial p/\partial x$. If $\partial p/\partial x$ is not very large such that the flow is laminar, the fluid will move horizontally with $v = w = 0$, which lead to $\partial u/\partial x = 0$ by using the mass conservation equation (see Equation 1.23). Considering that $\partial u/\partial t = 0$, $\partial v/\partial t = 0$, and $\partial w/\partial t = 0$, the Navier–Stokes equations (see Equations 1.21a–c) can be simplified as follows:

$$0 = -\frac{\partial p}{\partial x} + \mu \frac{\partial^2 u}{\partial y^2} \tag{1.24a}$$

$$0 = -\frac{\partial p}{\partial y} - \rho g \tag{1.24b}$$

$$0 = -\frac{\partial p}{\partial z} \tag{1.24c}$$

where $g_x = g_z = 0$ and $g_y = -g$ are used. Equations 1.24a–c can be easily integrated to give the flow velocity

$$u = \frac{1}{2\mu}\frac{\partial p}{\partial x} y^2 + c_1 y + c_2 \tag{1.25}$$

where c_1 and c_2 are the constants of integration. Their values can be determined as $c_1 = 0$ and $c_2 = -h^2/(2\mu)\cdot(\partial p/\partial x)$ by using the stick boundary conditions, i.e., $u = 0$ at $y = \pm h$. Thus, the velocity u can be calculated as follows:

$$u = \frac{1}{2\mu}\frac{\partial p}{\partial x}\left(y^2 - h^2\right) \tag{1.26}$$

The velocity profile given in Equation 1.26 is parabolic with the maximum value at the centerline of the flow (Figure 1.9b)

$$u_{\max} = -\frac{h^2}{2\mu}\frac{\partial p}{\partial x} \tag{1.27}$$

The volumetric flow rate per unit length in the z-direction, q, can be obtained as follows:

$$q = \int_{-h}^{h} u\,dy = -\frac{2h^3}{3\mu}\frac{\partial p}{\partial x} \tag{1.28}$$

Equation 1.28 shows that the flow rate is proportional to the pressure gradient, inversely proportional to the viscosity, and strongly dependent on the distance between the two plates. The mean velocity $\bar{V}$ of the flow is given as

$$\bar{V} = \frac{q}{2h} = -\frac{h^2}{3\mu}\frac{\partial p}{\partial x} \tag{1.29}$$

which, in terms of the maximum velocity, is equal to $\bar{V} = 2u_{\max}/3$.

1.7.2 Pressure-driven flows in circular tubes

Flows in circular tubes are very popular and are commonly known as the Poiseuille flows. For such flows, it is convenient to use cylindrical polar coordinates. Consider a flow in a horizontal circular tube of radius R, as illustrated in Figure 1.10a. If the flow is laminar, $V_r = V_\theta = 0$ and the mass conservation equation in cylindrical coordinates (for incompressible flows)

$$\frac{1}{r}\frac{\partial (rV_r)}{\partial r} + \frac{1}{r}\frac{\partial V_\theta}{\partial \theta} + \frac{\partial V_z}{\partial z} = 0 \tag{1.30}$$

gives $\partial V_z/\partial z = 0$. If the flow is steady, the Navier–Stokes equations in cylindrical coordinates (Equations 1.22a–c) are reduced to

$$0 = -\rho g \sin\theta - \frac{\partial p}{\partial r} \tag{1.31a}$$

$$0 = -\rho g \cos\theta - \frac{1}{r}\frac{\partial p}{\partial \theta} \tag{1.31b}$$

$$0 = -\frac{\partial p}{\partial z} + \mu\left[\frac{1}{r}\frac{\partial}{\partial r}\left(r\frac{\partial V_z}{\partial r}\right)\right] \tag{1.31c}$$

where $g_r = -g\sin\theta$ and $g_\theta = -g\cos\theta$ are used and θ is measured from the horizontal plane, as shown in Figure 1.10a.

Integrating Equations 1.31a and b, the pressure is obtained as

$$p = -\rho g r \sin\theta + f_1(\theta, z) \tag{1.32}$$

and

$$p = -\rho g r \sin\theta + f_2(r, z) \tag{1.33}$$

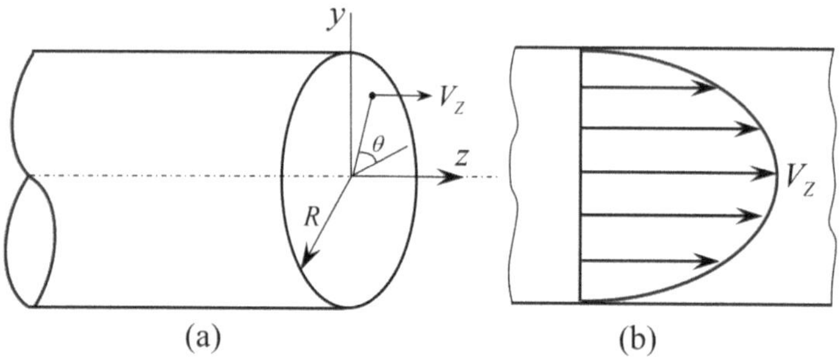

Figure 1.10 Pressure-driven flow in a circular tube: (a) notation and coordinate system and (b) parabolic velocity profile in a plane passing the centerline.

where $f_1(\theta,z)$ and $f_2(r,z)$ are two functions resulted from the integrations. Comparing Equations 1.32 and 1.33, it is easy to find that f_1 and f_2 are the same and are functions of z only, i.e., $f_1 = f_2 = f(z)$. Thus, $\partial p/\partial z$ is independent of r and θ, and can be treated as a constant when integrating Equation 1.31c with respect to r, which yields

$$V_z = \frac{1}{4\mu}\frac{\partial p}{\partial z}r^2 + c_1 \ln r + c_2 \tag{1.34}$$

Similarly, the two constants, $c_1 = 0$ and $c_2 = -R^2/(4\mu)\cdot(\partial p/\partial z)$, can be determined using the boundary conditions: V_z is finite at $r = 0$ (centerline of the tube) and $V_z = 0$ at $r = R$ (tube wall). Thus, the velocity distribution is given by

$$V_z = \frac{1}{4\mu}\frac{\partial p}{\partial z}\left(r^2 - R^2\right) \tag{1.35}$$

which is parabolic in any plane going through the tube centerline, as illustrated in Figure 1.10b.

Thus, the volumetric flow rate Q can be calculated by integrating the velocity

$$Q = \int_0^R V_z 2\pi r\,dr = -\frac{\pi R^4}{8\mu}\frac{\partial p}{\partial z} \tag{1.36}$$

and the mean velocity is given by

$$\bar{V} = \frac{Q}{\pi R^2} = -\frac{R^2}{8\mu}\frac{\partial p}{\partial z} \tag{1.37}$$

The maximum velocity takes place at the centerline of the tube. It is equal to

$$V_{max} = -\frac{R^2}{4\mu}\frac{\partial p}{\partial z} = 2\bar{V} \tag{1.38}$$

1.8 Electroosmotic flows

Pressure-driven flows in ducts represent a class of important flows in a variety of applications. Fluid flows can also be generated by other external forces, such as electrostatic forces. Electroosmotic flows are one of the most popular flows caused by electrostatic forces with the presence of an external electric field. They are the flows of electrolyte solutions caused by the motion of charged particles (e.g., ions) driven by an external electric field E. Electroosmotic flows usually take place near a charged solid surface, which affects the distribution of the charges in the electrolyte solution. To understand electroosmotic flows, some basic concepts need to be revisited, which will be discussed in the following sections.

1.8.1 Electric potential

It is known that two charged particles interact with each other and the force between them, i.e., the electrostatic force or Coulomb force, is given by

$$F_{\mathrm{E}} = \frac{qq'}{4\pi\dot{\varepsilon}r^2} \tag{1.39}$$

where q and q' are the charges carried by the two particles, $\dot{\varepsilon} = \varepsilon_{\mathrm{r}}\varepsilon_0$ is the electric permittivity of the medium (ε_{r} is the dielectric constant, or relative permittivity, of the medium and $\varepsilon_0 = 8.85\times10^{-12}\ \mathrm{C}^2/\left(\mathrm{Nm}^2\right)$ is the electric permittivity of a vacuum), and r is the distance between the particles. Equation 1.39 is called Coulomb's law. F_{E} is attractive or negative if the two charges are of opposite signs and repulsive or positive otherwise.

A charged particle generates an electric filed, which is a vector field defined as the Coulomb force acting on a unit charge. If the charge carried by the particle is q and the size of the particle is negligible, i.e., point charge, the electric field E at any location generated by the charge is given by

$$E = \frac{q}{4\pi\dot{\varepsilon}r^2} \tag{1.40}$$

where r is the distance from the point charge.

The electric field generated by a uniformly charged, infinite flat plane can be derived by considering the differential electric field dE_z due to an annulus, as shown in Figure 1.11. Consider a point, A, near the plane (Figure 1.11), where

$$dE_z = \int \frac{\cos\theta}{4\pi\dot{\varepsilon}\left(z^2+r^2\right)}dq = \frac{1}{4\pi\dot{\varepsilon}}\int_0^{2\pi} \frac{z}{\left(z^2+r^2\right)^{3/2}}\, r\, dr\rho_{\mathrm{e}}\, d\beta = \frac{\rho_{\mathrm{e}}}{2\dot{\varepsilon}}\frac{z}{\left(z^2+r^2\right)^{3/2}}rdr \tag{1.41}$$

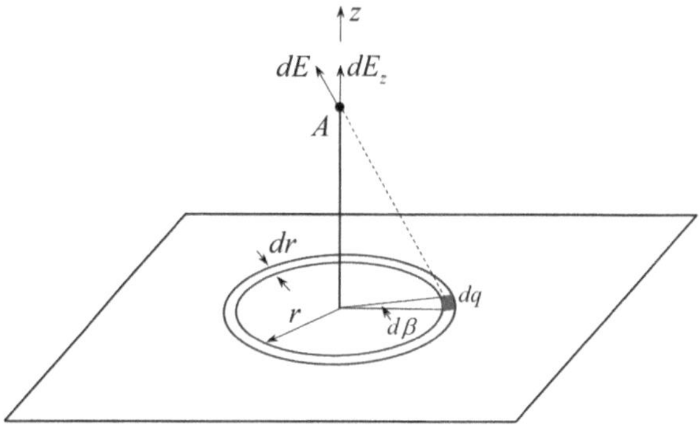

Figure 1.11 Determination of electric field caused by a uniformly charged flat surface.

where ρ_e is the charge density of the plane. Integrating Equation 1.41 yields

$$E = \frac{z\rho_e}{2\dot{\varepsilon}} \int_0^\infty \frac{r}{\left(z^2 + r^2\right)^{3/2}} dr = \frac{\rho_e}{2\dot{\varepsilon}} \tag{1.42}$$

An electric field is associated with an electric potential, ϕ, which is related to E through

$$E = -\frac{\partial \phi}{\partial r} \tag{1.43}$$

Thus, the electric potential at a location is the work required to move a unit charge from infinity to the location,

$$\phi = -\int_\infty^r E\,dr \tag{1.44}$$

The SI unit of the electric potential is N·m/C, which is equal to 1 volt, 1 V.

For a point charge q, the electric potential is

$$\phi = \frac{q}{4\pi\dot{\varepsilon}r} \tag{1.45}$$

and that for an infinite charged plane is given by

$$\phi = -\frac{\rho_e r}{2\dot{\varepsilon}} \tag{1.46}$$

Electric potential is an important quantity that is necessary in studying the motion of ions and electroosmotic flows under external electric fields, as will be shown later.

1.8.2 Electric double layer and zeta potential

When an electrolyte solution is in contact with a charged surface, the electrostatic forces between the surface and ions in the electrolyte will redistribute the ions in the vicinity of the surface. This occurs in electroosmotic flows in channels. Practically, most channel surfaces are negatively charged during the fabrication, such as silicon channels (Conlisk, 2013). In such a case, cations in the solution, which are called counterions, will collect around the surface under attractive electrostatic forces. Based on their motions, the ions near the surface can be divided into two layers. Right next to the surface, a layer of high-concentration cations are attracted by the surface. Cations in this layer are immobile due to the strong electrostatic forces from the surface. This layer is called the Stern layer, as illustrated in Figure 1.12. The thickness of the Stern layer is of order of ionic diameter. Outside the Stern layer is the diffuse layer, where the concentration of cations follows the Boltzmann distribution and all the ions (anions and cations) are mobile. The Stern layer and the diffuse layer form the so-called electric double layer (EDL). It is the motion of ions in the EDL that generates electroosmotic flows.

The charges in the surface and ions in the solution result in an electric potential in the electrolyte. The electric potential at the edge of the Stern layer (interface between the Stern layer and diffuse layer) is called zeta potential, which is designated by ζ.

1.8.3 Poisson's equation

Applying Gauss's theorem in calculus to electric field yields

$$\int_{\forall} \nabla \cdot \dot{\varepsilon} \mathbf{E} d\forall = \int_{\forall} \rho_e d\forall \tag{1.47}$$

where $\mathbf{E}$ is the electric field vector, ρ_e is the volume charge density, and $\forall$ denotes volume. Thus, Equation 1.47 can be written as

$$\nabla \cdot \mathbf{E} = \frac{\rho_e}{\dot{\varepsilon}} \tag{1.48}$$

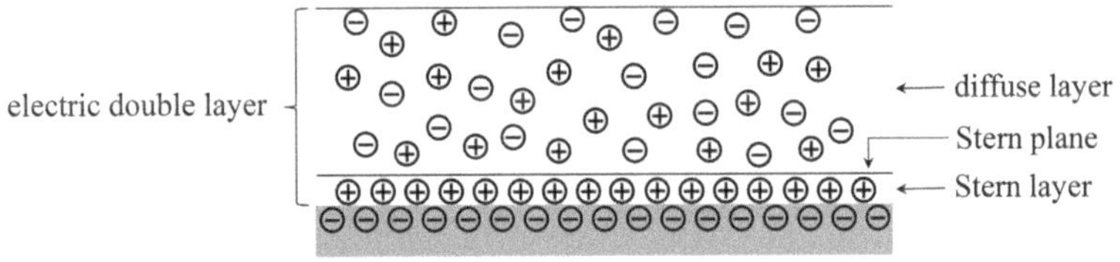

Figure 1.12 Schematic of electric double layer.

Substituting Equation 1.43 into Equation 1.48 gives

$$\nabla^2\phi = -\frac{\rho_e}{\dot{\varepsilon}} \tag{1.49}$$

This equation is called Poisson's equation.

1.8.4 Electroosmotic flows between two parallel charged plates

For an electrolyte solution between two parallel charged plates, the motion of ions caused by an external electric field drives the solution to move in the electric field direction if the surfaces are negatively charged, as depicted in Figure 1.13. Hence, the velocity field is strongly dependent on the ion and electric potential distributions. First, let us consider the case without an external electric field. In the steady state, the ion concentration and electric potential ϕ only vary in the y-direction (perpendicular to the plates) and can be described by the Nernst–Planck equation as follows:

$$\nabla \cdot \left(D_i \frac{dn_i}{dy} + \frac{D_i z_i n_i e}{k_B T} \frac{d\phi}{dy} \right) = 0 \tag{1.50}$$

where D_i, z_i, and n_i are the diffusion coefficient, valence, and number density of the ith species, respectively, and $e = 1.602 \times 10^{-19}$ C is the elementary charge. Thus, Equation 1.50 is coupled with Poisson's equation (see Equation 1.49) through charge density ρ_e

$$\rho_e = \sum_i z_i e n_i \tag{1.51}$$

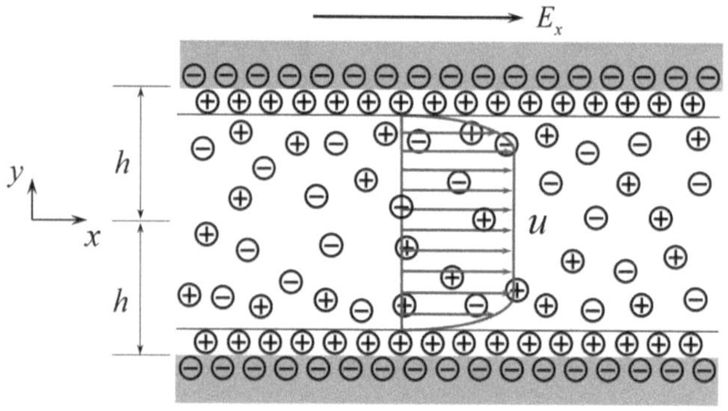

Figure 1.13 Velocity distribution in electroosmotic flows.

Considering that the net flux through any plane parallel to the plates is zero, Equation 1.50 can be simplified to

$$\frac{1}{n_i}\frac{dn_i}{dy}+\frac{z_i e}{k_B T}\frac{d\phi}{dy}=0 \tag{1.52}$$

By integrating Equation 1.52, the ion concentration n_i can be determined as follows:

$$n_i = n_{i,0}\exp\left(-\frac{z_i e\phi}{k_B T}\right) \tag{1.53}$$

where $n_{i,0}$ is the concentration at a location with $\phi = 0$ (far away from the plates).

For a binary, symmetric electrolyte ($z_1 = z_2 = z$, which is commonly referred to as $z{:}z$ electrolyte), Equations 1.49, 1.51, and 1.53 can be simplified to

$$\frac{d^2\phi^*}{dy^2}-\frac{2ze^2 n_0}{\dot{\varepsilon} k_B T}\sinh\left(z\phi^*\right)=0 \tag{1.54}$$

where $\phi^* = e\phi/(k_B T)$. By linearizing the hyperbolic sine term and replacing ϕ^* by ϕ, Equation 1.54 is reduced to

$$\frac{d^2\phi}{dy^2}-\kappa^2\phi=0 \tag{1.55}$$

where $\kappa=\sqrt{2z^2e^2n_0/(\dot{\varepsilon} k_B T)}$ is the Debye–Hückel parameter. $\lambda_D = 1/\kappa$ is a characteristic length for the thickness of the EDL and is commonly known as the Debye length. It is seen that λ_D depends on n_0 and it can vary from nanometers to micrometers. For electrolytes that are not simple $z{:}z$ electrolytes, the Debye length is given by

$$\lambda_D=\sqrt{\frac{\dot{\varepsilon} k_B T}{e^2\sum_i z_i^2 n_{i,0}}} \tag{1.56}$$

Equation 1.55 can be determined by using the boundary conditions: $\phi = \zeta$ at the Stern plane and $\phi = 0$ far away from the plates ($y \to \infty$), which lead to

$$\phi = \zeta \exp(-\kappa y) \tag{1.57}$$

Equation 1.57 is only valid for relatively low ζ and large distance between the plates, i.e., macro-systems without overlap in the EDLs of the plates. In nanochannels or for high ζ values, the error caused by Equation 1.57 could be significant.

The ions in the EDL move under electrostatic forces in the presence of an external electric field E_x in the x-direction (Figure 1.13). For pure electroosmotic flows (no pressure gradient in the x-direction), the motion of the electrolyte solution in the x-direction can be described by the following Navier–Stokes equation:

$$0 = \mu \frac{d^2 u}{dy^2} + \rho_e E_x \tag{1.58}$$

By using Poisson's equation, Equations 1.49 and 1.58 can be written as

$$\mu \frac{d^2 u}{dy^2} - \dot{\varepsilon} E_x \frac{d^2 \phi}{dy^2} = 0 \tag{1.59}$$

Integrating Equation 1.59 twice yields

$$u = \frac{\dot{\varepsilon}}{\mu} E_x (\phi - \zeta) \tag{1.60}$$

where boundary conditions, $\phi = \zeta$ and $u = 0$ at $y = 0$, $\phi = 0$, $du/dy = 0$, and $d\phi/dy = 0$ as $y \to \infty$, are used. Substituting Equation 1.57 into Equation 1.60 gives

$$u = \mu_{eo} E_x \left[1 - \exp(-\kappa y) \right] \tag{1.61}$$

where $\mu_{eo} = -\dot{\varepsilon}\zeta/\mu$ is denoted as the electroosmotic mobility. The velocity outside the EDL ($y \to \infty$) is then given by

$$u = \mu_{eo} E_x \tag{1.62}$$

which is a constant. Therefore, for macroscopic electroosmotic flows shown in Figure 1.13, the velocity u increases exponentially from zero at the Stern plane to $\mu_{eo} E_x$ at the edge of EDL, showing a plug flow-velocity profile, which is different from the parabolic velocity distribution in pressure-driven flows (Figure 1.9).

1.8.5 Electric current density and conductivity

In electroosmotic flows, an electric current may be generated due to the motion of charges in channels. For electroosmotic flows between two parallel plates (Figure 1.13), the total current per unit width I is calculated as

$$I = \int_{-h}^{h} J_x \, dy \tag{1.63}$$

where J_x is the electric current density given by

$$J_x = e \sum_i z_i \left(\frac{D_i z_i n_i e}{k_B T} E_x + n_i u \right) \tag{1.64}$$

In Equation 1.64, the mass transport due to the concentration gradient of a species is not considered, which is insignificant in most cases except for certain biological systems. Combining Equations 1.60, 1.63, and 1.64, the total current is expressed as follows:

$$I = E_x \frac{e^2}{k_B T} \sum_i D_i z_i^2 \int_{-h}^{h} n_i \, dy + E_x \frac{\dot{\varepsilon}}{\mu} \sum_i \int_{-h}^{h} (\phi - \zeta) e z_i n_i \, dy \tag{1.65}$$

By using Poisson's equation, Equations 1.49 and 1.51, I can be written as

$$I = 2h\bar{\sigma}_e E_x - E_x \frac{\dot{\varepsilon}^2}{\mu} \int_{-h}^{h} (\phi - \zeta) \frac{d^2\phi}{dy^2} dy \tag{1.66}$$

where

$$\bar{\sigma}_e = \frac{1}{2h} \frac{e^2}{k_B T} \sum_i D_i z_i^2 \int_{-h}^{h} n_i \, dy \tag{1.67}$$

is the average electric conductivity based on the definition of electric conductivity σ_e

$$\sigma_e = \frac{1}{2h} \frac{e^2}{k_B T} \sum_i D_i z_i^2 n_i \tag{1.68}$$

Integrating Equation 1.66 yields

$$I = 2h\bar{\sigma}_e E_x + E_x \frac{\dot{\varepsilon}^2}{\mu} \int_{-h}^{h} \left(\frac{d\phi}{dy} \right)^2 dy \tag{1.69}$$

1.8.6 Streaming potential and current

As introduced previously, electroosmotic flows are driven by external electric fields. When a charge-containing fluid (e.g., an electrolyte solution) flows in a channel driven by a non-electrostatic force, such as a pressure drop, gravitational force, or centrifugal force, the motion of the charges (typically ions) generates an electric potential difference between the upstream and downstream if there are no electrodes connecting the upstream and downstream, analogous to an open electric circuit. This potential is termed the streaming potential. In this case, an electric field E is induced by the streaming potential because the net current in the channel vanishes. Consider the flow of an electrolyte solution driven by a pressure gradient, dp/dx, as shown in Figure 1.14. The Navier–Stokes equation in the x-direction for this flow is given as follows:

$$0 = -\frac{dp}{dx} + \mu \frac{d^2 u}{dy^2} + \rho_e E_x \tag{1.70}$$

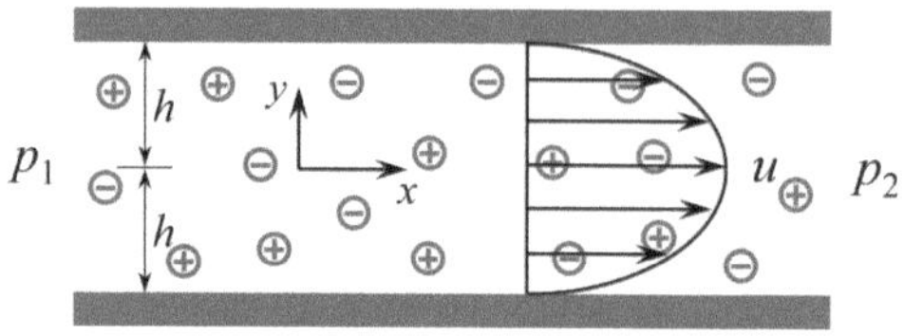

Figure 1.14 Flow of an electrolyte solution driven by a pressure drop (p_1 and p_2 are the upstream and downstream pressures, respectively).

The velocity distribution governed by Equation 1.70 actually is the combination of Equations 1.26 and 1.60, which are the solutions of Equations 1.24a and 1.58, respectively. Hence, u is given by

$$u = \frac{\dot{\varepsilon}}{\mu} E_x (\phi - \zeta) + \frac{1}{2\mu} \frac{dp}{dx} \left(y^2 - h^2 \right) \quad (1.71)$$

Since the net current vanishes in the channel, Equations 1.63, 1.64, and 1.71 lead to

$$E_x \frac{e^2}{k_B T} \sum_i D_i z_i^{\ 2} \int_{-h}^{h} n_i \, dy + E_x \frac{\dot{\varepsilon}}{\mu} \sum_i \int_{-h}^{h} (\phi - \zeta) e z_i n_i \, dy$$
$$+ \frac{1}{2\mu} \frac{dp}{dx} \sum_i \int_{-h}^{h} \left(y^2 - h^2 \right) e z_i n_i \, dy = 0 \quad (1.72)$$

Following the analysis from Equations 1.65 to 1.69, it is obtained that

$$2h\bar{\sigma}_e E_x + E_x \frac{\dot{\varepsilon}^2}{\mu} \int_{-h}^{h} \left(\frac{d\phi}{dy} \right)^2 dy + \frac{\dot{\varepsilon}}{\mu} \frac{dp}{dx} \int_{-h}^{h} y \frac{d\phi}{dy} dy = 0 \quad (1.73)$$

Thus, we have

$$E_x = -2h \frac{\dfrac{\dot{\varepsilon}}{\mu} \dfrac{dp}{dx} \left(\zeta - \bar{\phi} \right)}{2h\bar{\sigma}_e + \dfrac{\dot{\varepsilon}^2}{\mu} \displaystyle\int_{-h}^{h} \left(\frac{d\phi}{dy} \right)^2 dy} \quad (1.74)$$

where the conditions $\phi = \zeta$ at $y = \pm h$ are used and $\bar{\phi}$ is the mean potential defined as $\bar{\phi} = \int_{-h}^{h} \phi dy \Big/ 2h$.

In macroscale systems, the thickness of the EDL is thin as compared with the channel size. In this case, the potential can be treated as a constant (zero at any cross-section) in the channel and the electric field induced by the streaming potential is given by

$$E_x = -\frac{\dot{\varepsilon}\zeta}{\mu\bar{\sigma}_e}\frac{dp}{dx} \tag{1.75}$$

Equation 1.75 can be employed to measure the zeta potential ζ. If the length of the channel is L and the pressure difference between the upstream and downstream reservoirs is Δp, then $dp/dx = -\Delta p/L$ and ζ can be expressed as

$$\zeta = \frac{\mu\bar{\sigma}_e}{\dot{\varepsilon}}\frac{\Delta\phi}{\Delta p} \tag{1.76}$$

where $E_x = -d\phi/dx = \Delta\phi/L$ is used and $\Delta\phi$ is the potential difference between the upstream and downstream, which can be measured in experiments.

If the upstream and downstream reservoirs of the flow channel are connected through two electrodes to form a resistanceless loop, as depicted in Figure 1.15, the electric potential difference between the upstream and downstream will disappear and there is no electric field in the channel. The electric current I through the loop due to the motion of charges/ions in the channel is referred to as the streaming current. Following the analysis for the streaming potential, the streaming current can be obtained as

$$I = -\frac{\dot{\varepsilon}}{\mu}\frac{dp}{dx}\int_{-h}^{h} y\frac{d\phi}{dy}dy = -\frac{\dot{\varepsilon}}{\mu}\frac{dp}{dx}2h\left(\zeta - \bar{\phi}\right) \tag{1.77}$$

Again, for large channels, the electric potential ϕ can be treated as a constant (it can be chosen to be zero) and

$$I = \frac{2h\dot{\varepsilon}\zeta\Delta p}{\mu L} \tag{1.78}$$

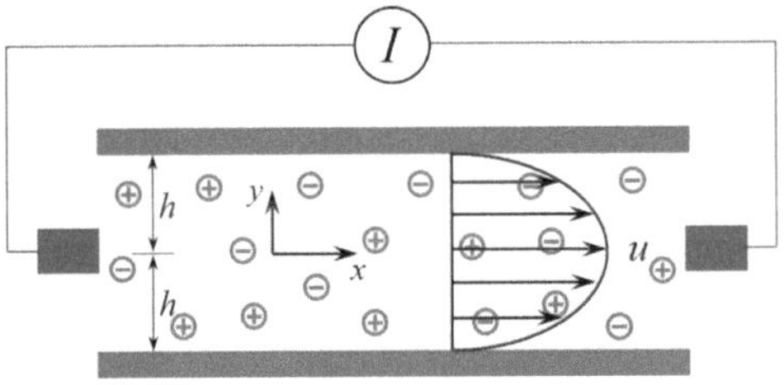

Figure 1.15 Streaming current in a pressure-driven flow.

Since I and Δp can be measured in experiments, Equation 1.78 offers an alternative approach to determine the zeta potential

$$\zeta = \frac{\mu I L}{2h\dot{\varepsilon}\Delta p} \tag{1.79}$$

References

Bird, R.B., Stewart, W.E., and Lightfoot, E.N. 2002. *Transport Phenomena*. New York: John Wiley & Sons.

Conlisk, A.T. 2013. *Essentials of Micro- and Nanofluidics*. Singapore: Cambridge University Press.

Lees, F.P. and Sarram, P. 1971. Diffusion coefficient of water in some organic liquids. *J. Chem. Eng. Data* 16:41–44.

Li, Z.G. and Drazer, G. 2007. Separation of suspended particles by arrays of obstacles in microfluidic devices. *Phys. Rev. Lett.* 98:050602.

Liu, C. and Li, Z.G. 2009. Flow regimes and parameter dependence in nanochannel flows. *Phys. Rev. E* 80:036302.

Lu, H.M. and Jiang, Q. 2005. Size-dependent surface tension and Tolman's length of droplets. *Langmuir* 21:779–781.

Munson, B.R., Young, D.F., and Okiishi, T.H. 2006. *Fundamentals of Fluid Mechanics*. Singapore: John Wiley & Sons.

Reif, F. 1985. *Fundamentals of Statistical and Thermal Physics*. Singapore: McGraw-Hill.

Sholl, D.S. 2006. Understanding macroscopic diffusion of adsorbed molecules in crystalline nanoporous materials via atomistic simulations. *Acc. Chem. Res.* 39:403–411.

Squires, T.M. and Quake, S.R. 2005. Microfluidics: Fluid physics at the nanoliter scale. *Rev. Mod. Phys.* 77:977–1026.

Stokes, G.G. 1851. On the effect of the internal friction of fluids in motion. *Proc. Cambridge Philos. Soc.* 9:8–106.

Todd, B.D., Evans, D.J., and Daivis, P.J. 1995. Pressure tensor for inhomogeneous fluids. *Phys. Rev. E* 52:1627–1638.

Chapter 2 Introduction to nanofluidics

2.1 What is nanofluidics

Nanofluidics is a subject that studies the static and dynamic behaviors of simple or complex fluids in confined structures with at least one dimension below 100 nm ($1 \text{ nm} = 1 \times 10^{-9} \text{ m}$). In terms of flow systems, the only difference between nanofluidics and the classic fluid mechanics is the size of the space available for a fluid flow. However, the small confinement in nanoscale flows can greatly affect the dynamics of a fluid and make nanofluidics quite different from the classic fluid mechanics.

A quantity that marks the difference between nanofluidics and the classic fluid mechanics is the surface-area-to-volume ratio (SAVR), which is the fluid–solid interfacial surface area divided by the volume of the fluid. The value of SAVR for a nanosystem could be several orders of magnitude larger than that of a macroscale system. For instance, $\text{SAVR} = 4 \times 10^{-2} \text{ m}^{-1}$ for a channel with square cross-sectional area equal to 1 cm^2. However, if the height or the width of the channel is reduced to 100 nm, $\text{SAVR} \approx 2 \times 10^{7} \text{ m}^{-1}$. The large value of SAVR for a nanosystem places the surface in an important position. In a macro-system, the role of the surface lies in the boundary condition. In a nanofluidic system, surface effects are significant and go beyond the boundary condition. They affect not only the flow velocity at the interface but also the fluid properties, which make nanofluidics more complex than the classic fluid mechanics. Consequently, the classic fluid mechanics theories (e.g., the Navier–Stokes equation) may be invalid for nanofluidics. Fortunately, the small confinement at the nanoscale can lead to new flow phenomena, which bring about novel applications of fluid flows and benefit many research areas, such as biology, medicine, chemistry, and engineering.

At the nanoscale, a surface affects the static and dynamic behaviors of a fluid through molecular forces, which involve multi-body interactions and make the theoretical analysis of nanofluidics challenging. For nanoscale flows, an accurate theoretical description of fluid dynamics is unavailable. The major tools for understanding nanofluidics are molecular dynamics (MD) simulations and physical experiments, which will be discussed in Chapters 3 and 4.

2.2 The role of confinements

To understand flows in a nanofluidic system, the role of the surface has to be well studied before taking advantage of the confinement to achieve various flow scenarios for diverse applications. Usually, a surface influences the motion of a fluid through molecular interactions. The most common forces are the van der Waals and electrostatic forces. The van der Waals force is a short-range force, which is generally weaker than electrostatic forces. It is important in systems with a dimension below ~50 nm (Liu and Li, 2011). Electrostatic forces can be considered as long-range forces that can reach tens of diameters of an atom. Electrostatic forces are essential in charged systems.

Theoretically, a molecular force is usually described by a molecular interaction potential, U. For the van der Waals force, the potential is position dependent and the force F is given as

$$F = -\frac{\partial U(r)}{\partial r} \tag{2.1}$$

where r is the distance between interacting molecules/atoms. Equation 2.1 also applies to the calculation of electrostatic forces, for which, the corresponding potential depends on the charges carried by the interacting molecules, in addition to the distance r. The unit of the potential U is energy. For the interaction between a fluid and a solid surface, the potential U is strongly associated with the free energy of the surface, i.e., surface energy (see Chapter 1) and the molecular species of the fluid, as will be discussed in Chapter 3.

For a fluid confined in a nanochannel, the distribution of fluid molecules is greatly affected by the fluid–surface interactions. In a channel of high surface energy, the molecular interaction between the fluid and the surface is strong and is largely attractive, which can lead to fluid adsorption. Figure 2.1a shows the distribution of fluid molecules in a 4-nm slit channel formed by two parallel, hydrophilic (high surface energy) walls. It is seen that some fluid molecules (about two layers) are adsorbed on the walls, resulting in a high fluid density near the walls, as will be shown in the next section. As the surface energy is reduced, the phenomenon of fluid adsorption is weakened. At an intermediate surface energy, the fluid distribution in the channel is somewhat uniform, as depicted in Figure 2.1b. If the walls are hydrophobic (low surface energy), the interaction forces between the fluid molecules and the walls at the interfaces, on average, tend to be mainly repulsive, leading to the depletion of fluid molecules in the vicinity of the walls (Figure 2.1c). Such surface effects are clearer when the total number of fluid molecules is reduced by 50%, as shown in Figure 2.1d–f.

If a fluid flow in a nanochannel is driven by a pressure gradient or an external force F_E, the fluid–wall intermolecular force F_{fw} will compete with the external force. If $F_E >> F_{fw}$, then the external force can break the fluid–wall interaction and

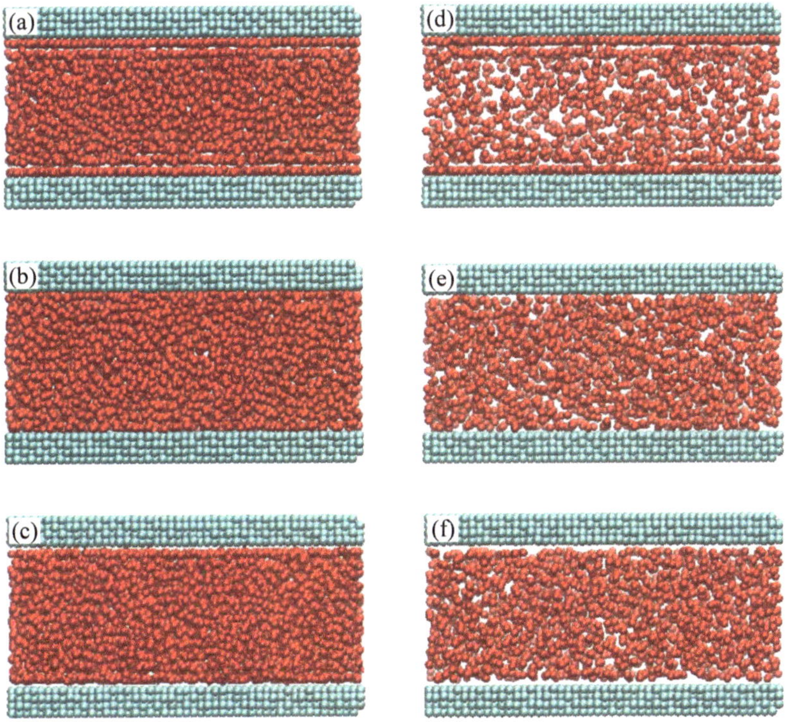

Figure 2.1 Distribution of fluid molecules in a 4-nm slit channel formed by two parallel walls. (a–c) The number density of the fluid is 20 nm^{-3}. (d–f) The number density of the fluid is 10 nm^{-3}. (a, d) Channel walls are of high surface energy. (b, e) Intermediate surface energy. (c–f) Low surface energy.

cause a relative motion between the fluid and the wall, leading to a velocity slip at the interface and the assumption of the stick boundary condition falls apart. Such a case tends to take place in low surface energy channels (e.g., hydrophobic channels), where F_{fw} is relatively small, as indicated by the velocity profile of an Ar flow in a 4-nm slit channel (low surface energy case) in Figure 2.2. If $F_E << F_{fw}$, the fluid–wall interaction is relatively strong and the external force cannot cause flow slip at the interface (Figure 2.2). This situation usually occurs in high surface energy channels (e.g., hydrophilic channels). In this case, the stick boundary condition appears to be valid. However, if the surface energy of the channel is sufficiently high, fluid adsorption on the channel wall will occur and negative velocity slip may take place, as will be discussed later.

For electroosmotic flows, other than the van der Waals force, the electrostatic forces between surface charges and a fluid containing charged particles (e.g., ions) also greatly influence the flow dynamics by affecting the particle and chemical

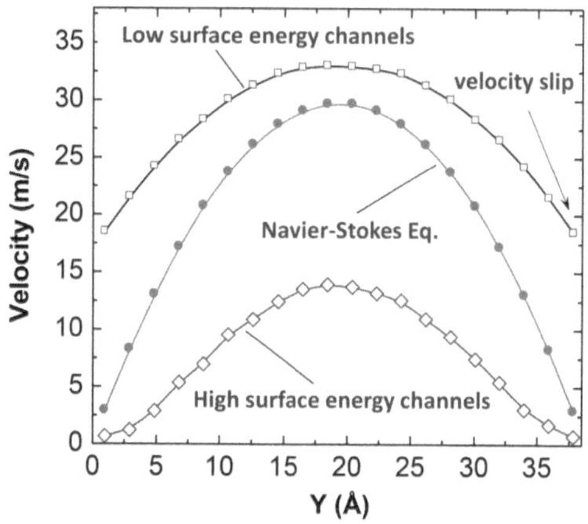

Figure 2.2 Velocity distribution of a fluid in a 4-nm slit channel driven by an external force. The *y*-direction is the direction perpendicular to the channel walls. Solid symbols are the predictions of the Navier–Stokes equation.

potential distributions. Studies have shown that the ion concentration distribution in nanochannels can deviate from the prediction of the Poisson–Boltzmann theory near the channel wall due to the finite size of ions and ion–surface interactions (Freund, 2002; Qiao and Aluru, 2003). The applicability of the Poisson–Boltzmann equation for nanoscale electroosmotic flows depends on the interplay of EDLs and the ion concentration. If the Debye length, i.e., the characteristic thickness of EDL, is comparable to or larger than the channel size, the interaction of EDLs, which is not considered in the classic electroosmotic theory, plays an important role and causes the Poisson–Boltzmann equation to fall apart (Wang and Chen, 2008). On the other hand, even if the interplay of EDLs is insignificant, the accuracy of the Poisson–Boltzmann equation is affected by the ion concentration. Numerical and theoretical analyses have shown that the Poisson–Boltzmann theory predicts the ion distribution well if the ion concentration is below 1 M (Wang and Chen, 2008). At high ion concentrations, the classic theory underestimates the ion density distribution next to the channel wall, as illustrated in Figure 2.3. The failure of the classic Poisson–Boltzmann equation at the nanoscale has also been reported in other MD simulations (Cui and Cochran, 2004; Dufreche et al., 2005).

In summary, the fluctuations of fluid properties and/or ion distribution are the direct consequence of fluid–surface intermolecular interactions. Molecular interactions, on one hand, may cause the classic theories to break down at the nanoscale and, on the other hand, can offer new opportunities to have diverse flow fashions and new flow phenomena, which cannot be observed in micro- and macroscale systems, as will be discussed in detail in Chapter 6.

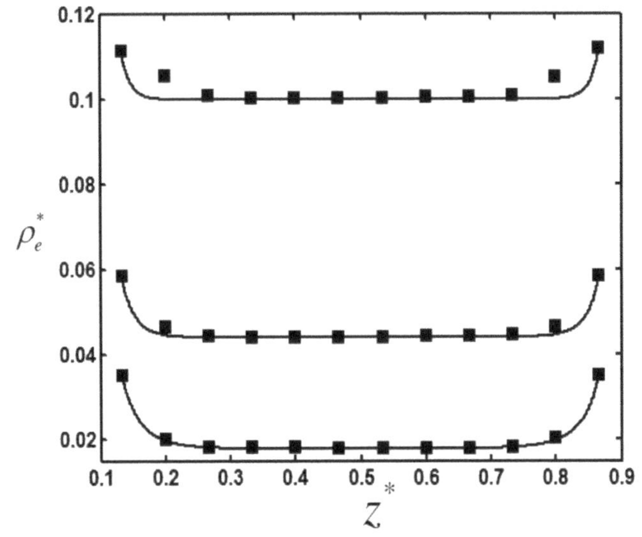

Figure 2.3 Normalized ion density distribution in a 4.98-nm slit channel at different ion concentrations. z^* is the relative position in the direction perpendicular to the channel walls. Symbols and solid lines are numerical results and the predictions of the Poisson–Boltzmann theory, respectively. The ion concentrations from bottom to top are about 1, 2, and 4 M. (Reprinted from Wang and Chen, 2008, with permission from Global Science Press.)

2.3 Physical properties of confined fluids

As demonstrated in Section 2.2, the molecular interaction between the surface and the fluid in a nanoscale flow greatly affects the distribution of the fluid molecules. This causes significant fluctuations in fluid properties (e.g., density and viscosity) next to the surface, and consequently affects the flow dynamics. Figure 2.4 shows the density ρ of a fluid confined in a nanoscale space between two parallel plates (Liu and Li, 2011). In Figure 2.4, $\varepsilon_{\mathrm{fw}}$ denotes the fluid–wall binding energy, which measures the strength of the molecular interaction between the fluid and the channel surface (this will be discussed in Chapter 3). It is an important parameter for controlling the surface energy or surface wettability. It is seen that the density varies greatly near the plates, especially in small nanochannels, where the surface effects are significant (Figure 2.4a). In relatively large channels (Figure 2.4b), the density is roughly constant around the center of the channel. It is also seen that the density distribution depends on the surface energy. As the surface energy ($\varepsilon_{\mathrm{fw}}$) increases, the fluid density next to the channel surfaces increases and is much larger than the mean density. This is because the channel walls tend to absorb the fluid molecules as the attractive force between the fluid and the walls increases, as illustrated in Figure 2.1a and d. Consequently, the density around the centerline decreases with increasing surface energy. In hydrophobic nanochannels, the density close to the channel wall is quite low due to the dominant repulsive molecular forces, as shown in Figure 2.5, which is the case in Figure 2.1f.

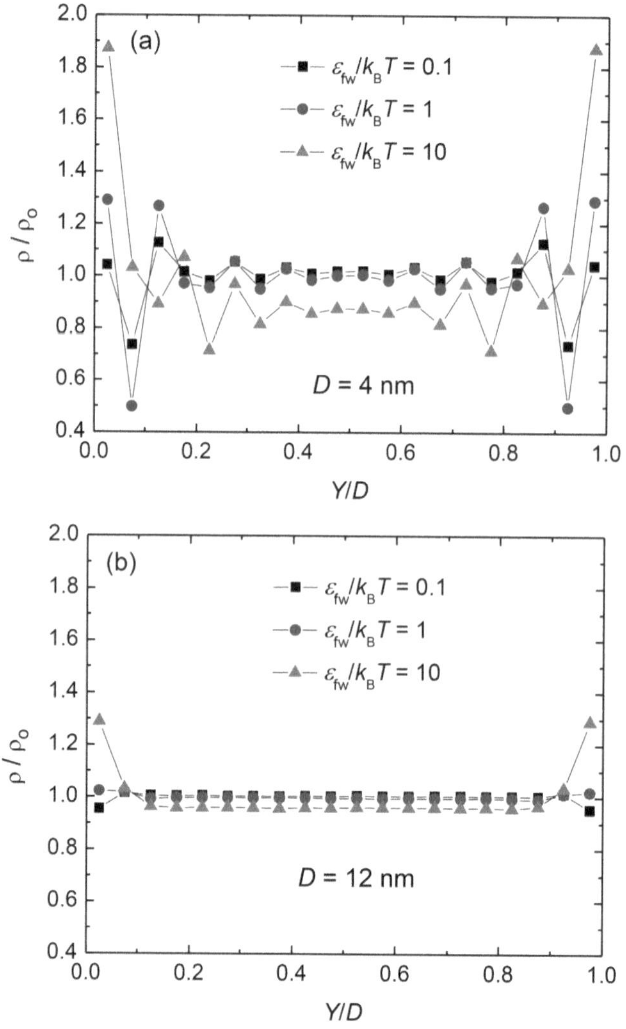

Figure 2.4 Density distribution of a fluid between two parallel plates. ρ and ρ_0 are the local and mean densities of the fluid. Y and D are the position coordinate and distance between the two plates, respectively. ε_{fw} is the binding energy between the fluid molecules and wall atoms. (a) $D = 4$ nm and (b) $D = 12$ nm. (Adapted from Liu and Li, 2011.)

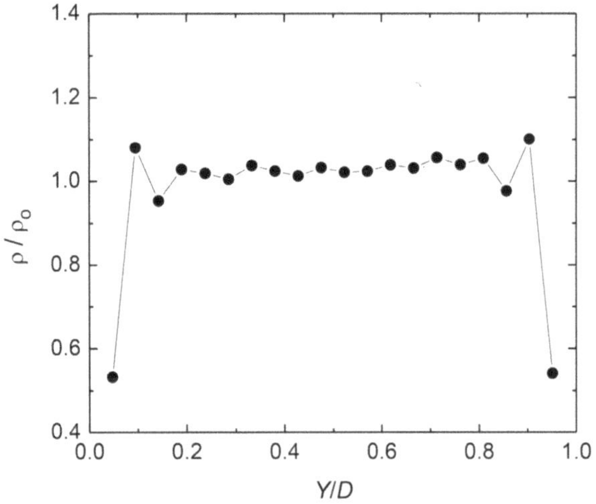

Figure 2.5 Density distribution for the case in Figure 2.1f.

Since the viscosity of a fluid is strongly related to the density, the viscosity distribution of a fluid between two parallel plates should follow similar fashion to the density shown in Figure 2.4 (Akhmatskaya et al., 1997; Gao et al., 2015). Figure 2.6 shows the viscosity distribution of water in a 1.6-nm slit channel. It is

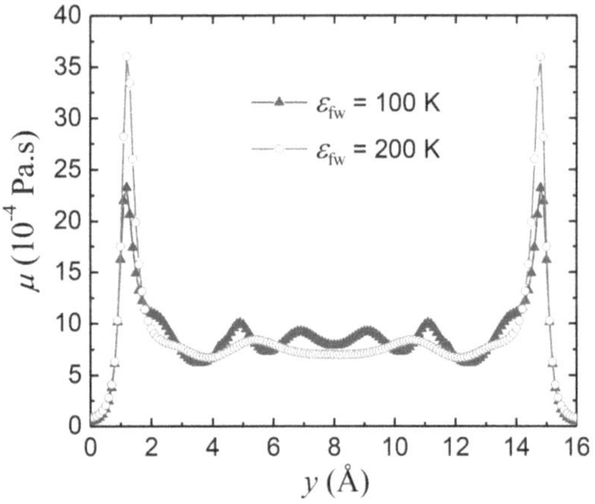

Figure 2.6 Distribution of water viscosity between two parallel plates (the distance between the plates in the *y*-direction, perpendicular to the plates, is 1.6 nm). ε_{fw} is the fluid–wall binding energy (1 K = 1.38 × 10^{-23} J). The contact angle of water is $\theta \approx 110^\circ$ for $\varepsilon_{fw} = 100$ K (or 0.0086 eV).

seen that the variation of viscosity is similar to that of density. The viscosity fluctuates greatly next to the surfaces and strongly depends on the surface energy (Figure 2.6).

2.4 Flow slip in nanochannels

The fluid–surface interaction can also change the boundary condition. In a large-scale system, the stick or nonslip boundary condition is usually assumed such that the velocity of the fluid at a solid surface is the same as the velocity of the surface, as demonstrated in Figure 2.7a. In a nanosystem, the velocity of a fluid at a fluid–solid interface depends on the surface properties and the external driving force. The stick boundary condition may or may not be valid. In certain conditions, the velocity of the fluid could be different from that of the surface, i.e., slip takes place at the interface. In this case, a quantity termed slip length, L_s, is usually used to characterize a slip boundary condition. Slip length is the length from the surface to the point where the extrapolated velocity vanishes in the direction perpendicular to the flow, as illustrated in Figure 2.7b. Given the coordinate in Figure 2.7b, the slip length can be expressed as follows:

$$L_s = u_w \Big/ \left.\frac{\partial u}{\partial z}\right|_{z=0} \tag{2.2}$$

where u_w is the flow velocity at the wall/surface and $\partial u/\partial z|_{z=0}$ is the velocity gradient at the wall, which is also termed shear rate and is commonly denoted by $\dot{\gamma}$.

The slip length depends on several parameters, including the surface and fluid properties, the external driving force, and temperature. It can be positive or negative. Strong molecular forces between the fluid and the surface may lead to fluid

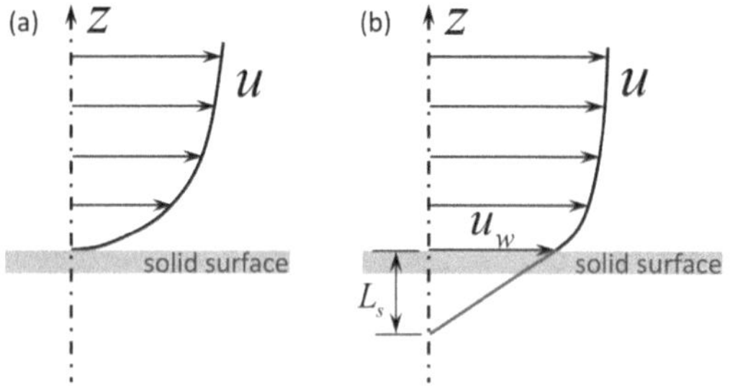

Figure 2.7 Boundary conditions for fluid flows moving over a planar surface. u is the fluid velocity parallel to the surface and u_w is the fluid velocity at the surface. (a) Stick boundary condition, where $u_w = 0$. (b) Slip boundary condition with a finite u_w. L_s is defined as the velocity slip length.

adsorption on the surface and result in negative velocity slip. This usually takes place on wetting surfaces and the definition of negative slip length varies in different situations (Priezjev, 2009). If the fluid–surface molecular forces are weak, positive velocity slip may occur, which will enhance the flow rate (Figure 2.7b). Unless otherwise stated, positive velocity slip is normally assumed.

Therefore, nanoscale confinements affect the flow dynamics through molecular interactions, which cause variations in fluid properties and boundary conditions and consequently may make the classic fluid mechanics theories invalid for nanofluidics.

2.5 Validity of the classic theories

The equation of motion for fluid flows, i.e., the Navier–Stokes equation, is fundamentally Newton's second law, which is valid regardless of the size (scale) of a flow system. In this sense, the general idea of flow description of the classic theories works for nanoscale flows. To apply the Navier–Stokes equation to nanofluidics, however, the fluctuations of fluid properties have to be considered. This actually is quite challenging because the variations of fluid properties depend on the surface properties (e.g., wettability) as well as other parameters, such as temperature. A universal model that can accurately describe fluid property changes is unavailable. This is one reason why the Navier–Stokes equation can hardly be used to predict nanoscale flows.

In addition to the change of fluid properties, the boundary condition is another issue that makes the classic theories fall apart for nanoscale flows. As discussed in the previous section, a velocity slip may occur at an interface and the slip length depends on surface properties, temperature, and external parameters. Under similar conditions, a hydrophobic surface usually enhances velocity slip and increases the flow rate. A large external driving force or a high temperature also promotes velocity slip. Unfortunately, how these parameters quantitatively affect the velocity slip is unclear. This poses a challenge for applying the classic theories to nanoscale flows.

Figure 2.8 shows the flow rate ratio Q/Q_{NS} of liquid argon flows in nanochannels formed by two parallel silver plates as a function of channel height D (Liu and Li, 2011). Q is the volumetric flow rate obtained through MD simulations and Q_{NS} is the flow rate predicted by the Navier–Stokes equation. It is seen that Q/Q_{NS} deviates greatly from unity in small nanochannels. This is because the surface effects are significant due to high SAVR values. As the channel height D increases, the surface effects become unimportant, and Q/Q_{NS} tends to converge to unity. In addition to the channel size, the surface energy is also critical. For large $\varepsilon_{\mathrm{fw}}/k_{\mathrm{B}}T$ values, $Q/Q_{\mathrm{NS}} < 1$ because the fluid–surface interaction is strong (surface energy

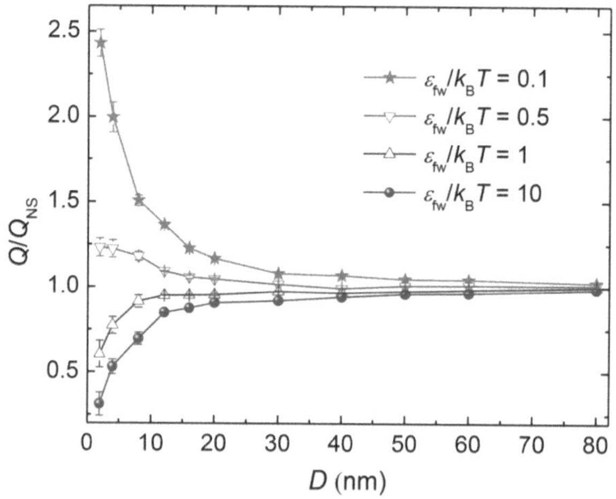

Figure 2.8 Flow rate ratio Q/Q_{NS} of liquid argon flows as a function of channel height D between two parallel silver plates. ε_{fw} is the binding energy between the fluid molecules and wall atoms. (Adapted from Liu and Li, 2011.)

is high), which leads to fluid adsorption on the channel wall and negative slip lengths. If ε_{fw}/k_BT is small, velocity slip occurs and the flow rate is enhanced.

The slip boundary condition and fluid property variations, which make the Navier–Stokes equation breakdown at the nanoscale, also challenge the validity of the classic electrokinetic theory in nanosystems. For instance, to understand nanoscale electroosmotic flows using the classic electrokinetic theory, fluid properties, such as viscosity, and ion distribution are required, which are affected by different factors (Bocquet and Charlaix, 2009). Figure 2.9 shows the electroosmotic mobility μ_{eo} of electroosmotic flows in micro- and nanochannels (glass) as a function of the Debye length λ_D and κh ($\lambda_D = 1/\kappa$, see Chapter 1), where h is half of the channel height (Haywood et al., 2014). It is seen that the predictions of the classic theory (lines in Figure 2.9) agree well with experiments in microchannels and relatively large nanochannels. However, in small nanochannels, i.e., $h = 27$ nm, theoretical predictions deviate from the experiments, especially at large Debye lengths. The discrepancy could be caused by the nonuniform fluid properties and ion distribution in nanochannels. The latter cannot be predicted by the Poisson–Boltzmann equation (Qiao and Aluru, 2003). Another reason could be the error in zeta potential (Bocquet and Charlaix, 2009), which depends on the channel size and charge concentration. The dependence of the electroosmotic velocity on channel size and ion concentration has also been observed in polydimethylsiloxane (PDMS) nanochannels (Peng and Li, 2016).

It must be noted that some efforts have been made to predict nanoscale flows using the classic theories. Some researchers argue that the classic theories, such

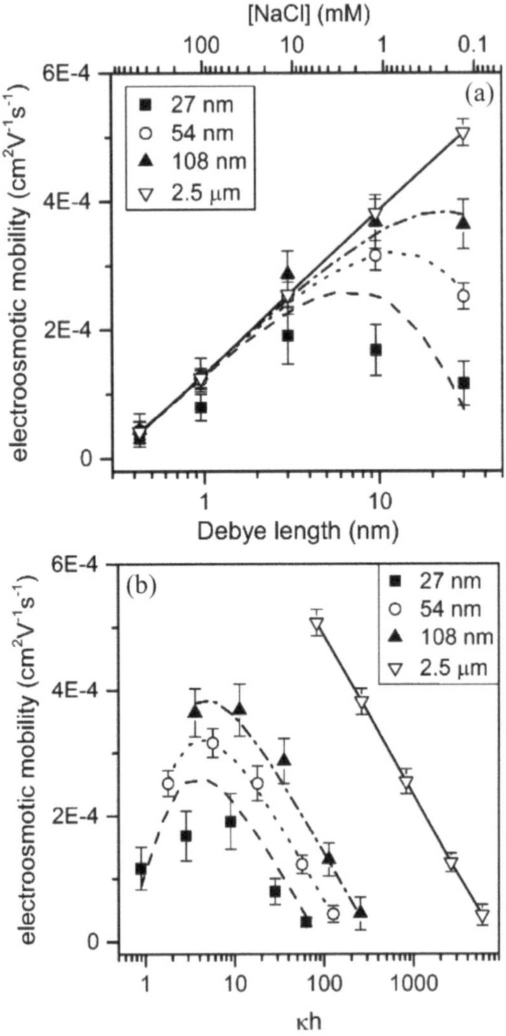

Figure 2.9 Average electroosmotic mobility for NaCl solutions as a function of Debye length (a) and κh (b) in channels of different heights. h is half of the channel height. Symbols are experimental data and lines are theoretical predictions. (Reprinted from Haywood et al., 2014.)

as the Navier–Stokes equation and electrokinetic theory, work well in predicting nanoscale flows if some essential parameters (e.g., slip length and zeta potential) can be accurately obtained. This may be true in cases, where the surface effects can be modeled by certain quantities. However, the classic theories fundamentally do not represent a precise description of nanofluidics because they do not consider molecular interactions, which are critical at the nanoscale. Even though they happen to offer a satisfactory prediction, the classic theories fail in explaining new flow phenomena that are unique to nanofluidics. For instance, in a macroscale fixed structure (no moving parts), a fluidic diode, where fluid flows are only allowed in a specific direction but blocked in the opposite direction, cannot be achieved using Newtonian fluids regardless of the value of Re. At the nanoscale, however, fluidic diodes can be realized without using moving parts for simple fluids, as will be discussed in Chapter 6. This is attributed to molecular interactions, which are not included in and cannot be explained by the classic theories.

Therefore, the classic theories may work well for nanofuidics in certain cases if the fluid properties and boundary conditions are properly considered. However, the absence of molecular interactions and the lack of accurate models for quantitative descriptions of fluid properties and boundary conditions make the classic theories fundamentally invalid for nanofluidics.

2.6 Flow characteristics at the nanoscale

In a macroscale system, it is known that a flow can be laminar, transitional, or turbulent, depending on the Re number. In laminar flows, where Re is relatively small, fluid particles move in different layers without clear lateral mixing such that the velocity and the momentum transfer perpendicular to the flow direction can be neglected, as illustrated in Figure 2.10a. At a sufficiently high Re, the inertial effect dominates over the viscous effect and the flow becomes turbulent. In turbulent flows, fluid particles do not move in layers but undergo intensive mixing such that the flow is primarily characterized by eddies and swirls (Figure 2.10b). The momentum transfer becomes strong in all the directions due to the mixing of fluid particles. The velocity and pressure at a point change with time and the flow analysis is quite challenging.

The concepts of laminar and turbulent flows also apply to most microscale flows. At the nanoscale, however, it is necessary to look into the motion of individual fluid molecules instead of the motion of a collection of fluid molecules. In this case, the diffusion of molecules, which is ignored in macroscale systems, becomes important as fluid molecules can travel a scale of nanometers quickly through diffusion. For example, the self-diffusion coefficient of water D_s at room temperature is about $2.57 \times 10^{-9}\ \mathrm{m^2/s}$ (Wang, 1965), which, according to Equation 1.9, shows that the mean distance traveled by a water molecule per second in a direction is given by

$$d \approx \sqrt{2D_s} = 77\ \mu\mathrm{m} \tag{2.3}$$

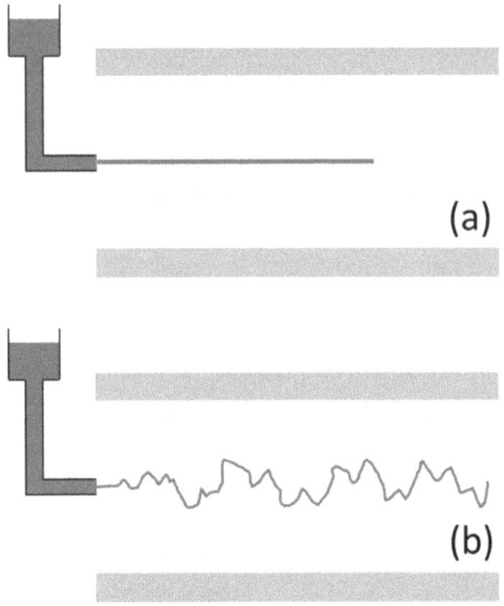

Figure 2.10 Motion of dye (lines) injected into a flow between two parallel plates. (a) Laminar flow with dye moving on a horizontal line. (b) Turbulent flow, where dye molecules move chaotically.

which is much larger than the dimension of nanoconfinements. On the other hand, the streaming velocity of nanoscale fluid flows is typically of order of micrometer per second or less (Li et al., 2015), indicating that the diffusion of fluid molecules is an important transport mechanism at the nanoscale.

As molecular diffusion is comparable to or even more important than the translational motion of a fluid at the nanoscale, it is uncommon to classify nanoscale flows into laminar or turbulent flows. At the nanoscale, to describe flow dynamics, statistical results (or mean values) of a quantity is more meaningful than the instantaneous values. Such statistically results are termed the ensemble average, which will be discussed in Chapter 3.

2.7 Why nanofluidics

Whether a research area can attract great interest and become an independent scientific subject largely depends on the potential of observing new phenomena, developing new theories, and/or finding new applications. Microfluidics gained popularity in late 1980s due to its tremendous applications in many areas, as well as the advances in microfabrication technologies. Nanofluidics, which offers new flow phenomena and brings about new applications, is following the footsteps of microfluidics and emerging as a new research area.

A remarkable label of nanofluidics is small size, which is in line with the development of many research areas, such as the miniaturization of electronic devices and lab-on-a-chip (LOC) technologies. After being integrated into a chip, nanofluidic systems can provide various laboratory processes with great performance. Depending on the applications, integrated nanofluidic systems can be used for chemical reactions, sensing, bioanalyses, medical diagnosis, and so on. For chemical and biological analyses, nanofluidics systems can achieve accurate and fast analyses without consuming too much reagents. As sensors, nanofluidic systems can greatly improve the sensing sensitivity. The small scale of nanofluidic systems can also help lower the cost of relevant devices with the advances in nanofabrication techniques. It is worth mentioning that integrated nanofluidic systems for medical analysis play an important role in improving worldwide public health, especially in developing and undeveloped countries, where low-cost, portable medical kits with a long shelf life are preferred.

Furthermore, the involvement of surface effects in nanofluidics offers new flow phenomena, which lead to novel applications. For example, flows driven by a temperature gradient in a nanochannel depend greatly on the surface properties. In high surface energy nanochannels (e.g., hydrophilic surface), fluids move from low temperature to high temperature, as illustrated in the top panel of Figure 2.11. In nanochannels of low surface energy, fluids migrate from high temperature to low temperature (middle panel of Figure 2.11). If a nanochannel is fabricated with heterogeneous surface properties, i.e., half of the channel surface is of high surface energy while the other half is of low surface energy, fluids can be pumped by symmetric temperature gradients, as illustrated in the bottom panel of Figure 2.11 (Liu and Li, 2010). Such a pumping mechanism, which cannot be realized in macroscale systems, can be employed for device cooling using the waste heat generated by the device, instead of using external pumps to drive the fluid. In addition, the pressure needed to drive a fluid into a nanochannel, which is called infiltration pressure, also depends on the channel properties. The surface property-dependent infiltration pressure can be used to develop flow rectifiers, where the flow rate is direction dependent, without using moving parts. Flow rectifiers can be used to control flows or sieve ions, which have important applications in seawater desalination. These examples and many others will be discussed in detail in later chapters.

It must be noted that nanofluidics has some inherent disadvantages. One is the large flow resistance due to the small confinement. This may limit the applications of nanofluidics. The other is the requirement of fabrication technologies. The fabrication of very small nanopores can be challenging, which can increase the cost of fluidic devices. Therefore, the wide applications of nanofluidics need novel nanofabrication techniques. Nevertheless, the rich flow phenomena offered by and the tremendous applications of nanoscale flows, with the advances in nanofabrication technologies, are gaining increasing attention and making nanofluidics an attractive scientific field.

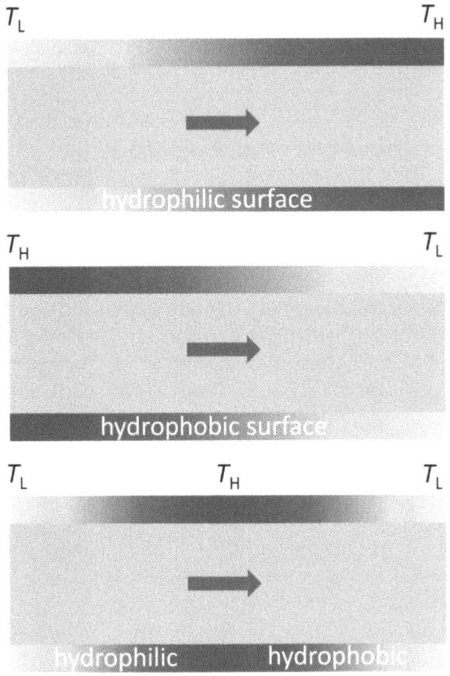

Figure 2.11 Flows driven by temperature gradients in nanochannels. T_H and T_L denote high and low temperatures, respectively. Top panel: Fluid moves from low to high temperature in a hydrophilic channel. Middle panel: Fluid is driven from high to low temperature in a hydrophobic channel. Bottom panel: Fluid flow is driven by symmetric temperature gradients.

References

Akhmatskaya, E., Todd, B.D., Daivis, P.J., Evans, D.J., Gubbins, K.E., and Pozhar, L.A. 1997. A study of viscosity inhomogeneity in porous media. *J. Chem. Phys.* 106:4684–4695.

Bocquet, L. and Charlaix, E. 2009. Nanofluidics, from bulk to interfaces. *Chem. Soc. Rev.* 39:1073–1095.

Cui, S.T. and Cochran, H.D. 2004. Electroosmotic flow in nanoscale parallel-plate channels: Molecular simulation study and comparison with classical Poisson–Boltzmann theory. *Mol. Simulat.* 30:259–266.

Dufreche, J. F., Marry, V., Malikova, N., and Turq, P. 2005. Molecular hydrodynamics for electroosmosis in clays: From Kubo to Smoluchowski, *J. Mol. Liquid.* 118:145–153.

Freund, J.B. 2002. Electro-osmosis in a nanometer-scale channel studied by atomistic simulation, *J. Chem. Phys.* 116:2194–2200.

Gao, X., Zhao, T.S., and Li, Z.G. 2015. Controlling flow direction in nanochannels by electric field strength. *Phys. Rev. E* 92:023017.

Haywood, D.G., Harms, Z.D., and Jacobson, S.C. 2014. Electroosmotic flow in nanofluidic channels. *Anal. Chem.* 86:11174–11180.

Li, L., Mo, J.W., and Li, Z.G. 2015. Nanofluidic diode for simple fluids without moving parts. *Phys. Rev. Lett.* 115:134503.
Liu, C. and Li, Z.G. 2010. Molecular dynamics simulation of composite nanochannels as nanopumps driven by symmetric temperature gradients. *Phys. Rev. Lett.* 105:174501.
Liu, C. and Li, Z.G. 2011. On the validity of the Navier–Stokes equation for nanoscale liquid flows: The role of channel size. *AIP Adv.* 1:032108.
Peng, R. and Li, D.Q. 2016. Electroosmotic flow in single PDMS nanochannels. *Nanoscale* 8:12237–12246.
Priezjev, N.V. 2009. Shear rate threshold for the boundary slip in dense polymer films. *Phys. Rev. E* 80:031608.
Qiao, R. and Aluru, N.R. 2003. Ion concentrations and velocity profiles in nanochannel electroosmotic flows. *J. Chem. Phys.* 118:4692–4701.
Wang, J.H. 1965. Self-diffusion coefficients of water. *J. Phys. Chem.* 69:4412–4412.
Wang, M.R. and Chen, S.Y. 2008. On applicability of Poisson–Boltzmann equation for micro-and nanoscale electroosmotic flows. *Comm. Comp. Phys.* 3:1087–1099.

Chapter 3 Molecular dynamics simulation

3.1 Introduction

Theoretical and experimental analyses are the major approaches for scientific studies in most research areas. One objective of a theoretical analysis is to find the solutions of certain governing equations, such as the Navier–Stokes and Poisson–Boltzmann equations. For simple fluid flows (e.g., Poiseuille's flow), the analytical solution of the Navier–Stokes equation can be obtained. However, in many cases, flows are rather complex and it is difficult to theoretically solve the governing equations. This leads to the development of various numerical methods, which employ different strategies in interpreting and solving the governing equations. Numerical methods or simulations usually involve intensive calculations and can only be performed using computers. Describing fluid flows through numerically solving the governing equations is an important approach and has been developed as a branch of fluid mechanics, which is called computational fluid dynamics (CFD). As most of CFD methods numerically solve the Navier–Stokes equation, they can accurately describe macroscale flows and have been widely used for various flow analyses. For nanoscale flows, as discussed in Chapter 2, the Navier–Stokes equation may become invalid and the traditional CFD method is not regarded as a quantitatively reliable approach for catching nanoscale flow physics. The most popular numerical method for nanofluidics is molecular dynamics (MD) simulation, which considers multi-body interactions of fluid molecules and solid surface atoms. Some other numerical methods, such as Monte Carlo simulation, can also be used for nanoscale flow analyses, especially for gas flows. In this chapter, only the method of MD simulation is introduced.

MD simulation is a numerical method that deals with the motions of a large number of particles, such as molecules/atoms and charges, by solving Newton's equation of motion based on inter- and/or intramolecular interactions. In an MD simulation, the forces in all the directions on each particle are first computed using empirical potentials, which depend on the distance between interacting particles. The forces are then used to calculate the acceleration, velocity, and position of the particle through Newton's equation of motion. The potentials, forces, velocities, and positions of all the particles are the quantities that an MD simulation can provide, which can be further analyzed to obtain necessary properties of a flow.

3.2 Molecular interaction potentials

3.2.1 Lennard-Jones and Coulomb's potentials

An MD system contains a large number of particles and the motions of these particles are obtained by solving Newton's equation of motion, which, for a particle, *i*, is given by

$$\mathbf{F}_i = m_i \ddot{\mathbf{r}}_i \tag{3.1}$$

where **F** and **r** are the force and position vectors, m is the mass of the particle, and $\ddot{\mathbf{r}}_i = \partial^2 \mathbf{r}/\partial t^2$ is the acceleration of the particle. The force acting on the particle, $\mathbf{F}_i$, is obtained as the spatial derivative of a molecular interaction potential $U(r)$,

$$\mathbf{F}_i = -\frac{\partial U(r)}{\partial r} \sum_{j,j \neq i} \frac{\mathbf{r}_{ij}}{r_{ij}} \tag{3.2}$$

where r_{ij} is the distance between interacting particles i and j and $\boldsymbol{r}_{ij}$ is the position vector from particle i to j. Therefore, in MD simulations, it is the potentials that determine the motions of molecules/atoms. However, it is extremely challenging to probe the interactions among molecules and acquire accurate descriptions for molecular interaction potentials. Most potential functions employed in MD simulations are empirical. These potentials contain empirical parameters, which are usually determined by matching certain physical properties of a substance computed from simulations to well-accepted values.

Potential functions for different materials could be very different. A simple and popular potential is the Lennard-Jones (LJ) potential, which describes the van der Waals interactions of atoms. The LJ potential can model the attractive force when two atoms are far away and the repulsive force when they are close to each other. The most commonly used LJ potential is the LJ 12-6 potential, which is given as

$$U_{\mathrm{LJ}}(r) = 4\varepsilon\left[\left(\frac{\sigma}{r}\right)^{12} - \left(\frac{\sigma}{r}\right)^{6}\right] \tag{3.3}$$

where ε and σ are the binding energy and collision diameter. These two parameters physically represent the well depth of the potential and the distance where the potential vanishes, as illustrated in Figure 3.1. In some research areas, σ is considered as the diameter of an atom. The LJ potential reaches the minimum at a distance r^*, where the force is zero. If $r > r^*$, the r^{-6} term in Equation 3.3 dominates and the force between two interacting atoms is attractive. If $r < r^*$, the r^{-12} term becomes important and the force is repulsive.

Since molecular forces go to zero as $r \to \infty$, potentials are usually truncated at a cutoff distance, r_c, beyond which the forces are not considered. A typical value for

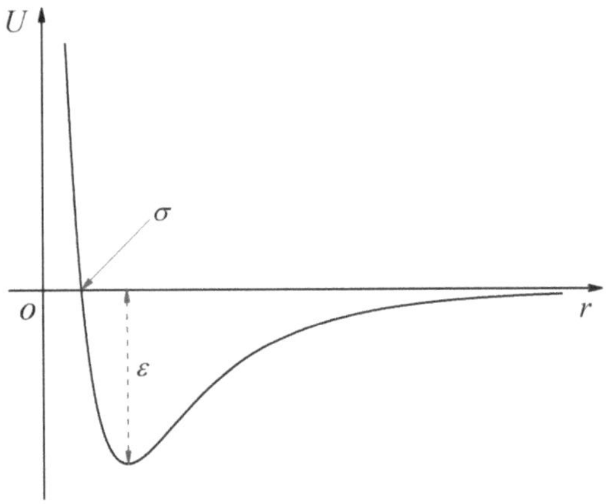

Figure 3.1 The Lennard-Jones potential.

r_c is $r_c = 2.5\sigma$. The truncation of a potential generates a step change in the force and causes errors in force calculation. To avoid error accumulation, the potential can be shifted to make the force at r_c continuous. The shifted LJ potential can be written as

$$U_{LJ}^{shifted}(r) = \begin{cases} U_{LJ}(r) - U_{LJ}(r_c) - (r - r_c)\left.\dfrac{\partial U}{\partial r}\right|_{r_c}, & r \le r_c \\ 0, & r > r_c \end{cases} \tag{3.4}$$

This shifted potential and the corresponding force are equal to zero at $r = r_c$. The error caused by the potential truncation can be significantly decreased through increasing the cutoff distance. If $r_c > 4\sigma$, numerical errors due to the potential truncation can be neglected. However, if r_c is large, the simulation becomes numerically expensive as the number of molecular interactions included in calculating the force on a particle scales with $r_c^{\,3}$.

The LJ potential was initially developed to model noble gases (Lennard-Jones, 1924). Furthermore, it was widely used for other gases, liquids, and even solids. The LJ parameters, ε and σ, for various fluids have been extensively studied in the literature (Mourits and Rummens, 1977; Hippler et al., 1983) and can be easily adopted in simulations. It must be noted that the LJ parameters in the literature are the self-interaction parameters, i.e., they are valid for atomic interactions between the same species. For interactions between different species, for example, an oxygen and a nitrogen molecule, the LJ parameters can be determined by the Loretz–Berthelot combining rules as

$$\sigma_{\alpha\beta} = (\sigma_{\alpha\alpha} + \sigma_{\beta\beta})/2,\ \varepsilon_{\alpha\beta} = \sqrt{\varepsilon_{\alpha\alpha}\varepsilon_{\beta\beta}} \tag{3.5}$$

where the subscripts α and β denote different species. For fluid–solid interactions, the LJ parameters, ε_{fs} and σ_{fs}, are strongly related to the surface energy or the wettability of the solid. By simply varying ε_{fs}, a solid surface can be tuned from hydrophilic to hydrophobic (Mo et al., 2015; Li et al., 2016). A large ε_{fs} corresponds to a high surface energy and the surface tends to be wetting (hydrophilic). To generate a superhydrophobic surface, both ε_{fs} and σ_{fs} may need to be controlled properly. A refined LJ potential for tuning the wettability of a solid surface is given by

$$U(r) = 4a\varepsilon_{fs}\left[\left(\frac{\sigma_{fs}}{r}\right)^{12} - b\left(\frac{\sigma_{fs}}{r}\right)^{6}\right] \tag{3.6}$$

where a and b are parameters. If $a = b = 1$, Equation 3.6 is reduced to the regular LJ potential (see Equation 3.3). If the values of a and b are reduced, the contact angle of the fluid on the solid surface is increased and the surface can be tuned to be superhydrophobic (Nagayama et al., 2006).

The LJ potential is simple and numerically inexpensive because it only considers two-body interactions. Although it can reasonably approximate interactions between neutral molecules/atoms of many materials, complex potentials have been developed to improve the accuracy of atomic interactions. The general expression of such potentials includes two-body and three-body interactions (higher-order interactions are usually not considered in simulations due to their minor contribution to forces and high time consumption) and can be simplified as

$$U(r) = \sum_{i}\sum_{j>i} U_2(r_{ij}) + \sum_{i}\sum_{j>i}\sum_{k>j>i} U_3(r_{ij}, r_{jk}, r_{ki}) \tag{3.7}$$

Compared with the LJ potential, this multi-body interaction potential is more accurate in describing the properties of materials, but the three-body interaction term makes simulations time-consuming. The two- and three-body interaction terms in Equation 3.7 may take different forms for different materials.

The LJ potential and Equation 3.7 are for neutral particles. If there are charges in a system, electrostatic forces need to be considered through the Coulomb potential

$$U_{\text{Coulomb}}(r) = \frac{q_1 q_2}{4\pi\varepsilon_0 r} \tag{3.8}$$

where q_1 and q_2 are the charges carried by the interacting particles, and ε_0 is the permittivity of free space. Compared with the LJ and multi-body interaction potentials, the Coulomb potential can reach a relatively long distance, as indicated by the r^{-1} dependence in Equation 3.8. Therefore, the Coulomb potential is considered as a long-range potential, while the LJ and multi-body potentials, which account for the van der Waals force, are short-range potentials.

3.2.2 Water models

For certain popular fluids and solids, efforts have been made to develop specific potentials to accurately describe their physical properties. For water, several potentials have been proposed, which can reproduce the physical properties of bulk water, such as viscosity and diffusion coefficient, very well. Relatively simple and widely used water models are rigid models, where the intramolecular interactions between the oxygen and the hydrogen atoms of a water molecule are not considered such that the intramolecular O–H bond length and H–O–H bond angle are fixed. The representatives of such rigid models are the transferable intermolecular potential 3 point (TIP3P) (Jorgensen et al., 1983) and the extended simple point charge (SPC/E) (Berendsen et al., 1987) water models. The latter is a modified version of the simple point charge (SPC) model (Berendsen et al., 1981). They combine the LJ and the Coulomb potentials to describe the van der Waals and electrostatic interactions. These two models for the interaction between two water molecules a and b take the form of

$$U_{ab}(r) = 4\varepsilon_{OO}\left[\left(\frac{\sigma_{OO}}{r_{OO}}\right)^{12} - \left(\frac{\sigma_{OO}}{r_{OO}}\right)^{6}\right] + \sum_{i\in a}\sum_{j\in b}\frac{k_c q_i q_j}{r_{ij}}, \quad i,j = 1,2,3 \tag{3.9}$$

where $k_c = 1389.5$ Å kJ/mol is the electrostatic constant, r_{OO} is the separation between the oxygen atoms of the two interacting water molecules, and the subscripts i and j represent the ith and jth atom (H or O) of water molecules a and b, respectively. The geometric and potential parameters for the two water models are listed in Table 3.1. Although the parameters for the two models are very close, they give quite different results in water diffusivity. The self-diffusion coefficient of water calculated with the SPC/E model is 2.5×10^{-9} m^2/s (Berendsen et al., 1987), which is in good agreement with experimental result, 2.3×10^{-9} m^2/s (Mills, 1973), while the TIP3P model predicts a large diffusion coefficient, ~5.7×10^{-9} m^2/s (Mills, 1973). In terms of water structure and self-diffusion coefficient, the SPC/E model appears to be the best, compared with the TIP3P and SPC water models (Mark and Nilsson, 2001).

3.2.3 Potentials for silicon

In a nanofluidic system, other than the fluid, the atomic dynamics of the solid surface may also need to be considered. Silicon nanochannels and carbon nanotubes (CNTs) are frequently used in experiments. In simulations, in addition to Si and CNTs, graphene and metals are also widely employed. The potentials of these materials will be introduced in this section.

The atomic dynamics of Si can be described by the Stillinger–Weber potential, which is a combination of two- and three-body interactions (Stillinger and Weber, 1985):

Table 3.1 Parameters for TIP3P and SPC/E Water Potentials

Parameters	TIP3P	SPC/E
O–H bond length (Å)	0.9572	1.0
H–O–H bond angle (deg.)	104.52	109.47
q_O (e)	−0.834	−0.8476
q_H (e)	0.417	0.4238
ε_{OO} (kJ/mol)	0.634	0.650
σ_{OO} (Å)	3.151	3.166

$$U = \sum_i \sum_{j>i} \varepsilon f_2\left(r_{ij}/\sigma\right) + \sum_i \sum_{j>i} \sum_{k>j>i} \varepsilon\left[h\left(r_{ij}, r_{ik}, \theta_{jik}\right) + h\left(r_{ji}, r_{jk}, \theta_{ijk}\right) + h\left(r_{ki}, r_{kj}, \theta_{ikj}\right)\right] \quad (3.10a)$$

where ε and σ have the same meanings as those in the LJ potential in Equation 3.3, the subscripts i, j, and k indicate the atoms of a system, θ_{jik} is the bond angle between bonds ij and ik, and

$$f_2(r) = \begin{cases} A\left(Br^{-p} - r^{-q}\right)\exp\left[(r-a)^{-1}\right], & r < a \\ 0, & r \geq a \end{cases} \quad (3.10b)$$

$$h\left(r_{ij}, r_{ik}, \theta_{jik}\right) = \lambda \exp\left[\gamma\left(r_{ij} - a\right)^{-1} + \gamma\left(r_{ik} - a\right)^{-1}\right] \times \left(\cos\theta_{jik} + \frac{1}{3}\right)^2 \quad (3.10c)$$

The parameters in this potential are given in Table 3.2.

Table 3.2 Parameters for the Stillinger–Weber Potential for Si

Parameters	Values
ε (eV)	2.16817
σ (Å)	2.0951
A	7.049556277
B	0.6022245584
p	4.0
q	0.0
a	1.8
λ	21.0
γ	1.2

Another potential that works well for Si is the Tersoff potential (Tersoff, 1988a), which is given by

$$U = \sum_{i}\sum_{j>i} f_C(r_{ij})\left[a_{ij}\exp(-\lambda_1 r_{ij}) + b_{ij}\exp(-\lambda_2 r_{ij})\right] \tag{3.11a}$$

with

$$f_C(r) = \begin{cases} 1, & r < R-D \\ \frac{1}{2} - \frac{1}{2}\sin\left[\frac{\pi}{2}(r-R)/D\right], & R-D < r < R+D \\ 0, & r > R+D \end{cases} \tag{3.11b}$$

$$a_{ij} = A\left(1+\alpha^n \eta_{ij}{}^n\right)^{-1/2n};\ \eta_{ij} = \sum_{k\neq i,j} f_C(r_{ik})\exp\left[\lambda_3{}^3(r_{ij}-r_{ik})^3\right] \tag{3.11c}$$

$$b_{ij} = -B\left(1+\beta^n \xi_{ij}{}^n\right)^{-1/2n};\ \xi_{ij} = \sum_{k\neq i,j} f_C(r_{ik})g(\theta_{ijk})\exp\left[\lambda_3{}^3(r_{ij}-r_{ik})^3\right] \tag{3.11d}$$

$$g(\theta) = 1 + c^2/d^2 - c^2/\left[d^2 + (h-\cos\theta)^2\right] \tag{3.11e}$$

where θ_{ijk} is the bond angle between bonds ij and ik. The constants in Equations 3.11a–e are provided in Table 3.3.

Both the Tersoff and Stillinger–Weber potentials can model Si structures very well if periodic boundary conditions (PBCs), which will be discussed later, are

Table 3.3 Parameters for the Tersoff Potential for Si and CNTs/Graphene

Parameters	**Si**	**CNTs/graphene**
R (Å)	2.85	1.95
D (Å)	0.15	0.15
A (eV)	1.8308×10^3	1.3936×10^3
B (eV)	4.7118×10^2	3.4674×10^2
λ_1 (Å^{-1})	2.4799	3.4879
λ_2 (Å^{-1})	1.7322	2.2119
λ_3 (Å^{-1})	1.7322	0.0
α	0.0	0.0
β	1.0999×10^{-6}	1.5724×10^{-7}
n	0.78734	0.72751
c	1.0039×10^5	3.8049×10^4
d	16.218	4.3484
h	−0.59826	−0.57058

used in simulations. However, if a free surface of Si is involved in a simulation, the temperature of the system has to be maintained sufficiently low (~<150 K) to generate a stable surface. At relatively high temperatures, these two potentials cannot offer a stable system with a free Si surface.

3.2.4 Potentials for CNTs and graphene

CNTs and graphene have the same atomic structure. Graphene is a two-dimensional, monolayer material consisting of a hexagonal lattice of sp^2-bonded carbon atoms. A CNT can be thought as a cylindrical structure by rolling a graphene sheet. Modeling CNTs or graphene is to describe the interactions of carbon atoms.

The Tersoff potential in Equation 3.11 has been shown reasonable in reproducing the dynamics of carbon materials with appropriate parameters, which are given in Table 3.3. For CNTs and graphene, the Tersoff potential is a relatively simple model that can offer quite accurate description of the thermal and structural properties of carbon (Tersoff, 1988b; Lindsay and Broido, 2010).

Brenner (1990) proposed a more complex empirical potential for hydrocarbons based on the Tersoff potential. This potential is commonly termed reactive bond order (REBO) potential or the Tersoff–Brenner potential, which contains attractive U_{ij}^{A} and repulsive U_{ij}^{R} contributions,

$$U_{\mathrm{REBO}} = \sum_{i}\sum_{j>i}\left(U_{ij}^{\mathrm{R}} + b_{ij}U_{ij}^{\mathrm{A}}\right) \tag{3.12a}$$

where

$$U_{ij}^{\mathrm{R}} = f_{ij}\left(r_{ij}\right)\left(1+\frac{Q}{r_{ij}}\right)A\exp\left(-\alpha r_{ij}\right) \text{ and } U_{ij}^{\mathrm{A}} = -f_{ij}\left(r_{ij}\right)\sum_{n=1}^{3}B_n\exp\left(-\beta_n r_{ij}\right) \tag{3.12b}$$

with

$$f_{ij}\left(r_{ij}\right)=\begin{cases} 1, & r_{ij} < R_{\min} \\ \dfrac{1}{2}\left\{1+\cos\left[\dfrac{\pi\left(r_{ij}-R_{\min}\right)}{R_{\max}-R_{\min}}\right]\right\}, & R_{\min} < r_{ij} < R_{\max} \\ 0, & r_{ij} > R_{\max} \end{cases} \tag{3.12c}$$

In Equation 3.12a, b_{ij} indicates the bond order for the interaction between atoms i and j, which is given by

$$b_{ij} = \frac{1}{2}\left(b_{ij}^{*} + b_{ji}^{*}\right) + \Pi_{ij} + b_{ji}^{\mathrm{DH}} \tag{3.12d}$$

where

$$b_{ij}^{*}=\left(1+\sum_{k\neq i,j}f_{ij}\left(r_{ij}\right)g_{ijk}\right)^{-1/2}\quad\text{with }g_{ijk}=\sum_{i=0}^{5}\lambda_i\cos^{i}\theta_{ijk}\tag{3.12e}$$

specifies the covalent bond interaction, Π_{ij} represents the bond order due to radical and conjugation effects, which can be ignored in cases, where radical energetics is unimportant, and b_{ji}^{DH} is a dihedral function accounting for a penalty for rotation around multiple bonds, given as

$$b_{ji}^{\mathrm{DH}}=\gamma\sum_{k,l\neq i,j}f_{ik}f_{jl}\left(1-\cos^{2}\Theta_{ijkl}\right)\tag{3.12f}$$

where $\Theta_{ijkl}=\mathbf{e}_{jik}\cdot\mathbf{e}_{ijl}$ is the dihedral angle of atoms i, j, k, and l, and $\mathbf{e}_{jik}$ is the unit vector normal to the triangle formed by atoms j, i, and k. For graphene, $\Theta_{ijkl}=0$ or π and the dihedral term vanishes. The parameters for the REBO potential are listed in Table 3.4.

Table 3.4 Parameters for the REBO and AIREBO Potentials

Parameters	REBO potential	AIREBO potential
A (eV)	1.0953544×10^{4}	1.0953544×10^{4}
B_1 (eV)	1.2388792×10^{4}	1.2388792×10^{4}
B_2 (eV)	1.7567406×10^{1}	1.7567406×10^{1}
B_3 (eV)	3.0714932×10^{1}	3.0714932×10^{1}
α (Å^{-1})	4.7465391	4.7465391
Q (Å)	3.1346029×10^{-1}	3.1346029×10^{-1}
β_1 (Å^{-1})	4.7204523	4.7204523
β_2 (Å^{-1})	1.4332132	1.4332132
β_3 (Å^{-1})	1.3826913	1.3826913
R_{mim} (Å)	1.7	1.7
R_{max} (Å)	2.0	2.0
λ_1 (Å^{-1})	2.4799	2.4799
λ_2 (Å^{-1})	1.7322	1.7322
λ_3 (Å^{-1})	1.7322	1.7322
γ	-8.09675×10^{-3}	-8.09675×10^{-3}
ε (eV)	—	2.84×10^{-3}
σ (Å)	—	3.4
$R_{\mathrm{min}}^{\mathrm{LJ}}$ (Å)	—	σ
$R_{\mathrm{max}}^{\mathrm{LJ}}$ (Å)	—	$2^{1/6}\sigma$
b_{min} (Å)	—	0.77
b_{max} (Å)	—	0.81
ε_{ijkl} (eV)	—	0.3079

The REBO potential was later modified to improve the energetic, elastic, and vibrational properties of hydrocarbons (Brenner et al., 2002). It works well in describing carbon–carbon interactions. Nevertheless, this potential does not include non-bonded (e.g., the van der Waals) and torsional interactions, which make the potential unsuitable for systems with strong intermolecular interactions. To overcome these drawbacks, an adaptive intermolecular reactive bond order (AIREBO) potential was developed by adding an LJ and a torsional interaction to the REBO model (Stuart et al., 2000):

$$U_{\text{AIREBO}} = U_{\text{REBO}} + U_{\text{LJ}} + U_{\text{Tors}} \tag{3.13a}$$

where

$$U_{\text{LJ}}\left(r_{ij}\right) = \left\{1 + f_{ij}'\left(r_{ij}\right)\left[f_{ij}''\left(\bar{b}_{ij}\right) - 1\right]\right\} C_{ij} 4\varepsilon\left[\left(\frac{\sigma}{r}\right)^{12} - \left(\frac{\sigma}{r}\right)^{6}\right] \tag{3.13b}$$

and

$$U_{\text{Tors}}\left(r_{ij}\right) = \frac{1}{2}\sum_{i}\sum_{j\neq i}\sum_{k\neq i,j}\sum_{l\neq i,j,k} f_{ij} f_{jk} f_{kl} \times \varepsilon_{ijkl}\left[\frac{256}{405}\cos^{10}\left(\frac{\Theta_{ijkl}}{2}\right) - \frac{1}{10}\right] \tag{3.13c}$$

In Equation 3.13b, $f_{ij}'\left(r_{ij}\right)$ and $f_{ij}''\left(\bar{b}_{ij}\right)$ are switching functions slightly different from Equation 3.12c and are given as follows:

$$f_{ij}'\left(r_{ij}\right) = \begin{cases} 1, & r_{ij} < R_{\min}^{\text{LJ}} \\ 1 - \left(\dfrac{r_{ij} - R_{\min}^{\text{LJ}}}{R_{\max}^{\text{LJ}} - R_{\min}^{\text{LJ}}}\right)^{2}\left[3 - 2\dfrac{r_{ij} - R_{\min}^{\text{LJ}}}{R_{\max}^{\text{LJ}} - R_{\min}^{\text{LJ}}}\right], & R_{\min}^{\text{LJ}} < r_{ij} < R_{\max}^{\text{LJ}} \\ 0, & r_{ij} > R_{\max}^{\text{LJ}} \end{cases} \tag{3.13d}$$

$$f_{ij}''\left(b_{ij}\right) = \begin{cases} 1, & b_{ij} < b_{\min} \\ 1 - \left(\dfrac{b_{ij} - b_{\min}}{b_{\max} - b_{ij}^{\min}}\right)^{2}\left[3 - 2\dfrac{b_{ij} - b_{\min}}{b_{\max} - b_{\min}}\right], & b_{\min} < b_{ij} < b_{\max} \\ 0, & b_{ij} > b_{\max} \end{cases} \tag{3.13e}$$

and $\bar{b}_{ij} = b_{ij}\,|_{r_{ij}=R_{\min}}$. In Equation 3.13b, C_{ij} works as a switch to turn on the LJ interaction and can be written as follows: $C_{ij} = 1 - \max\left\{f_{ij}, \left[f_{ik} f_{kj}, \forall k\right], \left[f_{ik} f_{kl} f_{lj}, \forall k, l\right]\right\}$, where the symbol $\forall$ means "for all the values of." The extra parameters in the AIREBO potential are provided in Table 3.4.

3.2.5 Potentials for metals

As nanochannels can also be fabricated using metals, the potential functions for metals are necessary in MD simulations. A popular potential for face-centered cubic (fcc) metals, such as Cu, Ag, Au, Ni, Pd, Pt, and their alloys, is the embedded-atom method (EAM) potential (Foiles et al., 1986). This potential has been extended to materials of different structures (Baskes, 1992), including body-centered cubic (bcc) materials, hexagonal close packed (hcp) materials, diamond, and even gaseous materials. The modified EAM (MEAM) potential includes an embedding function and a pair-interaction term,

$$U_{\mathrm{MEAM}} = \sum_i \left[F_i(\bar{\rho}_i) + \frac{1}{2}\sum_{j \neq i} \phi_{ij}(r_{ij}) \right] \tag{3.14}$$

where $\bar{\rho}_i$ is the background electron density. Although the MEAM potential has been widely employed for various metals, it is too complex. Details of this potential can be found in the literature (Baskes, 1992).

Another simple potential for transition metals and alloys is the tight-binding (TB) potential (Cleri and Rosato, 1993), which produces accurate physical properties of fcc transition metals (Li and Wang, 2005). This TB potential can be written as

$$U_{\mathrm{TB}} = \sum_i \left(U_{\mathrm{R}}{}^i + U_{\mathrm{B}}{}^i \right) \tag{3.15a}$$

where $U_{\mathrm{R}}{}^i$ is a repulsive interaction term and $U_{\mathrm{B}}{}^i$ represents the band energy of atom i. $U_{\mathrm{R}}{}^i$ and $U_{\mathrm{B}}{}^i$ are given by

$$U_{\mathrm{R}}{}^i = \sum_j A \exp\left[-p\left(\frac{r_{ij}}{r_0} - 1 \right) \right] \tag{3.15b}$$

and

$$U_{\mathrm{B}}{}^i = -\left\{ \sum_j \xi^2 \exp\left[-2q\left(\frac{r_{ij}}{r_0} - 1 \right) \right] \right\}^{1/2} \tag{3.15c}$$

where r_0 is the first-neighbor distance of the lattice structure. The other parameters of this potential for fcc transition metals can be found in Table 3.5.

3.3 Neighbor lists

The calculation of intermolecular forces is the most time-consuming part of an MD simulation. To calculate the force on a particle (a molecule or an atom), the distances between the particle and its neighboring particles, which lie within a distance r_L of the particle, are needed (Equation 3.2). To save time, a neighbor list for each particle is necessary in MD simulations. The neighbor list of a particle, say, particle i, stores all the particles, j, with $r_{ij} < r_L$. To compute the force on a particle, only the particles in its neighbor list are searched. The distance

Table 3.5 Parameters of the Tight-Binding Potential for fcc Transition Metals

Metals	A (eV)	ξ (eV)	p	q	r_0
Ni	0.0376	1.070	16.999	1.189	2.491
Cu	0.0855	1.224	10.960	2.278	2.556
Rh	0.0629	1.660	18.450	1.867	2.689
Pd	0.1746	1.718	10.867	3.742	2.749
Ag	0.1028	1.178	10.928	3.139	2.889
Ir	0.1156	2.289	16.980	2.691	2.715
Pt	0.2975	2.695	10.612	4.004	2.775
Au	0.2061	1.790	10.229	4.036	2.884
Al	0.1221	1.316	8.612	2.516	2.864
Pb	0.0980	0.914	9.576	3.648	3.501

r_L for setting up neighbor lists can be set to be the cutoff distance, $r_L = r_c$. If so, a neighbor list stores exactly all the neighbor particles, some of which, after one time step, may not be the neighbors of the same particle. This requires the update of the neighbor list at every time step, which is time-consuming. Usually, $r_L > r_c$, such that a neighbor list stores more particles than the "real" neighbors, as illustrated in Figure 3.2. In this case, the neighbor list is not required to be updated at every time step, but needs rebuilt after a certain number of time steps.

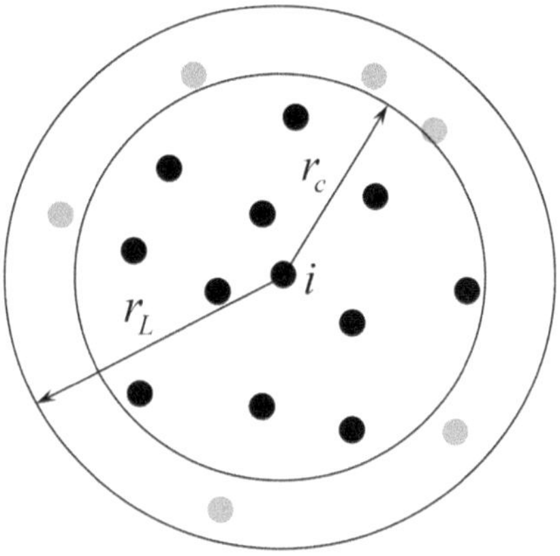

Figure 3.2 Particles for neighbor list construction. All the particles (black and gray circles) in the outer sphere, $r < r_L$, are stored in the neighbor list of particle i, but the particles between the inner and outer spheres (gray circles) are not considered for force calculation.

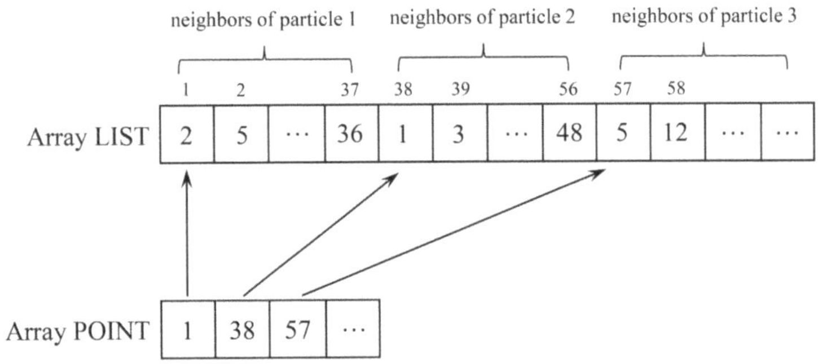

Figure 3.3 Two arrays, LIST and POINT, used in MD simulations to store and retrieve neighbors of all the particles.

There are two widely used ways to create a neighbor list. One is the Verlet neighbor list. This list requires two arrays: one array, say, LIST, stores the neighbors of all the particles sequentially and the other array, say POINT, gives the location of the first neighbor of all the particles in a system, as demonstrated in Figure 3.3.

The Verlet neighbor list is quite effective for small MD systems (containing less than 1,000 particles). For large systems, the Verlet neighbor list is long and can be a burden for data storage. This makes the simulation time-consuming. Another method that performs well for large systems is the cell index method (Allen and Tildesley, 1989). In this method, the simulation system is divided into many cells of equal size. The side length of these cells is larger than the cutoff distance. To compute the force on a particle in a cell, only particles in the neighboring cells are searched. A two-dimensional system is demonstrated in Figure 3.4, where the

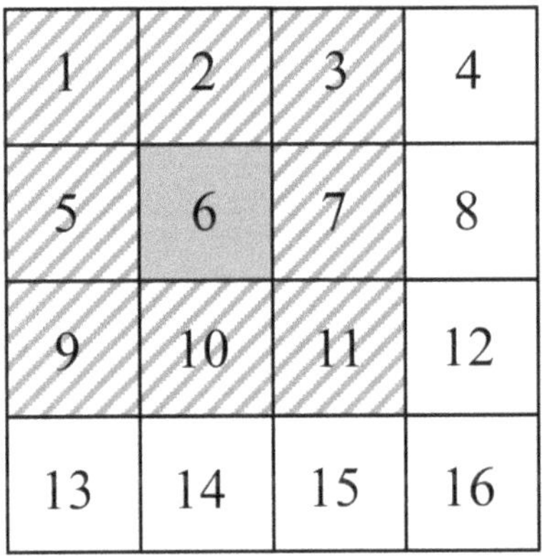

Figure 3.4 Cells of a simulation system for setting up the neighbor list. Particles in cells 1, 2, 3, 5, 7, 9, 10, and 11 are searched for creating the neighbor lists of particles in cell 6.

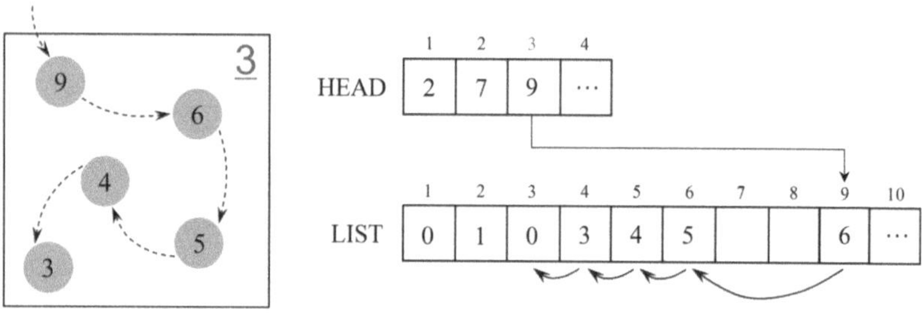

Figure 3.5 Neighbor list setup using the technique of linked-lists. Left panel: particles in cell 3. Right panel: an element in array HEAD indicates the "head" of a particle chain, which is followed by other particles in cell 3 stored in LIST.

simulation system is split into 16 cells. In each cell, there are many MD particles. To calculate the forces on the particles in a cell, say, cell 6, only particles in the neighboring cells, i.e., cells 1, 2, 3, 5, 7, 9, 10, and 11 (Figure 3.4), are checked for setting up the neighbor list. To create the neighbor list for such a cell structure, the technique of linked-lists is used (Allen and Tildesley, 1989). In this approach, particles in the same cell are linked in a "particle chain" with one of the particles as the "head" of the chain. Practically, it is similar to the setup of the Verlet list that two arrays, say, HEAD and LIST, are needed. The "head" particle of a cell is stored in array HEAD, which gives the position of the second particle of the chain in array LIST. In turn, the element of LIST for the second particle is the index of the third particle of the chain, and so on. A zero element of LIST denotes the end of a chain. This linked-list technique is illustrated in Figure 3.5. Let us take cell 3 in Figure 3.4, for example. Suppose cell 3 contains particles 9, 6, 5, 4, and 3. These particles form a chain with particle 9 being the "head" of the chain, which is stored as the third (corresponding to cell 3) element of array HEAD, as shown in Figure 3.5. The element "9" of HEAD gives the location of the second particle of the chain, particle 6, in array LIST. The element "6" of LIST indicates the position of the third particle of the chain, particle 5. Repeating such a tracking method, a zero element in LIST will be found, which is the end of the chain for cell 3.

3.4 Periodic boundary conditions

The size of an MD system in a direction usually varies from a few to tens of nanometers (some systems may reach a few hundreds of nanometers). To study the bulk properties of a fluid or the flow dynamics in a nanofluidic system with one or more dimensions being sufficiently long, PBCs are required. A PBC in a direction means that the simulation system, which can be viewed as the primary unit, duplicates itself in that direction periodically. In this sense, the simulation cell describes a system that is infinitely long in that direction. Figure 3.6a illustrates an MD system, where a fluid is confined in the y-direction by two parallel walls (solid circles). If this system is employed to simulate flows between two parallel plates, which are very long in the x-direction, a PBC can be applied in the x-direction. In this case, the left (AB) and right (CD) boundaries of the simulation system are basically the same. When a particle moves across the right boundary (CD) from the simulation domain, it will enter into the system from the left boundary with the same velocity, as indicated by the arrows in Figure 3.6a. In this sense, the MD system simulates a fluidic system that is infinite in the x-direction because it is composed of the MD system (primary unit) and its replicas (image units), as shown in Figure 3.6a. A two-dimensional case is demonstrated in Figure 3.6b, which can be used to model a system without confinements in the x- and y-directions.

PBCs affect the calculation of the distances of MD particles. If a PBC is used in a direction, the distance between two particles, i and j, in that direction should be the minimum distance between particles i and j as well as the image particles of j. For example, the distance between particles i and j in Figure 3.6b is d_2, instead of d_1. It must be noted that such distances should also be used for creating the neighbor list and calculating intermolecular forces.

3.5 Molecular dynamics algorithms

In MD simulations, the positions and velocities of all the particles are obtained by solving Newton's equation, Equation 3.1, which can be rewritten as

$$\mathbf{F}_i = m_i \frac{d\mathbf{v}_i}{dt} \text{ with } \mathbf{v}_i = \frac{d\mathbf{r}_i}{dt} \tag{3.16}$$

where $\mathbf{v}$ is the velocity vector.

Equation 3.16 can be solved using the straightforward finite difference method. The idea is to use the current ($t = t$) forces, positions, and velocities of MD particles to calculate the positions and velocities at a later time, $t = t + \Delta t$. A popular algorithm is the Verlet algorithm, where the position at time $t = t + \Delta t$ is obtained by combining Talyor's expansions for the position at times $t - \Delta t$ and $t + \Delta t$,

$$\mathbf{r}_i(t+\Delta t) = \mathbf{r}_i(t) + \mathbf{v}_i \Delta t + \frac{1}{2}\frac{d\mathbf{v}_i}{dt}\Delta t^2 + \frac{1}{3!}\frac{d^2\mathbf{v}_i}{dt^2}\Delta t^3 + O\left(\Delta t^4\right) \tag{3.17a}$$

and

$$\mathbf{r}_i(t-\Delta t) = \mathbf{r}_i(t) - \mathbf{v}_i \Delta t + \frac{1}{2}\frac{d\mathbf{v}_i}{dt}\Delta t^2 - \frac{1}{3!}\frac{d^2\mathbf{v}_i}{dt^2}\Delta t^3 + O\left(\Delta t^4\right) \tag{3.17b}$$

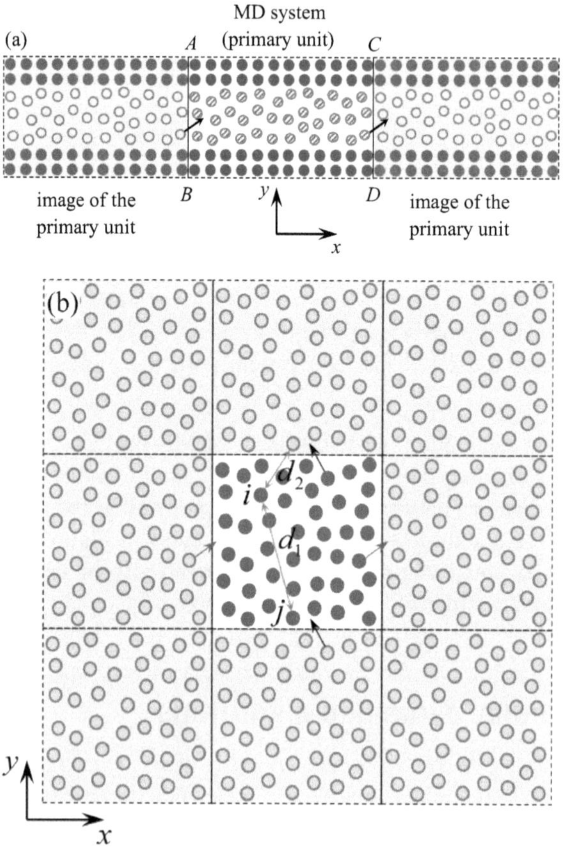

Figure 3.6 Periodic boundary condition. (a) A fluid confined by two parallel plates with a PBC in the *x*-direction. The unit in the middle represents an MD system (primary unit) and the other two units are the images of the MD system. Once a particle leaves the primary unit from the right boundary *CD*, the PBC requires that the particle enters the system from the left boundary with the same velocity, as indicated by the arrows. (b) A two-dimensional example with PBCs in both directions. The unit at the center is an MD system and the others are its image units. The distance between two particles *i* and *j* in the MD system is the minimum distance between particle *i* and particle *j* as well as the images of particle *j*, i.e., d_2.

Adding Equations 3.17a and 3.17b and omitting high-order terms yield

$$\mathbf{r}_i(t+\Delta t) = 2\mathbf{r}_i(t) - \mathbf{r}_i(t-\Delta t) + \frac{d\mathbf{v}_i}{dt}\Delta t^2 \tag{3.18}$$

The velocity at $t = t$ is simply obtained using the first-order central difference estimator,

$$\mathbf{v}_i(t) = \frac{\mathbf{r}_i(t+\Delta t) - \mathbf{r}_i(t-\Delta t)}{2\Delta t} \tag{3.19}$$

The omission of high-order terms in Equation 3.17 causes truncation errors. The error for positions is of order Δt^3 and that for velocities is of order Δt^2, which is the global error. The errors can be reduced by reducing the time step Δt. This, however, will make the algorithm time-consuming.

As a variant of the Verlet algorithm, the Beeman algorithm improves the global accuracy, where the position and velocity vectors are given by

$$\mathbf{r}_i(t+\Delta t) = \mathbf{r}_i(t) + \mathbf{v}_i(t)\Delta t + \left(\Delta t^2/6\right)\left[4\mathbf{a}_i(t) - \mathbf{a}_i(t-\Delta t)\right] \tag{3.20a}$$

$$\mathbf{v}_i(t+\Delta t) = \mathbf{v}_i(t) + \left(\Delta t/6\right)\left[2\mathbf{a}_i(t+\Delta t) + 5\mathbf{a}_i(t) - \mathbf{a}_i(t-\Delta t)\right] \tag{3.20b}$$

where $\mathbf{a}_i$ is the acceleration of particle i. The global error of Beeman's algorithm is of order Δt^3, which is more accurate than the Verlet algorithm.

There are other algorithms, such as predictor–corrector algorithm. The general predictor–corrector algorithm includes prediction, evaluation, and correction steps. First, $\mathbf{r}(t+\Delta t)$, $\mathbf{v}(t+\Delta t)$, and $\mathbf{a}(t+\Delta t)$ are predicted from the current position $\mathbf{r}(t)$, velocity $\mathbf{v}(t)$, and acceleration $\mathbf{a}(t)$. Then the force on each MD particle $\mathbf{F}$ and the acceleration are evaluated from the predicted position. Finally, correct the predicted position, velocity, and acceleration using the new acceleration. There are different options for the prediction and correction steps. Therefore, this method offers great flexibility in solving Newton's equation. In MD simulations, the most widely used predictor–corrector algorithms are those developed by Gear (1971). In these algorithms, positions and their derivatives at time $t = t + \Delta t$ are predicted first based on their values at $t = t$,

$$\mathbf{r}_{i,\mathrm{P}}(t+\Delta t) = \mathbf{r}_i(t) + \dot{\mathbf{r}}_i(t)\Delta t + \ddot{\mathbf{r}}_i(t)\frac{\Delta t^2}{2!} + \mathbf{r}_i^{(\mathrm{iii})}(t)\frac{\Delta t^3}{3!} + \mathbf{r}_i^{(\mathrm{iv})}(t)\frac{\Delta t^4}{4!} + \mathbf{r}_i^{(\mathrm{v})}(t)\frac{\Delta t^5}{5!} \tag{3.21a}$$

$$\dot{\mathbf{r}}_{i,\mathrm{P}}(t+\Delta t) = \dot{\mathbf{r}}_i(t) + \ddot{\mathbf{r}}_i(t)\Delta t + \mathbf{r}_i^{(\mathrm{iii})}(t)\frac{\Delta t^2}{2!} + \mathbf{r}_i^{(\mathrm{iv})}(t)\frac{\Delta t^3}{3!} + \mathbf{r}_i^{(\mathrm{v})}(t)\frac{\Delta t^4}{4!} \tag{3.21b}$$

$$\ddot{\mathbf{r}}_{i,\mathrm{P}}(t+\Delta t) = \ddot{\mathbf{r}}_i(t) + \mathbf{r}_i^{(\mathrm{iii})}(t)\Delta t + \mathbf{r}_i^{(\mathrm{iv})}(t)\frac{\Delta t^2}{2!} + \mathbf{r}_i^{(\mathrm{v})}(t)\frac{\Delta t^3}{3!} \tag{3.21c}$$

$$\mathbf{r}_{i,\mathrm{P}}^{(\mathrm{iii})}(t+\Delta t) = \mathbf{r}_i^{(\mathrm{iii})}(t) + \mathbf{r}_i^{(\mathrm{iv})}(t)\Delta t + \mathbf{r}_i^{(\mathrm{v})}(t)\frac{\Delta t^2}{2!} \tag{3.21d}$$

$$\mathbf{r}_{i,\mathrm{P}}^{(\mathrm{iv})}(t+\Delta t)=\mathbf{r}_i^{(\mathrm{iv})}(t)+\mathbf{r}_i^{(\mathrm{v})}(t)\Delta t \tag{3.21e}$$

$$\mathbf{r}_{i,\mathrm{P}}^{(\mathrm{v})}(t+\Delta t)=\mathbf{r}_i^{(\mathrm{v})}(t) \tag{3.21f}$$

where $\dot{\mathbf{r}}_i$, $\ddot{\mathbf{r}}_i$, $\mathbf{r}_i^{(\mathrm{iii})}$, $\mathbf{r}_i^{(\mathrm{iv})}$, and $\mathbf{r}_i^{(\mathrm{v})}$ are the first- to fifth-order derivatives of the position vector of particle i. Then the force $\mathbf{F}_i(t+\Delta t)$ on each particle at $t+\Delta t$ is evaluated using the predicted positions, $\mathbf{r}_{i,\mathrm{P}}(t+\Delta t)$, through Equation 3.2. The forces at $t+\Delta t$, $\mathbf{F}_i(t+\Delta t)$, lead to updated accelerations, $\ddot{\mathbf{r}}_i(t+\Delta t)$, and the difference between the updated and the predicted accelerations, $\Delta\ddot{\mathbf{r}}_i=\ddot{\mathbf{r}}_i(t+\Delta t)-\ddot{\mathbf{r}}_{i,\mathrm{P}}(t+\Delta t)$, is used to correct the predicted positions and their derivatives in Equation 3.21,

$$\mathbf{r}_i(t+\Delta t)=\mathbf{r}_{i,\mathrm{P}}(t+\Delta t)+\alpha_0\Delta\ddot{\mathbf{r}}_i\frac{\Delta t^2}{2} \tag{3.22a}$$

$$\dot{\mathbf{r}}_i(t+\Delta t)=\dot{\mathbf{r}}_{i,\mathrm{P}}(t+\Delta t)+\alpha_1\Delta\ddot{\mathbf{r}}_i\frac{\Delta t}{2} \tag{3.22b}$$

$$\ddot{\mathbf{r}}_i(t+\Delta t)=\ddot{\mathbf{r}}_{i,\mathrm{P}}(t+\Delta t)+\alpha_2\Delta\ddot{\mathbf{r}}_i \tag{3.22c}$$

$$\mathbf{r}_i^{(\mathrm{iii})}(t+\Delta t)=\mathbf{r}_{i,\mathrm{P}}^{(\mathrm{iii})}(t+\Delta t)+\alpha_3\Delta\ddot{\mathbf{r}}_i\frac{3}{\Delta t} \tag{3.22d}$$

$$\mathbf{r}_i^{(\mathrm{iv})}(t+\Delta t)=\mathbf{r}_{i,\mathrm{P}}^{(\mathrm{iv})}(t+\Delta t)+\alpha_4\Delta\ddot{\mathbf{r}}_i\frac{12}{\Delta t^2} \tag{3.22e}$$

$$\mathbf{r}_i^{(\mathrm{v})}(t+\Delta t)=\mathbf{r}_{i,\mathrm{P}}^{(\mathrm{v})}(t+\Delta t)+\alpha_5\Delta\ddot{\mathbf{r}}_i\frac{60}{\Delta t^3} \tag{3.22f}$$

The values of corrector coefficient α_k for the highest order of position derivatives, q, are given in Table 3.6. Compared with the Verlet and Beeman algorithms, Gear's method is more complex and time-consuming. However, if rotational motions of molecules are involved, Gear's algorithm appears to be more convenient (Allen and Tildesley, 1989).

Finally, it is worth mentioning that the time step Δt in all the algorithms has to be sufficiently small to guarantee the stability of the algorithms. In most MD simulations, $\Delta t \sim 1$ fs is appropriate. Small Δt can make simulations time-consuming, while large Δt may cause large numerical errors or instability. More detailed information about the time step is given in Section 3.8.2.

3.6 Units in MD simulations

As the size and mass of a molecule or an atom are much smaller than those of a macroscale object, the units typically used in a macroscale system (e.g., meter for

Table 3.6 Corrector Coefficient α_k in Equation 3.22 for Different q Values

α_k	$q = 3$	$q = 4$	$q = 5$
α_0	$\frac{1}{6}$	$\frac{19}{120}$	$\frac{3}{16}$
α_1	$\frac{5}{6}$	$\frac{3}{4}$	$\frac{251}{360}$
α_2	1	1	1
α_3	$\frac{1}{3}$	$\frac{1}{2}$	$\frac{11}{18}$
α_4	—	$\frac{1}{12}$	$\frac{1}{6}$
α_5	—	—	$\frac{1}{60}$

length and kilogram for mass) are inappropriate for MD simulations. If length (L), mass (M), and energy (E) are chosen to be the primary units, 1 Å, the mass of a fluid molecule or a solid atom, and 1 eV can be set as their unit values in MD simulations. All the other units can be obtained as a combination of these three primary units. For example, if water flows in a silver (Ag) nanochannel are studied through MD simulations and the mass of an Ag atom is used as the unit value of mass, then the unit value of time, $t_{\text{MD}} = 1$, in the simulations is equal to 0.1057 ps, which is given by $t \doteq L\sqrt{M/E} = 1\times10^{-10}\ (\text{m})\sqrt{0.107878/6.023\times10^{23}\ (\text{kg})/1.602177\times10^{-19}\ (\text{J})} =$ 0.1057 ps (the symbol $\doteq$ is used to denote "unit"). The values of the other variables in this case are summarized in Table 3.7.

3.7 Equilibria and ensembles

Literally, equilibrium means "equal balance." It describes a state of a system. If a system does not experience volume expansion or contraction, the system is in mechanical equilibrium. Thermal equilibrium is assumed if the temperature of a system is constant. If chemical reactions are involved in a system and all the reactions have taken place, the system is considered in chemical equilibrium. Once all these equilibria hold, a system is considered in thermodynamic equilibrium.

If an MD system is investigated after the system reaches the thermodynamic equilibrium, the MD simulation is defined as an equilibrium MD (EMD) simulation. Since many studies are performed in equilibrium states, MD simulations usually mean EMD simulations. However, if an MD system is considered under external forces or perturbations such that an equilibrium state does not hold, the MD simulation is referred to as a nonequilibrium MD (NEMD) simulation.

Table 3.7 Values of Various Variables in MD Simulations (1 Å, 1 eV, and the Mass of an Ag Atom are Chosen as the Units of Length, Energy, and Mass, respectively)

Variables	Dimension	MD Value	SI Unit Value
Length	L	1.0	1×10^{-10} m
Mass	M	1.0	1.7911×10^{-25} kg
Energy	E	1.0	1.6022×10^{-19} J
Time	$L\sqrt{M/E}$	1.0	1.057×10^{-13} s
Volume	L^3	1.0	1×10^{-30} m^3
Force	E/L	1.0	1.6022×10^{-9} N
Pressure	E/L^3	1.0	1.6022×10^{11} Pa
Temperature	E/k_B	1.0	1.1605×10^{4} K
Velocity	$\sqrt{E/M}$	1.0	9.4579×10^{2} m^2/s

"Ensemble" is a statistical mechanics term. An ensemble is a collection of systems, which are thermodynamically similar to a system of interest. The information of the systems in an ensemble is used to obtain the mean properties of the system under consideration. Depending on how a system interacts with its environment, the corresponding ensemble can be a microcanonical, canonical, or grand canonical ensemble.

For an isolated system, there is no mass and heat transfer between the system and its environment. The number of particles N, volume $\mathcal{V}$, and total energy E of the system are constant. The corresponding ensemble, i.e., an assembly of similar isolated systems, is called a microcanonical ensemble. Since N, $\mathcal{V}$, and E are constant for isolated systems, a microcanonical ensemble is also called an (N, $\mathcal{V}$, E) ensemble.

A closed, isothermal system, which can be maintained by bringing a closed system in contact with a heat reservoir, has constant number of particles N and temperature T. A collection of such systems forms a canonical ensemble. If the volume $\mathcal{V}$ is assumed to be constant, a canonical ensemble is also referred to as an (N, $\mathcal{V}$, T) ensemble.

An open, isothermal system is characterized by a constant chemical potential ς and temperature T. An example is a fluid in contact with its vapor at a constant temperature. The corresponding ensemble is termed grand canonical ensemble or (ς, $\mathcal{V}$, T) ensemble.

Based on the definition of equilibria and statistical ensembles, an EMD simulation can be performed in an (N, $\mathcal{V}$, E) ensemble if the system is isolated. If the temperature of an MD system is controlled by a thermostat, the simulation is conducted in an (N, $\mathcal{V}$, T) ensemble.

3.8 Major components of MD simulations

With the basic information introduced in the previous sections, now it is ready to start MD simulations. A simple MD simulation should contain at least three parts: initialization, equilibration, and production. Initialization is responsible for placing MD particles and assigning velocities to the particles. In equilibration, an MD system is relaxed to reach an appropriate state ready for data collection, which is conducted in the production part.

3.8.1 Initialization

A typical nanofluidic system contains a fluid confined by solid walls. An example is shown in Figure 3.7, where water is confined by two parallel plates. For the solid walls, a crystalline structure is usually employed, depending on the wall material (if amorphous structures are necessary, additional simulations are needed, which will not be discussed here). The most frequently used crystalline structures are fcc, bcc, hcp, and diamond cubic structures. Once the solid material and the corresponding structure are chosen, solid atoms are placed on the lattice sites of the crystalline structure. The lattice constant, i.e., the dimension of the unit cell of the crystalline structure, should be consistent with the bulk density of the material. If a PBC is used in a direction, the length of the wall in that direction should be multiples of the lattice constant. Otherwise, when atoms go across the periodic boundary, certain molecular distances r_{ij} may be too short, which will generate extremely large molecular forces and destroy the MD system.

Fluid molecules can be placed randomly in the space confined by the walls if the minimum distance among fluid molecules can be controlled. Nevertheless, to avoid the occurrence of unphysical intermolecular forces due to small molecular distances, fluid molecules can also be positioned on certain lattice sites. The initial positions of fluid molecules, after equilibration, will not affect the flow properties in the production part. Figure 3.8 shows an initial state of the system in Figure 3.7.

After setting the positions of MD particles, both solid atoms and fluid molecules, a velocity is to be assigned to each particle. Velocities can be generated by a random number generator such that the velocities follow the Maxwellian distribution. The magnitudes of the velocities are determined by the temperature of the system through

$$\frac{3}{2}Nk_{\mathrm{B}}T = \frac{1}{2}\sum_{i=1}^{N} m_i v_i^{\,2} \tag{3.23}$$

where N is the total number of solid atoms or fluid molecules and v_i is the velocity of particle i in a specific direction, x-, y-, or z-directions. Practically, it is

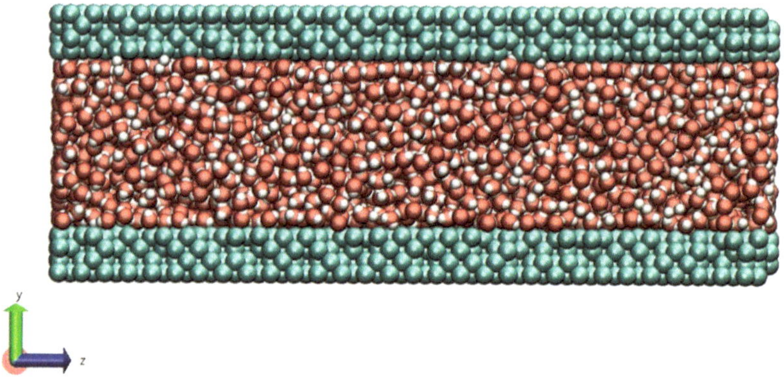

Figure 3.7 An example of an MD system: water confined between two parallel silver plates. Cyan, carmine, and white particles are silver, oxygen, and hydrogen atoms, respectively.

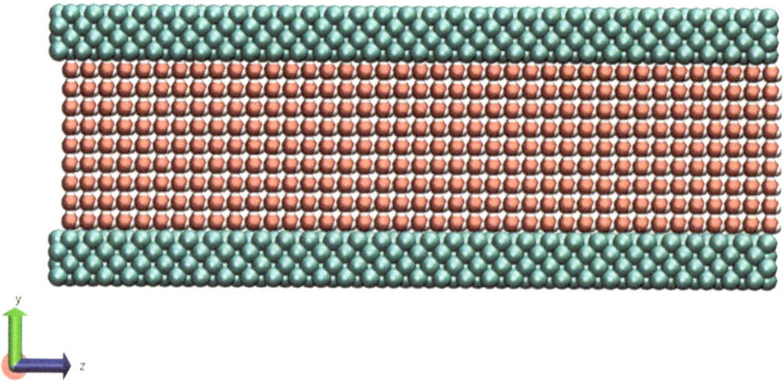

Figure 3.8 An initial state of the MD system in Figure 3.7. Silver atoms are on fcc lattice sites and water molecules are on cubic lattice sites.

performed by giving random velocities to all the fluid molecules and then scaling the velocities to satisfy Equation 3.23. As N is finite and the velocities are generated by a random number generator, the mean velocity of the wall atoms or the fluid molecules usually is not equal to zero. This will cause the fluid molecules or the wall atoms to drift in a specific direction, which is undesirable. Therefore, the last step of velocity initialization is to remove the net momentum of fluid molecules and wall atoms,

$$v_i^{\text{new}} = v_i - \frac{1}{N}\sum_{i=1}^{N} v_i \tag{3.24}$$

3.8.2 Equilibration

After the initialization, it is necessary to let the system relax to reach equilibrium. This can be performed by just running the simulation for a short time, ranging from tens of to hundreds of picoseconds, depending on the size of the system. If the initial state of the system significantly deviates from equilibrium, the velocities of the wall atoms and fluid molecules may need to be scaled frequently at the beginning stage of equilibration to maintain the temperature of the system at a desired value. During equilibration, fluid molecules depart from their initial positions, move under intermolecular forces, and finally reach a state, where the temperature, pressure, and total energy fluctuate about constant values. Figure 3.9 plots the kinetic, potential, and total energies of the system in Figure 3.7 as a function of time. It can be seen that the system reaches equilibrium after ~50 ps.

For an NEMD simulation, where the system may be subject to external forces or there are heat transfer processes, equilibration usually takes longer time as the external loads need to be applied to the system gradually, after which further relaxation is also required until a steady state is reached.

It must be noted that the time step Δt in MD simulations plays an important role. It affects the stability of the MD algorithm and determines the real simulation time. The stability requires Δt to be sufficiently small, while a large Δt is favored to shorten the simulation time. Therefore, there is a tradeoff between the stability of the system and the simulation time. The process of equilibration can be used to find a suitable time step by trial and error. Initially, a reasonable time step, say, $\Delta t \sim 1$ fs, can be used. In equilibration, the total energy of the system is monitored. If the total energy converges to a constant quickly, the time step can be

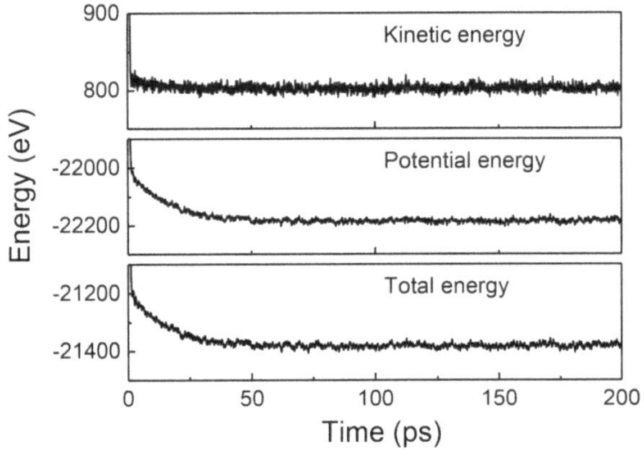

Figure 3.9 Kinetic, potential, and total energies as a function of time for the MD system in Figure 3.7.

increased. Repeating equilibration with increasing time step till the total energy diverges can determine the upper limit of the time step Δt^*. A proper time step should be considerably smaller than Δt^* to prevent numerical instabilities. For water flows in metal nanochannels at room temperature, $\Delta t \sim 1$ fs is quite safe. The time step is also strongly related to the temperature of the system. A small Δt should be chosen if the temperature is relatively high. This is because fluid molecules move fast at a high temperature. If Δt is large, molecules may approach each other closely, which can cause unphysically large intermolecular forces and consequently destroy the MD system.

3.8.3 Production

Once an MD system reaches an equilibrium or a steady state, the simulation enters the production part, where data collection and property calculations are performed. In the production part, the system is supposed to reach a situation of interest and the data are reliable. In many cases, particle positions and velocities need to be saved for further analyses after the simulation is completed. They can be used to compute static and dynamic properties of the fluid, which will be discussed in the next section. Some properties, such as temperature and pressure, can be computed instantaneously for monitoring the system. Depending on what problems are investigated, the physical time of the production part may vary from a few to hundreds of nanoseconds. For example, if the dynamics of solid nanoparticles in a fluid is of interest and the diffusion coefficient of the particles is to be determined, the simulation usually should be run for a very long time period. Sometimes, a series of simulations starting from different initial conditions are required. If a nanoscale Poiseuillie flow is studied and the velocity profile is the major concern, the simulation can complete very quickly since only the velocities of the fluid molecules in a very short time period are needed.

3.9 Property calculation

Studying nanoscale flows through MD simulations requires the determination of flow characteristics, such as velocity profile, flow resistance, and the thermodynamics properties of fluids (e.g., temperature and pressure). An MD simulation can directly provide the potential energy, kinetic energy, positions, and velocities of fluid molecules. These quantities can be used to compute various flow quantities.

3.9.1 Temperature

By using the thermal velocities of molecules/atoms, the temperature of an MD system (or a part of the system) can be calculated as follows:

$$T = \frac{1}{(3N - N_0)k_B} \sum_{i=1}^{N} m_i v_i^2 \tag{3.25}$$

where N is the total number of molecules involved and N_0 is the total number of internal and global constraints. For a monatomic fluid at rest, i.e., no net momentum in all the directions, $N_0 = 3$. If a system has a large number of molecules/atoms, $N \gg N_0$, the temperature can be computed using Equation 3.23,

$$T = \frac{1}{3Nk_B} \sum_{i=1}^{N} m_i v_i^2 \tag{3.26}$$

In an (N, $\mathcal{V}$, T) system, the temperature is controlled by a thermostat. The Brendsen thermostat is a simple temperature regulator, which scales the velocities of fluid molecules by a factor (Brendsen et al., 1984).

$$\chi = \sqrt{1 + \frac{\Delta t}{t^*}\left(\frac{T^*}{T} - 1\right)} \tag{3.27}$$

where T^* is the desired temperature and T is the current temperature, which can be computed using Equation 3.25, and t^* is a time constant. Although the value of t^* can be varied to tune the strength of the thermostat, $t^* = 0.4$ ps is considered as an appropriate value and is widely used. The Brendsen thermostat is an effective but strong temperature regulator as it changes the velocities directly. If a gentle thermostat is necessary, the Nosé–Hoover thermostat is a good choice. The Nosé–Hoover thermostat is also coherent with the canonical ensemble conditions. It modifies the equation of motion slightly,

$$m_i \frac{d\mathbf{v}_i}{dt} = \mathbf{F}_i - \zeta m_i \mathbf{v}_i \tag{3.28}$$

where ζ is a friction parameter governed by

$$\frac{d\zeta}{dt} = \frac{f k_B}{Q}\left(T - T^*\right) \tag{3.29}$$

where f is the number of degrees of freedom, i.e., $f = 3N - N_0$, and Q is the thermal inertial parameter, which can be optimized by trial and error. Compared with the Brendsen thermostat, the Nosé–Hoover thermostat is relatively time-consuming. For systems containing a large number of particles, the Brendsen thermostat works well and does not bring in significant errors (Liu and Li, 2010).

To simulate a fluid in motion, an external force is required to generate a fluid flow and the corresponding MD system is an NEMD system. In this case, the peculiar velocities, instead of the absolute velocities, of fluid molecules should be used in calculating the temperature T in Equations 3.25:

$$T = \frac{1}{3Nk_B} \sum_{i=1}^{N} m_i \left(v_i - \bar{V} \right)^2 \tag{3.30}$$

where $\bar{V}$ is the mean velocity of the flow. If the temperature is nonuniform in the flow field, fluid molecules should be divided into appropriate groups or bins and the temperature T should be computed for each bin based on the peculiar velocities of the fluid molecules in the bin,

$$T_k = \frac{1}{3N_k k_B} \sum_{i=1}^{N_k} m_i \left(v_i - \bar{V}_k \right)^2 \tag{3.31}$$

where N_k and $\bar{V}_k$ are the number of molecules and mean velocity of the kth bin, respectively.

A thermostat regulates the temperature by changing the velocities or kinetic energies of fluid molecules. This is equivalent to transferring energy to or taking energy away from the system. Therefore, thermostats can only be used in (N, $\mathcal{V}$, T) systems. In (N, $\mathcal{V}$, E) systems, thermostats cannot be applied because the total energy of the system is constant and energy transfer is not allowed. The temperature of an (N, $\mathcal{V}$, E) system can be obtained using Equation 3.25.

3.9.2 Pressure

By using the virial equation of state, the pressure of a fluid can be computed as follows:

$$P = nk_B T + \frac{1}{3\mathcal{V}} \sum_{i} \sum_{j>i} r_{ij} F_{ij} \tag{3.32}$$

where n and $\mathcal{V}$ are the number density and volume of the fluid, respectively. The two terms on the right-hand side of Equation 3.32 represent the kinetic and potential components of the pressure. It should be noted that Equation 3.32 is valid for systems with PBCs. For cases without PBCs (e.g., the pressure in certain volume or the pressure of a fluid confined by surfaces), the pressure evaluation should consider the intermolecular forces between fluid molecules and the surrounding molecules/atoms

$$P = nk_B T + \frac{1}{3\mathcal{V}} \left(\sum_{i} \sum_{j>i} r_{ij} F_{ij} + \sum_{i} \sum_{m} r_{im} F_{im} \right), \tag{3.33}$$

where i and $j \in \mathcal{V}$ and m represents a neighbor molecule of i outside of $\mathcal{V}$. Particle m can be a fluid molecule or a surface atom.

3.9.3 Time-correlation function

For a time-dependent function $A(t)$, the correlation function of A is defined as

$$C(\tau) = \langle A(t)A(t+\tau)\rangle \tag{3.34}$$

where $\langle\cdot\rangle$ denotes the ensemble average. A time-correlation function fundamentally measures how close a function is at $t = t$ to its value at a later time $t = t + \tau$. In MD simulations, force and velocity correlation functions are the most important quantities that are related to the transport properties of fluid molecules and suspension particles. If $F(t)$ is the force acting on a fluid molecule in an MD system in equilibrium, the correlation function of $F(t)$ is given as

$$C_{\mathrm{F}}(\tau) = \langle F(t)F(t+\tau)\rangle_0 \tag{3.35}$$

where the subscript "0" denotes equilibrium situation. It can be shown that the friction coefficient α, defined in Equation 1.11b, can be expressed as (Reif, 1985)

$$\alpha = \frac{1}{2k_{\mathrm{B}}T}\int_{-\infty}^{\infty}\langle F(0)F(\tau)\rangle_0 \, d\tau \tag{3.36}$$

It must be noted that Equation 3.36 is valid if Equation 1.11b holds, i.e., the drag force is a linear function of velocity, which is the case in most situations. In certain special cases, for instance, small nanoparticles moving in a rarefied gas at an extremely fast speed, the drag force may depend on higher-order terms of velocity, e.g., $F_{\mathrm{D}} \propto V^2$, and Equation 3.36 may not be accurate.

The velocity correlation function is related to the MSD through (Reif, 1985)

$$\langle \Delta r^2(t)\rangle = 2\int_0^t \langle v(0)v(s)\rangle (t-s)ds \tag{3.37}$$

The diffusion coefficient of fluid molecules can then be obtained using Equation 1.9.

3.9.4 Stress tensor

The stress tensor **J** is a 3×3 matrix for three-dimensional systems. It directly relates to the pressure of a system. Each element of **J** contains a kinetic and a potential contribution

$$J_{\alpha\beta} = m\sum_i v_i^{\alpha} v_i^{\beta} + \sum_i \sum_{j>i} r_{ij}^{\beta} F_{ij}^{\alpha}, \ \alpha,\beta = 1,2,3 \tag{3.38}$$

where α and β denote different directions in the Cartesian coordinates. Using Equations 3.23 and 3.32, it is obtained that

$$P = \frac{1}{3V} \sum_{\alpha=1}^{3} J_{\alpha\alpha} \tag{3.39}$$

The stress tensor can also be used to compute the viscosity of a fluid. According to the Green–Kubo relation, the viscosity of a homogeneous fluid in MD simulations can be determined as follows:

$$\mu = \frac{1}{V k_B T} \int_0^{\infty} \langle J_{xy}(0) J_{xy}(s) \rangle ds \tag{3.40}$$

Again, this equation is only valid for homogeneous systems subject to PBCs. If PBCs are not employed, the viscosity can be obtained based on its definition, Equation 1.2, by calculating shearing stresses in a flow.

3.10 Introduction to LAMMPS

The Large-scale Atomic/Molecular Massively Parallel Simulator, or LAMMPS, is a software package developed by Sandia National Laboratories. It is a free-to-use and open-source code that can conduct atomic, mesoscale, and Monte Carlo simulations. Predominantly, LAMMPS has been widely used to perform large-scale MD simulations in nanoscience and nanotechnologies, which involve most of the scientific fields, including chemistry, physics, biology, materials science, and mechanics. LAMMPS is integrated with rich force fields, which are capable of modeling a wide range of gases, liquids, and solids. LAMMPS can be used to simulate the dynamics of a variety of particles, such as atoms, coarse-grained particles, polymers, biomolecules, and granular materials. The latest version of LAMMPS is written in C++ and can be easily modified or extended with new constraints, force fields, and boundary conditions to meet user's requirements.

A notable feature of LAMMPS is the diverse force fields that can accommodate a great number of particles and models. The force fields integrated in LAMMPS range from simple pairwise potentials to complex molecular interactions for biological calculations. Almost all the potentials introduced in this Chapter can be found in LAMMPS. LAMMPS can model relatively simple, nanoscale water and other fluid flows. It can also study the dynamics of polymers and biomolecules (e.g., DNA and proteins) with the well-known CHARMM (chemistry at Harvard macromolecular mechanics), COMPASS (condensed-phase optimized molecular potentials for atomistic simulation studies), AMBER (assisted model building with energy refinement), and OPLS (optimized potential for liquid simulations) force fields. In addition to MD simulations, LAMMPS can also perform mesoscale simulations, such as DPD (dissipative particle dynamics) and DSMC (direct simulation Monte Carlo).

Another important attribute of LAMMPS is its good parallel performance. LAMMPS can run on a single processor but more generally in parallel using distributed-memory message-passing paradigm. It adopts spatial-decomposition of the simulation domain for parallelism. The efficient parallel performance of LAMMPS makes it possible to simulate systems with up to millions or billions of particles.

LAMMPS can perform $(N, \mathcal{V}, E)$, $(N, \mathcal{V}, T)$, (N, P, T) (isothermal–isobaric), and (N, P, H) (isoenthalpic–isobaric) ensembles. It can selectively apply thermostat to atoms/molecules in specific groups or regions in one to three dimensions. It can also perform both EMD and NEMD simulations. In addition, LAMMPS has other specialized but nontypical MD features, such as QM/MM (quantum mechanics/ molecular mechanics) coupling, targeted and steered MD, and DSMC. Detailed information about LAMMPS, including user's manual, tutorials, commands, and examples, can be found on LAMMPS web site: http://lammps.sandia.gov/.

LAMMPS is strong in building up diverse systems and tracking the motions of atoms/molecules, which are essential in computing various properties. However, LAMMPS is weak in pre- and post-processing. Users may need to employ other tools or develop independent codes for data analysis.

To perform MD simulations with LAMMPS, the first step is to download and install LAMMPS (http://lammps.sandia.gov/download.html#download). Both Microsoft Windows and Linux OS versions are available. Before setting up and running a system, users are strongly suggested to read through LAMMPS User's Manual to be familiar with LAMMPS commands (http://lammps.sandia.gov/doc/ Section_commands.html). The second step is to develop input files. A typical LAMMPS input file contains four parts: initialization, atom definition, settings, and running. Initialization defines basic parameters, such as units, dimension, boundary, and atom type. Atom definition is a step to create atoms on specific lattice sites. Atoms can also be read from a data file. In settings, parameters in the force field, constraints, ensembles, integrator, and output options are set. Finally, the system is executed. Herein, the input file of a Poiseuille flow in the examples of LAMMPS package is used for demonstration (a line starting with "#" is a comment line and the words after "/" are explanations).

```
# 2-d LJ Poiseuille flow simulation

dimension      2          /define a two-dimensional system
boundary       p s p      /set boundary conditions in the
                          x, y, z directions, p denotes
                          PBC and s denotes non-periodic
                          "shrink-wrapped" boundary.
atom_style     atomic     /set atom style
```

```
neighbor       0.3 bin      /select the style of
                            neighbor list
neigh_modify   delay 5      /set neighbor list
                            update frequency

# create geometry

lattice      hex 0.7      /set the lattice constant
region     box block 0 20 0 10 -0.25 0.25   /define a region
create_box    3 box        /create simulation box based on
                           the region
create_atoms 1 box         /create atoms in the box
mass         1 1.0         /set atom mass for atom type 1
mass         2 1.0         /set atom mass for atom type 2
mass         3 1.0         /set atom mass for atom type 3

# LJ potential

pair_style   lj/cut 1.12246       /set pairwise potential
                                  and the cutoff distance
pair_coeff   * * 1.0 1.0 1.12246  /set LJ parameters

# define groups

region       1 block INF INF INF 1.25 INF INF  /define block
                                               region 1
group        lower region 1                    /define atoms
                                               in block
                                               region 1 as
                                               group lower
region       2 block INF INF 8.75 INF INF INF  /define block
                                               region 2
group        upper region 2    /define atoms in block region
                               2 as group upper
group        boundary union lower upper     /group lower and
                                            group upper are
                                            united as group
                                            boundary
group        flow subtract all boundary   /define group flow
set          group lower type 2       /atom type in group
                                      lower is set to type 2
set          group upper type 3       /atom type in group
                                      upper is set to type 3

# initial velocities

compute      mobile flow temp   /compute the temperature of
                                group flow
```

```
velocity   flow create 1.0 482748 temp mobile  /velocity
                                                initialization
                                                for atoms in
                                                group flow
fix        1 all nve          /set a fix to employ NVE
                              integrator
fix        2 flow temp/rescale 200 1.0 1.0 0.02 1.0 /temperature
                                                 control by
                                                 scaling atom
                                                 velocities
fix_modify   2 temp mobile     /use the temperature calculated
                               for thermostat

# Poiseuille flow

velocity     boundary set 0.0 0.0 0.0     /set the initial
                                           velocity of group
                                           boundary to be zero
fix        3 lower setforce 0.0 0.0 0.0   /set a fix that
                                          removes the force on
                                          group lower
fix        4 upper setforce 0.0 NULL 0.0  /set a fix that
                                          removes the force on
                                          group upper in the x
                                          and z
                                          directions
fix        5 upper aveforce 0.0 -1.0 0.0  /set a fix that adds
                                          a force on group
                                          upper in the y
                                          direction
fix        6 flow addforce 0.5 0.0 0.0     /set a fix that
                                           adds a force on
                                           group flow in the x
                                           direction
fix        7 all enforce2d      /zero out z-direction velocity
                                and force

# Run

timestep 0.003            /set the time step
thermos  500              /set the frequency to output
                          thermodynamic data
thermo_modify   temp mobile       /specify the temperature
dump       1 all atom 500 dump.flow       /output atom
                                           positions to file
                                           "dump.flow"
run        10000         / set the total simulation steps and
                         execute the simulation
```

If the above input file is successfully executed, the following output will appear:

```
LAMMPS (17 Jun 2013)
Lattice spacing in x,y,z = 1.28436 2.22457 1.28436
Created orthogonal box = (0 0 -0.321089) to (25.6871
22.2457 0.321089)
   1 by 1 by 1 MPI processor grid
Created 420 atoms
60 atoms in group lower
60 atoms in group upper
120 atoms in group boundary
300 atoms in group flow
Setting atom values ...
   60 settings made for type
Setting atom values ...
   60 settings made for type
WARNING: Temperature for thermo pressure is not for group
all (../thermo.cpp:436)
Setting up run ...
Memory usage per processor = 2.07994 Mbytes
Step Temp E_pair E_mol TotEng Press Volume
   0         1     0       0 0.71190476  0.52314537  571.54286
 500 1.0990751 -0.36396768 0 0.41846911  2.522517  575.55806
1000        1  -0.34609791 0 0.36580685  1.9736175  584.67179
1500 1.2112861  -0.39062708  0 0.47169327 1.521881  587.24778
..........................
..........................
 8500      1.2245982 -0.73461969    0 0.13717758   2.4482542
506.2679
 9000          1 -0.697338          0 0.014566762  2.4757604
503.86953
 9500      1.1882    -0.66622491    0 0.17966032   2.3295348
505.67684
 10000     1 -0.70364194            0 0.008262826  2.218568
507.64681
Loop time of 0.221781 on 1 procs for 10000 steps with 420
atoms

Pair  time (%) = 0.0959458 (43.2615)
Neigh time (%) = 0.0247238 (11.1478)
Comm  time (%) = 0.00443673 (2.0005)
Outpt time (%) = 0.00601673 (2.71292)
Other time (%) = 0.0906579 (40.8772)

Nlocal:    420 ave 420 max 420 min
Histogram: 1 0 0 0 0 0 0 0 0 0
Nghost:    50 ave 50 max 50 min
```

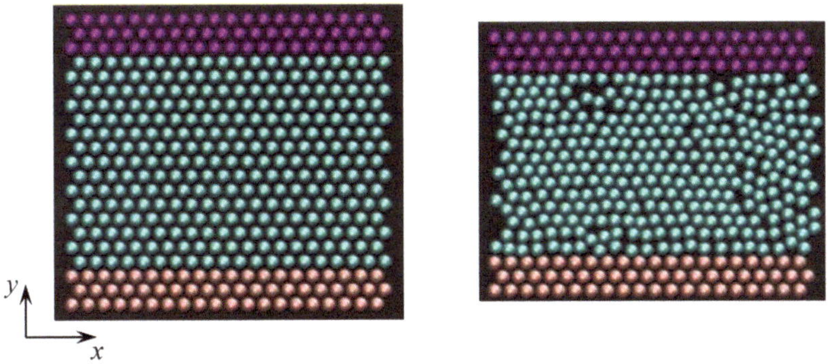

Figure 3.10 Snapshots of a 2D-Poiseuille flow system. Left panel: initial state. Right panel: final state. Notes: As an external force is added to the upper wall in the *y*-direction to control the pressure of the system, the channel heights are different for the initial and final states.

```
Histogram: 1 0 0 0 0 0 0 0 0 0
Neighs:    1116 ave 1116 max 1116 min
Histogram: 1 0 0 0 0 0 0 0 0 0

Total # of neighbors = 1116
Ave neighs/atom = 2.65714
Neighbor list builds = 563
Dangerous builds = 0
```

With the data in file "dump.flow", the Poiseuille flow system can be visualized using a visualization software package, such as VMD, as shown in Figure 3.10.

References

Allen, M.P. and Tildesley, D.J. 1989. *Computer Simulation of Liquids*. Oxford: Oxford University Press.

Baskes, M.I. 1992. Modified embedded-atom potentials for cubic materials and impurities. *Phys. Rev. B* 46:2727–2742.

Berendsen, H.J.C., Postma, J.P.M., van Gunsterrn, W.F., and Hermans, J. 1981. *Intermolecular Forces,* pp. 331. Edited by Pullman, B. Dordrecht: Reidel.

Brendsen, H.J.C. and Van Gunsteren, W.F. 1984. Molecular liquids: Dynamics and interaction. Edited by Barnes, A.J., Orville-Thomas, W.J., and Yarwood, J. *NATO ASI series C135*. New York: Reidel.

Berendsen, H.J.C., Grigera, J.R., and Straatsma, T.P. 1987. The missing term in effective pair potentials. *J. Phys. Chem.* 91:6269–6271.

Brenner, D.W. 1990. Empirical potential for hydrocarbons for use in simulating the chemical vapor deposition of diamond films. *Phys. Rev. B* 42:9458–9471.

Brenner, D.W., Shenderova, O.A., Harrison, J.A., Stuart, S.J., Ni, B., and Sinnott, S.B. 2002. A second-generation reactive empirical bond order (REBO) potential energy expression for hydrocarbons. *J. Phys.: Condens. Matter* 14:783–802.

Cleri, F. and Rosato, V. 1993. Tight-binding potentials for transition metals and alloys. *Phys. Rev. B* 48:22–33.

Foiles, S.M., Baskes, M.I., and Daw, M.S. 1986. Embedded-atom-method functions for the fcc metals Cu, Ag, Au, Ni, Pd, Pt, and their alloys. *Phys. Rev. B* 33:7983–7991.

Gear, C.W. 1971. *Numerical Initial Value Problems in Ordinary Differential equations*. Englewood Cliffs: Prentice-Hall.

Hippler, H., Troe, J., and Wendelken, H.J. 1983. Collisional deactivation of vibrationally highly excited polyatomic molecules. II. Direct observations for excited toluene. *J. Chem. Phys.* 78:6709–6717.

Jorgensen, W.L., Chandrasekhar, J., Madura, J.D., Impey, R.W., and Klein, M.L. 1983. Comparison of simple potential functions for simulating liquid water. *J. Chem. Phys.* 79:926–935.

Lennard-Jones, J.E. 1924. On the determination of molecular fields. *Proc. R. Soc. Lond. A* 106:463–477.

Mark, P. and Nilsson, L. 2001. Structure and dynamics of the TIP3P, SPC, and SPC/E water models at 298 K. *J. Phys. Chem.* 105:9954–9960.

Li, Z.G. and Wang, H. 2005. Gas-nanoparticle scattering: A molecular view of momentum accommodation function. *Phys. Rev. Lett.* 95:014502.

Li, C., Huang, J.Z., and Li, Z.G. 2016. A relation for nanodroplet diffusion on smooth surfaces. *Sci. Rep.* 6:26488.

Lindsay, L. and Broido, D.A. 2010. Optimized Tersoff and Brenner empirical potential parameters for lattice dynamics and phonon thermal transport in carbon nanotubes and graphene. *Phys. Rev. B* 81:205441.

Liu, C. and Li, Z.G. 2010. Surface effects on nanoscale Poiseuille flows under large driving force. *J. Chem. Phys.* 132:024507.

Mills, R. 1973. Self-diffusion in normal and heavy water in the range 1°–45°. *J. Phys. Chem.* 77:685–688.

Mo, J.W., Li, L., Zhou, J.F., Xu, D.Y., Huang, B.L., and Li, Z.G. 2015. Fluid infiltration pressure for hydrophobic nanochannels. *Phys. Rev. E* 91:033022.

Mourits, F.M. and Rummens, F.H.A. 1977. A critical evaluation of Lennard-Jones and Stockmayer potential parameters and of some correlation methods. *Can. J. Chem.* 55:3007–3020.

Nagayama, G., Tsuruta, T., and Cheng, P. 2006. Molecular dynamics simulation on bubble formation in a nanochannel. *Int. J. Heat Mass Transfer* 49:4437–4443.

Reif, F. 1985. *Fundamentals of Statistical and Thermal Physics*. Singapore: McGral-Hill.

Stillinger, F.H. and Weber, T.A. 1985. Computer simulation of local order in condensed phases of silicon. *Phys. Rev. B* 31:5262–5271.

Stuart, S.J., Tutein, A.B., and Harrison, J.A. 2000. A reactive potential for hydrocarbons with intermolecular interactions. *J. Chem. Phys.* 112:6472–6486.

Tersoff, J. 1988a. Expirical interatomic potential for silicon with improved elastic properties. *Phys. Rev. B* 38:9902–9905.

Tersoff, J. 1988b. Expirical interatomic potential for carbon, with applications to amorphous carbon. *Phys. Rev. Lett.* 61:2879–2882.

Chapter 4 Nanofabrication

4.1 Introduction

Alternatives to numerical approaches, physical experiments, if repeatable, offer reliable and, in many cases, insightful investigations of nanoscale fluid flows through direct or indirect measurements of specific properties. For system setup and parameter measurements, there is no significant difference in experiments at the macro- and nanoscales. In terms of fabrication techniques and flow investigation methods, however, nanoscale flow experiments do have special characteristics as compared with macroscale experiments. Experiments for macroscale flows can be conducted on a prototype, i.e., a physical system, or a model, which is similar to the prototype but may have different dimensions. For nanoscale flows, unfortunately, investigations usually cannot be conducted in a large-scale model because the surface effects in large-scale systems become insignificant and the key flow physics for nanoscale flows may be lost. Therefore, nanoscale flow experiments can be mainly performed in nanofluidic systems. This requires the fabrication of nanoscale flow channels and fluidic devices, which is more challenging than that of macroscale systems. Most of the fabrication techniques for microelectromechanical systems (MEMS) can be employed for nanostructure fabrication. Nanochannels can be etched in a substrate or constructed on a surface. These can also be formed by using advanced nanomaterials, such as CNTs. Due to the way that nanochannels are fabricated, nanochannel fabrication techniques are categorized into bulk nanomachining, surface nanomachining, nanoimprint lithography (NIL), nanolithography, and nanomaterial-based fabrication. Except for the last method, nanomaterial-based fabrication, nanochannels are fabricated on a substrate, which needs to be well-cleaned before the fabrication. In the following sections, wafer cleaning and substrate preparation are discussed first, which are followed by the introduction of different fabrication techniques.

4.2 Substrate preparation

Due to small dimensions, nanostructures can be easily contaminated by dust particles, organic residues, and metal ions. Before a fabrication process is started, the substrate should be cleaned to remove these common contaminants. The cleaning method varies for different materials and structures of the substrate. A standard cleaning procedure for silicon wafers is the so-called RCA clean, which was developed by Werner Kern at RCA laboratories (Kern, 1993). It is also called the standard clean (SC) method.

The RCA method consists of two levels: RCA-1 and RCA-2. RCA-1 (or SC-1) is to remove organic contaminants. A silicon wafer is first degreased with acetone and isopropyl alcohol. Then the wafer is soaked in the RCA-1 solution, which is a mixture of 5:1:1 deionized (DI) water, ammonium hydroxide (NH_4OH), and hydrogen peroxide (H_2O_2), at 75°C for 15 min. Finally, the wafer is flushed by DI water and dried using a wafer drier or a blow-drying system. As RCA-1 oxidizes silicon and leaves an oxide layer on the wafer surface, the silicon wafer can be cleaned using a 1:50 solution of HF and H_2O at room temperature to remove the oxide layer if a pure silicon surface is needed.

A succeeding process, RCA-2 (or SC-2), is required for removing ionic contaminants. In this process, the silicon wafer is immersed in a 6:1:1 solution of DI water, hydrochloric acid (HCl), and H_2O_2 for 10 min. This is then followed by several times of DI water rinses of the wafer. After the RCA clean, the substrate is ready for further fabrication activities.

4.3 Bulk nanomachining

Bulk nanomachining is a widely used technique for fabricating nanochannels. In this method, nanochannels are fabricated in a substrate through four steps: substrate preparation, pattern definition via photolithography, channel etching, and bonding.

4.3.1 Pattern definition

The process of pattern definition is to transfer a desired nanopattern from a mask onto a substrate through photolithography. The desired nanopattern is first designed on a mask, which typically is a glass or quartz plate, through wet chemical etching. In most cases, the size of the pattern on the mask is the same as that transferred onto the substrate. It can also be larger than the final pattern on the substrate if a reduction imaging method is used in defining the pattern. The pattern on the mask is usually printed on the substrate through photolithography, which typically is ultraviolet (UV) light radiation. Therefore, the substrate needs to be coated with a layer of photoresist material before the pattern definition. Photoresist materials are usually polymers, whose chemical structure can be easily changed upon the exposure to UV light. Depending on how the molecular structure and the solubility of the photoresist are varied by UV light, photoresist materials can be referred to as either positive or negative. For positive resist materials, the UV light radiation energy can break the bonds along polymer backbones and lower their molecular weight, which increases the solubility of the resist material in developer solutions. In this case, when the photoresist is radiated by UV light through the mask, the radiated area, which has the same pattern as that on the mask, is removed in the developer solution and the pattern on the mask is transferred to the substrate, as illustrated in Figure 4.1a.

For negative photoresist, the energy of UV light can increase the molecular weight of the photoresist polymer and make the radiated area resistant to the developer solution. The unexposed area of the photoresist will be dissolved by the developer solution, as shown in Figure 4.1b. Positive photoresist is more frequently employed in nanofabrication compared with negative photoresist.

The wavelength of UV light ranges from 10 to 400 nm and that from most laboratory UV light sources is about 250 nm. Therefore, patterns with characteristic dimensions larger than 250 nm can be well defined. If a characteristic dimension is smaller than half of the UV light wavelength, the corresponding pattern will be blurred.

4.3.2 Channel etching

Etching is an important step in nanochannel fabrication. It creates nanostructures by physically or chemically removing parts of the substrate material using an etchant. Based on the etchant used, etching processes are classified into two groups: wet etching and dry etching. Wet etching uses liquid etchant to react with parts of the substrate that are uncovered by the photoresist, as shown in Figure 4.2. In dry etching, nanostructures are etched on a substrate by a plasma or an etching gas. Therefore, the only difference between wet etching and dry

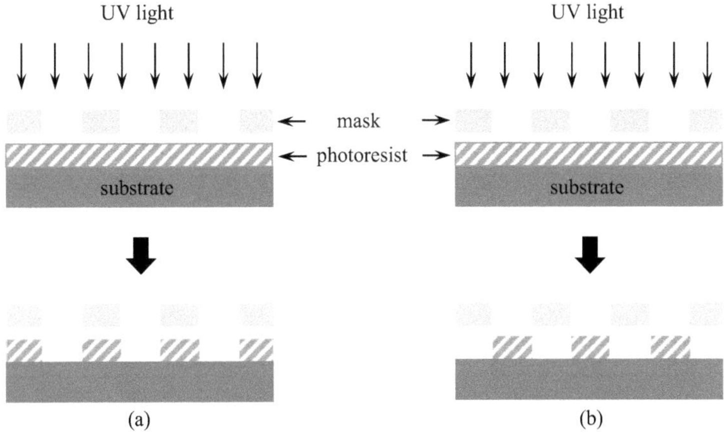

Figure 4.1 Pattern definition through photolithography: (a) positive photoresist and (b) negative photoresist.

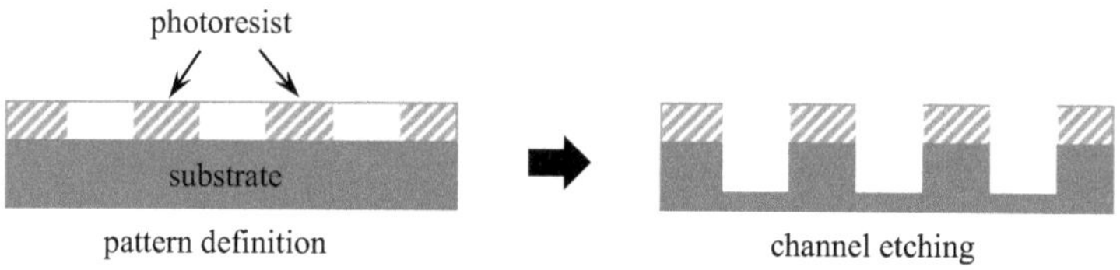

Figure 4.2 Schematic of channel etching.

etching is the etchant. Wet etching typically has higher etch rate and can accommodate more wafers per batch than dry etching.

Etching processes can be isotropic or anisotropic. In isotropic etching, the etch rate is the same in different directions, as illustrated in Figure 4.3. This makes it difficult to control the shape of the structure fabricated in the substrate. By choosing proper etchants and substrates, anisotropic etching, where the etch rate is direction dependent, can be achieved. This can help fabricate desired structures, such as sharp edges (Figure 4.3).

In wet etching, for silicon oxide (or glass) substrates, isotropic etching can be realized using hydrofluoric acid (HF) or buffered oxide etch. HF reacts with silicon oxide through the following reaction:

$$SiO_2 + 6HF \rightarrow H_2SiF_6 + 2H_2O \tag{4.1}$$

Most wet etchants are isotropic. However, for single-crystalline silicon wafers, the etching rates in different crystal orientations can be significantly different in certain alkaline solutions, such as KOH and NaOH. This can help reach anisotropic wet etching. For example, KOH shows an etch rate two orders of magnitude higher in the <100> crystal plane of Si than that in the <111> plane.

In contrast, most dry etching processes are anisotropic. In nanochannel fabrication, anisotropic dry etching is commonly used. The most widely used dry etching methods are reactive ion etching (RIE) and deep reactive ion etching (DRIE). RIE employs a radio-frequency (RF) excited plasma to remove substrate material. High-energy ions from a plasma bombard and cut the exposed part of

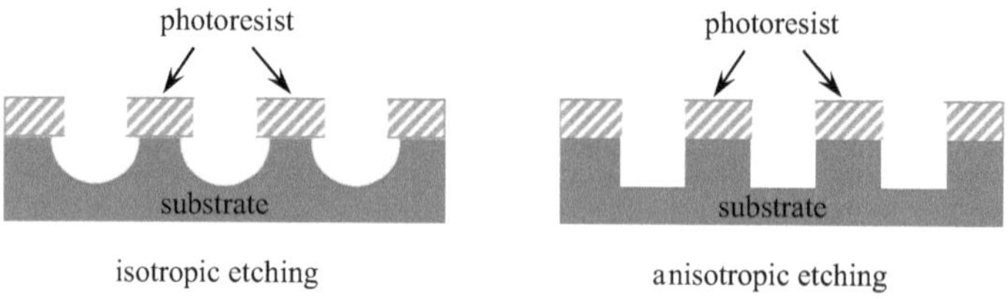

Figure 4.3 Isotropic and anisotropic etching.

the substrate. A popular RIE etchant for Si is sulfur hexafluoride (SF_6) and the reaction is given as follows:

$$Si + 2SF_6 \rightarrow SiF_4 + 2SF_4 \quad (4.2)$$

Compared with RIE, DRIE is highly anisotropic. It is capable of creating deep trenches with steep sidewalls and of high aspect ratio in a substrate. In DRIE, two RF power sources are used, which ensure the generation of high-density plasma ($>10^{11}\,cm^{-3}$) at low pressures ($<1\,Pa$), leading to a high etching anisotropy. Another reason that DRIE can achieve steep sidewalls is the Bosch etching technique, which is also known as the pulsed or time-multiplexed etching. The Bosch etching includes two alternating processes: one is the standard etching using plasma and the other is the protection of sidewalls through passivation deposition, which coats the sidewalls by a layer of Teflon-like substance using octafluorocyclobutane (C_4F_8). These two processes are performed for a short time alternatively. During etching, after the sidewalls and the bottom surface of a channel are deposited with a passivation layer, the standard etching process resumes, which removes the passivation layer on the bottom surface first through ion bombardments and then etches the substrate vertically downward, as shown in Figure 4.4. Channels of a desired depth can be fabricated by controlling the number of cycles of the two processes. Since DRIE is capable of offering fast and anisotropic etching, it can be used for the fabrication of microchannels and fluid reservoirs, which are essential in a nanofluidic system. For nanochannel fabrication, DRIE is less common because the roughness of the sidewalls can be a few to tens of nanometers (Quévy et al., 2002).

Finally, etchants for different substrates and the corresponding etch rates for wet and dry etchings are briefly summarized in Table 4.1.

4.3.3 Bonding

The last step of bulk nanomachining is bonding, which is basically the encapsulation of nanochannels or the whole nanofluidic device using a cover wafer. In addition to protecting nanostructures fabricated on the substrate from contamination, the bonding step seals nanochannels to form enclosed fluidic ducts. Bonding also helps form microchannels and fluid reservoirs, which are important parts of a nanofluidic device. There are three major bonding methods: adhesion bonding, fusion bonding, and anodic bonding.

Adhesion bonding is a very simple approach for chip bonding. It uses an adhesion layer as a glue to bond two wafers. The most widely used adhesion materials are SU-8 and benzocyclobutene (BCB), including photoresist, polyimide, and poly(methyl methacrylate) (PMMA) (Jaeger, 2002). Adhesion bonding can be performed at a relatively low temperature (<250°C), which makes the bonding

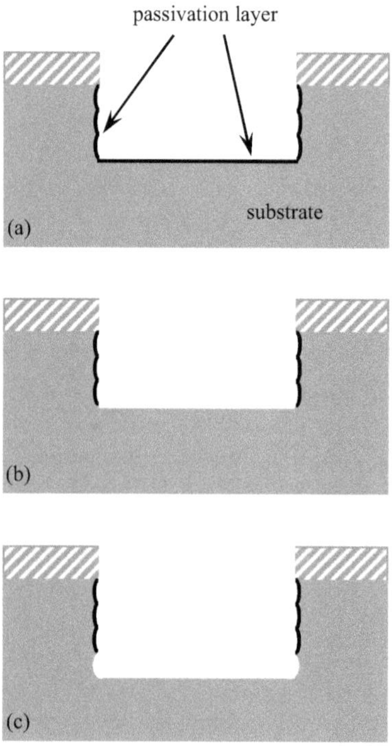

Figure 4.4 The Bosch process in deep reactive ion etching (the roughness of the sidewalls depends on the cycle times of the two processes). (a) Deposition of a passivation layer on the sidewalls and bottom surface of a channel. (b) Removal of the passivation layer on the bottom surface. (c) Etching of exposed substrate material.

method suitable for a wide range of materials, such as silicon, glass, polymers, and metals. However, the adhesion layer can extrude into and alter the dimension of the nanostructure fabricated on the substrate, and consequently may affect the accuracy of experiments. Another disadvantage of this bonding method is that most organic adhesion materials are permeable to gases and water, which results in poor seal. Furthermore, the adhesion material may react with some solvents and produce undesired products in applications involving chemical reactions. Finally, the bonding method may not work at high temperatures.

Fusion bonding is a physical process that connects two wafers through the chemical bonds between the two wafers by annealing them at a high temperature. Fusion bonding can achieve quite strong linking between two surfaces due to atomic diffusion and strong covalent bonds. As this method involves thermal treatment, the two bonding surfaces should have similar thermal expansion coefficients to avoid the generation of internal stresses when the wafers are cooled down to room temperature. Fusion bonding is mostly used for silicon–silicon and glass–glass

Table 4.1 Etchants and Etch Rates for Wet and Dry Etchings of Common Substrate Materials (Williams et al., 2003)

Substrate Material	Wet Etchants	Etch Rate (nm/min)	Dry Etchants	Etch Rate (nm/min)
Silicon	HNO_3:H_2O:NH_4F (126:60:5)	100–900	XeF_2 (0.35 Pa)	180–450
	KOH	670–1,100	$SF_6 + O_2$	1,500
Silicon dioxide	10:1 HF (10 H_2O:1 49% HF)	23–1,500	HF vapor	66–210
	KOH	6.7–38	$SF_6 + O_2$	10–73
Silicon nitride	HNO_3:H_2O:NH_4F (126:60:5)	0.23–66	HF vapor	1.0–1.9
	KOH	0.67	$SF_6 + O_2$	150–200
Aluminum	HNO_3:H_2O:NH_4F (126:60:5)	60	$SF_6 + O_2$	<2.8
	KOH	12,900		
Aluminum oxide	HNO_3:H_2O:NH_4F (126:60:5)	12–99	$SF_6 + O_2$	0.41–0.55
Gold	I_2:KI:H_2O (5:10:85)	660	$CF_4 + O_2$	8.3
Nickel	HNO_3:H_2O:NH_4F (126:60:5)	21	$SF_6 + O_2$	0.71
Photoresist (positive)	KOH	>17,900	$SF_6 + O_2$	180

bonding. For silicon–silicon bonding, the annealing temperature typically ranges from 600 to 1200°C. For glass–glass bonding, the annealing temperature is about 550°C. The drawback of fusion bonding is that the high annealing temperature may deform or even damage the nanostructures on the substrate.

Anodic bonding is also known as a field-assisted bonding or electrostatic sealing (Wallis and Pomerantz, 1969). This technique bonds wafers using a sufficiently strong electric field. It is commonly employed to bond a silicon substrate with a glass cover. The bonding is typically performed at a relatively high temperature (250–400°C) and a voltage of around 1000 V across the wafers. Under such conditions, oxygen ions diffuse from glass to silicon and Si–O bonds are formed at the interface, leading to a strong binding between the silicon substrate and the glass cover. The silicon substrate and the glass cover should be clean, smooth, and flat to ensure a good bonding. Anodic bonding can also be used for Si–Si bonding (Quenzer et al., 1996). However, silicon is nontransparent, which makes the observation of fluid flows impossible. This is why glass wafers are mostly used in anodic bonding for micro- and nanofluidic devices. One disadvantage of this technique is that the high electrostatic force generated by the high-bonding voltage may cause collapse of nanochannels.

4.4 Surface nanomachining

Instead of etching materials to create nanostructures in a substrate, nanochannels can also be constructed on a substrate. The method of surface nanomachining is such a technique. It employs a sacrificial layer, which is removed at a later step of the fabrication process, to form nanochannels on top of a substrate. Therefore, this method is also known as sacrificial layer releasing. The fabrication process is demonstrated in Figure 4.5. After a substrate is prepared, a sacrificial layer, such as silicon dioxide, is first deposited on the surface of the substrate (Figure 4.5a). The sacrificial layer is then covered by a capping layer through a deposition process, which is followed by etching holes in the capping layer (Figure 4.5b). The holes are used for removing the sacrificial layer. They can also serve as the inlets and outlets of the final fluidic system. Finally, the sacrificial layer is removed through selective etching (either wet or dry etching) or thermal degradation (Figure 4.5c). The remaining capping layer and the substrate form the walls of nanochannels and the space occupied by the sacrificial layer becomes flow channels. The size of nanochannels can be controlled by varying the dimensions of the sacrificial layer. To make the channel surface properties uniform, the substrate can be coated by the capping material before the deposition of the sacrificial layer (Figure 4.5d).

Polysilicon (Poly-Si) and silicon oxide are the most popular sacrificial layer materials due to their great thermal stability, mechanical rigidity, and process compatibility. Table 4.2 gives some combinations of sacrificial and capping layer materials, and the corresponding etchants.

Surface nanomachining is capable of fabricating complex nanostructures, such as nanochannels of heterogeneous surface properties (Li et al., 2015). The channel height in this method can also be easily controlled by changing the thickness of the sacrificial layer. If wet etching is employed for removing the sacrificial layer, however, residues at the interface of the etchant and the sacrificial layer may agglomerate and reduce the etching rate progressively (Abgrall and Gué, 2007).

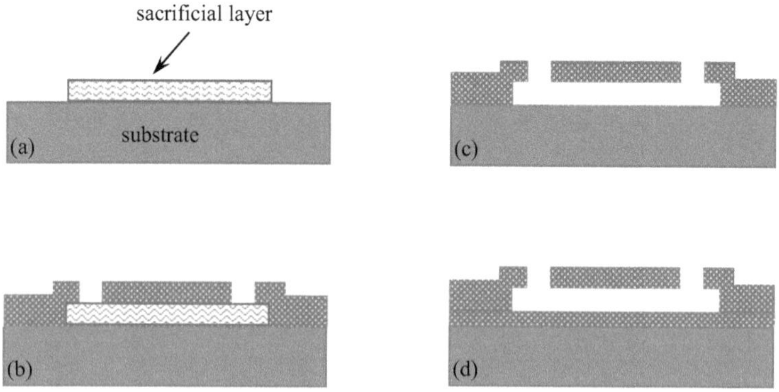

Figure 4.5 Schematic of surface nanomachining. (a) Sacrificial layer deposition. (b) Capping layer deposition. (c) Removal of sacrificial layer. (d) Channels with uniform surface properties.

Table 4.2 Materials as Sacrificial and Capping Layers and the Corresponding Etchants (Bagolini et al., 2002)

Sacrificial Layer Materials	Capping Layer Materials	Etchants
Poly-Si	SiC, SiN	TMAH, KOH
Poly-Si	SiO_2	TMAH
SiO_2	SiC, SiN, Poly-Si, TiN	HF
SiO_2	Ti, Al	73% HF
Polyimide	Al, SiC, SiN, Ti	Oxygen plasma

The etching rate of this method drops as the channel height is reduced. Therefore, this method can be time-consuming for nanochannel fabrication.

4.5 Nanoimprint lithography

NIL is also a common nanofabrication method, which uses a stiff mold to transfer nanostructures to an imprint resist through mechanical deformation and subsequent thermal or optical treatments. The imprint resist is typically a thermoplastic material, such as polymers, whose properties vary under a thermal or UV light cure. In an NIL process, the mold should be first patterned with a predefined structure using a high-resolution technique, such as e-beam lithography, as will be introduced later, and an imprint resist film needs to be deposited on a substrate, as shown in Figure 4.6a. Then the mold is pressed into the imprint resist (Figure 4.6b). This is followed by thermal or UV light treatments depending on the imprint resist used. After the solidification of the imprint resist, the mold is separated from the substrate, which leaves the desired nanostructure on the imprint resist (Figure 4.6c).

The nanopattern on the imprint resist can be further transferred to the substrate through bulk nanomachining technique discussed in Section 4.3, as illustrated in Figure 4.6d. It can also be sealed directly through solvent vapor sealing or melting reflow sealing (Cho et al., 2010; Xia et al., 2008) (Figure 4.6e). The former uses chemical vapor (e.g., toluene or methanol) to chemically melt the top part of the nanostructure to achieve self-sealing. The latter employs an UV laser pulse to melt a layer of the top surface to seal the nanostructure.

Nanochannels can also be formed using NIL without sealing. As shown in Figure 4.7, the height of the structure on the mold can be larger than the thickness of the resist film. After the mold is stamped into the resist, the gap between the resist and the mold form closed channels. The height of the channels can be adjusted by changing the size of the structure on the mold and the thickness of the resist.

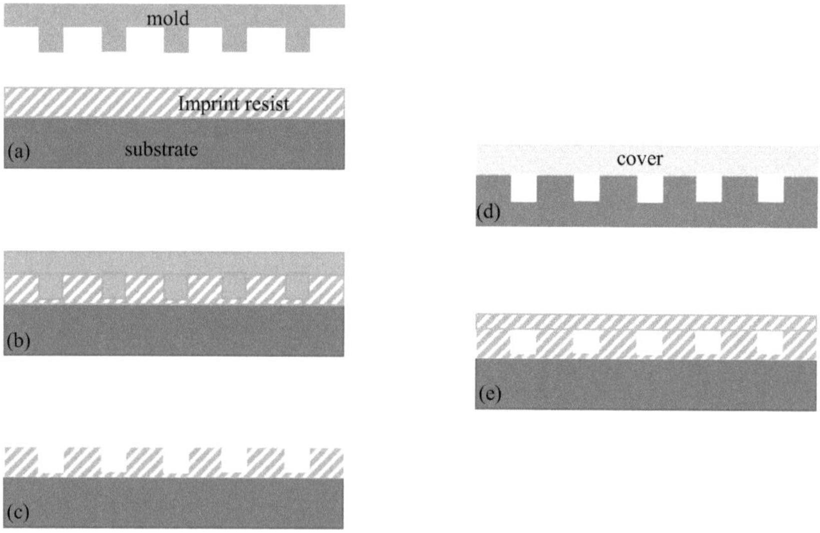

Figure 4.6 Schematic of nanoimprint lithography. (a) Preparation of the mold and imprint resist deposition on a substrate. (b) Pattern transfer from the mold to the imprint resist followed by thermal or UV light treatments (not shown in the figure). (c) Removal of the mold. (d) Pattern transfer to the substrate through further bulk nanomachining. (e) Direct sealing of nanostructures using solvent vapor sealing or melting reflow sealing techniques.

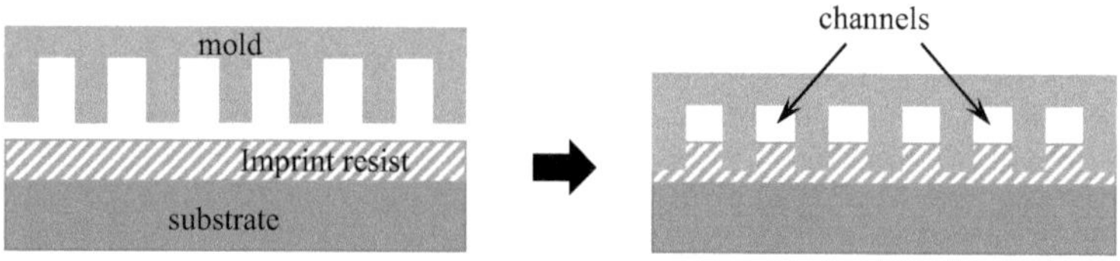

Figure 4.7 Self-sealed nanochannels through nanoimprint lithography.

Nanoimprint lithograph method is a low-cost fabrication technique because it does not require the traditional lithography and etching processes and the mold can be reused many times. It is also a fast method suitable for mass production of nanostructures. Furthermore, this method is capable of fabricating three-dimensional structures, which are stackable on the mold. The major disadvantage of this method is the defects caused by the mold. In some cases, the mold may not be sufficiently flat and air can be trapped during imprinting, which may lead to bubble defects. Mold wear is another problem that can reduce the lifetime of the mold.

4.6 Nanolithography

In the traditional photolithography techniques, the characteristic size of the fabricated structure is determined by the wavelength of the incident light. Since UV light is the most common light source and the wavelength of the widely used UV light is around 250 nm, the typical resolution of the traditional photolithography methods is ~100 nm for mass production. Although further improvement of the resolution can be achieved, it may require expensive light sources and instrument, such as extreme UV light (wavelength: 10–70 nm) and X-rays (wavelength: 10–70 nm) (Alexe et al., 2004). For the fabrication of nanochannels with cross-sectional area below 1,000 nm^2, the traditional photolithography methods face challenges. To meet the requirements of nanofluidic devices and improve the fabrication resolution, the techniques of electron beam lithography (EBL) and focused ion-beam lithography can be employed.

4.6.1 Electron beam lithography

EBL is a direct pattern-creation technology that uses a focused electron beam to write patterns in a resist, which is an electron-sensitive material, such as PMMA. Similar to UV light exposure in the traditional photolithography, an electron beam changes the solubility of the resist, for which either the exposed or non-exposed part is selectively removed during the developing process, leading to the generation of desired nanostructures with small lateral dimensions in the resist. The characteristic size of a nanopattern depends on the wavelength of electrons, which can be controlled by varying the energy of electrons. For instance, the wavelength of an electron is 1.23 nm if the electron energy is 1 eV. It is reduced to 0.87 nm if the energy is raised to 2 eV. In any case, the typical wavelength of electrons is much smaller than that of UV light and the characteristic pattern size can be down to 10 nm or even smaller (Yasin et al., 2001).

Therefore, the main advantage of EBL is the high resolution for patterning. However, this technique is not suitable for mass production of nanostructures because it consists of a series of sequential processes, which make the method time-consuming.

4.6.2 Focused ion-beam lithography

Similar to EBL, focused ion-beam lithography (FIBL) also draws nanopatterns on a substrate directly with a high resolution without using a mask and even without a resist. Compared with EBL, which uses electron beams, FIBL employs ion beams to create nanostructures. In this method, ions, which carry certain energy, are shot onto a surface. When they collide with the atoms of the surface, their energy is transferred to the surface atoms. If sufficient energy is transferred during the collisions, the surface atoms can overcome the bonding forces from neighboring atoms and escape from the surface. This leaves a nanoscale trench on the surface. Liquid metal ion sources are popularly employed in FIBL and gallium is the most widely used liquid metal ion source due to its good compatibility. FIBL can perform different nanoscale processes, including milling, swelling, implantation, deposition, and etching (Tseng, 2005). Milling can be conducted to fabricate nanochannels because the width of a milled grove can be controlled down to 5 nm (Yan et al., 2009).

Although FIBL works well in cutting materials at a very small scale, the high energy of ions may cause defects or even damage to a substrate. Therefore, the ion energy, beam current, and beam size should be properly controlled to reduce defects. Similar to EBL, FIBL is also time-consuming and not suitable for high-throughput fabrication.

4.7 Nanomaterial-based fabrication

For a nanofluidic device/chip, the major part is the fabrication of nanochannels, which can be constructed using the methods described in the previous sections. Nanochannels can also be formed by making use of the intrinsic structure of certain nanomaterials. In this section, anodized aluminum oxide (AAO) and CNTs will be introduced as two representative nanomaterials.

4.7.1 Anodized aluminum oxide

AAO is a honeycomb-like, nanoporous material with arrays of uniform and parallel nanopores, as shown in Figure 4.8 (Liang et al., 2002). AAO can be fabricated through anodizing aluminum in an acid electrolyte (O'Sullivan and Wood, 1970; Thompson and Wood, 1981). The initialization of pores is caused by the nonuniform distribution of electric field and the formation of pores is the consequence of the balance between the oxide growth on aluminum and the dissolution of aluminum at the pore bases (O'Sullivan and Wood, 1970). The pore diameter and

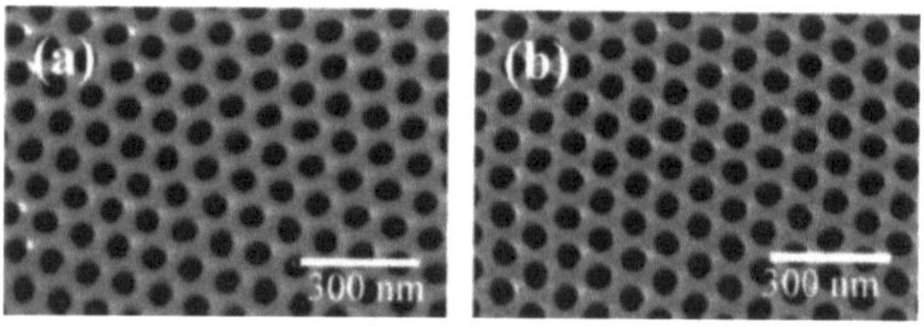

Figure 4.8 SEM image of an AAO membrane: (a) top view and (b) bottom view. (Reprinted from Liang et al., 2002, with the permission of AIP Publishing.)

pore–pore spacing can be controlled by varying the voltage during anodization (Crouse et al., 2000). The nanopores in AAO are typically positively charged and the pore properties are affected by the electrolyte solution. Due to the uniform and orderly nanopores, AAO can be directly used in nanofluidic systems, such as electroosmostic pumps (Wang and Smirnov, 2009). Another application of AAO is that a thin film of AAO can serve as an etch mask for creating nanostructures in a substrate through RIE or other etching techniques, as discussed in the method of bulk nanomachining (Section 4.3). Figure 4.9 shows a pattern transferred from AAO to a silicon substrate (Crouse et al., 2000) and Figure 4.10 illustrates the fabrication of cupper nanodots as resistive switching memory using AAO as the template mask (Han and Lee, 2016).

4.7.2 Carbon nanotubes

Carbon nanotubes (CNTs) are cylindrical structures of carbon atoms. A single-wall CNT can be formed by rolling a graphene sheet. CNTs exhibit extraordinary mechanical, thermal, and electrical properties and have been widely used in a variety of areas (Dai, 2002). The cylindrical structure of a CNT provides an ideal passage for nanoscale fluid flows. It has been shown that the smooth and hydrophobic surface of CNTs can lead to a large flow slip and enhance the flow rate (Sokhan et al., 2002). CNTs can be synthesized by vaporizing a solid-state carbon source, such as graphite, at a high temperature to render CNTs to grow on a cool surface. Such a process is employed in arc discharge and laser ablation methods. CNTs can also be synthesized through chemical vapor deposition (CVD), where a hydrocarbon gas is used as a source for carbon atoms and metal nanoparticles (e.g., nickel and ion) are used as catalysts to grow CNTs. More information about CNT synthesis can be found in the literature (Dai, 2002; Prasek et al., 2011).

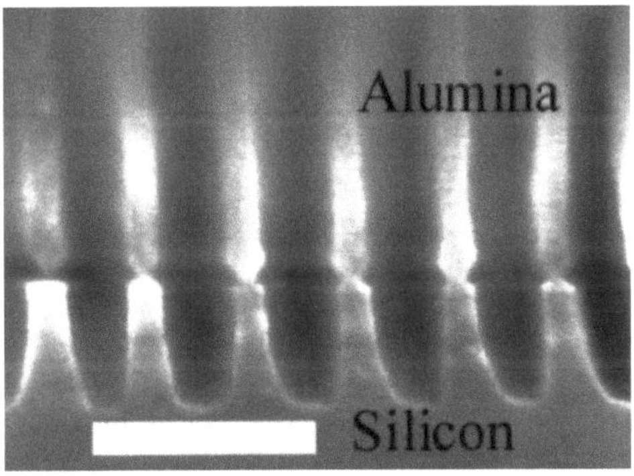

Figure 4.9 Pattern transfer from AAO to a silicon substrate. (Reprinted from Crouse et al., 2000, with the permission of AIP Publishing.)

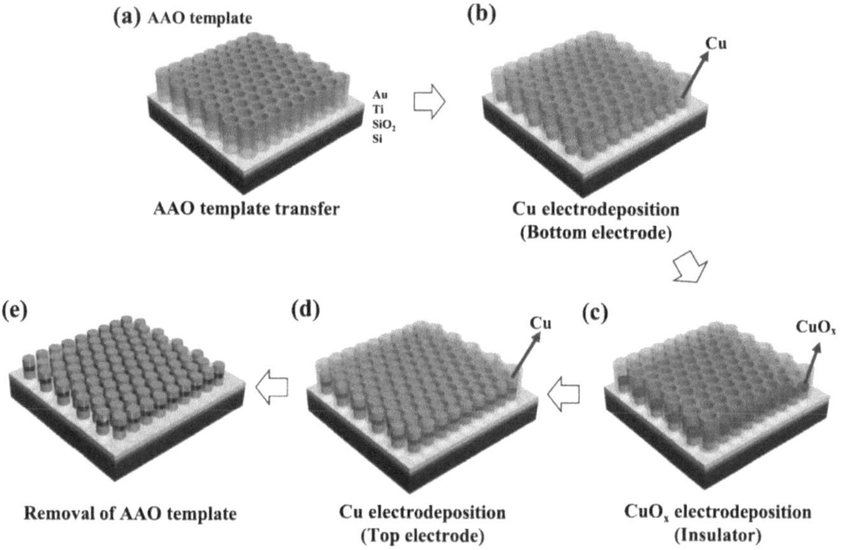

Figure 4.10 Illustration of nanoscale resistive switching memory using AAO as the mask. (a) AAO template transferred onto a substrate. (b) Cu deposition through AAO mask. (c) CuO_x deposition through electrochemical deposition (ECD). (d) Cu deposition through ECD. (e) Final resistive switching memory structure after the removal of AAO. (Reprinted from Han and Lee 2016.)

In the applications in nanofluidic devices, aligned CNTs can be fabricated into a membrane by coating them with a polymer, as sketched in Figure 4.11. The CNTs in the membrane serve as nanochannels, through which fluid flows can be generated by pressure drops or electric fields. Single CNT-based fluidic system can also be setup with the help of scanning electron microscope. An electroosmotic flow system using an isolated CNT is shown in Figure 4.12 (Pang et al., 2011). In such a system, the transport of ions and the electric current generated by ions can be studied.

4.7.3 Other nanotubes

Other nanotubes, such as ZnO, TiO_2, MnO_2, and SiO_2 nanotubes, can also be synthesized through a so-called sol–gel method. In this method, nanotubes of a certain material are prepared inside the pores of a nanopore membrane, which is immersed in a precursor sol and then followed by gelation (Zhang et al., 2000). The AAO porous material discussed in Section 4.7.1 can be employed as a nanopore membrane template for the fabrication of various nanotubes.

Herein, the traditional sol–gel method is introduced through elucidating the preparation of SiO_2 nanotubes. The gel is prepared by mixing a TEOS (tetraethyl

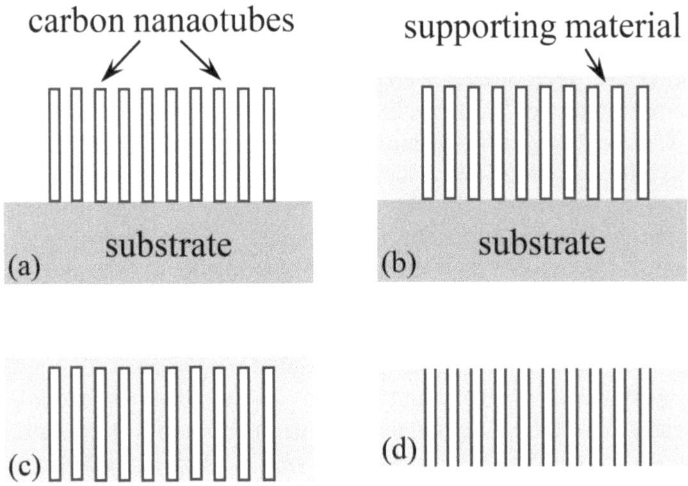

Figure 4.11 Fabrication of carbon nanotube-based membrane. (a) Synthesis of CNTs on a substrate (e.g., silicon or quartz). (b) Filling CNTs with a supporting material (e.g., polymer). (c) Removal of the substrate. (d) Removal of excess supporting material and opening of the CNTs.

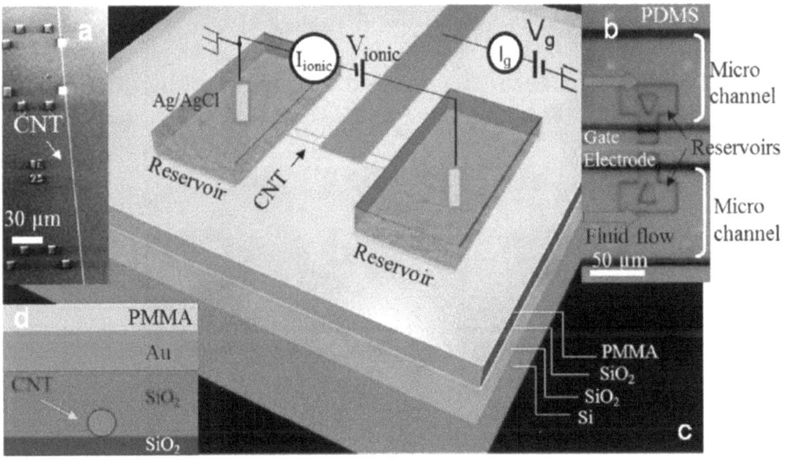

Figure 4.12 A single CNT-based electroosmotic flow system. (a) SEM image of a single-walled CNT (diameter is about 1.7 nm). (b) Optical microscope image of a device. (c) Schematic of the electroosmotic flow device, where a 10–30 μm long CNT connects two fluid reservoirs. Ag/AgCl electrodes are used to apply a voltage across the reservoirs. (d) Sketch of the cross-sectional view of the barrier region. (Reprinted with permission from Pang et al., 2011, Copyright (2011) American Chemical Society.)

orthosilicate) solution and an ethanol/HCl solution. The former can be made by adding 52-mL TEOS into 115-mL ethanol and the latter is a mixture of 115-mL ethanol, 18-mL water, and 0.27-mL HCl (Zhang et al., 2000). The gel is aged at room temperature for 15–30 days before a nanoporous AAO membrane is dipped into it. If the gel is aged for a shorter time (e.g., 2–15 days), this method can be used to synthesize nanowires (Zhang et al., 2000). The dipping time of an AAO template can range from a few seconds to one minute. The gel-containing AAO is then heated in air at 200°C for 1 day. Finally, the AAO template is dissolved by a chemical solution (H_3PO_4, CrO_3, and H_2O). Figure 4.13 shows a TEM image of SiO_2 nanotubes fabricated using this approach (Zhang et al., 2000). The inner surface roughness of the nanotubes can be improved by changing the temperature and increasing the viscosity of the gel. The latter can be achieved by aging the gel for a longer time. The diameter and length of the nanotubes are determined by the pore dimensions of the AAO template, while the thickness of the nanotubes can be controlled from about 3 to 10 nm by varying the dipping time.

A drawback of the traditional sol–gel method is the poor control over the nanotube thickness and morphology. These issues can be solved using a layer-by-layer deposition technique or the surface sol–gel (SSG) method. In a layer-by-layer deposition approach, molecular precursors are adsorbed on a growing surface as a layer at a time to achieve precise control of thickness. The SSG method also involves the chemisorption of molecular precursors, which is followed by a washing step to remove weakly bound molecules and a hydrolysis step. By repeating these steps, the nanotube thickness and morphology can be precisely controlled. The SSG method has been demonstrated to synthesize SiO_2 nanotubes with smooth inner surface (Kovtyukhova et al., 2003).

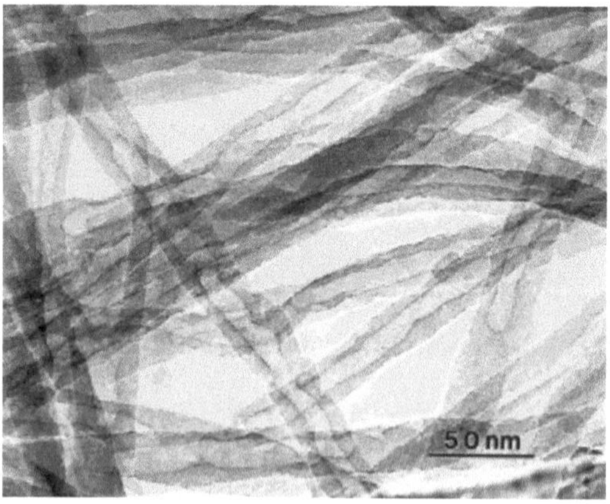

Figure 4.13 TEM image of SiO_2 nanotubes synthesized through the sol–gel method (the dipping time is 20 s and the aging time of the gel is 15 days). (Reprinted from Zhang et al. 2000.)

The SSG method for the fabrication of SiO_2 nanotubes is sketched in Figure 4.14. First, $SiCl_4$ molecules are chemisorbed on the hydrated pore surfaces of an AAO membrane. Weakly bound and unbound adsorbate molecules are then cleaned by washing the AAO pores with CCl_4. This is followed by the hydrolysis of $SiCl_4$ layer to generate SiO_2 nanotubes. Repeating these steps and after a certain number of cycles, SiO_2 nanotubes with a desired inner diameter can be synthesized. Finally, the AAO membrane is dissolved in 50% H_2SO_4 to obtain freestanding SiO_2 nanotubes. Figure 4.15 shows TEM images of SiO_2 nanotubes after different deposition cycles (Kovtyukhova et al., 2003). It is seen that the inner surfaces of

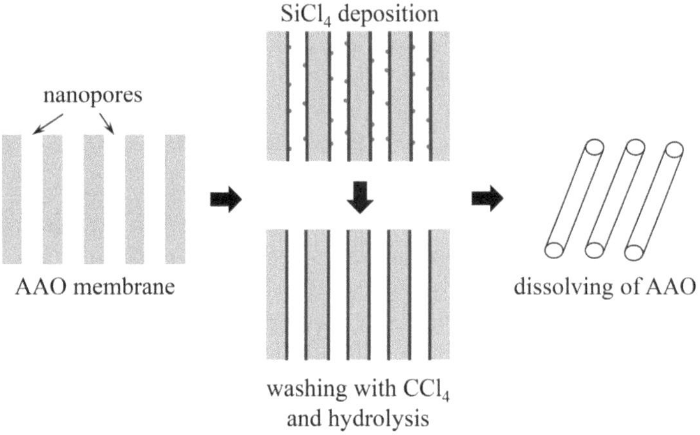

Figure 4.14 Schematic of SiO_2 nanotube synthesis using the SSG method.

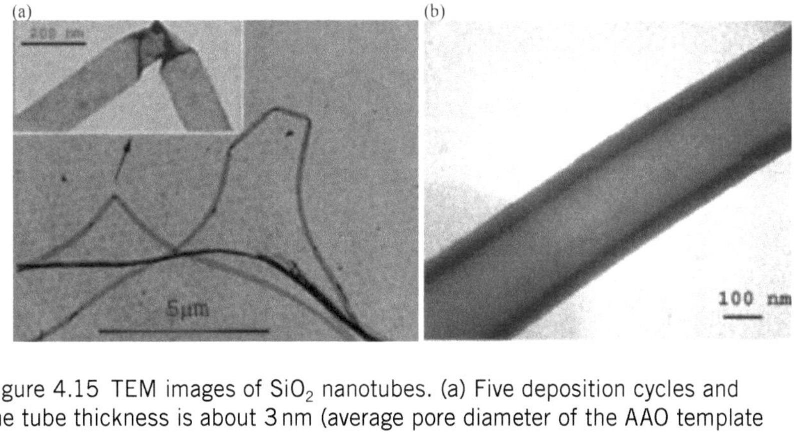

Figure 4.15 TEM images of SiO_2 nanotubes. (a) Five deposition cycles and the tube thickness is about 3 nm (average pore diameter of the AAO template is 90 nm). (b) Twenty deposition cycles and the tube thickness is about 25 nm (average pore diameter of the AAO template is 280 nm). (Reprinted from Kovtyukhova et al. 2003.)

the nanotubes are smooth and the inner diameter is quite uniform. The inner diameters of the nanotubes in Figure 4.15a and b are about 3 and 25 nm, respectively.

4.8 Materials for fabrication

Depending on the characteristics and the applications of a nanofluidic system and the fabrication method employed, a proper material for the fabrication of the system should be selected. Generally, silicon, silicon oxide (quartz and glass), silicon nitride, and polymers are some popular materials for nanofabrication. For fluidic devices, the mechanical, electrical, and optical properties of a material are of major concerns in material selection. The thermal properties of a material could also be critical as some fabrication techniques may involve high-temperature processes. As these properties vary for different materials, it is important to choose a suitable material that is compatible with the features of a device and the operation conditions.

Silicon is the most widely used material in nanofabrication. It is chemically stable, cheap, and easy to be integrated with other fluidic components. For pressure-driven fluidic systems, silicon is a good choice. For electroosmotic flows, however, the semi-conductive property of silicon can cause electrical leakage and affect the electric field distribution. If silicon is used in an electroosmotic flow system, the channel surface needs to be passivated by an insulation layer, for which silicon oxide and silicon nitride are good candidates because they have great mechanical and chemical properties to serve as channel walls, in addition to electrical insulation ability. Another issue of silicon is its nontransparency to visible light, which does not support flow observation. If videos of a flow are to be taken or the flow dynamics needs to be visually investigated, quartz or glass substrates can be used to replace silicon. If glass is used, the fabrication cannot involve a high-temperature process because glass deforms as the temperature becomes higher than ~600°C. Alternatively, the surfaces of silicon channels can be oxidized into silicon dioxide and the channels can be enclosed by a glass wafer, which accommodates flow observation.

Aluminum oxide is also a popular material for nanochannel fabrication, especially in the cases where channel surface properties need to be modified. If a channel surface is made of or coated with aluminum oxide, the surface is usually positively charged. This is very different from silicon surfaces, which are usually negatively charged. In electroosmotic flows, surface charges play an important role and aluminum oxide is frequently employed to change the surface charge. Furthermore, aluminum oxide exhibits hydrophobic characteristics while silicon and silicon dioxide are hydrophilic. These versatile surface properties of different materials can bring about diverse flow phenomena at the nanoscale. Heterogeneous nanochannels, half of which is silicon dioxide and the other half is aluminum oxide, have been fabricated to generate direction-dependent flows to achieve flow rectifications (Li et al., 2015).

Other than those inorganic materials, polymers also have unique properties that can benefit nanofluidic systems, such as low cost, transparency, and biocompatibility. In addition, their surface properties can be easily modified by functional groups or the cross-link density (Liu et al., 2003; Chen and Li, 2010). Polymers, such as PMMA, polydimethylsiloxane (PDMS), and polyethylene terephthalate (PET), have been used for fabricating nanofluidic devices (Duan et al., 2013). However, the mechanical strength of polymers cannot compete with that of inorganic materials. This may limit the applications of polymers in fluidic devices containing very small nanochannels as the driving pressure needed increases with decreasing channel size (Li et al., 2015).

4.9 Nanofluidic systems: examples

In this last section, two nanofluidic systems are introduced. The first system is fabricated through focused ion-beam milling, similar to FIBL discussed in Section 4.6.2. It is a flexible method for various substrates and the size of nanochannels can be controlled down to 5 nm. The second system contains an array of nanochannels with heterogeneous surface properties, which are constructed using the surface nanomaching method. The fluidic system is capable of rectifying fluid flows in different flow directions, including a special case, i.e., fluidic diode, where flows in one direction are completely blocked.

4.9.1 Fabrication of sub-5 nm nanochannels

In this example, nanochannels are fabricated in a quartz substrate using gallium ion beams (Menard and Ramsey, 2011). First, the substrate is coated with a 130-nm-thick chromium masking layer, as illustrated in Figure 4.16a. Chromium is chosen because its sputtering yield is suitable for the fabrication technique and it is easy to be removed through chemical etching. Nanochannels are then milled by rastering the Ga ion beam over a rectangular region, whose width is about the desired width of the channel (Figure 4.16b). The channel width can be adjusted by varying the beam current, beam probe size, and the thickness of the Cr layer. The depth of the channels depends on the number of times the beam is scanned.

Figure 4.17a and b shows 50- and 100-nm channels milled by a 50-pA-current and 20-nm-diameter (fwhm) beam. The corresponding channels in the quartz substrate are shown in Figure 4.17d and e, where the Cr layer is removed by chemical etchants (e.g., 15-g cerium sulfate, 1,215-mL H_2O, and 90-mL HNO_3; Geschke et al., 2004). It can be seen that the channels have roughly square cross-sections.

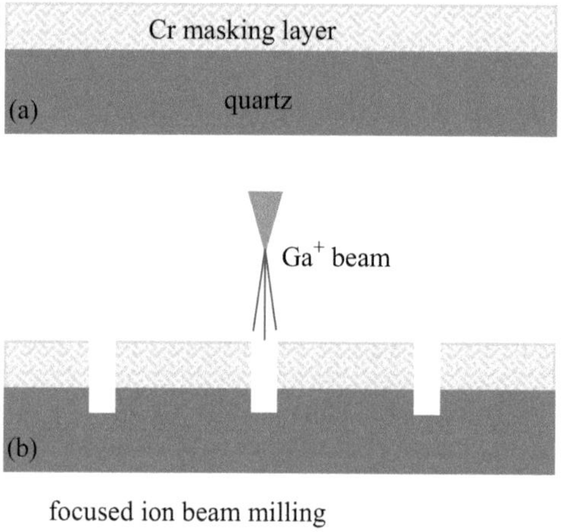

Figure 4.16 Schematic of focused ion-beam milling. (a) A quartz substrate coated with a Cr masking layer. (b) Milling of nanochannels using Ga ion beams.

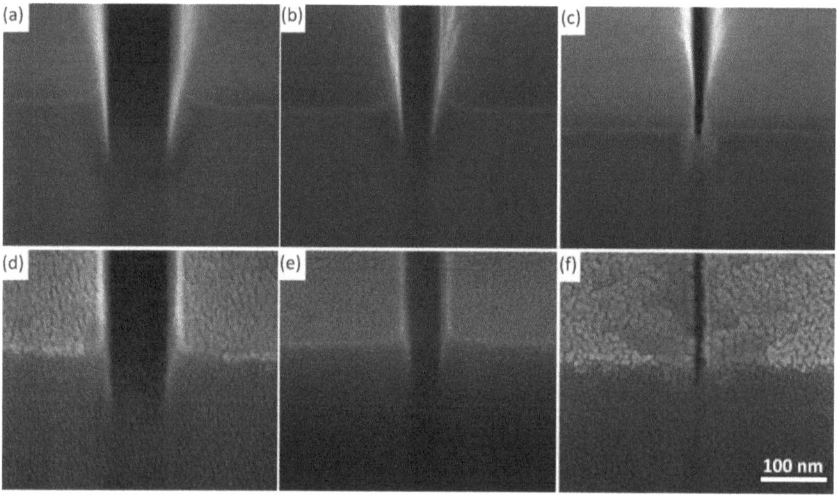

Figure 4.17 SEM images of nanochannels fabricated in a quartz substrate coated with a 130-nm-thick Cr masking film using the method of focused ion-beam milling. (a–c) with and (d–f) without a Cr layer. (a, d) cross section is about 100 nm × 100 nm; (b, e) cross section is about 50 nm × 50 nm; (c, f) cross section is about 12 nm × 12 nm. (Reprinted with permission from Menard and Ramsey (2011), Copyright (2011) American Chemical Society.)

If the ion beam is scanned along a line, channels of smaller widths can be created. Figure 4.17c and f shows a nanochannel of about 15 nm in width with and without the Cr layer (beam current: 1.5 pA; probe diameter: 7 nm). If the thickness of Cr layer is increased, the channel size can be further decreased. Figure 4.18 shows a nanochannel with a width smaller than 5 nm in a quartz substrate coated with a 300-nm Cr film.

To make a complete fluidic system, microchannels or fluid reservoirs connecting with the nanochannels need to be fabricated. This is conducted through the regular DRIE. After the removal of the Cr layer, the whole structures, including micro- and nanochannels, are sealed by a quartz plate through fusion bonding discussed in Section 4.3.3. The substrate and the cover plate are cleaned first and the bonding between the substrate and the cover is performed at 900°C for more than 10 h.

The quality of the fabrication is tested by filling the channels with a dye or electrically driving fluorescent DNA molecules through the channels, as demonstrated in Figure 4.19. The microscopic images of the channels indicate that the channels are free of defects and the surface roughness is small.

This method also works well for Si and other substrates. Figure 4.20 depicts nanohannels fabricated in Si substrates. More details about the fabrication method can be found in the literature (Menard and Ramsey, 2011).

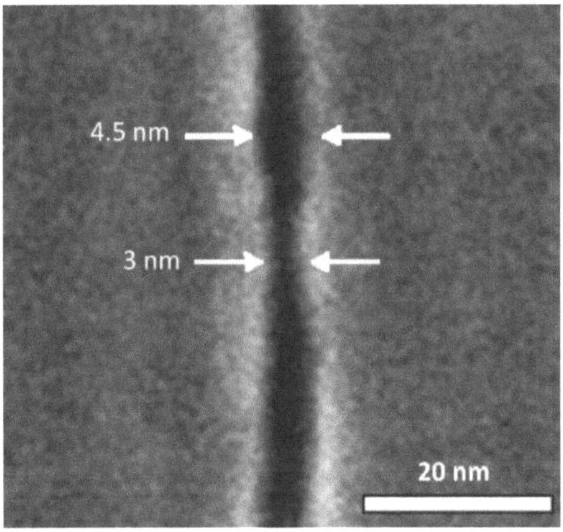

Figure 4.18 SEM image of a sub-5-nm channel in a quartz substrate coated with a 300-nm-thick Cr film. (Reprinted with permission from Menard and Ramsey (2011), Copyright (2011) American Chemical Society.)

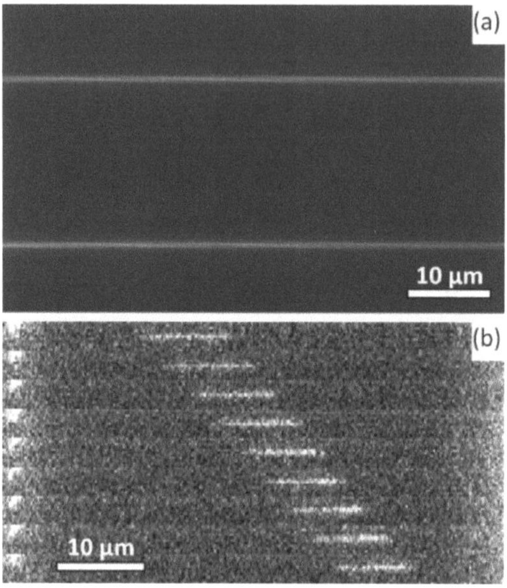

Figure 4.19 Testing of nanochannels. (a) SEM image of 30-nm channels filled with 2-mM fluorescein dye. (b) Frames showing the transport of a DNA molecule in a 50-nm channel. (Reprinted with permission from Menard and Ramsey (2011), Copyright (2011) American Chemical Society.)

4.9.2 Fabrication of heterogeneous nanochannels

Nanochannels with different surface energies can be used to achieve various functions, such as flow rectifications (Li et al., 2015) and thermally driven flows (Liu and Li, 2010). Heterogeneous channels can be fabricated using the surface nanomachining method. A nanofluidic system containing heterogeneous nanochannels is illustrated in Figure 4.21, where an array of nanochannels are connected to four reservoirs through two microchannels. The nanochannels have different surface properties. The left half of the channels is hydrophilic and the right half is hydrophobic.

The nanochannel fabrication starts with the hydrophobic part. The pattern of this part is first defined through the traditional photolithography on a silicon wafer coated with 1-μm-thick SiO_2 layer and then wet etched (H_2SO_4:H_2O_2 = 1:10). This is followed by growing a 5-nm-thick Al_2O_3 layer using plasma atomic layer deposition (Figure 4.22a). The Al_2O_3 layer is then modified by a low surface energy material (1H, 1H, 2H, 2H-Perfluorododecyltrichlorosilane, 97%, Sigma-Aldrich) through CVD at 60°C for 20 h. The modified Al_2O_3 layer is hydrophobic and will serve as the bottom surface of the nanochannels. Next, a 100-nm-thick silicon sacrificial layer is deposited on the substrate (Figure 4.22b). The thickness of

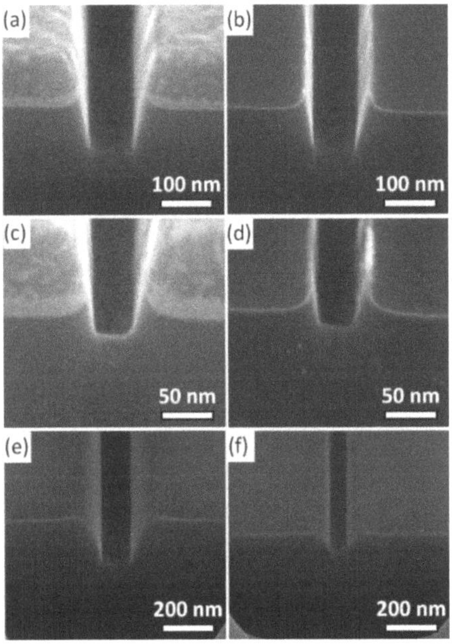

Figure 4.20 SEM images of nanochannels milled in Si substrates coated with a 150-nm-thick Al film. (a, c) with and (b, d) without the Al layer. The cross section is about 100 nm × 100 nm in (a, b) and 35 nm × 35 nm in (c, d). Channels in (e) and (f) are milled in Si directly and their sizes are 100 nm × 200 nm and 65 nm × 70 nm, respectively. (Reprinted with permission from Menard and Ramsey (2011), Copyright (2011) American Chemical Society.)

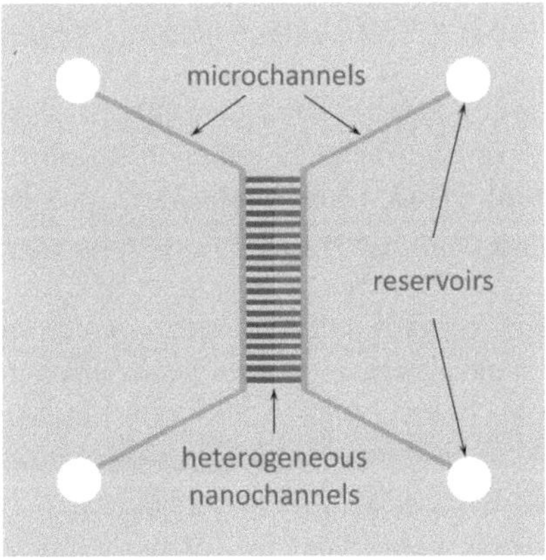

Figure 4.21 Schematic of a nanofluidic system containing heterogeneous nanochannels. (Reprinted with permission from Li et al., 2015, Copyright (2015) by American Physical Society.)

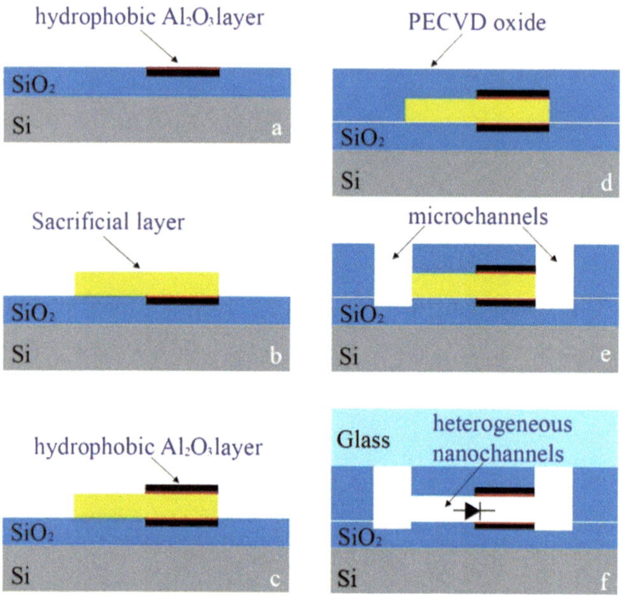

Figure 4.22 Fabrication of nanochannels with heterogeneous surface wettabilities. (a) Growth of a 5-nm-thick Al_2O_3 layer on SiO_2-coated silicon wafer through photolithography and wet etching. The Al_2O_3 layer is subsequently modified by a low surface energy material (red). (b) Deposition of silicon sacrificial layer. (c) Fabrication of another Al_2O_3 layer serving as the top wall of the hydrophobic part. (d) Wrapping with SiO_2 layer through plasma-enhanced CVD (PECVD). (e) Reactive ion etching of microchannels and reservoirs (not shown). (f) Anodic bonding of a glass cover and the removal of the sacrificial layer. (Reprinted with permission from Li et al., 2015, Copyright (2015) by American Physical Society.)

the sacrificial layer is the height of the nanochannels, which can be controlled by changing the deposition time. The same approach is used to construct the top hydrophobic surface (Al_2O_3 layer) on the sacrificial layer (Figure 4.22c). The substrate is then capped by a 5-μm-thick SiO_2 film using plasma-enhanced CVD (PECVD) (Figure 4.22d). Photolithography and RIE are conducted to fabricate the microchannels and through holes are etched as the reservoirs by deep RIE (Figure 4.22e). The nanostructures are enclosed through anodic bonding by a clean pyrex glass wafer below 380°C and 800 V. Finally, the sacrificial layer is removed by dry XeF_2 etching and heterogeneous nanochannels are fabricated (Figure 4.22f). In this fabrication, the hydrophobic part is Al_2O_3 and the hydrophilic part is SiO_2. Water contact angles for these two parts are 158° and 16°, respectively, as shown in Figure 4.23.

The top panel of Figure 4.24 shows an SEM image of the cross section of a nanochannel. The height and width of the channel are 97.8 nm and 10 μm, respectively.

Figure 4.23 Water contact angles for the hydrophobic and hydrophilic parts of the heterogeneous nanochannels. (a) The contact angle is 158° for an Al_2O_3 surface modified by perfluorododecyltrichlorosilane. (b) The contact angle is 16° for a SiO_2 surface. (Reprinted with permission from Li et al., 2015, Copyright (2015) by American Physical Society.)

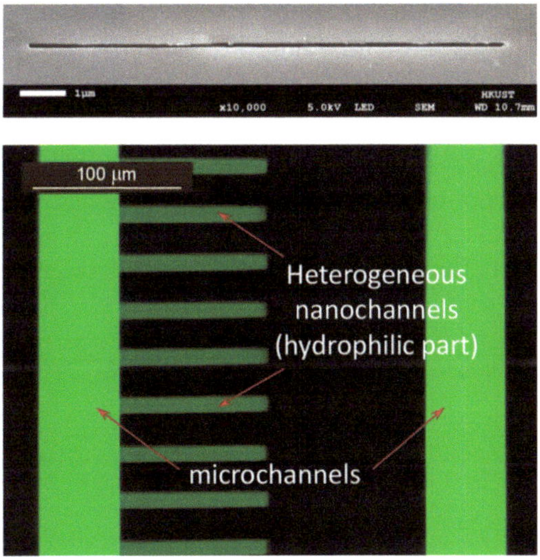

Figure 4.24 Nanochannel characterization. Top panel: SEM image showing the height (97.8 nm) and width (10 μm) of a nanochannel. Bottom panel: Microscopic image of a fluorescent flow through the fluidic system in the forward direction. (Reprinted with permission from Li et al., 2015, Copyright (2015) by American Physical Society.)

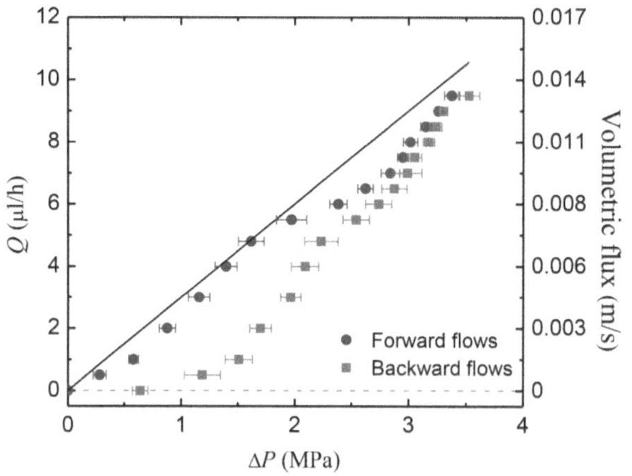

Figure 4.25 Total volumetric flow rate and flux through the fluidic system. The solid line is the prediction of the Navier–Stokes equation. Error bars denote the standard deviation of three experiments. (Adapted with permission from Li et al., 2015, Copyright (2015) by American Physical Society.)

The fluidic system is tested by flushing fluorescent dye (Rhodamine 6G)-laden DI water. A snapshot of a forward flow is shown in the bottom panel of Figure 4.24. It is seen that the flow (green) could go through nanochannels smoothly from the left to the right microchannel, indicating that the nanochannels are free of defects (the hydrophobic part is not transparent to fluorescence).

The fluidic system in Figure 4.21 can achieve different flow scenarios. Figure 4.25 depicts flow rates in the forward (from hydrophilic to hydrophobic) and backward directions as a function of pressure drop. It can be seen that the fluidic system acts as a fluidic diode for $\Delta P < 0.63$ MPa, where flows in the backward direction are completely blocked. The mechanism for the flow phenomenon will be discussed in Chapter 6.

References

Abgrall, P. and Gué, A.-M. 2007. Lab-on-chip technologies: Making a microfluidic network and coupling it into a complete microsystem—A review. *J. Micromech. Microeng.* 17:R15–R49.

Alexe, M., Harnagea, C., and Hesse, D. 2004. Non-conventional micro-and nanopatterning techniques for electroceramics. *J. Electroceramics* 12:69–88.

Bagolini, A., Pakula, L., Scholtes, T.L.M., Pham, H.T.M., French, P.J., and Sarro, P.M. 2002. Polyimide sacrificial layer and novel materials for post-processing surface micromachining. *J. Micromech. Microeng.* 12:385–389.

Chen, L.Q. and Li, Z.G. 2010. Bouncing droplets on nonsuperhydrophobic surfaces. *Phys. Rev. E* 82:016308.

Cho, Y.H., Park, J., Park, H., Cheng, X., Kim, B.J., and Han, A. 2010. Fabrication of high-aspect-ratio polymer nanochannels using a novel Si nanoimprint mold and solvent-assisted sealing. *Microfluid. Nanofluid.* 9:163–170.

Crouse, D., Lo, Y.H., Miller, A.E., and Crouse, M. 2000. Self-ordered pore structure of anodized aluminum on silicon and pattern transfer. *Appl. Phys. Lett.* 76:49–51.

Dai, H.J. 2002. Carbon nanotubes: Synthesis, integration, and properties. *Acc. Chem. Res.* 35:1035–1044.

Duan, C.H., Wang, W., and Xie, Q. 2013. Fabrication of nanofluidic devices. *Biomicrofluidics* 7:026501.

Geschke, O., Klank, H., and Telleman, P. 2004. *Microsystem Engineering of Lab-on-a-Chip Devices.* p. 168. Weiheim: Wiley-VCH.

Jaeger, R.C. 2002. *Lithography: Introduction to Microelectronic Fabrication.* Upper Saddle River: Prentice Hall.

Han, U.B. and Lee, J.S. 2016. Bottom-up synthesis of ordered metal/oxide/metal nanodots on substrates for nanoscale resistive switching memory. *Sci. Rep.* 6: 25537.

Kern, W. 1993. *Handbook of Semiconductor Cleaning Technology.* Park Ridge: Noyes Publishing.

Kovtyukhova, N.I., Mallouk, T.E., and Mayer, T.S. 2003. Templated surface sol-gel synthesis of SiO_2 nanotubes and SiO_2-insulated metal nanowires. *Adv. Mater.* 15:780–785.

Li, L., Mo, J.W., and Li, Z.G. 2015. Nanofluidic diode for simple fluids without moving parts. *Phys. Rev. Lett.* 115:134503.

Liang, J., Chik, H., Yin, A., and Xu, J. 2002. Two-dimensional lateral superlattices of nanostructures: Nonlithographic formation by anodic membrane template. *J. Appl. Phys.* 91:2544–2546.

Liu, C. and Li, Z.G. 2010. Molecular dynamics simulation of composite nanochannels as nanopumps driven by symmetric temperature gradients. *Phys. Rev. Lett.* 105:174501.

Liu, T.B., Burger, C., and Chu, B. 2003. Nanofabrication in polymer matrices. *Prog. Polym. Sci.* 28:5–26.

Menard, L.D. and Ramsey, J.M. 2011. Fabrication of sub-5 nm nanochannels in insulating substrates using focused ion beam milling. *Nano Lett.* 11:512–517.

O'Sullivan, J.P. and Wood, G.C. 1970. The morphology and mechanism of formation of porous anodic films on aluminum. *Proc. Roy. Soc. Lond. A.* 317:511:543.

Pang, P., He, J., Park, J.H., Krstic, P.S., and Lindsay, S. 2011. Origin of giant ionic currents in carbon nanotube channels. *ACS Nano.* 9:7277–7283.

Prasek, J., Drbohlavova, J., Chomoucka, J., Hubalek, J., Jasek, O., Adam, V., and Kizek, R. 2011. Methods for carbon nanotubes synthesis-review. *J. Mat. Chem.* 21:15872–15884.

Quenzer, H.J., Dell, C., and Wagner, B. 1996. Silicon-silicon anodic-bonding with intermediate glass layers using spin-on glasses. Pro. IEEE 9th Int. Workshop on Microelectromechanical Systems. pp. 272–276.

Quévy, E., Rarvais, B., Baskin, J.P., Buchaillot, L., Flandre, D., and Collard, D. 2002. A modified bosch-type process for precise surface micromachining of polysilicon. *J. Micromech. Microeng.* 12:328–333.

Sokhan, V.P., Nicholson, D., and Quirke, N. 2002. Fluid flow in nanopores: Accurate boundary conditions for carbon nanotubes. *J. Chem. Phys.* 117:8531–8539.

Thompson, G.E. and Wood, G.C. 1981. Porous anodic film formation on aluminum. *Nature* 290:230–232.

Tseng, A.A. 2005. Recent developments in nanofabrication using focused ion beams. *Small* 1:924–939.

Wallis, G. and Pomerantz, D.I. 1969. *Field assisted glass-metal sealing. J. Appl. Phys.* 40:3946–3949.

Wang, X. and Smirnov, S. 2009. Label-free DNA sensor based on surface charge modulated ionic conductance. *ACS Nano* 3:1004–1010.

Williams, K.R., Gupta, K., and Wasilik, M. 2003. Etch rates for micromachining processing II. *J. Microelectromech. Syst.* 12:761–778.

Xia, Q.F., Morton, K.J., Austin, R.H., and Chou, S.Y. 2008. Sub-10nm self-enclosed self-limited nanofluidic channel arrays. *Nano Lett.* 8:3830–3833.

Yan, R.X., Liang, W.J., Fan, R., and Yang, P.D. 2009. Nanofluidic diodes based on nanotube heterojunctions. *Nano Lett.* 9:3820–3825.

Yasin, S., Hasko, D.G., and Ahmed, H. 2001. Fabrication of<5nm width lines in poly(methylmethacrylate) resist using a water:isopropyl alcohol developer and ultrasonically-assisted development. *Appl. Phys. Lett.* 78:2760–2762.

Zhang, M., Bando, Y., and Wada, K. 2000. Silicon dioxide nanotubes prepared by anodic alumina as templates. *J. Mater. Res.* 15:387–392.

Chapter 5 Nanoscale capillarity

5.1 Introduction

Capillarity at the nanoscale is referred to as the filling of a fluid into a nanochannel. Similar to the capillarity at the macroscale discussed in Chapter 1, nanoscale capillarity also depends on the surface tension of the fluid and the surface energy of the nanochannel, which determine the contact angle θ of the fluid. If the channel surface energy is high such that $\theta < 90°$ (e.g., hydrophilic channels), the surface tension of the concave meniscus of the fluid will pull the fluid into the channel, as illustrated in Figure 5.1a. However, if $\theta > 90°$, a pressure is required to drive the fluid into the channel (Figure 5.1b), which is usually called fluid infiltration.

At the macroscale, the pressure difference for cylindrical tubes across the meniscus ΔP at the steady state is described by the Young–Laplace equation:

$$\Delta P = P_l - P_v = -\frac{2\gamma \cos\theta}{R} \tag{5.1}$$

where the subscripts l and v denote liquid and vapor phases and R is the tube radius. For rectangular channels, the pressure difference ΔP is given as follows:

$$\Delta P = -2\gamma \cos\theta \left(\frac{1}{H} + \frac{1}{W} \right) \tag{5.2}$$

where H and W are the channel height and width, respectively.

For the dynamic filling of a fluid into a hydrophilic, cylindrical channel, on the basis of Poiseuille's law, the length of the fluid in the channel is given by (Washburn, 1921)

$$L = \sqrt{\frac{R\gamma \cos\theta}{2\mu} t} \tag{5.3}$$

For rectangular channels, the length of the fluid is given by (Middleman, 1977)

$$L = \sqrt{\frac{4\gamma \cos\theta \left(1/H + 1/W \right) C_s}{\mu} t} \tag{5.4}$$

where C_s is a shape factor given by

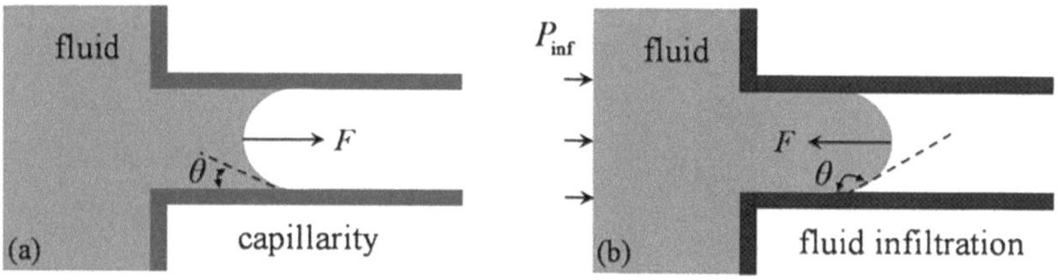

Figure 5.1 Capillarity of fluids in channels with different surface energies. (a) High surface energy channel with contact angle $\theta < 90°$. (b) Low surface energy channel with $\theta > 90°$.

$$C_s = \frac{H^2}{12}\left(1 - \frac{192H}{\pi^5 W}\sum_{k=1}^{\infty}\frac{1}{(2k-1)^5}\tanh\frac{(2k-1)\pi W}{2H}\right) \tag{5.5}$$

For two-dimensional channels, $H/W \ll 1$, $C_s \approx H^2/12$ and Equation 5.4 is reduced to

$$L = \sqrt{\frac{H\gamma\cos\theta}{3\mu}t} \tag{5.6}$$

From Equations 5.1 to 5.4, it can be seen that capillarity strongly depends on the physical properties of a fluid, i.e., surface tension γ and viscosity μ. As discussed in Chapter 2, the viscosity of a fluid fluctuates at the nanoscale, especially next to the channel surface. This may cause Equations 5.3 and 5.4 to be invalid for fluid filling in nanochannels. The applicability of Equations 5.1 and 5.2 to nanoscale capillarity is also questionable because molecular interactions, which are important at the nanoscale, are not considered in these equations. In the following sections, the size effects on the surface tension are discussed first, which is followed by nanoscale capillarity in both hydrophilic and hydrophobic nanochannels.

5.2 Size effects on fluid surface tension

For curved fluid surfaces, such as droplet surfaces and the meniscus of a liquid in a capillary, intensive investigations have been conducted to study the dependence of the fluid surface tension on the curvature of the fluid surface. Thermodynamics analysis has shown that the variation of the surface tension γ of a liquid caused by the curvature of the liquid surface is negligible if the characteristic size (e.g., diameter of a capillary) is above 100 nm (Guggenheim, 1940). If the characteristic size is below 100 nm, however, γ may change. For droplets, it has been suggested that γ varies when the radius R of a droplet is different from the equimolar radius

R_e (also known as the Gibbs adsorption radius, which depends on the density change from the liquid to vapor phase) (Tolman, 1949), and γ can be described by

$$\gamma = \gamma_0 \left(\frac{1}{1 + 2\delta / R} \right) \tag{5.7}$$

where γ_0 is the surface tension for a planar surface and $\delta = R_e - R$ is termed Tolman's length, which is of the order of angstrom to ~10 nm. For water ($\gamma_0 = 71.9$ mN/m), if the OH bond length is used as the Tolman length, $\delta = 0.96$ Å (Dewar et al., 1985), γ, as a function of droplet radius, can be calculated, as shown in Figure 5.2. Another surface tension model (Lu and Jiang, 2005) yields the following equation:

$$\gamma = \gamma_0 \left(1 - \frac{1}{4R/h - 1} \right) \exp\left(-\frac{2S_b}{3R_g} \frac{1}{4R/h - 1} \right) \tag{5.8}$$

where h is the atomic diameter, S_b is the solid–vapor transition entropy of crystals, and R_g is the gas constant. The surface tension of water predicted by Equation 5.8 is also plotted in Figure 5.2 ($h = 0.96$ Å and $S_b = 36.5$ J/(mol·K) are used). It is seen that γ approaches γ_0 as the droplet radius is larger than 4 nm. Compared with Equation 5.7, Equation 5.8 predicts a smaller deviation of γ from γ_0 for small droplets and $\gamma/\gamma_0 > 95\%$ for $R > 2$ nm.

The surface tension of water determined by Equation 5.1 in a capillary has also been measured through experiments (Sobolev et al., 2000). It is found that $\gamma/\gamma_0 \approx 1$

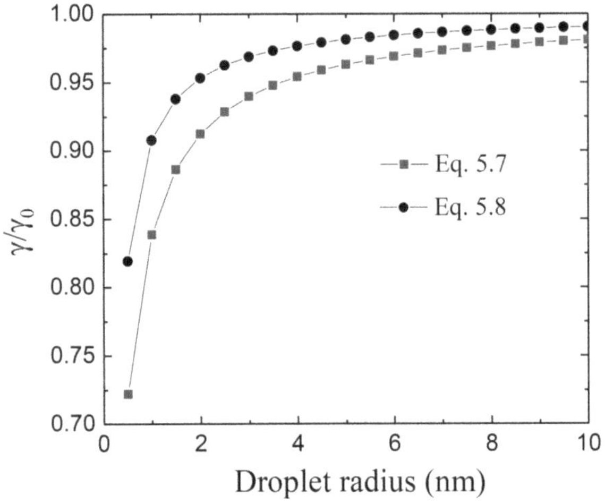

Figure 5.2 Surface tension of water droplets versus droplet radius.

for capillary radius down to 45 nm, which is consistent with Equations 5.7 and 5.8. Measurements of surface tension in capillaries with radius below 10 nm are unavailable due to the limitation in spatial resolution. Molecular dynamics simulations of water droplets also show that $\gamma/\gamma_0 \approx 1$ when the radius of water droplets is larger than 1 nm, although numerical approaches suffer from large numerical errors (Samsonov et al., 2003). Therefore, the surface tension of fluids, especially that of water, can be roughly considered as size independent at the nanoscale if the characteristic size is larger than 2 nm. The error caused by such a treatment is below 10%. If the characteristic size is below 2 nm, however, special attention should be paid to the variation of surface tension.

5.3 Capillary filling in hydrophilic nanochannels

The validity of Equation 5.1 for static capillarity has been confirmed by experiments in cylindrical quartz channels with radius down to 45 nm (Sobolev et al., 2000). For the dynamic filling of a fluid into a nanochannel, Equations 5.3 and 5.6 suggest that the relationship between the length L of the fluid in the channel and time depends on both the surface tension and the viscosity of the fluid. Even though γ can be treated independent of channel size, the fluctuation of fluid viscosity may make Equations 5.3 and 5.6 invalid in nanochannels.

Capillary filling of water in nanochannels has been investigated (Tas et al., 2004). The fluidic system contains an array of parallel nanochannels, which are in connection with two microchannels at the ends, as shown in Figure 5.3. The whole system is fabricated through micro- and nanomachining techniques in a silicon substrate. Nanochannels are created by wet etching and microchannels are prepared by RIE. The nanochannel length and width are 1 cm and 20 μm, respectively. The height of the nanochannels ranges from 50 to 150 nm. The surface of the nanochannels is thermally oxidized such that it is covered by a 20-nm-thick oxide layer, which makes the nanochannels hydrophilic. Rulers are also fabricated next to the nanochannels to measure the position of water meniscus in the nanochannels, as shown in Figure 5.4. The channels are enclosed by fusion bonding with a borofloat glass wafer. Access holes at the ends of the microchannels are powder-blasted in the glass cover.

In the experiments, a 7-μL demineralized water droplet is introduced into the access holes at the ends of a microchannel by pipetting. Subsequently, the droplet fills into the microchannel and then into nanochannels. The position of the water meniscus is recorded through video imaging with a frame rate of 25 images/s. Figure 5.4 shows the position of water meniscus at a time. The measurements are conducted up to a position, where bubbles form due to the enclosure of air. Figure 5.5 shows the position of water meniscus, x, as a function of $\sqrt{t}$ in nanochannels of different heights. The predictions of Equation 5.6 are also plotted

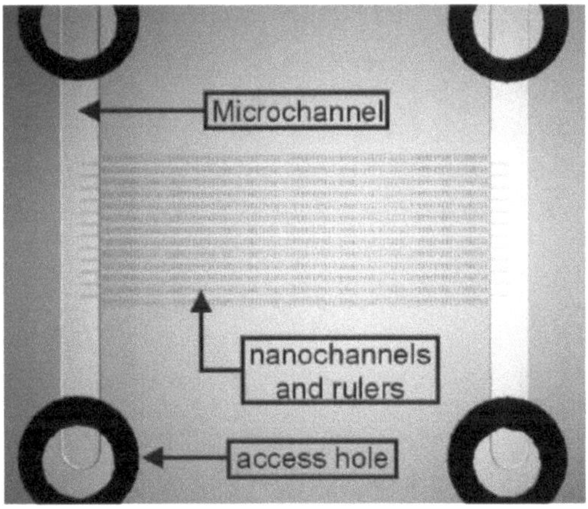

Figure 5.3 Fluidic system for studying the dynamic capillary filling of water in nanochannels. (Reprinted from Tas et al., 2004 with permission of AIP Publishing.)

Figure 5.4 Rulers fabricated next to the nanochannels and the capillary filling of water in 110-nm-height nanochannels. The distance between two large ticks is 100 μm. (Reprinted from Tas et al., 2004 with permission of AIP Publishing.)

(lines) with $\cos\theta = 0.94$ and $\gamma = 71.0\,\text{mN/m}$. It can be seen that the position of the moving water meniscus is a linear function of $\sqrt{t}$ regardless of the channel height. Such a linear relationship agrees well with Equation 5.6. However, the quantitative dependence of x on $\sqrt{t}$ depends on the channel height. In relatively large nanochannels (152 nm), the experimental results are close to theoretical predictions. As the channel size decreases, the discrepancy between experiments and theory increases, as shown in Figure 5.5.

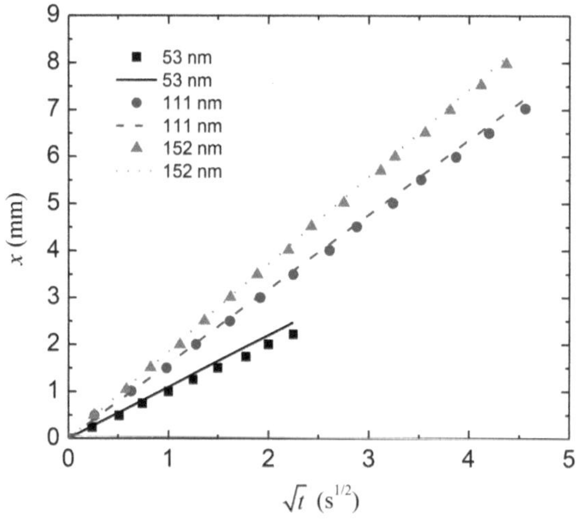

Figure 5.5 Position of water meniscus in nanochannels. Symbols are experimental data and lines are the predictions of Equation 5.6. (Data are extracted from Tas et al., 2004.)

If the surface tension and contact angle are assumed to be invariant, the increasing discrepancy between experiments and Equation 5.6 as the channel size decreases in Figure 5.5 should be attributed to the variation of water viscosity, which, as discussed previously, deviates from the bulk value in nanochannels. On the basis of Equation 5.6, the apparent viscosity μ_{ap} of water is obtained and the ratio of μ_{ap} to the bulk viscosity μ, μ_{ap}/μ, is given in Table 5.1. It must be noted that the apparent viscosity μ_{ap} obtained from the experimental data in Figure 5.5 can be viewed as the mean viscosity, which does not reflect the viscosity fluctuation in nanochannels. Nevertheless, the values of μ_{ap} indicate that the channel surface does affect fluid viscosity, especially in small nanoconfinements, which qualitatively confirms the numerical results of fluid density and viscosity distributions in nanochannels in Chapter 2 (Figures 2.4–2.6).

As the channel height is reduced further, the deviation of experiments from Washburn's law (Equation 5.6) increases significantly. Figure 5.6 depicts the length of DI water column in small nanochannels as a function of $\sqrt{t}$ (Haneveld et al., 2008). The nanochannels are fabricated in a silicon dioxide layer ($\theta < 10°$

Table 5.1 Ratio of Apparent Viscosity to the Bulk Viscosity of Water in Nanochannels[a]

Channel Height	**53 nm**	**111 nm**	**152 nm**
μ_{ap}/μ	1.22 ± 0.04	1.09 ± 0.04	1.06 ± 0.02

[a] Data extracted from Tas et al. (2004).

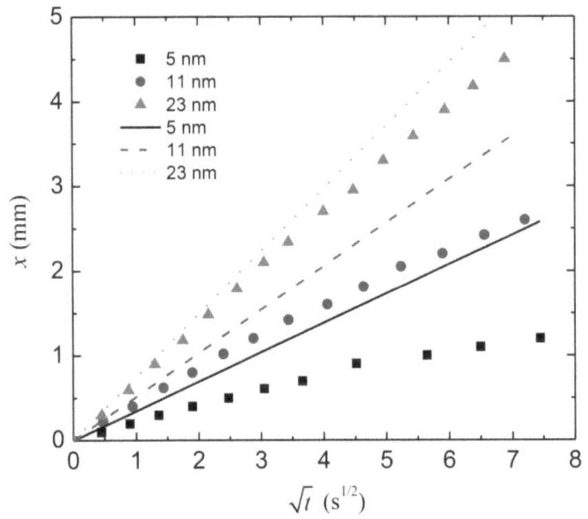

Figure 5.6 Position of water meniscus in small nanochannels. Symbols are experimental data and lines are the predictions of Equation 5.6. (Data are extracted from Haneveld et al., 2008.)

and $\cos\theta \approx 0.99$) on a silicon wafer. The length and width of the channels are 1 cm and 20 μm, respectively. The heights of the channels are 5, 11, and 23 nm. It can be seen that experimental data deviate greatly from Equation 5.6 in small nanochannels, in comparison with Figure 5.5. In such small channels, even though the contact angle θ may vary (Sobolev et al., 2000), the variation in θ could not account for the large discrepancy. If the difference between experiments and theory is assumed to be mainly caused by the changes in viscosity, the apparent viscosity is larger than the bulk value by a factor of 1.5–2 in 5-nm channels. As can be seen from Figure 5.6, the relationship between x and $\sqrt{t}$ becomes nonlinear as the channel height decreases. This defies Washburn's law. Possible reasons for the nonlinear behavior of dynamic filling could be the electroviscous effect and channel distortion (Haneveld et al., 2008). Furthermore, the friction due to the strong molecular interaction in small hydrophilic nanochannels could play an important role (Liu and Li, 2010). Currently, how these factors affect the capillary kinetics is not well understood. There could also be new flow physics underlying the unusual capillary behavior. All these require extensive numerical and experimental studies.

5.4 Capillary filling in hydrophobic nanochannels

For hydrophobic (or low surface energy) channels, the contact angle $\theta > 90°$ and the capillary filling of a fluid requires an infiltration pressure at the channel entrance, i.e., $\Delta P = P_l - P_v > 0$, as indicated by Equation 5.1 (see Figure 5.1b). For nanochannels, however, Equation 5.1 becomes questionable. Experiments of water intrusion into silica nanchannels have been performed (Helmy et al., 2005). Ordered mesoporous silicas with pore radius, $R_{pore} \approx 2 - 4$ nm, are prepared using the water porosimetry technique (Fadeev and Eroshenko, 1997). The channel surface is modified by monolayers of trimethylsilyl groups (TMS), for which the water contact angle is about 106°. A TEM image of the nanochannels is shown in Figure 5.7.

Figure 5.8 depicts the pressure difference ΔP for nanochannels of different radii. It can be seen that the infiltration pressure measured in experiments is much higher than the predictions of Equation 5.1. The best fit between the experimental data and Equation 5.1 gives a contact angle of ~120°, which is unreasonably high for TMS surfaces (Helmy et al., 2005). The significant discrepancy is speculated to be caused by the formation of a low-density fluid layer at the interface. This hypothesis is based on the measurement of water volume in the channels, which is smaller than the volume measured through nitrogen adsorption (Helmy et al., 2005). Such a low-density fluid layer reduces the volume of the channels available to water such that the effective radius is smaller than the physical radius of the channels, $R_{pore}^{eff} < R_{pore}$. The thickness of the low-density fluid layer h is estimated to range from 0.42 to 0.64 nm for channels of different radii. If $R_{pore}^{eff} = R_{pore} - h$ is employed, the experimental results are found to be in good agreement with the classic theory (Helmy et al., 2005).

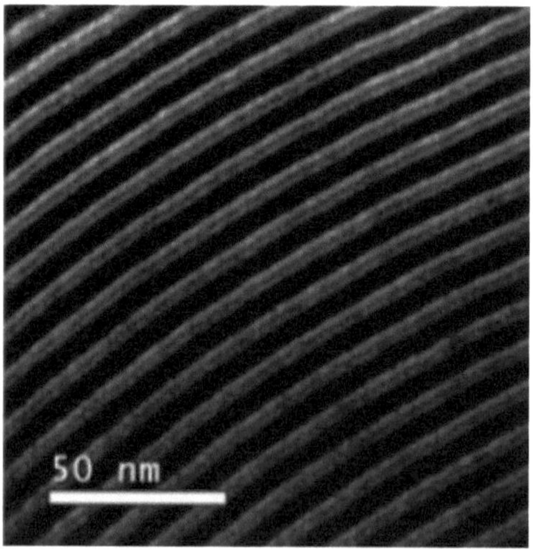

Figure 5.7 TEM image of ordered nanochannels in a mesoporous silica. (Reprinted with permission from Helmy et al., 2005, Copyright (2005) American Chemical Society.)

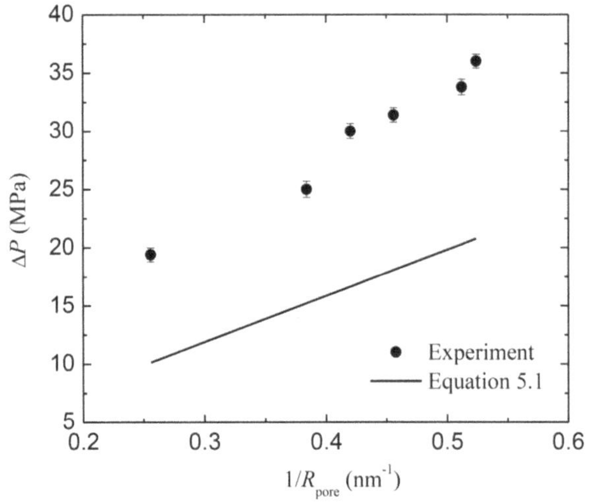

Figure 5.8 Water infiltration pressure for hydrophobic silica nanochannels of different radii R_{pore}. (Dara are extracted from Helmy et al., 2005.)

Although the existence of the low-density layer has not been experimentally confirmed, the explanation of the reduced channel size, i.e., $R_{pore}^{eff} < R_{pore}$, is consistent with numerical observations of fluid distribution in hydrophobic channels, as shown in Figures 2.1c and f, and 2.5. As discussed in Chapter 2, the molecular interactions between fluid molecules and wall atoms in low-surface energy channels are mainly repulsive, which generate a depletion layer next to the wall and reduce the effective channel size. Therefore, the size variation for hydrophobic nanochannels due to molecular interactions is a reason that causes Equation 5.1 to fall apart at the nanoscale.

In addition to the variation of channel size, a recent MD study shows that the fluid–surface molecular interactions generate an energy barrier at the entrance of hydrophobic nanochannels, which increases the infiltration pressure (Mo et al., 2015). The MD system contains a slit nanochannel and a reservoir. The channel height and length are 3 and 9.8 nm, respectively. Initially, a certain number of water molecules are filled in the reservoir and confined by two rigid walls, as illustrated in the top panel of Figure 5.9. The two parallel walls, which have four atomic layers, are constructed by truncating a rectangular portion of an fcc structure. To control the temperature of the system, the inner three layers of wall atoms are free to vibrate and the temperature is controlled by the Berendsen thermostat at the room temperature, $T = 300\,K$. Periodic boundary conditions (PBCs) are employed in the y- and z-directions for water in the reservoir and in the z-direction for the channel.

The SPC/E water model is used to describe water molecules. Each oxygen and hydrogen atom carries a point charge q equal to −0.8476 and 0.4238 e, respectively. The potential for any two interacting water molecules is given by Equation 3.9. The interactions among wall atoms are modeled by the tight-binding potential and the parameters for Ag are used (see Chapter 3). The water–surface interaction is described by the Lennard-Jones (LJ) potential: $U(r) = 4\varepsilon_{\text{fw}}[(\sigma_{\text{fw}}/r)^{12} - (\sigma_{\text{fw}}/r)^{6}]$ between oxygen and wall atoms with $\sigma_{\text{fw}} = 2.85$ Å. ε_{fw} is varied to change the water–surface interaction strength for considering the surface effects. The equations of motion are integrated using Beeman's algorithm with a time step equal to 1 fs.

The system is first relaxed for 20 ps. The right wall of the reservoir (top panel of Figure 5.9) is then removed and the system is relaxed for another 30 ps. Water is driven into the channel by moving the left rigid wall of the reservoir in the positive x-direction in a stepwise and quasi-static manner. The left wall is moved one step of 0.3 Å every 8 ps, such that the system has sufficiently long time to reach a new equilibrium state. A snapshot in this process is given in the bottom panel of Figure 5.9, where some water molecules are pushed into the nanochannel and the convex water meniscus indicates that the channel surface is hydrophobic.

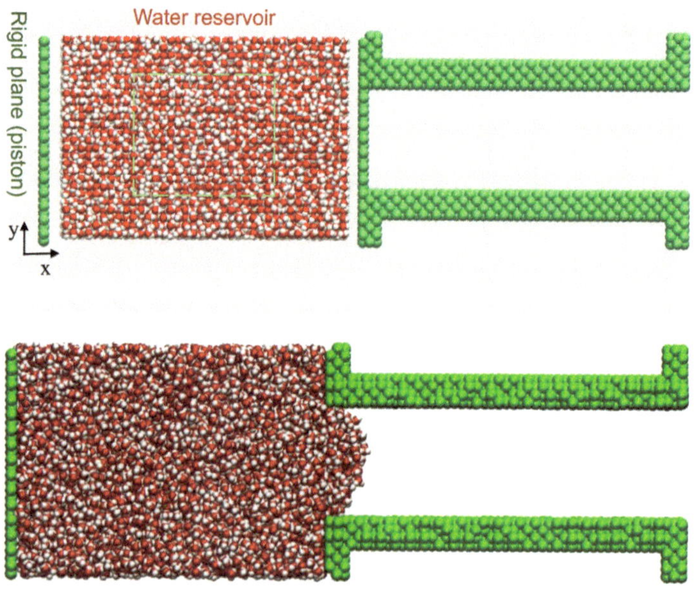

Figure 5.9 Top panel: Initial state of an MD system simulating water infiltration into a hydrophobic nanochannel. Bottom panel: Snapshot of the system during infiltration. (Top panel reprinted with permission from Mo et al., 2015, Copyright (2015) by the American Physical Society.)

The pressure p in the reservoir is computed using Equation 3.29 by sampling sufficient water molecules in a cuboid in the reservoir, as indicated by the dashed lines in the top panel of Figure 5.9. The contact angle θ of water is obtained by fitting the meniscus using a series of isopycnic lines of different densities, as demonstrated in Figure 5.10, where only the 300 kg/m^3 isopycnic line is shown. The contact angles for the isopycnic lines ranging from 200 to 600 kg/m^3 are obtained and the average value is considered as the water contact angle. Figure 5.11 shows the contact angle as ε_{fw} is varied. It can be seen that the contact angle decreases linearly with increasing ε_{fw} and the channel is hydrophobic when ε_{fw} is in the range from 10 K to 130 K (1 K = 1.38065×10^{-23} J).

The pressure p of the reservoir versus the number of water molecules infiltrated into the channel N_{inf} at T = 300 K for water–surface binding energy ε_{fw} = 30 K (θ = 140°) is plotted in Figure 5.12. It is clear that the infiltration is quite different from that in hydrophilic channels. For p < 10 MPa, water molecules do not infiltrate into the channel. Infiltration takes place when p becomes higher than 10 MPa and N_{inf} increases linearly with p for 10 < p < 45 MPa, which corresponds to the formation of meniscus. When p reaches about 45 MPa, it remains almost constant while the number of water molecules keeps increasing, which is the consequence of continuous water infiltration. The constant pressure, ~45 MPa, is denoted as the infiltration pressure p_{inf}, as illustrated in Figure 5.12.

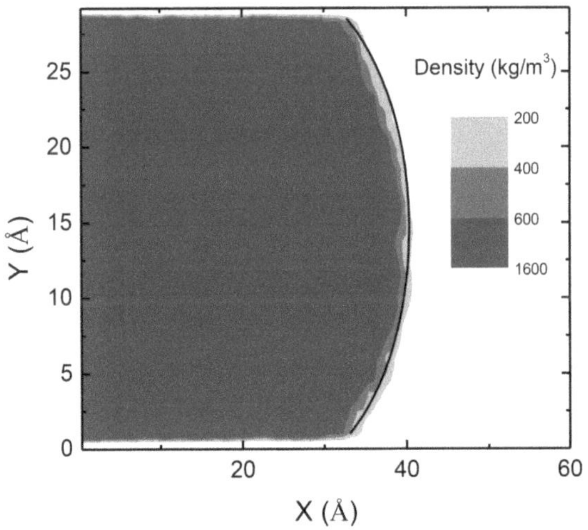

Figure 5.10 Density contour of water in the nanochannel in Figure 5.9 for ε_{fw} = 50 K. The solid line is the isopycnic line for density equal to 300 kg/m^3. (Reprinted with permission from Mo et al., 2015, Copyright (2015) by the American Physical Society.)

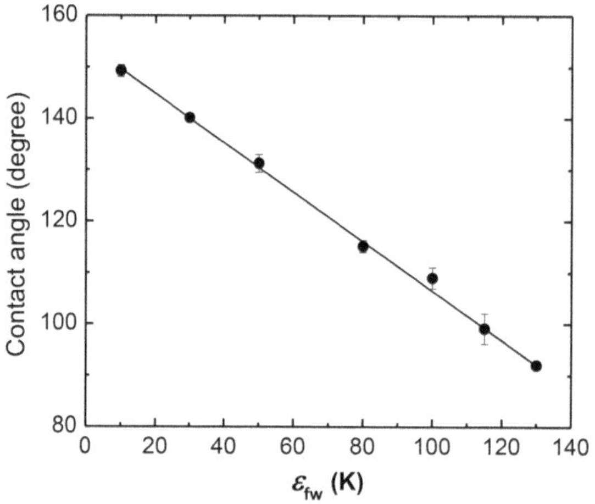

Figure 5.11 Water contact angle in the nanochannel. (Data are extracted from Mo et al., 2015.)

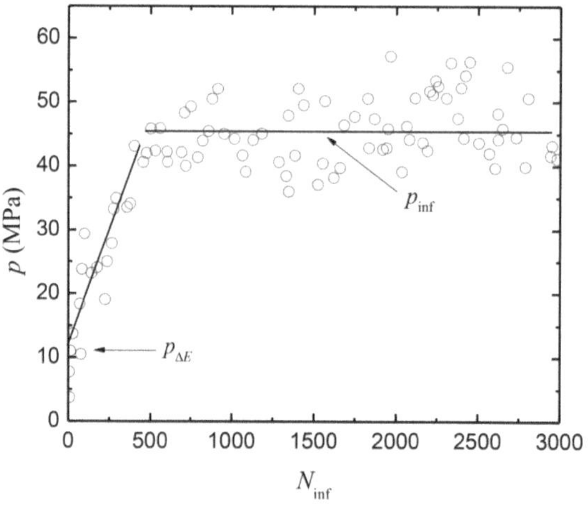

Figure 5.12 Number of water molecules infiltrated into the nanochannel (ε_{fw} = 30 K) as the pressure of the reservoir is varied. (Reprinted with permission from Mo et al., 2015, Copyright (2015) by the American Physical Society.)

The physical pictures for the linear ($10 < p < 45$ MPa) and continuous ($p \approx 45$ MPa) water infiltration in Figure 5.12 are clear. However, the threshold pressure, ~10 MPa, below which water infiltration does not occur, is not straightforward. Further investigations find that the threshold pressure is caused by the water–wall molecular interactions, which generate a potential energy barrier at the channel entrance. As shown in Figure 5.13a, the potential energy in most of the areas at the entrance is above 0.0 eV, which is higher than that of bulk water (–0.4 eV). This indicates that a pressure is needed to help water molecules overcome a potential energy barrier, 0.4 eV, to move into the channel. For the case in Figure 5.12, this pressure is about 10 MPa. As can be seen from Figure 5.13a and b, the mean potential at the channel entrance increases as ε_{fw} decreases, which suggests that the pressure required to drive a fluid into a hydrophobic channel will increase when the hydrophobicity of the channel is increased.

The threshold pressure in Figure 5.12 is named the entrance barrier pressure, $p_{\Delta E}$, which is about 25% of the infiltration pressure p_{inf} and cannot be neglected at the nanoscale. Therefore, for hydrophobic nanochannels, the infiltration pressure p_{inf} is composed of the entrance barrier pressure $p_{\Delta E}$ and the capillary pressure (Equation 5.1). By combining $p_{\Delta E}$, which is computed through MD simulations, and the predictions of Equation 5.1 ($\gamma = 71.6$ mN/m is used), the semi-theoretical infiltration pressures for different surface properties are obtained and compared with the MD results, as shown in Figure 5.14. It is seen that the capillary pressure given by Equation 5.1 is indeed smaller than the infiltration pressure p_{inf}. If it is combined with the entrance barrier pressure $p_{\Delta E}$, the total pressure is in good agreement with p_{inf} computed directly from MD simulations. Therefore, for a hydrophobic nanochannel, where the entrance barrier pressure $p_{\Delta E}$ is significant, the pressure drop in Equation 5.1 can be calculated as follows:

$$P_{inf} = P_l - P_v = p_{\Delta E} - \frac{2\gamma\cos\theta}{R} \tag{5.9}$$

where $p_{\Delta E}$, based on the linear fit of the data in Figures 5.11 and 5.14, is given as

$$p_{\Delta E} = 4.723 - 9.24\cos\theta \tag{5.10}$$

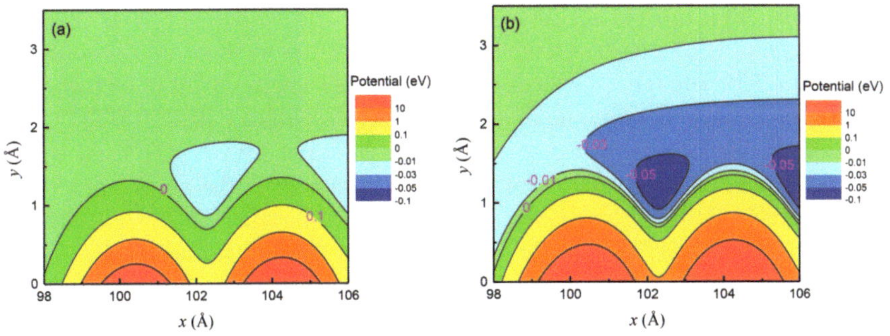

Figure 5.13 Potential distribution in an *xy* plane close to the bottom wall at the channel entrance (the channel starts at $x = 98$ Å): (a) $\varepsilon_{fw} = 30$ K and (b) $\varepsilon_{fw} = 130$ K. (Reprinted with permission from Mo et al., 2015, Copyright (2015) by the American Physical Society.)

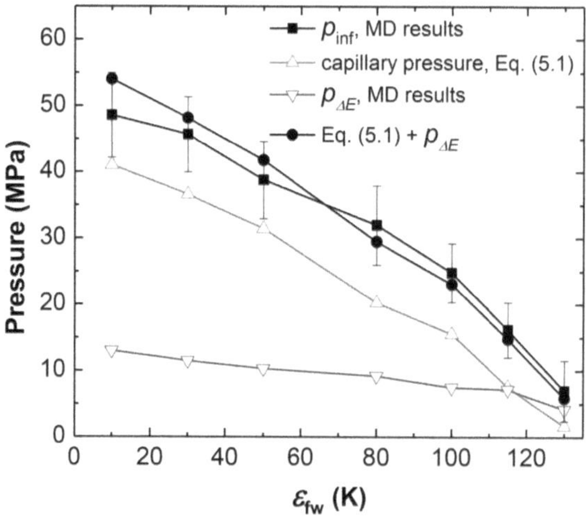

Figure 5.14 Entrance barrier pressure, capillary pressure, and infiltration pressure as a function of ε_{fw}. (Reprinted with permission from Mo et al., 2015, Copyright (2015) by the American Physical Society.)

It should be noted that Equation 5.10 is empirical and is only valid for $\theta \in (90°, 150°)$ and very small nanochannels ($H = 3$ nm). For channels of different heights, extensive studies should be conducted to obtain an empirical expression for $p_{\Delta E}$.

5.5 Other studies

There are other studies on nanoscale capillarity. Herein, an experimental investigation conducted in hydrophobic silica nanotubes with diameter ranging from 30 to 170 nm (Jayaraman et al., 2005) is introduced. The method of SSG is employed for the fabrication of silica nanotubes (details of the method are introduced in Section 4.7.3). The inner surfaces of the nanotubes are modified to be hydrophobic by covalently bonding them with octadecyltrimethoxysilane (OTS). This is achieved by dipping silica nanotubes into a solution of 5% (v/v) OTS and 16-mM acetic acid in ethanol/water (95% v/v) for half an hour at the room temperature, which is then followed by rinsing with ethanol and curing for 2 h at 150°C. An advantage of the OTS-modified silica surface is that the wettability of the surface can be changed by varying mole fraction of the solution components. For methanol–water mixtures, the contact angle θ on an OTS-coated silica surface can be tuned from 10° to 102° by altering the mole fraction of methanol, X_{me}, as shown in Table 5.2.

Capillarity is explored using the method of fluorescence recovery after photobleaching (FRAP). For a methanol–water mixture with a specific X_{me} value,

Table 5.2 Contact Angle of Methanol–Water Mixtures on an OTS Surface

X_{me}	0.0	0.05	0.13	0.31	0.57	1.0
θ(deg)	102	92	76	65	34	10

Source: From Jayaraman et al. (2005).

a group of experiments are conducted and the wettability of the nanotubes is obtained as the fraction of experiments, where FRAP is observed. Figure 5.15 depicts the wettability of the nanotubes as a function of X_{me}. As X_{me} is varied, it is seen that the wettability undergoes a transition from 0 (non-wetting) to 1 (wetting), which corresponds a change in the contact angle from $\theta < 90°$ to $\theta > 90°$. If the classic Young–Laplace equation, Equation 5.1, is valid for nanotubes, the contact angles in Table 5.2 indicate that the transition of the wettability should take place at a X_{me} value slightly larger than 0.05, $X_{me} \approx 0.06$. However, Figure 5.15 shows that the transition occurs at $X_{me} \approx 0.4 - 0.5$.

The clear discrepancy between experiments and the Young–Laplace equation suggests that the classic theory, Equation 5.1, breaks down at the nanoscale. The data in Figure 5.15 could be an experimental support to the entrance barrier pressure $p_{\Delta E}$ in Equation 5.9 found in MD simulations. According to the data in Figure 5.14 or Equation 5.10, $p_{\Delta E}$ is positive even if θ is slightly smaller than 90°. This suggests that the transition for $P_{inf} = P_l - P_v = p_{\Delta E} - \dfrac{2\gamma \cos\theta}{R}$ to change from negative to positive may take place at a contact angle smaller than 90°. This will make

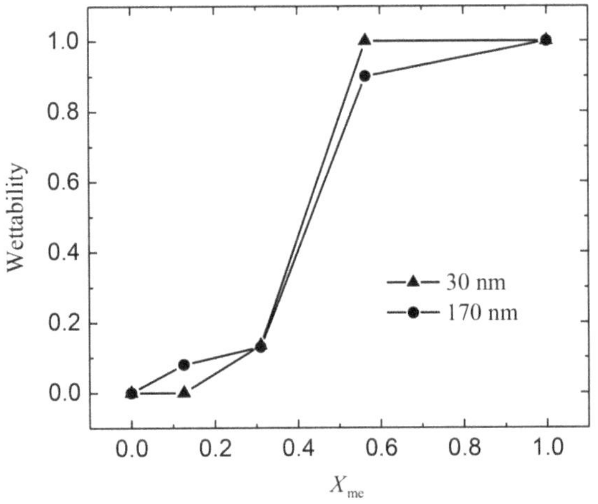

Figure 5.15 Wettability (fraction of experiments where FRAP is observed) for 30- and 170-nm-diameter channels as the mole fraction of methanol X_{me} of methanol–water mixtures is varied. (Data are extracted from Jayaraman et al., 2005.)

the transition of wettability shift to a relatively large X_{me} value in Table 5.2, which is consistent with the experimental data in Figure 5.15.

Another reason for the inconsistency between the experiments and the Young–Laplace equation might be the changes of the contact angle in nanoconfinements. In hydrophilic channels, variations of contact angle have been found (Sobolev et al., 2000). In hydrophobic channels, due to repulsive molecular forces, the apparent contact angle could be larger than that on a flat surface without confinement. This, however, is just a hypothesis and requires further exploration.

In addition to surface properties and channel size, temperature also greatly affects nanoscale capillarity as fluid surface tension and contact angle depend on temperature. For CNTs, MD simulations show that the infiltration pressure can be decreased by 30%–60% as the temperature is increased from 300 K to 360 K due to the reduction in the fluid surface tension and contact angle (Zhao et al., 2009; Xu et al., 2011). The experiments performed on water infiltration into hydrophobic zeolite with pore size less than 1 nm show that the infiltration pressure can be decreased from 94 to 56 MPa as the temperature is increased from 5°C to 85°C (Qiao et al., 2006). Although various studies have been carried out to understand how temperature affects nanoscale capillarity, a quantitative description of thermal effects is unavailable and requires extensive numerical and experimental investigations.

5.6 Summary remarks

Studies in the literature demonstrate that nanoscale capillarity is more complex than that at the macroscale. In a nanocapillary, even though the surface tension of a fluid can be treated as a constant, the molecular interactions and viscosity can significantly affect the static and dynamic capillary behaviors and make the classic theories invalid. Although great efforts have been made to explore nanoscale capillarity, extensive investigations are required to quantitatively examine how capillary size, surface wettability, and temperature affect capillarity and fluid infiltration in nanochannels. In particular, the coupled effects of these factors (e.g., how the coupling of channel size and temperature affects surface tension) on nanoscale capillarity are unclear and need to be systematically studied.

Nanoscale capillarity plays an essential role in a variety of areas, including drug delivery, nanofluidic device design, and molecular manipulation. A deep understanding of nanoscale capillarity can benefit many applications. For instance, high infiltration pressures for hydrophobic nanochannels pose a challenge to the applications of nanofluidics. How to reduce the infiltration pressure becomes a critical topic, which requires extensive exploration.

References

Dewar, M.J.S., Zoebisch, E.G., Healy, E.F., and Stewart, J.J.P. 1985 AM1: A new general purpose quantum mechanical molecular model. *J. Am. Chem. Soc.* 107:3902–3909.

Fadeev, A.Y. and Eroshenko, V.A. 1997. Study of penetration of water into hydrophobized porous silicas. *J. Colloid Interface Sci.* 187:275–282.

Guggenheim, E.A. 1940. The thermodynamics of interfaces in systems of several components. *Trans. Faraday Soc.* 35:397–412.

Haneveld, J., Tas, N.R., Brunets, N., Jansen, H.V., and Elwenspoek, M. 2008. Capillary filling of sub-10 nm nanochannels. *J. Appl. Phys.* 104:014309.

Helmy, R., Kazakevich, Y., Ni, C.Y., and Fadeev, A.Y. 2005. Wetting in hydrophobic nanochannels: A challenge of classical capillarity. *J. Am. Chem. Soc.* 127:12446–12447.

Jayaraman, K., Okamoto, K., Son, S.J., and Luckett, C. 2005. Observing capillarity in hydrophobic silica nanotubes. *J. Am. Chem. Soc.* 127:17385–17392.

Liu, C. and Li, Z.G. 2010. Surface effects on nanoscale Poiseuilli flows under large driving force. *J. Chem. Phys.* 132:024507.

Lu, H.M. and Jiang, Q. 2005. Size-dependent surface tension and Tolman's length of droplets. *Langmuir* 21:779–781.

Middleman, S. 1977. *Fundamentals of Polymer Processing.* New York: McGraw-Hill.

Mo, J.W., Li, L., Zhou, J.F., Xu, D.Y., Huang, B.L., and Li, Z.G. 2015. Fluid infiltration pressure for hydrophobic nanochannels. *Phys. Rev. E* 91:033022.

Qiao, Y., Punyamurtula, V.K., Han, A., Kong, X., and Surani, F.B. 2006. Temperature dependence of working pressure of a nanoporous liquid spring. *Appl. Phys. Lett.* 89:251905.

Samsonov, V.M., Bazulev, A.N., and Sdobnyakov, N.Y. 2003. Rusanov's liner formula for the surface tension of small objects. *Dok. Phys. Chem.* 389:83–85.

Sobolev, V.D., Churaev, N.V., Velarde, M.G., and Zorin, Z.M. 2000. Surface tension and dynamic contact angle of water in think quartz capillaries. *J. Colloid Interface Sci.* 222:51–54.

Tas, N.R., Haneveld, J., Jansen, H.V., Elwenspoek, M., and van der Berg, A. 2004. Capillary filling speed of water in nanochannels. *Appl. Phys. Lett.* 85:3274–3276.

Tolman, R.C. 1949. The effect of droplet size on surface tension. *J. Chem. Phys.* 17:333–337.

Washburn, E.D. 1921. The dynamics of capillary flow. *Phys. Rev.* 17:273.

Xu, B.X., Qiao, Y., Park, T., Tak, M., Zhou, Q., and Chen, X. 2011. A conceptual thermal actuation system driven by interface tension of nanofluids. *Energy Environ. Sci.* 4:3632–3639.

Zhao, J.B., Liu, L., Culligan, P.J., and Chen, X. 2009. Thermal effect on the dynamic infiltration of water into single-walled carbon nanotubes. *Phys. Rev. E* 80:061206.

Chapter 6 Nanoscale simple fluid flows

6.1 Introduction

This chapter deals with nanoscale flows of simple fluids driven by a pressure difference or an external force. The fluids involved are pure, simple fluids, which are different from those in electrokinetic flows. Nanoscale electrokinetic phenomena will be discussed in Chapter 8.

In the following sections, new flow phenomena for pressure driven flows at the nanoscale and the underlying mechanisms, regarding flow slip, flow regimes, and thermally driven flows, will be discussed.

6.2 Flow slip

6.2.1 Slip length

6.2.1.1 General summary Flow slip is one of the most popular topics of nanofluidics, especially in the early studies of nanoscale flows. This, perhaps, is because the boundary condition is an important issue in fluid mechanics, which is directly related to and, in many cases, determines the flow field. Another reason is that researchers expect to use the classic fluid mechanics to describe nanoscale flows by employing proper boundary conditions (e.g., flow slip).

Flow slip is commonly characterized by the slip length, L_s, which is illustrated in Figure 2.7. It is a fundamental problem in fluid mechanics as it can enhance flow rates in practical applications. Flow slip at the nanoscale has been observed in experiments and numerical simulations and the slip length has been directly or indirectly measured (or numerically calculated). Unfortunately, the results are inconsistent. Some work reports that L_s is a constant for a specific fluid–surface interaction, i.e., surface energy. This conclusion agrees well with the classic slip theories (Leal, 2007). Some other work, however, shows that L_s is shear rate dependent. In this case, it is interesting that three distinct scenarios for the dependence of L_s on the shear rate $\dot{\gamma}$ have been found. One is the linear dependence on the shear rate when the shear rate is larger than a critical value. The other two scenarios are nonlinear. In one of the two nonlinear cases, the slip length appears to be unbounded as the shear rate increases. For the other case, the slip length asymptotes to a constant value as the shear rate becomes sufficiently high.

These diverse slip behaviors qualitatively seem to be reasonable as they are supported by certain mechanisms. The quantitative discrepancy might be caused by the flow conditions, including surface properties and shear rates, because the surface energy and roughness as well as the shear rate affect the velocity at fluid–solid interfaces. In different studies, the surface wettabilities could be different, which may lead to apparently contradictive results. In addition, shear rates in experiments are usually below $10^5\,s^{-1}$, while in simulations they are typically in the range from 10^8 to $10^{11}\,s^{-1}$. This may cause the slip lengths measured in experiments to be different from those computed in simulations. Finally, artificial effects in simulations and unexpected errors in experiments, which are inevitable, may also lead to inconsistency in slip length determination.

Nevertheless, the different fashions of slip length mentioned above will be introduced first in the next three subsections. Then, the molecular mechanisms for different flow slip behaviors will be discussed.

6.2.1.2 Constant slip length A popular experimental approach for determining the slip length is to measure the hydrodynamic force acting on a curved surface as it moves toward another fixed surface in an aqueous medium. A schematic of this method is illustrated in Figure 6.1, where the moving curved surface is denoted by a sphere of radius R. As the sphere approaches the fixed, planar surface, the distance h between them decreases and the hydrodynamic force F_h acting on the sphere by the fluid between the two solid surfaces is measured. The relationship between F_h and h depends on the boundary conditions at the fluid–solid interfaces and can be used to determine the slip length.

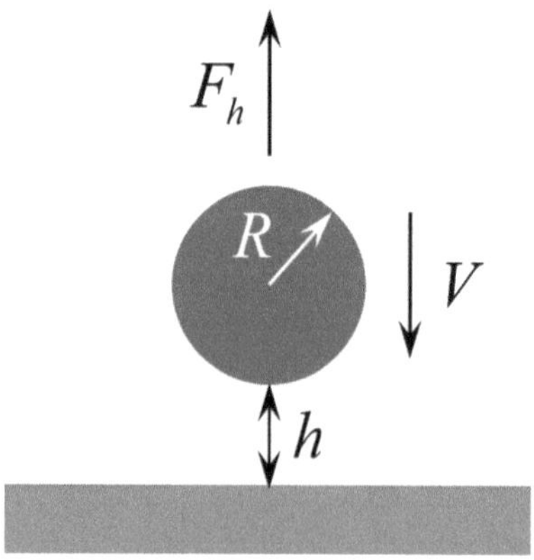

Figure 6.1 Representative experimental setup for the determination of slip length: a curved surface (sphere) interacts with a planar surface.

Theoretically, the hydrodynamic force F_h on a sphere for creeping (low Re) flows of Newtonian fluids for the case shown in Figure 6.1, under the assumption of the stick boundary condition, is given as follows:

$$F_h = 6\pi\mu RV\lambda \tag{6.1}$$

where $\lambda = g(h/R)$ is a function of ratio h/R (Li and Drazer, 2008). $\lambda \to 1$ for $h/R \gg 1$ and Equation 6.1 is reduced to the Stokes law (see Equation 1.11a). If there is velocity slip at the interface, Equation 6.1 is modified as (Vinogradova, 1995)

$$F_h = f^* \frac{6\pi\mu R^2 V}{h} \tag{6.2}$$

where f^* is a correction factor for velocity slip. If both the sphere and the fixed surface have the same slip length L_s, f^* is given by

$$f^* = \frac{h}{3L_s}\left[\left(1+\frac{h}{6L_s}\right)\ln\left(1+\frac{6L_s}{h}\right)-1\right] \tag{6.3}$$

If velocity slip takes place only at one surface, f^* reads

$$f^* = \frac{1}{4}\left\{1+\frac{6h}{4L_s}\left[\left(1+\frac{h}{4L_s}\right)\ln\left(1+\frac{4L_s}{h}\right)-1\right]\right\} \tag{6.4}$$

In experiments, F_h is measured as a function of h and the slip length can be obtained by fitting the $F_h - h$ curve using Equation 6.2. It should be noted that Figure 6.1 is just a representative setup. Different configurations may be used in experiments under the same principle. For instance, the curved surface can be fixed, while the flat surface moves with a constant velocity toward the curved surface or oscillates with a specific frequency.

Experiments on the basis of the above principle have been conducted to determine the slip length. In an experiment, a borosilicate glass particle of $R = 10\,\mu m$ is attached to an AFM cantilever and a mica surface moves up and down periodically below the glass particle with a NaCl solution in between (Bonaccurso et al., 2002). During the motion of the mica surface, the force on the glass particle is obtained based on the deflection of the AFM cantilever. Figure 6.2 depicts the hydrodynamic force as the separation h is varied for different velocities of the mica surface. It can be seen that the hydrodynamic force deviates from Equation 6.1 (non-slip case) as the separation decreases, $h < 50\,nm$, indicating that velocity

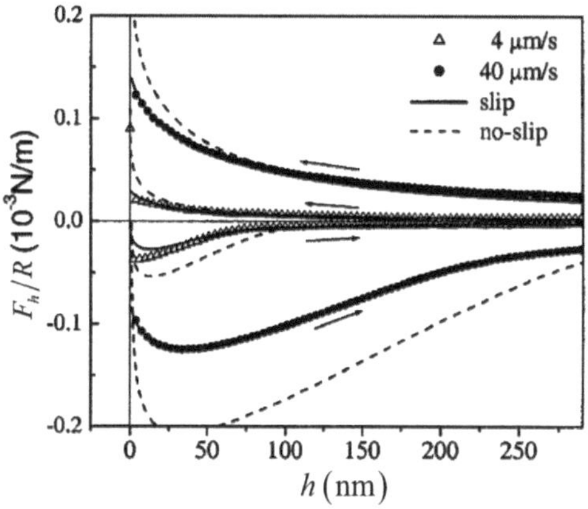

Figure 6.2 Normalized hydrodynamic force F_h/R as a function of distance h at different velocities, $V = 4$ and 40 μm/s. Arrows indicate approaching and retracting motions. Symbols are experimental data. Dashed lines are the predictions of Equation 6.1 with the stick boundary condition and solid lines represent Equation 6.2 with $L_s = 8.6$ nm. (Reprinted with permission from Bonaccurso et al., 2002, Copyright (2002) by the American Physical Society.)

slip occurs. By fitting the data in Figure 6.2 using Equation 6.2, the slip length is obtained as $L_s = 8 - 9$ nm, which is independent of the velocity and the shear rate.

As mica and glass surfaces are hydrophilic, the slippage on them is sort of surprising from the point of view of the classic fluid mechanics because the no-slip boundary condition for water on hydrophilic surfaces has been widely accepted. The molecular view of velocity slip actually supports the experimental results because velocity slip also depends on the competition between the fluid–surface interaction and the external force, as discussed in Section 2.2 and demonstrated in Figure 2.2. If the external force is larger than the fluid–surface interaction, or the shear rate is sufficiently high, velocity slip will take place. In the experiment in Figure 6.2, the maximum shear rate is estimated to be ~$10^4\,s^{-1}$, for which the observation of slippage appears to be reasonable.

In another experimental work, which employs a similar idea to that in Figure 6.1, the oscillating force $F(\omega)$ on the plane in response to the oscillation of a sphere is measured using a dynamic surface force apparatus and the sphere–plane distance was obtained by a capacitive sensor (Cottin-Bizonne et al., 2005). The independent measurements of the force and distance ensure an accurate determination of the complex force $G(\omega) = F(\omega)/h_0$, where h_0 is the amplitude of the vibrating distance between the sphere and the plane. Furthermore, this strategy does not depend on the fluid properties, such as viscosity and diffusion coefficient. This makes the experimental data reliable. For simple fluids and rigid surfaces, the imaginary part of $G(\omega)$ is the viscous damping due to the motion of the fluid, which is theoretically given as

$$G''(\omega) = 6\pi\mu\omega R^2 f^* / h \tag{6.5}$$

Experiments are performed for symmetric hydrophilic Pyrex surfaces (i.e., the surfaces of the sphere and the plane are the same) and asymmetric surfaces (a plain Pyrex sphere and an OTS-coated Pyrex plane) in water and dodecane. The reciprocal of the damping, $G''(\omega)^{-1}$, and the force are plotted in Figure 6.3 as a function of sphere–plane distance h for symmetric Pyrex surfaces with dodecane as the fluid. The linear dependence of $G''(\omega)^{-1}$ on h indicates a no-slip boundary condition (the viscosity determined from the experimental data is $\mu = 1.34 \pm 0.07$ mPa, which agrees well with the bulk viscosity of dodecane, 1.29 mPa at 27°C).

When water is confined between the sphere and plane, velocity slip is not observed for the symmetric plain Pyrex surfaces. However, for the asymmetric configuration, flow slip is found. Figure 6.4 shows $G''(\omega)^{-1}$ as the sphere–plane distance is varied. The nonlinear relationship between $G''(\omega)^{-1}$ and h at small h values for the asymmetric surfaces suggests the occurrence of velocity slip. A comparison between the experimental data and Equations 6.2 and 6.4 (slip only occurred on the OTS-coated plane) gives a slip length $L_s = 19 \pm 2$ nm. Therefore, velocity slip

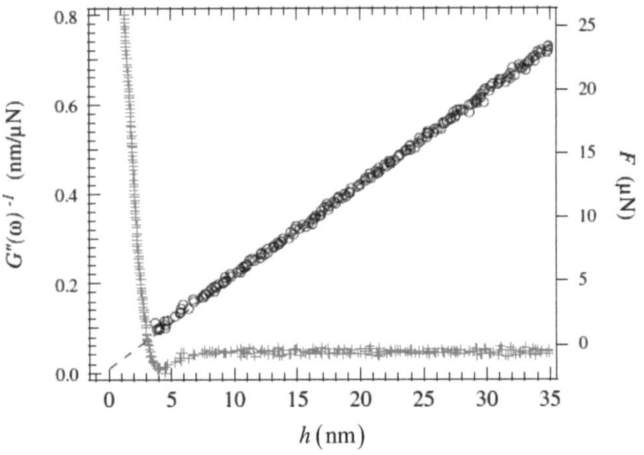

Figure 6.3 Hydrodynamic responses of a plane due to the oscillation of a sphere normal to the plane (open circles: inverse of the damping $G''(\omega)^{-1}$ and crosses: hydrodynamic force on the plane). The surfaces of the sphere and plane are plain Pyrex. h is the distance between the sphere and the plane. The dashed line is the linear fit of the data. (Reprinted with permission from Cottin-Bizonne et al., 2005, Copyright (2005) by the American Physical Society).

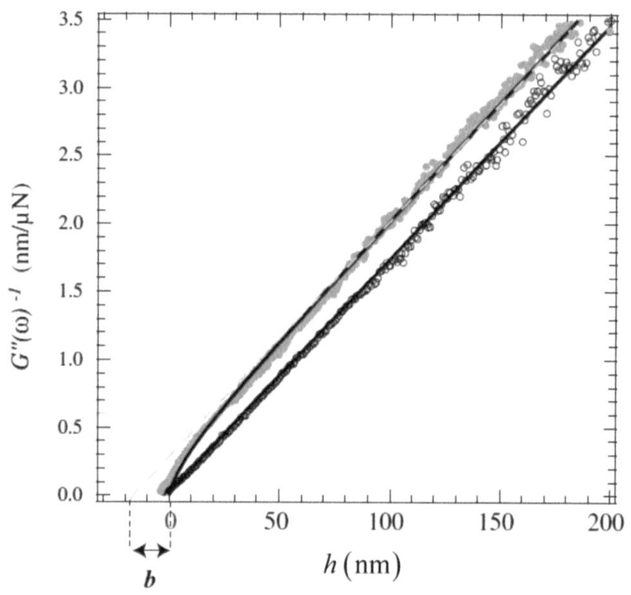

Figure 6.4 Inverse of the damping, $G''(\omega)^{-1}$, versus the sphere–plane distance with water being the fluid. Open circles: symmetric surfaces (both the sphere and plane are plain Pyrex). Gray solid circles: asymmetric Pyrex/OTS-coated surfaces. The solid lines represent the theoretical predictions for the stick boundary condition or a slip length of 19 nm. The dashed line is a linear fit to the data with slip. (Reprinted with permission from Cottin-Bizonne et al., 2005, Copyright (2005) by the American Physical Society.)

is only observed for hydrophobic surfaces and the slip length is independent of the shear rate (Cottin-Bizonne et al., 2005).

The experiments in Figures 6.3 and 6.4 are well designed with independent determinations of hydrodynamic forces and sphere–plane distances. However, it should be noted that the conclusion of constant slip length is only valid under the experimental conditions, i.e., shear rates up to $5 \times 10^3 s^{-1}$. This is also the case for the experiments in Figure 6.2. As the shear rate increases, the slip length may become shear rate dependent, as will be discussed later.

Other experiments using similar or different strategies with nanoscale *h* are also conducted and the slip length is of order 10 nm (Vinogradova and Yakubov, 2003; Joly et al., 2006; Maali et al., 2008). This value is smaller than those measured at the microscale, which range from 100 nm to 1 μm (Pit et al., 2000; Tretheway and Meinhart, 2002; Joseph and Tabeling, 2005).

In addition to experiments, constant slip lengths have also been found in numerical studies for shear rates up to $10^{10} s^{-1}$. Slip lengths for water flows in slit-like nanochannels with different surface properties have been explored through MD simulations (Sendner et al., 2009). To study slip lengths, shear flows are frequently employed. A typical MD system of a shear flow between two moving parallel plates and a possible velocity distribution with velocity slip are shown in Figure 6.5. As discussed in Section 3.2.1, the surface energy can be tuned by changing the fluid–surface interaction binding energy ε, which affects the contact angle of the fluid. For water flows generated by two moving diamond plates, the contact angle has been studied (Sendner et al., 2009). Figure 6.6 shows the contact angle as a function of the C–O (diamond–water) interaction binding energy, ε_{CO}, where the dependence of the contact angle is θ on the binding energy ε_{CO} is similar to that in Figure 5.11. Hydrophobic surfaces can be created by terminating the diamond plates with hydrogen atoms because oxygen (water)-hydrogen binding energy is relatively weak. The channel surfaces become hydrophilic if some carbon atoms of the diamond surfaces are replaced by C–O–H groups. For hydrophobic channels, different surface roughnesses are considered, as shown in Figure 6.7. For hydrophilic channels, the surface wettability is varied by changing the fraction of OH groups. Two examples are illustrated in Figure 6.8.

In hydrophilic channels, water adsorption on the channel surfaces is observed due to the large water–surface binding energy, as depicted by the water density and velocity profiles in Figure 6.9. This leads to a slightly negative slip length, $L_s \approx -0.3$ nm, which is insignificant and the stick boundary condition can be considered valid. In this case, the viscosity of water next to the channel surfaces

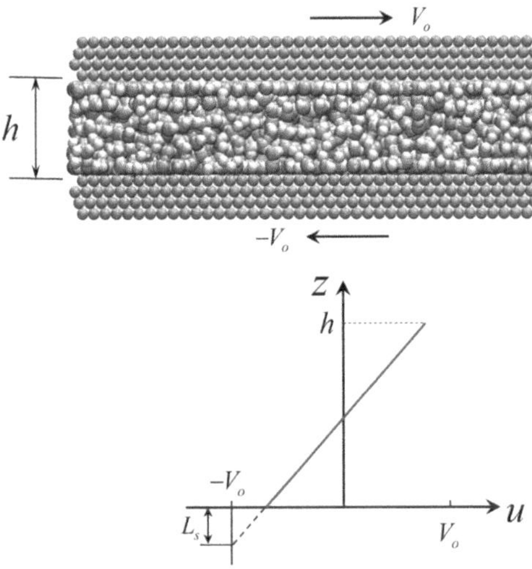

Figure 6.5 Numerical study of slip length in nanochannels. Top panel: molecular dynamics simulation system with channel height *h*. The motion of the two plates generates a shear flow. Bottom panel: a possible velocity profile with velocity slip.

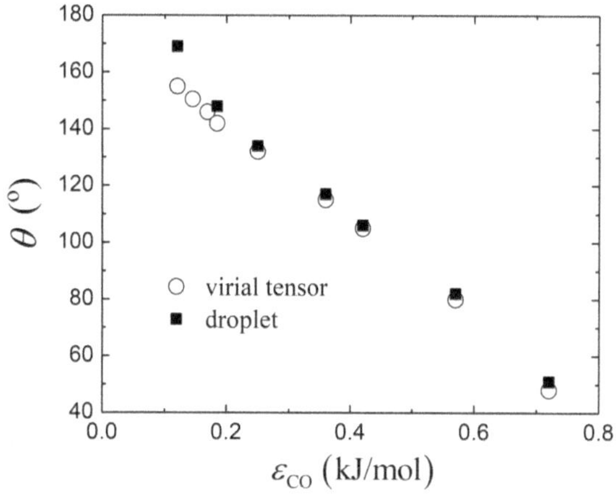

Figure 6.6 Contact angle of water as a function of the diamond–water binding energy ε_{CO}. Circles are the results calculated from the virial tensor and squares represent the contact angles obtained by simulating a droplet on surfaces. (Data are extracted from Sendner et al., 2009.)

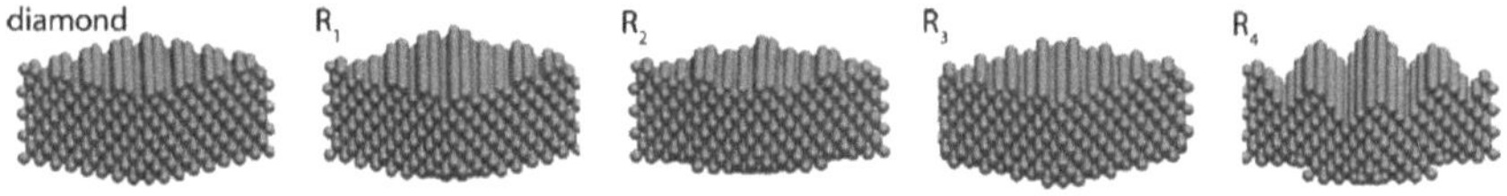

Figure 6.7 Hydrophobic diamond surfaces with different surface roughnesses. (Reprinted with permission from Sendner et al., 2009, Copyright (2009) American Chemical Society.)

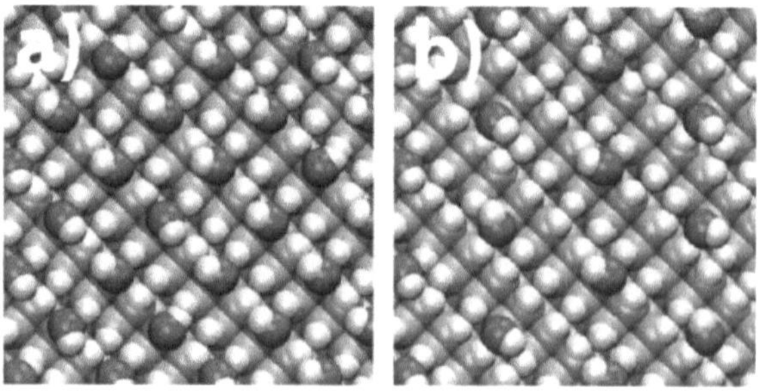

Figure 6.8 Hydrophilic diamond surfaces modified with OH groups. Surface fraction of OH groups is $x_{OH} = 1/4$ (a) and $x_{OH} = 1/8$ (b). (Reprinted with permission from Sendner et al., 2009, Copyright (2009) American Chemical Society.)

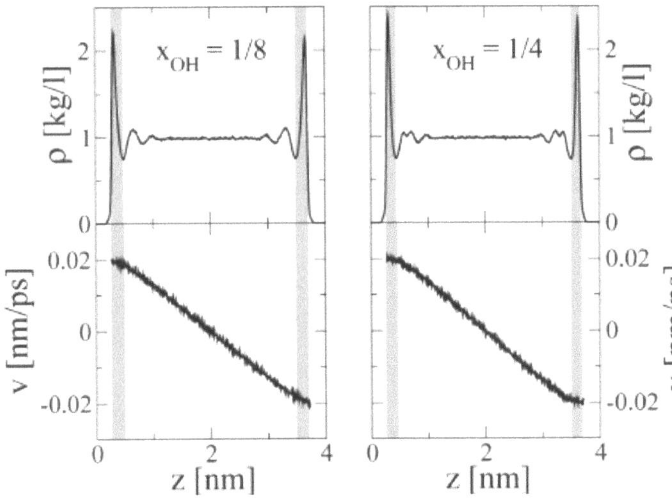

Figure 6.9 Water density and velocity profiles in hydrophilic channels with different surface fraction of OH groups, $x_{OH} = 1/8$ (left) and $x_{OH} = 1/4$ (right). (Reprinted with permission from Sendner et al., 2009, Copyright (2009) American Chemical Society.)

becomes higher than the bulk viscosity due to water adsorption (see Chapter 2). Table 6.1 shows the shear rates, viscosity, and slip length for the surfaces in Figure 6.8.

If the channel is hydrophobic, however, flow slip is found. Figure 6.10 shows the slip length as a function of shear rate $\dot{\gamma}$. It can be seen that the slip length $L_s = 2-3$ nm and remains almost unchanged for $\dot{\gamma} < 10^{10}$ s^{-1} for channel height ranging from 2 to 8 nm. When $\dot{\gamma} > 10^{10}$ s^{-1}, the slip length tends to increase slightly (Sendner et al., 2009). The slip length depends on the surface energy (or contact angle), which can be characterized by the fluid–surface binding energy ε_{CO}, and surface roughness. Figure 6.11 shows L_s as ε_{CO} is varied for different surface structures in Figure 6.7. It can be seen that the slip length increases linearly with $1/\varepsilon_{CO}^2$ and decreases as the surface roughness increases. The linear

Table 6.1 Shear Rate and Viscosity in the Bulk ($\dot{\gamma}_0$, μ_0) and Interfacial ($\dot{\gamma}_s$, μ_s) Regions and Slip Length for Water Flows in Hydrophilic Nanochannels

x_{OH}	$\dot{\gamma}_0$(ns^{-1})	$\dot{\gamma}_s$(ns^{-1})	μ_s/μ_0	L_s (nm)
1/4	13.4	3.6±0.8	3.7±0.9	−0.32±0.01
1/8	12.8	5.9±1.0	2.2±0.4	−0.29±0.01

Source: Sendner et al. (2009).

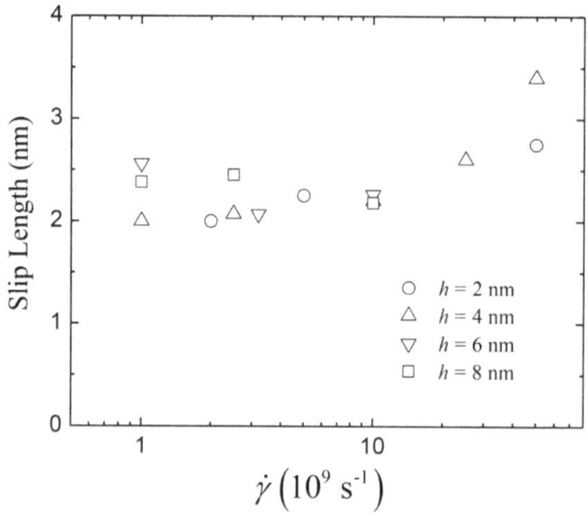

Figure 6.10 Slip length versus shear rate in hydrophobic channels with channel height *h* ranging from 2 to 8 nm. (Data are extracted from Sendner et al., 2009.)

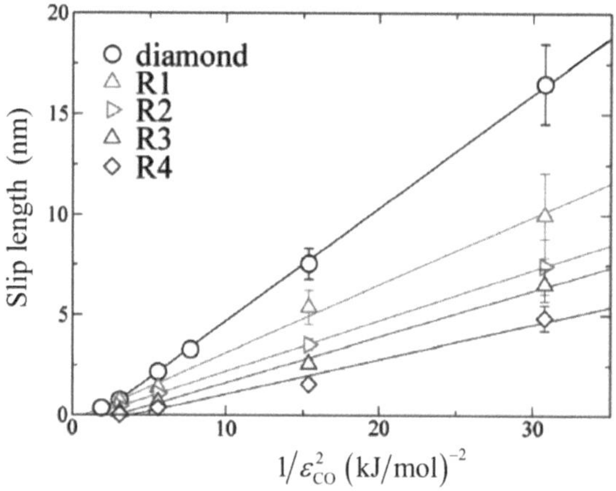

Figure 6.11 Slip length as a function of $1/\varepsilon_{\mathrm{CO}}^2$ for surface structures shown in Figure 6.7. Solid lines are linear fits to the data. (Reprinted with permission from Sendner et al., 2009, Copyright (2009) American Chemical Society.)

relationship between L_s and $1/\varepsilon_{\mathrm{CO}}^2$ is an important finding, which is helpful in slip length predictions, as will be discussed later.

Slip lengths can also be predicted through semi-theoretical approaches. First, a general relationship between the slip length L_s and certain variables is derived using the classic fluid mechanics treatment. The general relationship contains unknown parameters, which require numerical methods to determine. A sound approach is developed by considering the equation of motion of a fluid layer next to the channel walls (Hansen et al., 2011). Consider a horizontal flow (x-direction) of velocity $u(y)$ between two parallel walls, as shown in Figure 6.12a. Flows can be generated by an external force F_e, which is equivalent to a pressure drop, or the motion of the walls. For the latter, if the velocity of the walls is denoted as u_w, the relative velocity between the fluid and the walls is defined as $\delta u = u_w - u$, and the flow is assumed to be laminar and continuous, the slip length L_s can be written as follows (Figure 6.12b):

$$L_s = \frac{\delta u}{\dot{\gamma}} \tag{6.6}$$

As the shear rate $\dot{\gamma} = \partial u / \partial y \,|_{y=0}$ is related to the fluid viscosity through Equation 1.2, the slip length in Equation 6.6 can be rewritten as follows:

$$L_s = \mu \frac{\delta u}{\tau} = \frac{\mu}{\xi_w} \tag{6.7}$$

where τ is the shearing stress and $\xi_w = \tau / \delta u$ is the friction coefficient per unit area at the wall, which is unknown.

At the nanoscale, Equation 6.7 theoretically holds but the size effect makes the physical meaning of ξ_w vague because the interface is sort of dynamic due to the thermal motions of wall atoms and fluid molecules. A similar expression for the slip length in nanochannels can be obtained by considering a fluid slab of thickness δy at the interfaces, as shown by the orange, shaded areas in Figure 6.13a. Let us consider the fluid slab next to the bottom wall. The equation of motion for the fluid slab reads

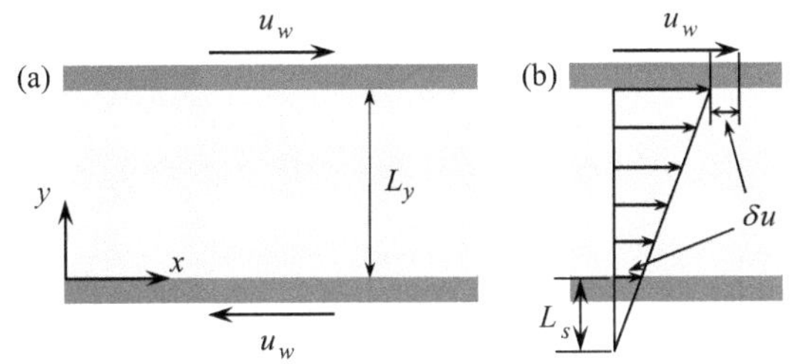

Figure 6.12 Schematic of a fluid flow between two parallel walls. (a) Coordinate system and notation. The flow is driven by the horizontal motion of the walls. (b) Velocity distribution and slip length for a case with moving upper wall and fixed bottom wall.

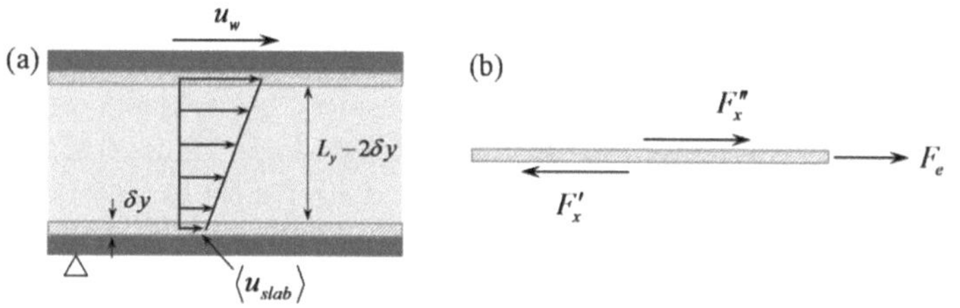

Figure 6.13 Fluid slabs next to the channel walls for analyzing the slip length. (a) Size and positions of the fluid slabs in a flow caused by the moving upper wall (the bottom wall is fixed). (b) Free body diagram of the fluid slab next to the bottom wall.

$$m\frac{du_{\text{slab}}}{dt} = F_x' + F_x'' + mF_{\text{e}} \tag{6.8}$$

where m and u_{slab} are the mass and center-of-mass velocity of the fluid slab, F_x' and F_x'' are the forces acting on the slab by the wall and interior fluid, respectively, and F_{e} is the external force per unit mass, as depicted in Figure 6.13b. F_x' is nothing but the shearing force at the interface, which depends on the relative velocity between the wall and the fluid slab $\Delta u'$. If $\Delta u'$ is sufficiently small, a linear constitutive equation assumes (Hansen et al., 2011)

$$F_x'(t) = -\int_0^t \zeta(t-t')\Delta u'(t')dt' + F_r'(t) \tag{6.9}$$

where ζ is a friction kernel and $F_r'(t)$ is a random force with zero mean, $\langle F_r'(t)\rangle = 0$. By taking average of both sides of Equation 6.9, it is obtained that

$$\langle F_x'(t)\rangle = -\zeta_0\langle \Delta u'\rangle \tag{6.10}$$

where ζ_0 is the zero-frequency friction coefficient. The force due to the relatively motion between the slab and the interior fluid, F_x'', is given by

$$F_x''(t) = A\tau\big|_{y=\delta y} = A\mu\frac{\partial u}{\partial y}\bigg|_{y=\delta y} \tag{6.11}$$

where A is the interfacial area between the slab and the interior fluid.

For a Couette or shear flow (a flow caused by the motion of the walls), as shown in Figure 6.13, the Navier–Stokes equation in the flow direction is simplified as

$$\frac{\partial^2 u}{\partial y^2} = 0 \tag{6.12}$$

If the upper wall moves with u_w while the bottom wall is fixed and the two walls are identical, the boundary conditions for Equation 6.12 are as follows:

$$\langle u_{\text{slab}} \rangle = \frac{1}{\delta y}\int_0^{\delta y} u\,dy \quad \text{and} \quad u_w = \langle u_{\text{slab}} \rangle + \frac{1}{\delta y}\int_{L_y-\delta y}^{L_y} u\,dy \tag{6.13}$$

where $\langle u_{\text{slab}} \rangle$ is the mean velocity of the fluid slab next to the bottom wall. It is also the velocity difference between the upper wall and the fluid slab close to it, as illustrated in Figure 6.12b. The boundary conditions in Equation 6.13 lead to the velocity profile (Hansen et al., 2011)

$$u = \frac{u_w - 2\langle u_{\text{slab}} \rangle}{L_y - \delta y}\left(y - \frac{\delta y}{2}\right) + \langle u_{\text{slab}} \rangle \tag{6.14}$$

Equation 6.14 gives $u = \langle u_{\text{slab}} \rangle$ at $y = \delta y/2$ and $u = u_w - \langle u_{\text{slab}} \rangle$ at $y = L_y - \delta y/2$ as $\langle u_{\text{slab}} \rangle$ is the center-of-mass velocity of the fluid slab next to the bottom wall. As the velocity distribution is linear, the slope of the velocity in Equation 6.14 gives the shear rate $\dot{\gamma}_s$, i.e.,

$$\dot{\gamma} = \frac{u_w - 2\langle u_{\text{slab}} \rangle}{L_y - \delta y} \tag{6.15}$$

In the steady state with $F_e = 0$, Equation 6.8, by taking average on both sides, becomes $\langle F_x' \rangle + \langle F_x'' \rangle = 0$, which can be written as

$$-\zeta_0 \langle u_{\text{slab}} \rangle + A\mu \frac{u_w - 2\langle u_{\text{slab}} \rangle}{L_y - \delta y} = 0 \tag{6.16}$$

where Equation 6.15 is used. From Equation 6.16, $\langle u_{\text{slab}} \rangle$ is solved as

$$\langle u_{\text{slab}} \rangle = \frac{\mu u_w}{\xi_0 (L_y - \delta y) + 2\mu} \tag{6.17}$$

where $\xi_0 = \zeta_0 / A$ is the friction coefficient per unit area.

Based on the definition of slip length, $L_s = u(0) / \partial u/\partial y|_{y=0}$, L_s can be obtained by combining Equations 6.14 and 6.17

$$L_s = \frac{\mu}{\xi_0} - \frac{\delta y}{2} \tag{6.18}$$

It should be noted that Equation 6.18 is reduced to Equation 6.7 as the width of the fluid slab $\delta y \to 0$. ξ_0 in Equation 6.18 represents the mean friction coefficient per unit area between a thin fluid layer (the fluid slab in Figure 6.13) and the wall, which practically caries a clearer physical meaning than ξ_w in Equation 6.7 at the nanoscale.

If the flow is driven by an external force, the Navier–Stokes equation is simplified as

$$\frac{\partial^2 u}{\partial y^2} = -\frac{\rho F_e}{\mu} \tag{6.19}$$

where F_e is the same as that in Equation 6.8, i.e., force per unit mass. Using the same boundary conditions for $\langle u_{slab} \rangle$ given in Equation 6.13, the solution of Equation 6.19 is obtained as (Hansen et al., 2011)

$$u = \frac{\rho F_e}{12\mu}\left[6y(L_y - y) + \delta y(2\delta y - 3L_y)\right] + \langle u_{slab} \rangle \tag{6.20}$$

Similarly, if a fluid slab is considered (Figure 6.13), at the steady state, Equation 6.8 becomes $\langle F_x' \rangle + \langle F_x'' \rangle + mF_e = 0$, which, in combining with Equations 6.10 and 6.11, yields

$$-\zeta_0 \langle u_{slab} \rangle + \frac{A\rho F_e}{2}(L_y - 2\delta y) + mF_e = 0 \tag{6.21}$$

and $\langle u_{slab} \rangle$ is given by

$$\langle u_{slab} \rangle = \frac{\rho F_e L_y}{2\xi_0} \tag{6.22}$$

If the density ρ is assumed to be constant, the slip length reads

$$L_s = \frac{\mu}{\xi_0} - \delta y\left(\frac{1}{2} - \frac{\delta y}{3L_y}\right) \tag{6.23}$$

This result, as expected, reduces to Equation 6.7 as $\delta y \to 0$. It should be noted that the slip length in Equation 6.23 is slightly different from that given by Equation 6.18 and the difference is $\delta y^2/3L_y$, which is negligible for small δy (Hansen et al., 2011).

Equations 6.18 and 6.23 are theoretical predictions for the slip length. Although they are rigorous, the thickness of the fluid slab δy in the analyses appears to be arbitrary. Even though the terms containing δy can be neglected for sufficiently small δy, the friction coefficient ξ_0 is unknown. ξ_0 should depend on the surface and fluid properties. It cannot be suggested by the classic fluid mechanics theory and has to be determined through numerical approaches, such as MD simulations.

In numerical simulations, the friction coefficient can be determined by calculating the velocity–force correlation function $C_{uF'}(t)$ and the velocity autocorrelation function $C_{uu}(t)$ (Hansen et al., 2011). For Poiseuille's flows, $u_w = 0$, Equation 6.9 becomes

$$F_x'(t) = -\int_0^t \zeta(t-t')u_{\text{slab}}(t')dt' + F_r'(t) \tag{6.24}$$

Multiplying both sides of Equation 6.24 with $u_{\text{slab}}(0)$ and taking ensemble average yield

$$C_{uF'}(t) = -\int_0^t \zeta(t-t')C_{uu}(t')\,dt' \tag{6.25}$$

where $C_{uF'}(t) = \left\langle u_{\text{slab}}(0)F_x'(t)\right\rangle$ and $C_{uu}(t) = \langle u_{\text{slab}}(0)u_{\text{slab}}(t)\rangle$. Equation 6.25 can be converted to a more convenient form through a Laplace transformation

$$\tilde{f}(s) = \int_0^\infty f(t)\exp(-st)\,dt \tag{6.26}$$

which yields

$$\tilde{C}_{uF'}(s) = -\tilde{\zeta}(s)\tilde{C}_{uu}(s) \tag{6.27}$$

As $\tilde{C}_{uF'}$ and $\tilde{C}_{uu}(s)$ can be easily obtained in numerical simulations, Equation 6.27 can be employed to estimate $\tilde{\zeta}(s)$ and consequently the friction kernel ζ_0. However, numerical fluctuations could be significant and may lead to large errors in the determination of the friction coefficient (Hansen et al., 2011). If the friction kernel is written as an n-term Maxwellian memory function (Evans and Morriss, 1990), $\zeta(t) = \sum_{i=1}^{n} B_i e^{-\lambda_i t}$, $\tilde{\zeta}(s)$ is then given by

$$\tilde{\zeta}(s) = \sum_{i=1}^{n} \frac{B_i}{s+\lambda_i} \tag{6.28}$$

and

$$\tilde{C}_{uF'}(s) = -\sum_{i=1}^{n} \frac{B_i \tilde{C}_{uu}(s)}{s+\lambda_i} \tag{6.29}$$

By fitting the $\tilde{C}_{uF'} - \tilde{C}_{uu}(s)$ curve obtained in numerical simulations, parameters B_i and λ_i can be determined, which can be used to compute the friction kernel through (Hansen et al., 2011)

$$\zeta_0 = \int_0^{\infty} \sum_{i=1}^{n} B_i e^{-\lambda_i t}\, dt = \sum_{i=1}^{n} \frac{B_i}{\lambda_i} \tag{6.30}$$

Using this method, MD simulations for flows between two parallel walls are performed (Travis et al., 1997; Hansen et al., 2011). Four systems employing the Weeks–Chandler–Andersen (WCA) and LJ potentials for fluid–fluid and fluid–wall interactions are studied, as given in Table 6.2. The WCA potential is pure repulsive and generates a relatively hydrophobic surface by using a small cutoff distance (Hansen et al., 2011).

In MD simulations, the velocities of fluid molecules and the fluid–wall interaction forces can be easily obtained. Therefore, the correlation functions in Equation 6.27 can be computed for a fluid slab of any thickness. This leads to the determination of ζ_0. Figure 6.14a plots ζ_0 as a function of δy for the systems in Table 6.2 and Figure 6.14b depicts ζ_0 for system 1 in Table 6.2. Based on the values of ζ_0, semi-empirical predictions of the slip length can be achieved using either Equation 6.18 or Equation 6.23. The slip length can also be determined directly from the velocity profile through MD simulations (see Figure 2.7). Figure 6.15 shows the slip lengths predicted by Equations 6.18 and 6.23 as well as those obtained from NEMD simulations of Couette and Poiseuille flows. It can be seen that the semi-empirical predictions are in good agreement with MD simulations. Although the slip length depends on the flow system, it is constant for a given configuration. This perhaps is because the shear rates considered in the MD simulations are relatively low (Hansen et al., 2011).

Table 6.2 Molecular Dynamics Simulation Systems with Different Combinations of Fluid–Fluid (U_{ff}) and Fluid–Wall (U_{fw}) Potentials. The Units Are Dimensionless MD Units (see Chapter 3)

System	U_{ff}	U_{fw}	ρ_{bulk}	ρ_{wall}	T	μ
1	WCA	WCA	0.51	0.80	0.80	0.43
2	WCA	LJ	0.44	0.80	0.80	0.31
3	LJ	LJ	0.79	0.90	0.73	1.94
4	WCA	WCA	0.70	0.80	0.73	1.03

Source: Hansen et al. (2011).

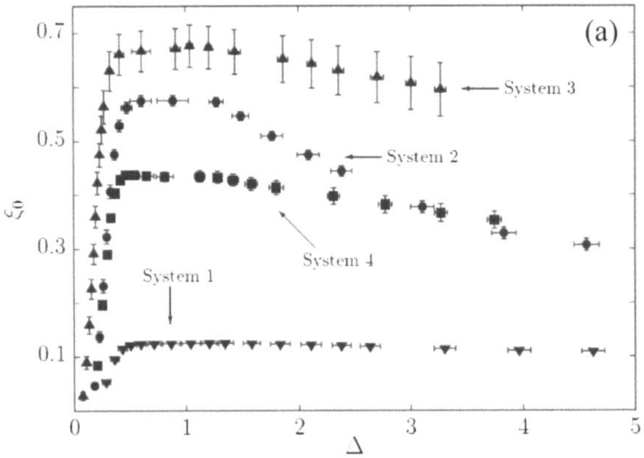

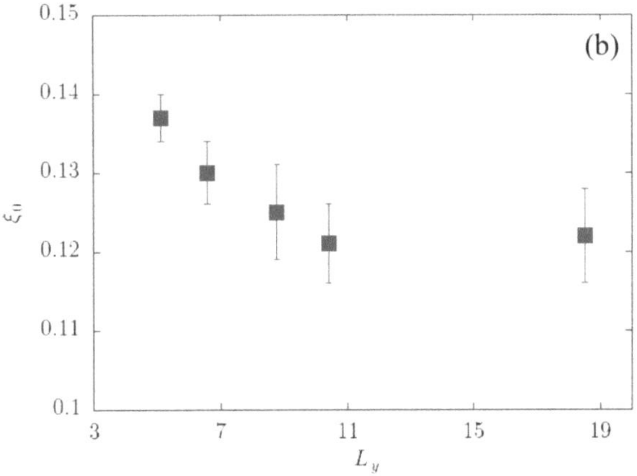

Figure 6.14 Friction coefficient ξ_0. (a) ξ_0 as a function of the thickness of the fluid slab Δ for systems given in Table 6.2. (b) ξ_0 versus channel height L_y for system 1 with $\Delta = 1.2$. All the units are dimensionless MD units. (Reprinted with permission from Hansen et al., 2011, Copyright (2011) by the American Physical Society.)

Nevertheless, constant slip length might be valid under certain conditions, such as low shear rates and proper surface properties. As the shear rate is increased or the surface properties (e.g., surface energy and surface roughness) are changed, the slip length may not remain constant, as discussed in the next subsections.

6.2.1.3 Shear-dependent slip length-unbounded Other than constant slip lengths, shear-dependent slip lengths are also found in experimental and numerical studies. Experiments based on the idea illustrated in Figure 6.1 have been conducted with sucrose solutions as the fluid. The experimental setup is depicted in Figure 6.16, where a flat surface moves toward a sphere. Both the flat surface and the sphere are coated with a layer of gold, which is treated with a mixture

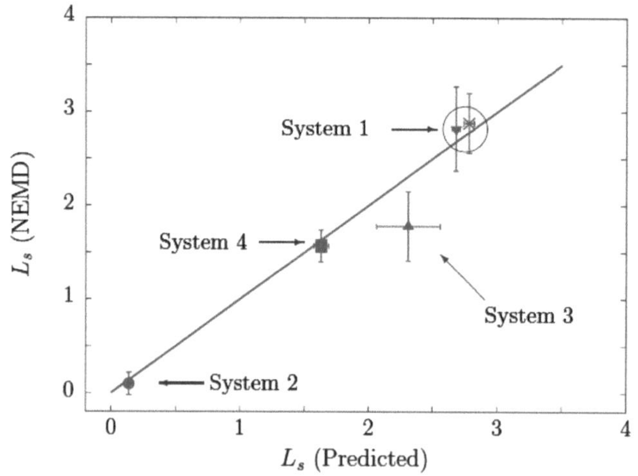

Figure 6.15 Slip length obtained from NEMD simulations and predicted by Equations 6.18 and 6.23. Circle, square, and triangles are Poiseuille flows with $L_s = 7.7$. Cross represents Couette flows with $L_s = 21.8$. The thickness of the fluid slab is 1.5σ (σ is the collision diameter of the LJ potential). (Reprinted with permission from Hansen et al., 2011, Copyright (2011) by the American Physical Society.)

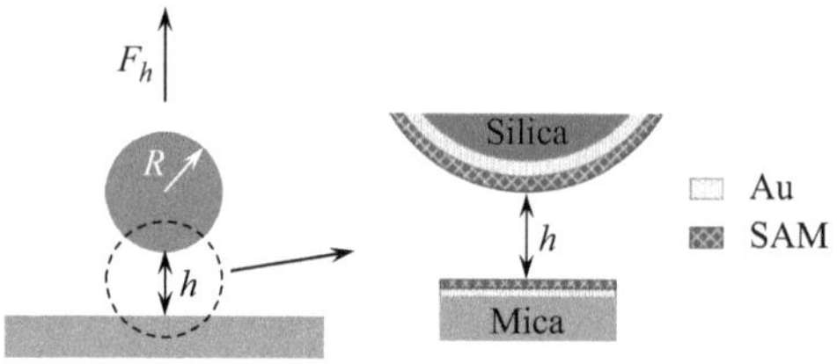

Figure 6.16 Sketch of a sphere–plane system for the determination of slip length. The surfaces of the sphere and the plane are coated with gold and SAM.

of 10-mM 11-mercapto-1-undecanol and 10-mM 1-dodecanethiol (20% dodecanethiol) in ethanol to chemically bond the gold coating with a self-assembled monolayer (SAM) (Craig et al., 2001). The motion of the flat surface is controlled by a piezoelectric transducer and the force on the sphere is measured by an AFM through the deflection of the AFM cantilever. The slip length L_s is determined based on the value of f^* given in Equation 6.3.

At a relatively low velocity of the flat surface (2.4 μm/s) and intermediate viscosity for the sucrose solution (27.0 mPa·s), the hydrodynamic force on the sphere agrees well with the predictions of Equation 6.1, indicating the no-slip boundary condition, as shown in Figure 6.17a. As the approaching velocity of the flat

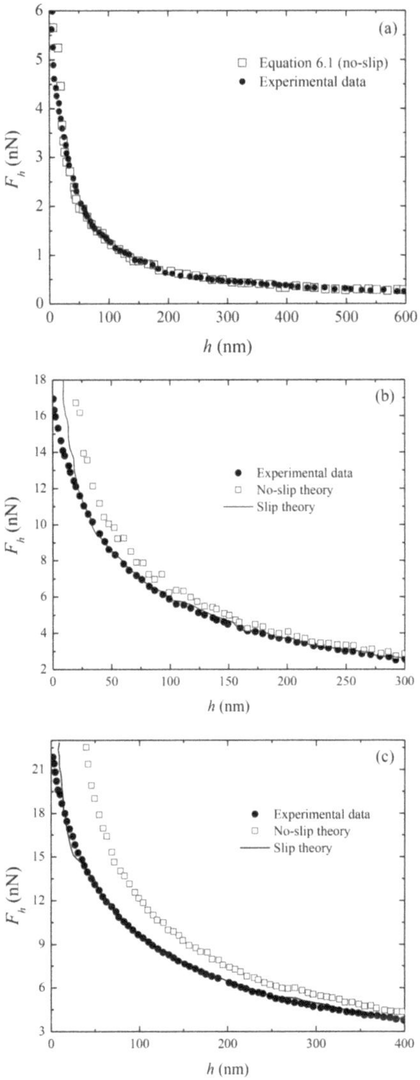

Figure 6.17 Hydrodynamic force in sucrose solutions. (a) The approaching velocity is 2.4 μm/s and the viscosity of the sucrose solution is 27.0 mPa·s. Dots are experimental data and squares are the predictions of Equation 6.1. (b) The approaching velocity is 21.6 μm/s and the viscosity of the sucrose solution is 19.2 mPa·s. Dots are experimental data, squares are the predictions of Equation 6.1, and the solid line represents Equation 6.2. (c) The approaching velocity is 21.6 μm/s and the viscosity of the sucrose solution is 38.9 mPa·s. Dots are experimental data, squares are the predictions of Equation 6.1, and the solid line represents Equation 6.2. (Data are extracted from Craig et al., 2001.)

surface is increased (21.6 μm/s), the hydrodynamic force becomes smaller than that given by Equation 6.1 due to the flow slip at the interfaces, as depicted in Figure 6.17b, where the viscosity of the sucrose solution is 19.2 mPa·s and the slip length is obtained as $L_s = 5\,\text{nm}$. The slip length also depends on the fluid viscosity. In Figure 6.17c, the velocity of the flat surface is kept at 21.6 μm/s while the viscosity of the sucrose solution is increased to 38.9 mPa·s, for which the slip length is obtained as $L_s = 12\,\text{nm}$ by fitting the hydrodynamic force using Equation 6.2 (Craig et al., 2001).

Figure 6.18 plots the slip length as the velocity of the flat surface is varied with different sucrose viscosities. It can be seen that the slip length increases as the approaching velocity is increased, regardless of the viscosity of the sucrose solution. The qualitative dependence of the slip length on the viscosity in Figure 6.18 is somewhat surprising. The slip length increases as the viscosity increases, which is different from constant slip lengths found in MD simulations (Figure 6.15). The reason for the inconsistency is unclear. It could be caused by the non-Newtonian behavior of the sucrose solution at a high sucrose concentration. Nevertheless, the slip length in Figure 6.18 generally shows monotonous increases with increasing shear rate. Although practically it cannot diverge to infinity, the slip length does not appear to converge to a value as the shear rate is increased.

Similar experiments are conducted with tetradecane and water as the fluids between octadecyltriethoxysiloxane (OTE)-coated mica surfaces (Zhu and Granick, 2001). On an OTE surface, the contact angles of water and tetradecane

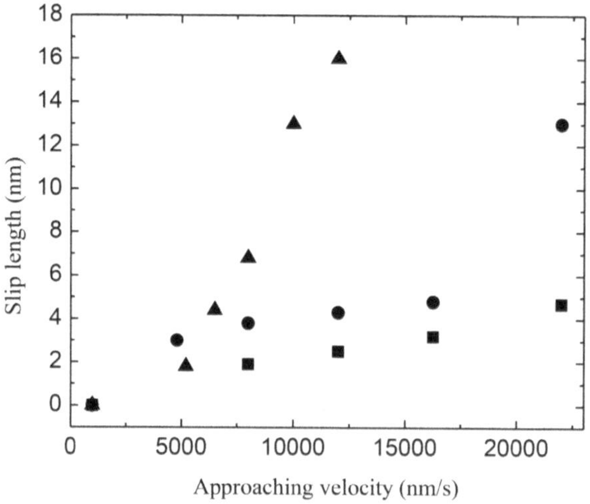

Figure 6.18 Slip length as a function of the approaching velocity. The viscosities of the sucrose solutions are 19.2 mPa·s (square), 38.9 mPa·s (circle), and 80.3 mPa·s (triangles). (Data are extracted from Craig et al., 2001).

are about 110°C and 44°C, respectively. The experimental setup is the same as that in Figure 6.16. In the experiments, the flat surface oscillates and generates an oscillatory hydrodynamic force. The peak velocity of the flat surface is denoted as v_{peak}. Based on the hydrodynamic forces measured, the correction factor f^* in Equation 6.3 is obtained.

Figure 6.19 shows f^* as a function of distance h for DI water between OTE surfaces for different peak velocities v_{peak}. It can be seen that the stick boundary condition holds, i.e., $f^* = 1$, for $h > 300$ nm regardless of the magnitude of the peak velocity. When h is below 300 nm, f^* starts to decrease and its value depends on the peak velocity. At a given distance, where flow slip takes place, the value of f^* at a relatively high peak velocity is smaller than that at a low peak velocity, implying that the flow slip and the slip length increases with increasing shear rate. Figure 6.20 plots the slip length as the peak velocity is varied for water and tetradecane between different surfaces (Zhu and Granick, 2001), where the slip lengths follow similar fashions for different cases as the peak velocity is increased. It is zero (no slip) at sufficiently small v_{peak}. As v_{peak} is higher than a critical value, the value of L_s increases exponentially. It can also be seen that the slip length increases as the surface wettability is decreased, i.e., non-wetting surfaces (large contact angle) promote flow slip.

It is worth mentioning that the slip lengths in Figure 6.20 are of micrometer scale, which are much larger than those in Figures 6.2, 6.4, and 6.18, where the

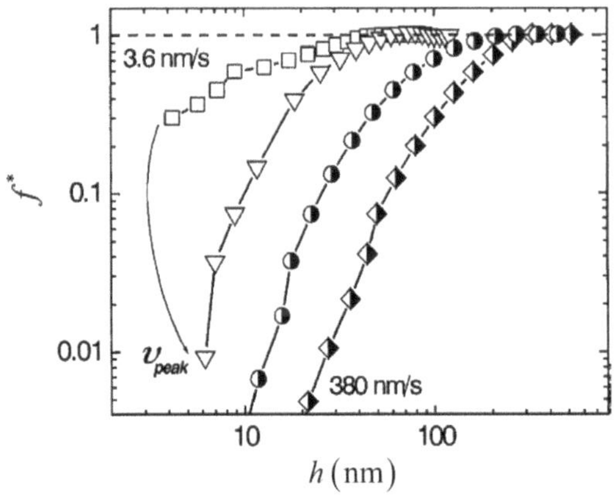

Figure 6.19 Correction factor for velocity slip versus separation h for DI water between OTE-coated mica surfaces. The peak velocities for squares, triangles, circles, and diamonds are 3.6, 40, 100, and 380 nm/s, respectively. The frequencies of the oscillatory plane are 1 Hz for squares and triangles and 10 Hz for circles and diamonds. (Reprinted with permission from Zhu and Granick, 2001, Copyright (2001) by the American Physical Society.)

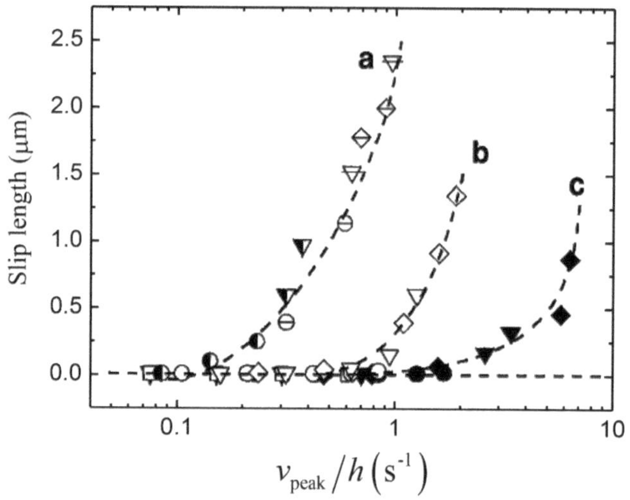

Figure 6.20 Slip length as a function of v_{peak}/h for (a) DI water between OTE surfaces (cross and semi-filled symbols, contact angle is about 110°C, (b) tetradecane between OTE surfaces (open symbols, contact angle is about 44°C), and (c) tetradecane containing 0.2% hexadecylamine between mica surfaces (filled symbols, contact angle is about 12°C). Shear rate depends on h and ranges from 10^3 and 10^4 s^{-1}. (Reprinted with permission from Zhu and Granick, 2001, Copyright (2001) by the American Physical Society.)

experimental strategies are similar. The large slip lengths in Figure 6.20 might be caused by systematic errors in the experiments, such as the errors in hydrodynamic force, velocity, and distance measurements. Nevertheless, the general observation of the monotonic increase of the slip length with increasing shear rate is consistent with many studies.

Shear-rate-dependent slip length has also been reported using MD simulations (Priezjev, 2007; Yong and Zhang, 2013). Figure 6.21 depicts an MD system, where an LJ fluid is confined by two parallel, rigid walls (wall atoms are fixed). The fluid moves in the x-direction under an external force, F_x, which acts on all the fluid molecules. Both F_x and the fluid–wall binding energy ε_{fw} are varied to consider the effects of the shear rate and the surface energy on the slip. The LJ parameters, binding energy ε and collision diameter σ, for the interactions among fluid molecules are used to express the other units.

Figure 6.22 shows the velocity profiles at different external forces and fluid–wall binding energies (only half of the velocity profiles are shown due to their symmetry with respect to $z = 0$ plane). Other than the slip at the interfaces, $z = \pm 11.8\sigma$, the velocity distributions can be well-fitted by parabolas. At a given fluid–wall

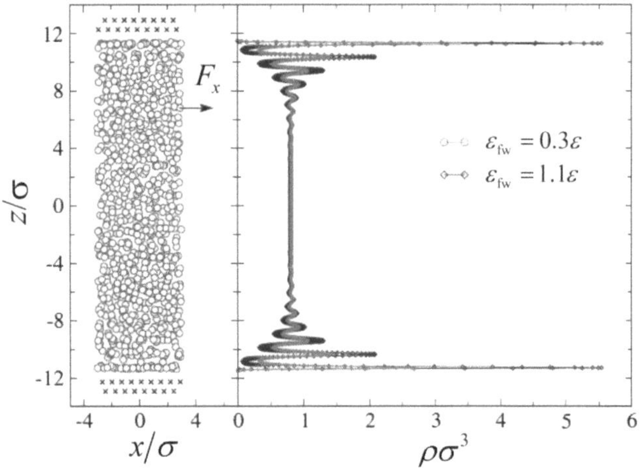

Figure 6.21 An MD system (left, partially shown) and density profiles (right). Circles and crosses in the system are fluid molecules and wall atoms, respectively. The density distributions are for $\varepsilon_{fw}/\varepsilon = 0.3$ (circles) and $\varepsilon_{fw}/\varepsilon = 1.1$ (diamonds) with $F_x = 0.001\,\varepsilon/\sigma$. (Reprinted with permission from Priezjev, 2007, Copyright (2007) by the American Physical Society.)

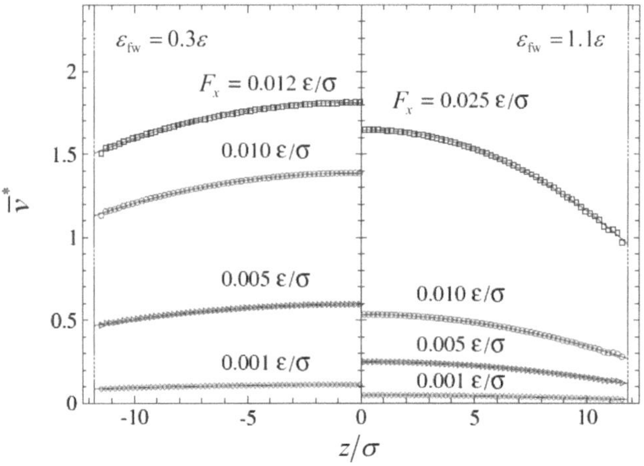

Figure 6.22 Velocity profiles at different external forces and fluid–wall binding energies. Dashed lines are the locations of the interfaces. $\bar{v}^* = \bar{v} \cdot \tau/\sigma$ is the reduced mean velocity with $\tau = \sqrt{m\sigma^2/\varepsilon}$ being the unit time in MD simulations. (Reprinted with permission from Priezjev, 2007, Copyright (2007) by the American Physical Society.)

binding energy, the slip velocity increases with increasing external force. For the same external force, the slip velocity decreases when the fluid–wall binding is strengthened.

By extrapolating the velocity profile to zero, the slip length is obtained, which is shown in Figure 6.23 as a function of the external force and the shear rate. The dependence of L_s on F_x and $\dot{\gamma}$ at different fluid–wall binding energies suggests that the slip length undergoes a transition as ε_{fw} is increased. For relatively large ε_{fw} values, $\varepsilon_{fw}/\varepsilon > 0.7$, the slip length increases linearly with increasing shear rate. For $0.3 < \varepsilon_{fw}/\varepsilon < 0.7$, the $L_s - \dot{\gamma}$ relationship becomes nonlinear, which appears to be a transition. As ε_{fw} is further reduced and the surface energy becomes sufficiently low ($\varepsilon_{fw}/\varepsilon = 0.3$), L_s follows a power-law dependence on $\dot{\gamma}$, which is in good agreement with other MD studies (Thompson and Troian, 1997) and can be modeled by

$$L_s = L_s^0\left(1 - \dot{\gamma}/\dot{\gamma}_0\right)^{-0.5} \tag{6.31}$$

where L_s^0 and $\dot{\gamma}_0$ are fitting parameters ($L_s^0 = 19.5\sigma$ and $\dot{\gamma}_0 = 0.093/\tau$ for $\varepsilon_{fw}/\varepsilon = 0.3$ in Figure 6.23).

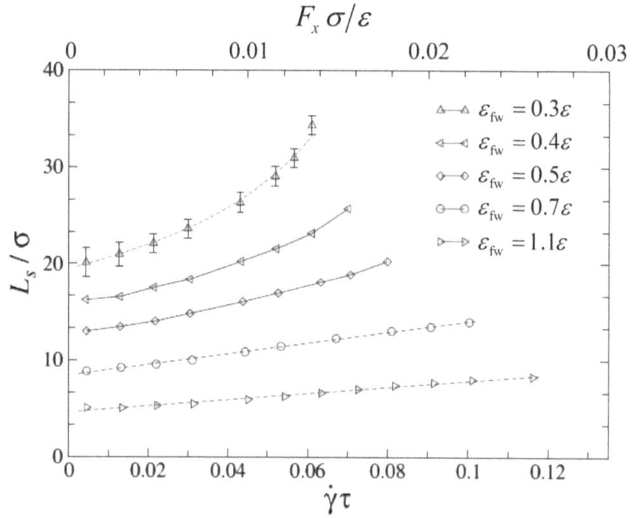

Figure 6.23 Slip length as a function of shear rate and external force at different values of fluid–wall binding energies. Lines are the best fits of the data. For $\varepsilon_{fw}/\varepsilon = 0.3$, the dashed line is the fit of the data using Equation 6.31 with $L_s^0 = 19.5\sigma$ and $\dot{\gamma}_0 = 0.093/\tau$. (Reprinted with permission from Priezjev, 2007, Copyright (2007) by the American Physical Society.)

Some other MD simulations also find that the slip length is shear rate dependent and unbounded (Yong and Zhang, 2013). A system similar to that in Figure 6.21 is employed and the slip length is determined as the surface rigidity is varied under different shear rates. It is found that there is a dramatic increase in the slip length as the shear rate is raised. At the low-shear-rate regime, the slip length increases monotonically with increasing shear rate, as depicted in Figure 6.24a. When the

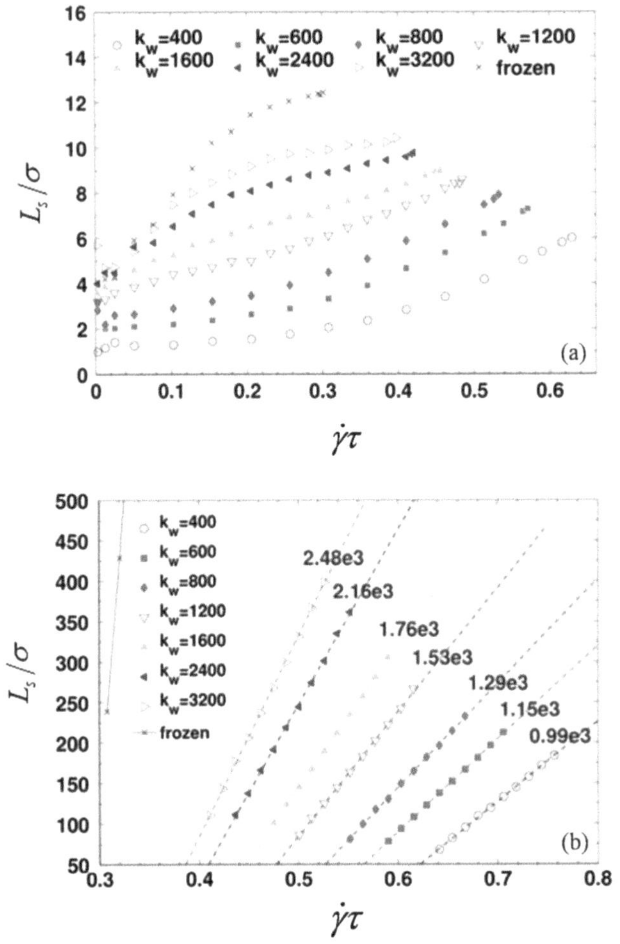

Figure 6.24 Slip length as a function of shear rate with different wall rigidities. k_w is the spring constant used to confine the wall atoms. The fluid–wall binding energy $\varepsilon_{fw} = 1.0\varepsilon$ for all the cases, for which the walls are hydrophilic. (a) Low shear rates $(0 < \dot{\gamma}\tau < 0.65)$ and (b) high shear rates $(0.3 < \dot{\gamma}\tau < 0.8)$. Dashed lines in (b) are linear fits to the data and the numbers denote the slopes of the lines. (Reprinted by permission from Springer Nature, Yong and Zhang, 2013, Copyright (2013).)

shear rate reaches a critical value, the slip length rises abruptly and increases linearly and unboundedly with increasing shear rate (Figure 6.24b).

Although the detailed dependence of the slip length on the shear rate in these studies are different, the slip length keeps increasing as the shear rate is increased. Such a behavior can only hold at relatively low shear rates as the slip length cannot go to infinity. In MD simulations, the shear rates are usually high and the fast increase of the slip length as the shear rate is increased might be caused by artificial effects, as will be discussed later. Practically, the slip length should approach an upper bound when the shear rate becomes sufficiently high.

6.2.1.4 Shear-dependent slip length-bounded At high shear rates, bounded slip lengths are found in theoretical and numerical analyses (Martini et al., 2008). Motions of *n*-decane molecules sheared by two parallel plates (Couette flows) are numerically studied with different wall models. The *n*-decane molecules are described by a united atom model, i.e., each molecule contains ten monomers connected by rigid bonds, which are allowed to bend and rotate. Both rigid (wall atoms are fixed) and flexible walls are considered. In the simulations, the walls move in opposite directions with velocities ranging from 3.2 to 1,000 m/s. The slip length is obtained as $L_s = u_w/\dot{\gamma}$, where u_w is the velocity difference between the fluid and the walls, i.e., the slip velocity at the interfaces. The shear rate $\dot{\gamma}$ is obtained by fitting the velocity profile. Figure 6.25 shows the slip length as the wall speed is changed. For rigid walls, it can be seen that the slip length increases exponentially with increasing wall speed. This

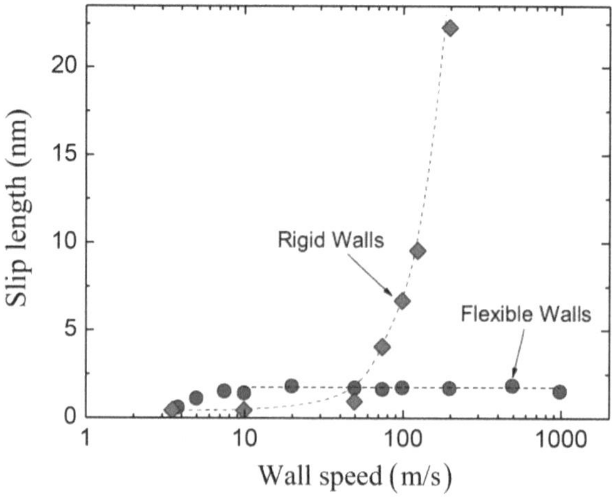

Figure 6.25 Slip length as a function of wall speed for rigid and flexible walls. Lines are a guide to the eye. (Reprinted with permission from Martini et al., 2008, Copyright (2008) by the American Physical Society.)

unbounded slip length is qualitatively similar to the MD studies in Figures 6.23 and 6.24. If the walls are flexible, however, the slip length converges to a constant value when the wall speed is higher than 10 m/s. The bounded slip length is supported by the continuum theory with the consideration of the potential near the walls and the compressibility of the fluid, for which the equation of motion is written as follows:

$$-\nabla p + \mu\nabla^2\mathbf{V} + \frac{\mu}{3}\nabla(\nabla\cdot\mathbf{V}) - \rho\nabla\phi = 0 \tag{6.32}$$

where $\mathbf{V} = \mathbf{V}(u,v)$ is the velocity vector and ϕ is the potential field. At the walls, the boundary condition is set as

$$\mu\frac{\partial u}{\partial y} = f_w u \tag{6.33}$$

where f_w is the friction coefficient per unit area at the walls. For a potential field given by (Braun and Kivshar, 1998)

$$\phi = \phi_0 e^{-ky}\cos(my)\left[1.1 + b\sin(mx)\right] \tag{6.34}$$

where ϕ_0, k, m, and b are parameters, Equation 6.32 is numerically solved for different wall speeds and values of the friction coefficient f_w. Figure 6.26 shows

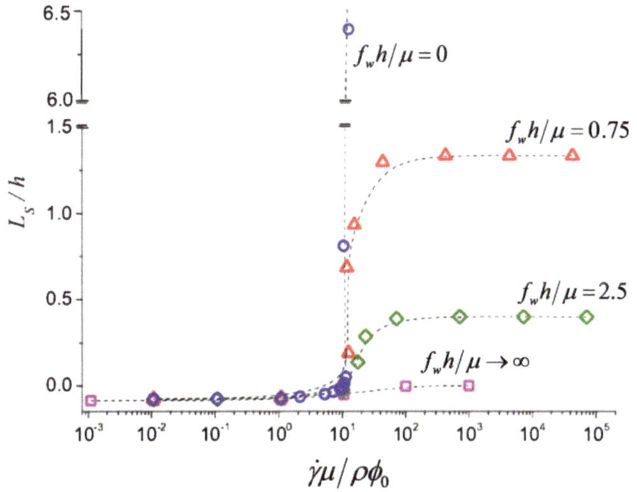

Figure 6.26 Slip length predicted from the continuum theory as a function of normalized shear rate. h is the distance between the parallel walls. (Reprinted with permission from Martini et al., 2008, Copyright (2008) by the American Physical Society.)

the slip length as the shear rate (or wall velocity) is varied. It can be seen that the dependence of L_s on $\dot{\gamma}$ is affected by the friction coefficient. If the flow is frictionless at the interface, i.e., $f_w = 0$, the slip length diverges and goes to infinity as the shear rate is higher than a critical value. However, if there is an interfacial friction, the slip length approaches to a constant value, which decreases as the friction coefficient is increased.

It is worth mentioning that the fashion of the slip length in Figure 6.26 is largely determined by the boundary condition in Equation 6.33, which has a strong physical background. This will be discussed in Section 6.2.2 on the mechanisms of flow slip.

Bounded slip lengths have also been confirmed in experiments, which are performed in a nanofluidic chip consisting of 100 parallel, 50-μm-spaced nanochannels (ten channels in the center part were 500-μm-spaced for fluorescent flow observation) (Li et al., 2014). The sketch of the fluidic system is illustrated in Figure 6.27a. In this fluidic system, nanochannels are connected at the ends by two microchannels, which are in connection with four reservoirs. The nanochannels are constructed by the photolithography technique, as discussed in Chapter 4. They are fabricated on a silicon substrate through RIE with an etching rate of 2.5 μm/min, which could guarantee a good control for the channel depth with small surface roughness. The mirochannels are etched using the DRIE technique. An oxide layer is then developed on both sides of the silicon wafer to protect the micro/nanochannels during the fabrication of fluid reservoirs through DRIE. Before bonding the channels, a 300-nm-thick wet thermal oxide layer is uniformly grown on the silicon wafer in a furnace at 1,000°C to prevent the nanochannels from collapsing during the final bonding. The fluidic system is closed through anodic bonding with a clean glass wafer under 380°C and 800 V. Figure 6.27b shows a fabricated chip. The depth, width, and length of the nanochannels are set

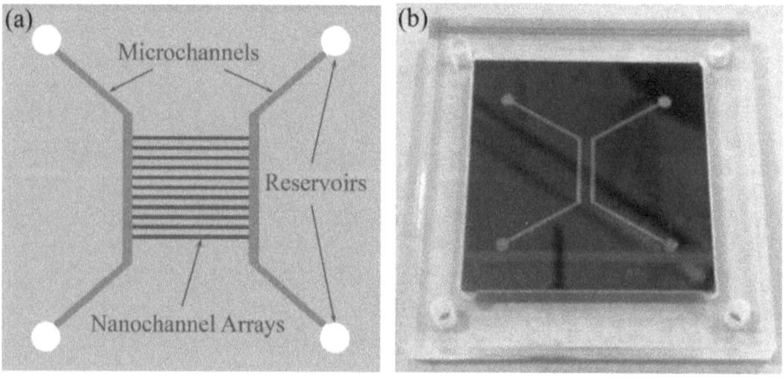

Figure 6.27 Schematic (a) and fabricated chip (b) of a nanofluidic system for flow slip investigation. (Reprinted with permission from Li et al., 2014, Copyright (2014) by the American Physical Society.)

as 100 nm, 100 μm, and 1.5 mm, respectively, and those for the microchannels are 10 mm, 500 μm, and 40 μm. After the fabrication, the nanochannels are checked by an AFM before the final bonding. The channel height and surface roughness are demonstrated in Figure 6.28, which shows that the channel height is 105.6 nm and the surface roughness is less than 1.1 nm.

The quality of the nanofluidic system is tested by fluorescent dye-laden DI water flows. A snapshot of a flow is shown in Figure 6.29, where the smooth flow (light gray parts) indicates that the channels are free of defects. Polytetrafluoroethylenc

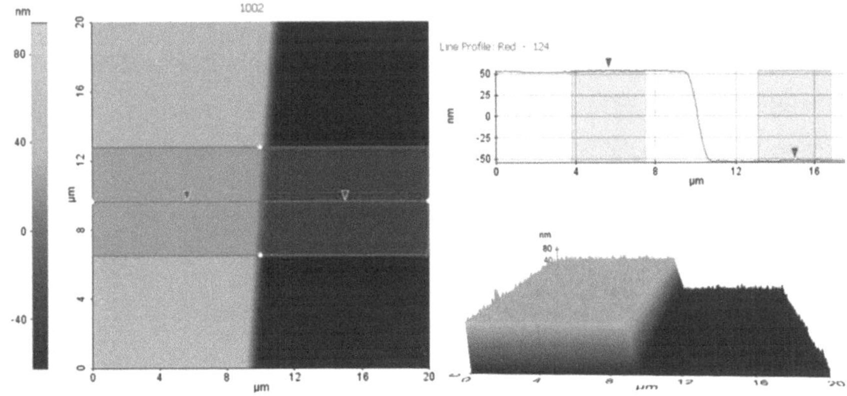

Figure 6.28 AFM characterization of nanochannels. Channel height is 105.6 nm with surface roughness less than 1.1 nm. (Reprinted with permission from Li et al., 2014, Copyright (2014) by the American Physical Society.)

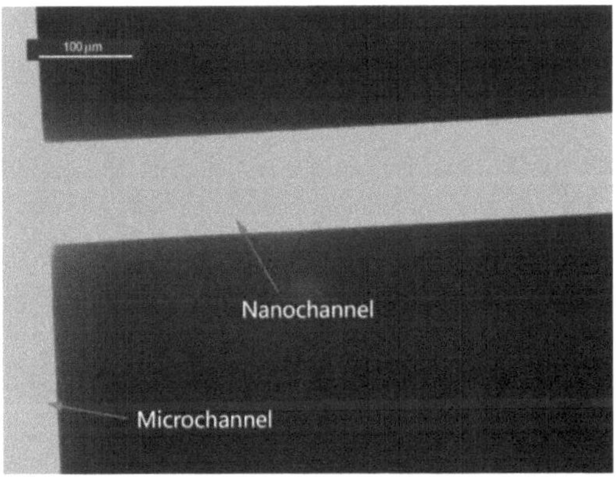

Figure 6.29 An image of a fluorescent flow in the nanofluidic system in Figure 6.27. (Reprinted with permission from Li et al., 2014, Copyright (2014) by the American Physical Society.)

(PTFE) tubings are used for connecting different parts and the joints are sealed with 2-mm-diameter rubber O-rings. Deionized water is driven by a stainless steel syringe mounted on an ultra-high linear force syringe pump, which controls the flow rate with an accuracy of 0.001 μL/h. Pressure transducers are used for the pressure measurement of the reservoirs. Valves are employed to make sure that air is removed from the system before experiments. The whole experimental setup is shown in Figure 6.30.

In the experiments, the flow rates are controlled by the syringe pump and the pressures of the upstream and downstream reservoirs, P_1 and P_3 (Figure 6.30) are measured (P_3 is 1 atm because the downstream is open to air). As the flow resistance in microchannels is significantly smaller than that in nanochannels, the pressure difference, $P_1 - P_3$, can be approximately treated as the pressure drop, Δp, over the nanochannels.

Figure 6.31 shows the flow rate Q in each nanochannel as a function of Δp. It is found that the flow rate experiences a transition between two linear regimes as Δp is varied. The transition occurs in the pressure drop range of $2.8 < \Delta p < 3.2$ MPa. The dependence of Q on Δp in Figure 6.31 is different from the classic Poiseuille law for channel flows under the stick boundary condition,

$$Q = \frac{h^3 w}{12\mu} \frac{\Delta p}{l_c} \tag{6.35}$$

where h, w, and l_c are the channel height, width, and length, respectively. Equation 6.35 suggests a linear relationship between Q and Δp, which is also depicted in

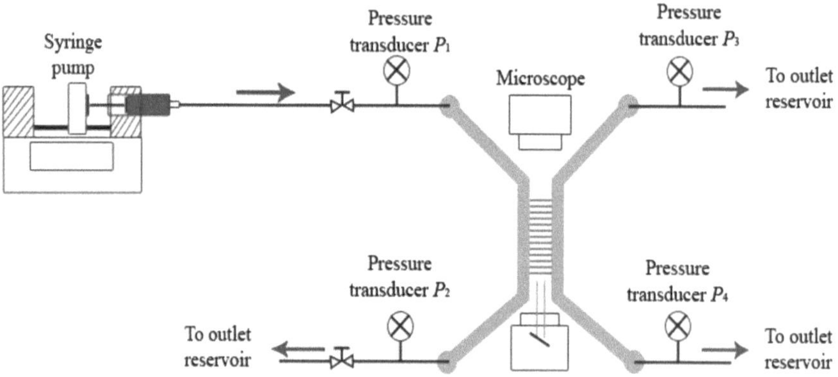

Figure 6.30 Experimental setup of the fluidic system. DI water is driven by a syringe pump with a desired flow rate. Arrows indicate the flow directions. Pressures of the reservoirs are measured by pressure transducers. (Reprinted with permission from Li et al., 2014, Copyright (2014) by the American Physical Society.)

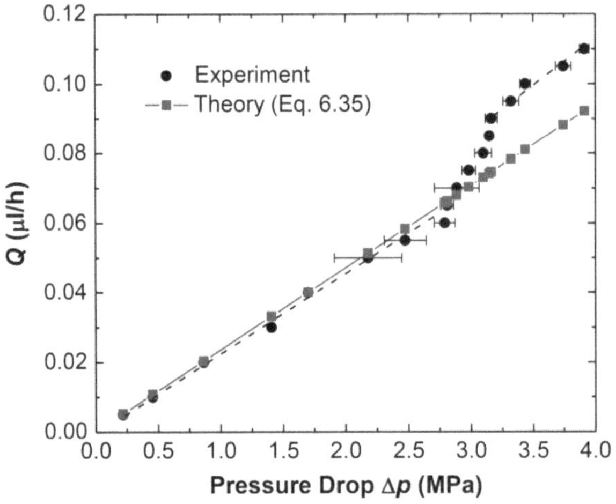

Figure 6.31 Flow rate as a function of pressure drop. Error bars represent the standard deviation of three experiments. Dashed lines are a guide to the eye. (Reprinted with permission from Li et al., 2014, Copyright (2014) by the American Physical Society.)

Figure 6.31. At low-pressure drops ($\Delta p < 2.8$ MPa), the experimental results are in good agreement with Equation 6.35. The slightly lower experimental flow rates might be the consequence of the relatively large flow resistance because the nanochannels are hydrophilic (Liu and Li, 2011). When $\Delta p > 2.8$ MPa, the experimental data deviate from Equation 6.35 and switch to another linear regime for $\Delta p > 3.2$ MPa, where the experimental results are about 20% larger than the predictions of Equation 6.35.

The flow enhancement after the transition is most probably caused by the velocity slip as the fluid velocity profile is parabolic (Todd et al., 1995; Liu and Li, 2011) and the viscosity variation is negligible in channels of ~ 100 nm (Akhmatskaya et al., 1997). If the velocity slip is considered, the flow rate in Equation 6.35 can be calculated as follows:

$$Q_{\text{slip}} = \frac{h^3 w}{12\mu} \frac{\Delta p}{l_c} \left(1 + \frac{6L_s}{h}\right) \tag{6.36}$$

By fitting the experimental data using Equation 6.36, the slip length L_s is obtained, which is plotted in Figure 6.32 as a function of shear rate $\dot{\gamma} = h\Delta p/(2\mu l_c)$. It can be seen that the stick boundary condition is only valid at low shear rates, for $\dot{\gamma} < 1.0 \times 10^5$ s^{-1} (the slip length herein actually is slightly negative due to fluid adsorption, as discussed in Chapter 2). For 1.0×10^5 s^{-1} $< \dot{\gamma} < 1.1 \times 10^5$ s^{-1}, the slip length increases linearly with increasing shear rate. For $\dot{\gamma} > 1.1 \times 10^5$s^{-1}, the slip

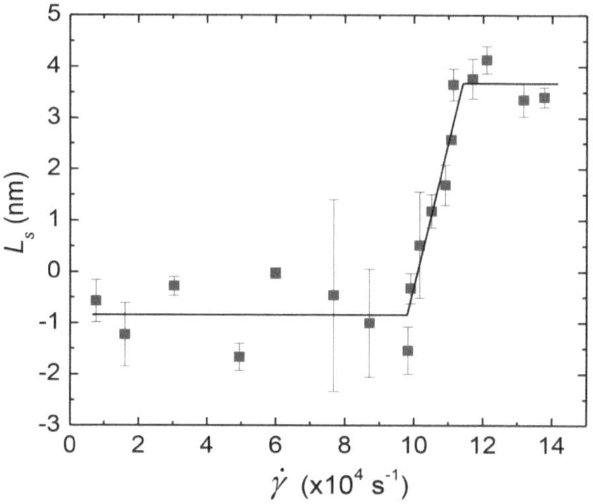

Figure 6.32 Slip length as a function of shear rate. Solid lines are a guide to the eye. (Reprinted with permission from Li et al., 2014, Copyright (2014) by the American Physical Society.)

length roughly remains constant with an average value of 3.66 nm. The slip length in Figure 6.32 is similar to that in Figure 6.26 and is the first experimental evidence for shear-dependent and bounded slip length in nanochannels.

The constant slip length at higher shear rates in Figure 6.32 can be explained by considering fluid molecules in a thin layer next to the channel wall. The forces acting on the fluid layer are the external force associated with the pressure difference, the shear force from the interior fluid, and the frictional force from the channel wall, as illustrated in Figure 6.33. Since the shear force is the intermolecular force, which is usually much larger than the external force, the equation of motion for the thin fluid layer can be written as follows (Martini et al., 2008):

$$m\ddot{x} = \alpha(V - \dot{x}) - \beta\dot{x}, \tag{6.37}$$

where m, x, $\dot{x}$, and $\ddot{x}$ are the mass, position, velocity, and acceleration of the fluid layer, respectively. The terms $\alpha(V - \dot{x})$ and $\beta\dot{x}$ denote the shear and frictional forces. Here, α and β are the shear and friction coefficients, V is the velocity of the interior fluid next to the thin fluid layer given by $V = \dot{\gamma}(L_s + d)$, where d is a constant representing the mean spacing of fluid layers in the direction normal to the wall surface and is about a few angstroms. In the steady state, the two force terms in Equation 6.37 are the same and the slip velocity u_s is given by $u_s = \alpha V/(\alpha + \beta)$. Based on the definition of slip length, $L_s = u_s/\dot{\gamma}$, and using $\dot{\gamma} = (V - u_s)/d$, it is obtained that

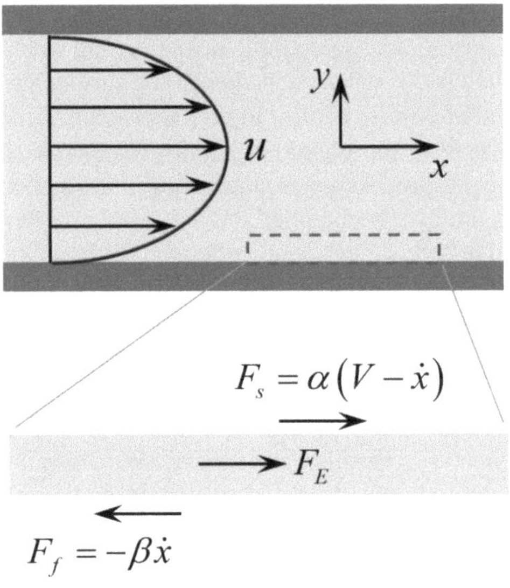

Figure 6.33 Forces acting on a thin fluid layer next to the channel wall. F_E, F_s, and F_f are the external, shearing, and frictional forces, respectively.

$$L_s = \frac{\alpha}{\beta} d, \tag{6.38}$$

which is a constant. It should be noted that α and β depend on the materials of the fluid and channel. At a high shear rate, $u_s \approx V$, which leads to $\alpha \gg \beta$. Therefore, L_s could range from a few to tens of nanometers. The experimental results in Figure 6.32 indicates that $\alpha \sim 10\beta$. It should be noted that Equation 6.38 is only valid under the assumption that the shear force and friction in Equation 6.37 are linearly related to velocity, which is valid if the shear rate is not extremely high.

6.2.2 Mechanisms

The classic fluid mechanics theory cannot explain the distinct behavior of slip length discussed in the previous section. The underlying mechanisms can be probed by examining nanoscale molecular interactions. As the boundary condition is mainly determined by the motion of interfacial fluid molecules, the forces acting on the fluid molecules near a channel wall and how they affect the dynamics of these molecules may offer a physical picture about the velocity slip. Usually, the intermolecular force between a fluid molecule and a solid atom is stronger than that between two fluid molecules. Therefore, the molecular forces applied on the fluid molecules near a channel wall by the wall atoms play an essential role in flow slip.

As introduced in Chapter 3, the fluid–wall intermolecular interaction is governed by the interaction potential between fluid molecules and wall atoms. The potential distribution next to a solid surface depends on the surface properties, such as the surface energy and atomic structure. Figure 6.34 depicts the potential distribution next to a copper wall for a liquid argon flow (Liu et al., 2010). It can be seen that the potential contains a series of potential wells and maxima (both local and global). The wells are of low potential and are the places fluid molecules tend to stay. The potential difference between a potential well and its adjacent maximum forms a potential barrier, ΔE, which determines the probability of a fluid molecule escaping from the potential well. Precisely, the probability p_r depends on the value of $\Delta E/k_B T$ in a way that $p_r \sim \exp\left(-\Delta E/k_B T\right)$ (Li, et al., 2014). For a given fluid, a surface of high surface energy generates a potential with high potential barriers and tends to adsorb fluid molecules. Figure 6.35 shows that a layer of fluid molecules (red spheres) are firmly trapped by a nanochannel wall (green spheres) of high surface energy. It must be noted that the potential in Figure 6.34 is quite

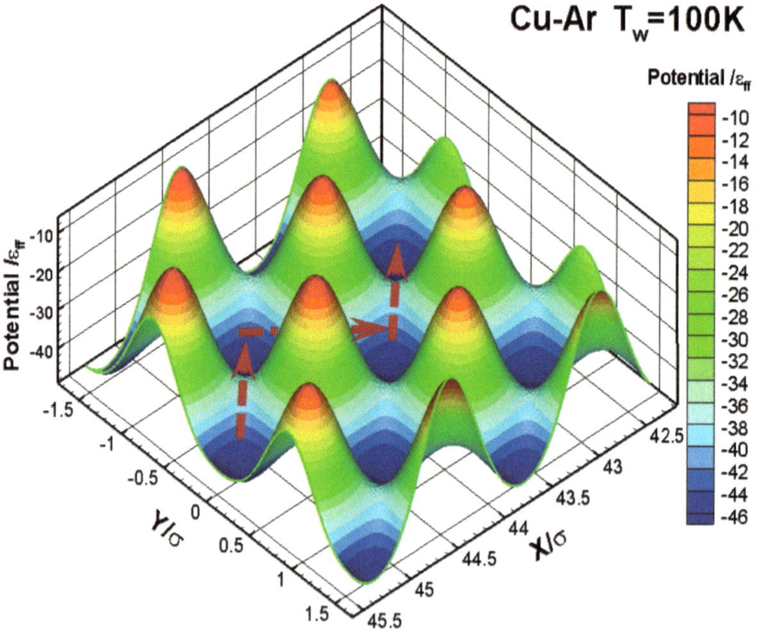

Figure 6.34 Three-dimensional potential distribution in an *xy* plane near and parallel to a copper wall for a liquid argon flow. The potential is normalized by the fluid–fluid binding energy ε_{ff}. *x* and *y* coordinates are scaled by the collision diameter σ. T_W is the wall temperature. Arrows show a possible hopping path of fluid molecules. (Reprinted from Liu et al., 2010, with the permission of AIP Publishing.)

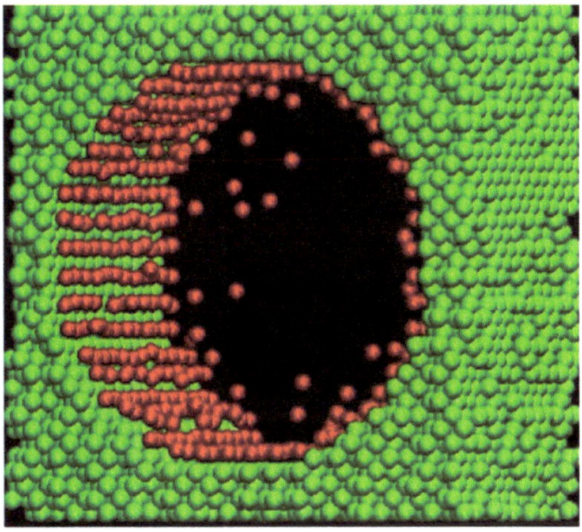

Figure 6.35 Adsorption of fluid molecules (red) on the interior surface of a high surface energy channel.

representative and the potential distributions near a wall for most nanoscale flows follow a similar fashion (Yong and Zhang, 2013).

Other than the potential, the temperature, which controls the thermal motion of fluid molecules, also affects the dynamics of fluid molecules when they move around at an interface. Furthermore, the external force or pressure drop, which drives the flow, helps fluid molecules overcome the potential barrier and enhances the mobility of fluid molecules. At the molecular scale, the ratio, $F_e\sigma/\varepsilon$ (ε and σ are the LJ parameters and can be viewed as the characteristic energy and length), can be used as a measure for the contribution of the external force. These three effects are the major factors that govern the dynamics of interfacial fluid molecules.

Therefore, two dimensionless numbers, $\Delta E/k_BT$ and $F_e\sigma/\varepsilon$, play important roles in flow slip. They have opposite effects on flow slip. A small $\Delta E/k_BT$ and a high $F_e\sigma/\varepsilon$ promote flow slip, while a high $\Delta E/k_BT$ and a small $F_e\sigma/\varepsilon$ usually lead to fluid adsorption, for which the no-slip boundary condition tends to be valid or a negative slip length may occur.

In high surface energy channels (e.g., hydrophilic channels), $\Delta E/k_BT \gg 1$, the wall attraction is strong and the thermal effect barely assists fluid molecules to move out from potential wells. In this case, if the external force is small (small $F_e\sigma/\varepsilon$), a layer of fluid molecules can be adsorbed by the wall and the stick boundary condition holds (there could be a negative slip length if the fluid–wall interaction is very strong). When the external force is sufficiently high (large $F_e\sigma/\varepsilon$), it is capable of assisting the admolecules to overcome potential barriers and consequently leads to a relatively motion between the admolecules and the wall, which is nothing but flow slip. This is why a critical shear rate is found for the slip in Figures 6.20, 6.24, 6.25, and 6.32. The arrows in Figure 6.34 show how admolecules move under a sufficiently large external force. They follow a zigzag path, where the potential is relatively low. If the external force is further increased, the slip velocity should increase. Before the slip velocity grows large, the friction at the interface is roughly a linear function of the slip velocity. However, as the slip velocity becomes sufficiently high, the friction between the fluid molecules and the channel wall will cause the temperature at the interface to increase (Li, 2009). This will enhance the thermal motions of fluid molecules and wall atoms, and consequently increases the friction, leading to a bounded slip length, as observed in Figures 6.25 and 6.32. The increased friction due to heating could serve as a support to the boundary condition in Equation 6.33.

In low surface energy channels, $\Delta E/k_B T \ll 1$, the fluid–wall interaction is relatively weak and fluid adsorption does not take place. In this case, fluid molecules can easily hop from one potential well to another and a small external force can lead to a large slip velocity. This explains why flow slip can be observed regardless of the magnitude of the shear rate (Figure 6.23). Furthermore, the friction at the interface is small due to the weak molecular forces, which does not generate heat and increase the temperature significantly even if at high shear rates. This is why the slip length keeps increasing as the shear rate is increased, as shown previously.

Therefore, whether the slip length is a constant or increases with increasing shear rate depends on the flow conditions, including surface properties (e.g., surface roughness and surface energy), shear rate range, and temperature, which are coupled and make flow slip a non-trivial issue.

6.2.3 Remarks on theoretical predictions

The micro-mechanisms of flow slip discussed in the previous section indicate that the slip length is influenced by the energy barrier ΔE, temperature, and external force. These factors make the prediction of slip length challenging. Even though the analysis from Equation 6.6 to 6.23 is rigorous, the slip length predictions of Equations 6.18 and 6.23 involve the friction coefficient ξ_0, which requires numerical approaches to provide its value. In this sense, the macroscale framework for the analysis of slip length is valid but the applications of Equations 6.18 and 6.23 are not straightforward. This remark also applies to Equation 6.38, where the shear and friction coefficients, α and β, are unavailable.

For numerical results, although rich data can be obtained through extensive MD or other simulations, the quantitative predictions face a few problems. The beauty of MD simulations lies in the qualitative description of flow phenomena instead of providing quantitative results. This is because the potentials employed in MD simulations are empirical and may not be reliable in offering accurate values for the fluid properties and flow velocities. Furthermore, there are artifacts in MD simulations caused by unrealistic treatments, such as rigid walls (wall atoms are fixed), and extremely large velocities. In MD simulations, the fluid velocity usually is quite high, from tens to hundreds of meters per second. This is to mask the large numerical error to ensure a reasonable result. The extremely high velocity is unpractical and can cause a series of issues, including temperature increase and inaccurate fluid property reproduction. Finally, the modeling of slip length on the basis of a large amount of simulations may contain fitting parameters (e.g., Equation 6.31), which vary for different situations and cause inconvenience in practical applications.

In summary, the prediction of slip length is not an easy problem. The complex nature due to the coupling of various parameters poses challenges in giving a simple universal law governing the slip length. This actually is not unique for flow slip. Many other issues, such as flow scenarios, also suffer from the influence of parameters at different scales. Nevertheless, extensive studies are needed for detailed investigations of flow slip.

6.3 Flow regimes

6.3.1 Dimensionless numbers

In Section 6.2, it is shown that surface properties greatly affect the motion of fluid molecules at an interface. Surface effects can propagate further through hydrodynamic interactions among fluid molecules and consequently may influence the flow scenarios.

For surface properties, surface energy is an important one, which, macroscopically, is related to the contact angle of a fluid on a surface. Microscopically, surface energy is described by the molecular binding energy, ε, between fluid molecules and surface atoms. ε can be viewed as a microscopic parameter and its effects on the dynamics of a fluid are not considered in the classic Navier–Stokes equation. In the previous section, the role of ε in flow slip is discussed. In this section, how it couples with other parameters and influences the flow behavior are discussed.

To understand how the surface energy affects the flow dynamics, it is necessary to find out the major parameters and the associated dimensionless numbers that control the flow at the nanoscale. As discussed in Chapter 2, the surface energy plays an important role in fluid adsorption. If the fluid–wall binding energy, ε, is used to describe fluid adsorption on a surface, it can be stated that a strong ε promotes fluid adsorption. This is because a strong fluid–wall binding energy can generate a low potential area next to a wall, as illustrated in Figure 6.36. The low

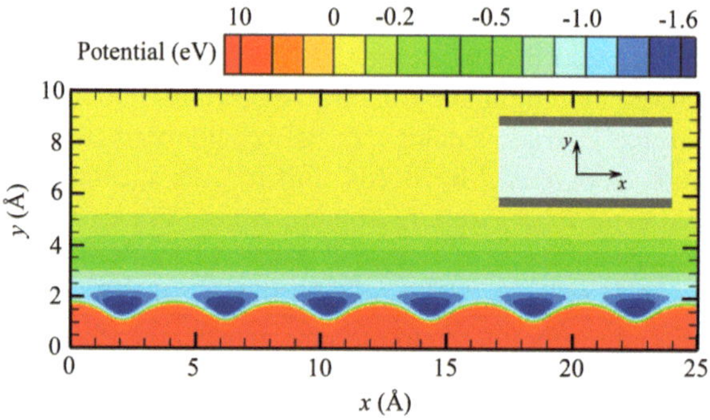

Figure 6.36 Potential distribution next to the bottom wall of a nanochannel flow between two parallel walls. The flow system is depicted in the inset and the bottom wall is located at $y = 0$. The fluid–wall binding energy $\varepsilon_{fw} = 600$ K. Blue areas are low potential areas, where fluid molecules are trapped by the wall.

potential can trap fluid molecules having relatively low kinetic energies and lead to fluid adsorption. To overcome the wall attraction, one way is to increase the kinetic energy of the admolecules. Once the kinetic energy of the admolecules becomes sufficiently high, they can escape from low potential areas and fluid adsorption is hindered. This usually requires a relatively high temperature T. Therefore, the fluid–wall binding energy ε_{fw} and $k_B T$ form a pair of competing parameters and their ratio, $\varepsilon_{fw}/k_B T$, can be used as a simple index for fluid adsorption (Li and Wang, 2005; Li and Hong, 2007). A large value of $\varepsilon_{fw}/k_B T$ can cause fluid adsorption.

Another parameter that plays a similar role to temperature is the fluid–fluid binding energy, ε_{ff}, which measures the strength of the molecular interactions among of fluid molecules. ε_{ff} is directly related to the contact angle of a fluid on a solid surface. At an intermediate $\varepsilon_{fw}/k_B T$ value, a large $\varepsilon_{fw}/\varepsilon_{ff}$ promotes fluid adsorption. In this sense, the product of $\varepsilon_{fw}/k_B T$ and $\varepsilon_{fw}/\varepsilon_{ff}$, i.e., $\varepsilon_{fw}^2/(k_B T \varepsilon_{ff})$, is also a dimensionless number that can characterize fluid adsorption.

If a flow is driven by an external force F_e, the external force can also help fluid molecules overcome the attraction from the surface atoms, as demonstrated in Figure 6.37. If the mean attractive force is characterized by $\varepsilon_{fw}/\sigma_{fw}$, where σ_{fw} is the LJ collision diameter and can be viewed as the characteristic length of fluid–wall interactions, such that $\bar{F}_{fw} \sim \varepsilon_{fw}/\sigma_{fw}$, the ratio $\varepsilon_{fw}/(F_e \sigma_{fw})$ can also offer a qualitative description of fluid adsorption.

As both $\varepsilon_{fw}/k_B T$ and $\varepsilon_{fw}/(F_e \sigma_{fw})$ are dimensionless numbers and qualitatively offer similar description for fluid adsorption, their combination, $\varepsilon_{fw}^2/(k_B T \cdot F_e \cdot \sigma_{fw})$, presents a comprehensive dimensionless number, which is sensitive to fluid–wall interactions and could be a reasonable dimensionless number for nanoscale flow

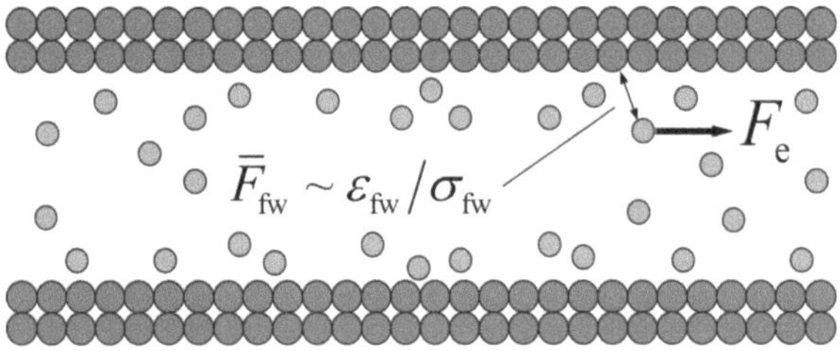

Figure 6.37 Competition between the external force F_e and the fluid–wall intermolecular force $\bar{F}_{fw} \sim \varepsilon_{fw}/\sigma_{fw}$.

Table 6.3 Dimensionless Numbers Characterizing Fluid–wall Interactions

Dimensionless Number	**Value**	**Phenomenon**
$\varepsilon_{fw}/\varepsilon_{ff}$	Large	Promotes fluid adsorption
ε_{fw}/k_BT	Large	Promotes fluid adsorption
$\varepsilon_{fw}^2/(k_BT\varepsilon_{ff})$	$\gg 1$	Fluid adsorption
$\varepsilon_{fw}/(F_e\sigma_{fw})$	Large	Promotes fluid adsorption
$\varepsilon_{fw}^2/(k_BT \cdot F_e \cdot \sigma_{fw})$	$\gg 1$	Fluid adsorption

characterization. Table 6.3 summarizes different dimensionless numbers and their relationships to fluid adsorption.

6.3.2 Flow scenarios

In this section, nanoscale flows between two parallel walls, i.e., Poiseuille flows, are investigated through NEMD simulations. The attention is on how the aforementioned parameters, ε_{ff}, ε_{fw}, T, and F_e, affect the flow rate, and how to characterize nanoscale flows using appropriate dimensionless numbers.

The Poiseuille flow system is shown in Figure 6.38. A fluid is confined by two parallel planar walls. The fluid is modeled by the LJ potential,

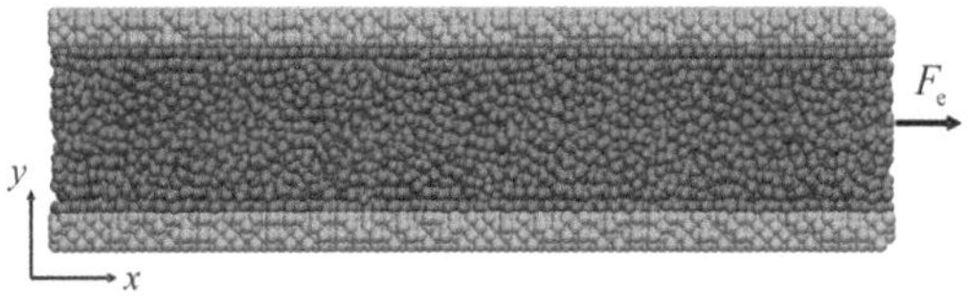

Figure 6.38 Structure of the Poiseuille flow system. Light and dark particles are wall atoms and fluid molecules, respectively.

$U(r) = 4\varepsilon[(\sigma/r)^{12} - (\sigma/r)^6]$, and the parameters for liquid Ar ($\varepsilon_{ff} = 114$ K) and He ($\varepsilon_{ff} = 10$ K) are used. The walls are described by the tight-binding potential (see Chapter 3 for detail) such that the thermal vibrations of the wall atoms are considered. The flexibility the walls is important for fluid–wall interactions as it affects the energy accommodation of the walls and the dynamics of fluid molecules at the interfaces (Li and Wang, 2005; Li and Hong, 2007). The LJ potential is also employed to calculate the interactions between the fluid molecules and wall atoms. The fluid–wall binding energy ε_{fw} is varied in a wide range to examine the effects of surface energy on the flow rate. Each wall contains four layers of atoms and the atoms in the outmost layer are fixed to prevent the walls from moving when fluid molecules move under an external force. Atoms in the other three inner layers are free to vibrate. The temperature of the walls is controlled at a specific value by applying the Berendsen thermostat to the vibrating wall atoms. The cutoff distance of the potentials is set at 10.21 Å, which is sufficiently large and greatly reduces the numerical error. Newton's equations are integrated with Beeman's leapfrog algorithm with time step equal to 1 fs. Different channel heights H (the distance between the two parallel walls in the y-direction), from 2 to 4 nm are considered. The lengths of the system in the x- and z-directions are 19.8 and 4.9 nm. An external force is applied to each fluid molecule in the x-direction and the PBCs are employed in the x- and z-directions.

The flux Q of liquid Ar in channels of $H = 3.7$ nm are computed by changing the fluid–wall binding energy under different temperatures and external forces. The ratio, Q/Q_{NS}, where Q_{NS} is the flux predicted by the Navier–Stokes equation at 1 atm, is depicted in Figure 6.39 as a function of $\varepsilon_{fw}^2/(k_B T \cdot F_e \cdot \sigma_{fw})$ (Liu and Li, 2009). It is easy to find that there is a critical value, $\varepsilon_{fw}^2/(k_B T \cdot F_e \cdot \sigma_{fw}) \approx 7$, in terms of the dependence of the flux ratio on the temperature. This critical $\varepsilon_{fw}^2/(k_B T \cdot F_e \cdot \sigma_{fw})$ value divides the flow into two regimes. For small $\varepsilon_{fw}^2/(k_B T \cdot F_e \cdot \sigma_{fw})$ values (below the critical value), let us denote this regime as regime I, the flux decreases as the temperature increases for a given value of $\varepsilon_{fw}^2/(k_B T \cdot F_e \cdot \sigma_{fw})$. However, the flux increases with increasing temperature when $\varepsilon_{fw}^2/(kT \cdot F_e \cdot \sigma_{fw})$ is larger than the critical value (regime II).

The critical value, $\varepsilon_{fw}^2/(k_B T \cdot F_e \cdot \sigma_{fw}) \approx 7$, in Figure 6.39 is a division point of strong and weak fluid–wall interactions. In regime I, the surface energy is relatively low and fluid adsorption is not favored, as illustrated by the velocity and density profiles in Figure 6.40. In this case, fluid molecules near the channel walls can move freely and experience a series of collisions with the walls without being trapped. As the temperature is increased, the thermal motion of the fluid molecules is strengthened. This enhances the collision frequency between the fluid molecules and the walls, and consequently increases the friction at the interface. This explains why the flux in this regime decreases with increasing temperature. In regime II, the surface

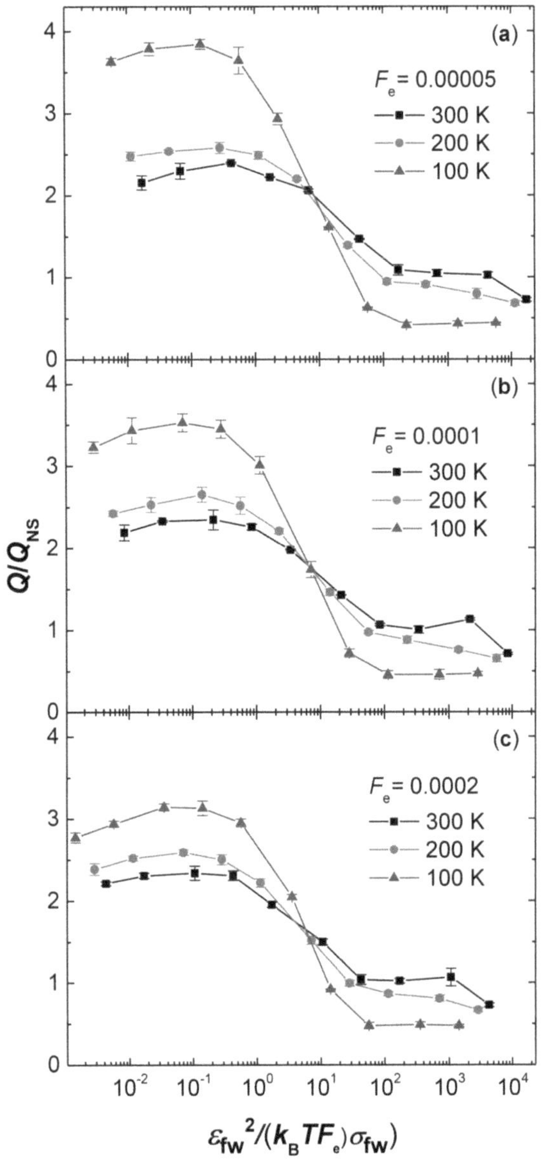

Figure 6.39 Flux ratio, Q/Q_{NS}, for liquid Ar flows as a function of $\varepsilon_{fw}^2/(k_B T \cdot F_e \cdot \sigma_{fw})$. The channel height is 3.7 nm. The pressure drops over the channel corresponding to the external forces F_e are (a) 33 MPa, (b) 66 MPa, and (c) 132 MPa. (Reprinted with permission from Liu and Li, 2009, Copyright (2009) by the American Physical Society.)

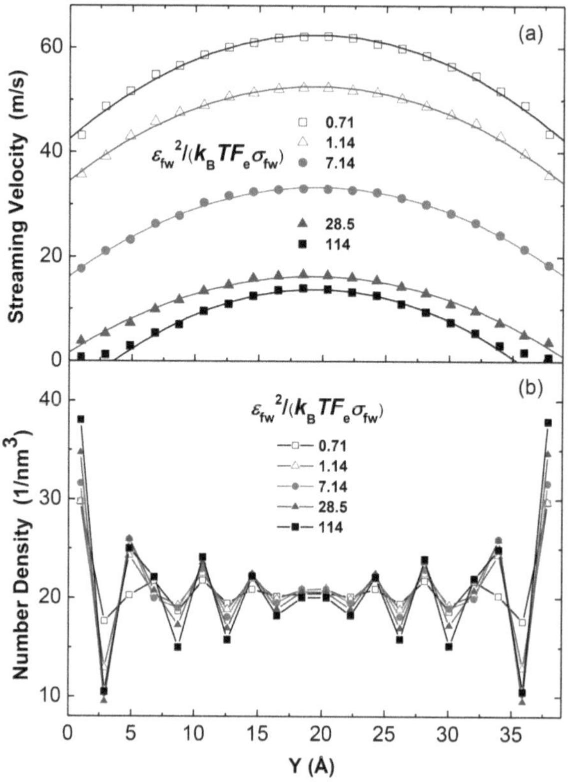

Figure 6.40 Velocity (a) and density (b) distributions for the case of $T = 100$ K and $F_e = 1 \times 10^{-4}$ in Figure 6.39. Solid lines in (a) are parabolic fits to the data. (Reprinted with permission from Liu and Li, 2009, Copyright (2009) by the American Physical Society.)

effect becomes important and fluid adsorption tends to occur as $\varepsilon_{fw}^2/(k_BT \cdot F_e \cdot \sigma_{fw})$ increases, as indicated by the density change next to the walls in Figure 6.40. For a strong fluid–wall binding energy ε_{fw}, a layer of fluid molecules can be adsorbed by the walls. This greatly reduces the number of moving molecules and leads to a low flow rate/flux, as shown in Figure 6.39. In this regime, however, if the temperature is raised, the kinetic energy of the admolecules is increased, which helps them overcome the attraction of the walls and join the main flow stream. This is why the flux increases with increasing temperature. Therefore, the roles of temperature in these two regimes are different. In regime I, a high temperature increases the flow friction at the interface and reduces the flow rate. In regime II, a high temperature enhances the flow rate through freeing admolecules from the wall attraction and increasing the number of fluid molecules in the flow.

For Ar, the fluid–fluid self-binding energy $\varepsilon_{\mathrm{ff}}/k_{\mathrm{B}}T \sim 1$, which is a typical value for common fluids, such as water. For some light species, however, $\varepsilon_{\mathrm{ff}}$ can be very small and $\varepsilon_{\mathrm{ff}}/k_{\mathrm{B}}T \ll 1$. As an alternative fluid, liquid He is studied. For He, $\varepsilon_{\mathrm{ff}} = 10$ K and $\varepsilon_{\mathrm{ff}}/k_{\mathrm{B}}T \sim 0.1$ (Hippler et al., 1983). Figure 6.41 shows the flux ratio for He as $\varepsilon_{\mathrm{fw}}^2/(k_{\mathrm{B}}T \cdot F_{\mathrm{e}} \cdot \sigma_{\mathrm{fw}})$ is varied. It is found that the flow scenarios for He

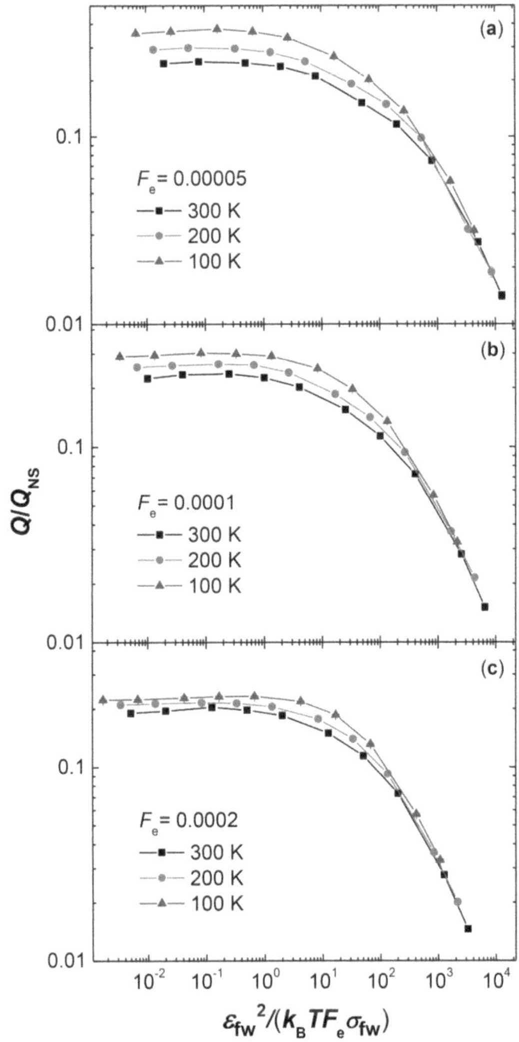

Figure 6.41 Flux ratio, Q/Q_{NS}, for liquid He flows as a function of $\varepsilon_{\mathrm{fw}}^2/(k_{\mathrm{B}}T \cdot F_{\mathrm{e}} \cdot \sigma_{\mathrm{fw}})$. The channel height is 3.7 nm and the pressure drops over the channel corresponding to the external forces F_{e} are (a) 30 MPa, (b) 59 MPa, and (c) 120 MPa. (Reprinted with permission from Liu and Li, 2009, Copyright (2009) by the American Physical Society.)

are different from those of liquid Ar. Instead of having two flow regimes, the He flows fall into three different regimes. At small $\varepsilon_{fw}^2/(k_BT \cdot F_e \cdot \sigma_{fw})$ values, the dependence of the flux on temperature is similar to that for Ar, i.e., the flux decreases with increasing temperature. Furthermore, the flux is largely independent of ε_{fw}, which indicates that in this regime the temperature effect dominates over the surface effect. For large $\varepsilon_{fw}^2/(k_BT \cdot F_e \cdot \sigma_{fw})$, all the flux curves collapse onto a single curve and the flux becomes independent of k_BT for a given value of $\varepsilon_{fw}^2/(k_BT \cdot F_e \cdot \sigma_{fw})$. In this regime, the flux is insensitive to temperature and the wall effect (ε_{fw}) is more important than temperature (k_BT). Between the two regimes is the transition regime, where the flux changes from ε_{fw} independent to k_BT independent.

For the Ar and He flows in Figures 6.39 and 6.41, the only difference is the value of ε_{ff}. A careful comparison between the fluxes for Ar and He flows indicates that the role of the fluid–fluid self-binding energy ε_{ff} lies in the regime of strong surface effect, where $\varepsilon_{fw}^2/(k_BT \cdot F_e \cdot \sigma_{fw})$ is large and fluid adsorption occurs. If ε_{ff} is comparable to the temperature, $\varepsilon_{ff}/k_BT \sim 1$, (the case for Ar), ε_{ff} can enhance the temperature effect, as indicated by the temperature dependent flux in Figure 6.39. If $\varepsilon_{ff}/k_BT \ll 1$, the fluid–fluid interaction is too weak to affect the dynamics of fluid molecules and the temperature plays a minor role when the surface effect is strong. This is why the flux becomes temperature independent at large $\varepsilon_{fw}^2/(k_BT \cdot F_e \cdot \sigma_{fw})$ in Figure 6.41.

In Figures 6.39 and 6.41, the channel height is 3.7 nm. As the channel height is not included in $\varepsilon_{fw}^2/(k_BT \cdot F_e \cdot \sigma_{fw})$, it is necessary to explore the role of channel size in affecting the flow regimes. Figures 6.42 and 6.43 show the flux ratio of liquid Ar and He in 1.8 and 2.8 nm channels. It is clear that the flow regimes observed in Figures 6.39 and 6.41 remain unchanged. However, it is also easy to find that the temperature effect diminishes as the channel height increases. This indicates that the flows may become temperature independent when the channel size is sufficiently large, where the classic theories are valid and the flow regimes in Figures 6.39 and 6.41 may disappear, as discussed in Chapter 2 (see Figure 2.8).

The flow regimes in Figures 6.39 and 6.41 can be qualitatively summarized in the $\varepsilon_{ff}/k_BT - \varepsilon_{fw}^2/(k_BT \cdot F_e \cdot \sigma_{fw})$ coordinates, as shown in Figure 6.44. For a weak fluid–fluid binding energy, i.e., a small ε_{ff}/k_BT (e.g., He), there are three flow regimes for the flux. As $\varepsilon_{fw}^2/(k_BT \cdot F_e \cdot \sigma_{fw})$ increases, the flux experiences a transition from ε_{fw} independent (T dominates) to temperature independent (ε_{fw} dominates). As ε_{ff}/k_BT is raised, for $\varepsilon_{ff}/kT \sim 1$, flows can be divided into weak and strong fluid–wall interaction regimes. In the weak interaction regime (small $\varepsilon_{fw}^2/(k_BT \cdot F_e \cdot \sigma_{fw})$), the flux decreases as temperature is raised, while it increases with increasing temperature in the strong interaction regime.

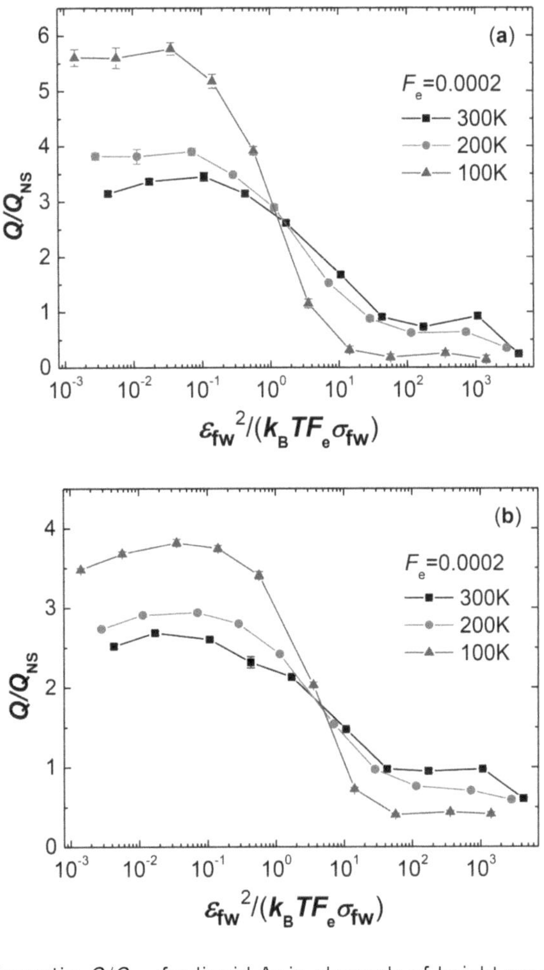

Figure 6.42 Flux ratio, Q/Q_{NS}, for liquid Ar in channels of height equal to (a) 1.8 nm and (b) 2.8 nm. (Reprinted with permission from Liu and Li, 2009, Copyright (2009) by the American Physical Society.)

6.3.3 Indication of flow regimes: fluid pump without moving parts

The flow regimes in Figure 6.39 show distinct temperature dependences for the flow rate in nanochannels with different surface energies. As the flow rate is related to the mobility of fluid molecules, it is indicated that a high temperature can enhance the mobility of fluid molecules in high surface energy channels, while

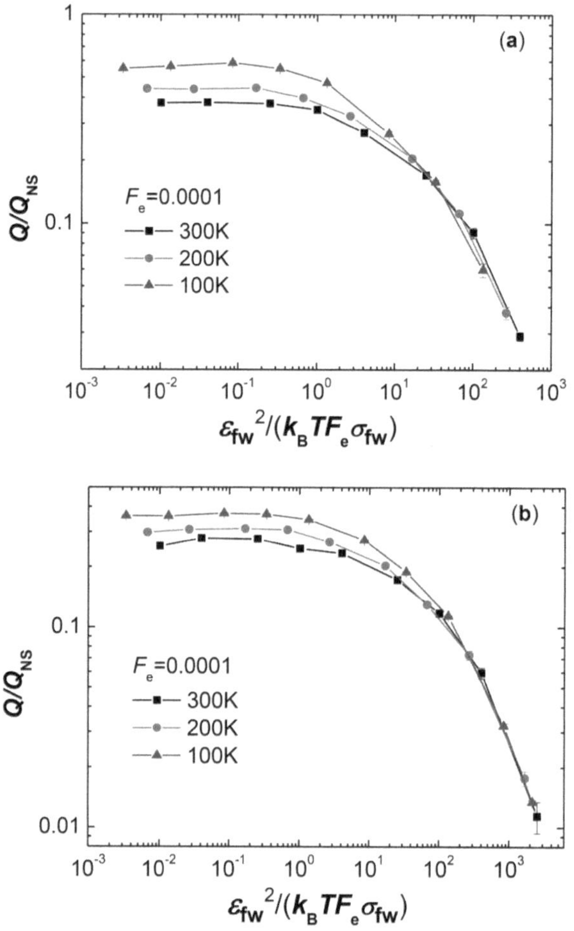

Figure 6.43 Flux ratio, Q/Q_{NS}, for liquid He in channels of height equal to (a) 1.8 nm and (b) 2.8 nm. (Reprinted with permission from Liu and Li, 2009, Copyright (2009) by the American Physical Society.)

the mobility of fluid molecules is reduced when the temperature is raised in low surface energy channels. The indication of these interesting phenomena can be significant. In a nanochannel flow system, if a temperature gradient is applied to the channel wall, the mobility of the fluid molecules will be different at different locations along the channel. The mobility gradient can act as an internal driving force and pump the fluid molecules to move from a low mobility area to a high mobility area. Namely, fluid molecules will move from high to low temperature in channels of low surface energy, while they will move from low to high temperature in high surface energy channels, as shown in the top panel of Figure 6.45. If so,

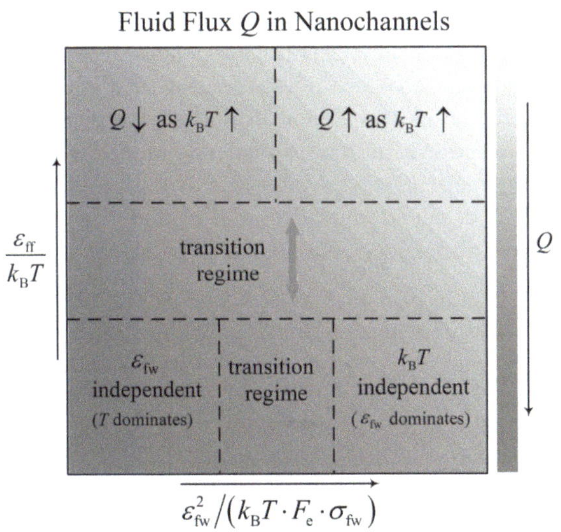

Figure 6.44 Flow regimes in $\varepsilon_{ff}/k_BT - \varepsilon_{fw}^2/(k_BT \cdot F_e \cdot \sigma_{fw})$ coordinates in terms of mass flux. (Reprinted with permission from Liu and Li, 2009, Copyright (2009) by the American Physical Society.)

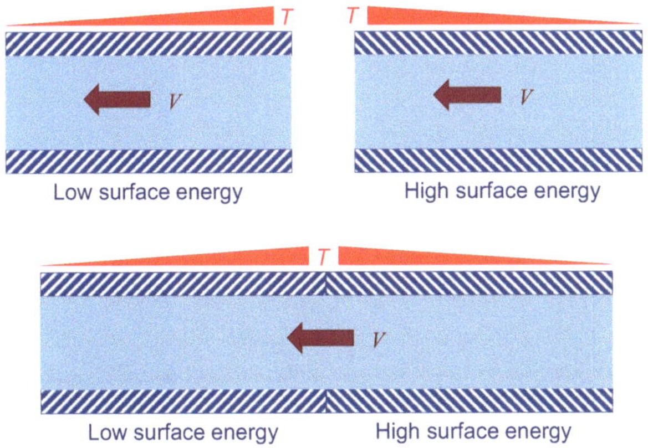

Figure 6.45 Schematic of flows in nanochannels of different surface energies. Arrows indicate the flow direction.

a fluid can be pumped by a symmetric temperature gradient in a nanochannel of heterogeneous surface properties, as illustrated in the bottom panel of Figure 6.45.

If the idea in Figure 6.45 works, the application of such flows driven by symmetric temperature gradients can be tremendous. First of all, the flow system does not involve any mechanical pumps, i.e., it is a moving part-free pumping system.

This is important for microscopic fluidic systems as the fabrication of small-scale pumps is challenging and they usually have reliability problems. Furthermore, the flow system can be used for chip-level cooling. In a microelectronic device, a chip generates heat and its temperature is higher than the ambient temperature. Therefore, the chip and the environment form a symmetric temperature gradient, which can be used to circulate a coolant to dissipate the heat from the chip. Such a cooling strategy does not consume extra energy and is in line with the miniaturization roadmap of the next generation electronic devices.

To test the idea in Figure 6.45, an MD simulation system containing a nanochannel of heterogeneous surface energies is constructed, as shown in Figure 6.46 (Liu and Li, 2010). The channel is formed by two parallel walls. A liquid is filled in between. There are two fluid reservoirs at the ends of the channel. The surface energies of the channel are controlled by the fluid–wall binding energy ε_{fw} such that the left and the right half of the channel are of low and high energies, respectively. The lengths of the system are 16 and 5 nm in the x- and z-directions, and the length of the wall is 8 nm. Again, the wall is modeled by the tight-binding potential and the parameters for silver (Ag) are used. The fluid–fluid and fluid–wall interactions are described by the LJ potential. The values of ε_{ff} and σ_{ff} are set to be 9.8×10^{-3} eV and 3.47 Å and the fluid number density $\rho_0 = 15.38\ 1/\text{nm}^3$. The temperatures at the middle and ends of the walls, T_H and T_L, are maintained at desired values by controlling the kinetic energy of wall atoms lying in a 5-Å-wide slit around the midline and ends of the walls through the Berendsen thermostat.

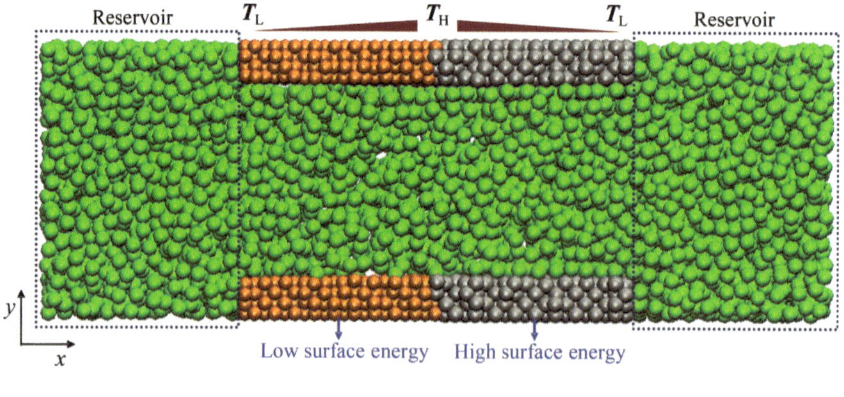

Figure 6.46 A composite nanochannel fluidic system driven by a symmetric temperature gradient. Green spheres are fluid molecules. The left half of the channel wall (orange) is of low surface energy and the right half (gray) is of high surface energy. The temperature in the middle of the channel T_H is higher than that at the ends of the channel T_L. (Reprinted with permission from Liu and Li, 2010, Copyright (2010) by the American Physical Society.)

Periodic boundary conditions are applied in the x- and z-directions. Other simulation details are similar to those discussed in the previous section.

The flow flux in 4-nm channels with different surface energy combinations is studied under various temperature gradients. The low temperatures T_L at the channel ends are maintained at 100 K and the high temperature T_H at the center is varied from 300 K to 750 K. The fluid–wall binding energy for the right half of the channel is fixed at $\varepsilon_{fw(R)} = 500$ K and that for the left half $\varepsilon_{fw(L)}$ is changed from 10 K to 200 K. Figure 6.47 shows the flow flux as a function of $\varepsilon_{fw(L)}$ for different temperature gradients (the reduced flux is the flux scaled by $\sqrt{eV/m}$, where m is the mass of a fluid molecule). It can be seen that fluid flows are generated by symmetric temperature gradients and the flux increases as the surface energy heterogeneity is strengthened. For a given surface energy combination, a large temperature gradient enhances the flux.

To understand the mechanisms of the flows, the density, temperature, and pressure distributions in the channel are obtained, as shown in Figure 6.48. The pressure is calculated using Equation 3.32, which contains a kinetic (ρkT) and a potential $\left(\sum_i \sum_{j>i} r_{ij}F_{ij} \big/ 3V\right)$ components. The contributions of these two components to the pressure change are different in the left and right parts of the channel. It can be seen that the density and pressure distributions are similar. They are generally higher in the right part of the channel than those in the left part. The temperature

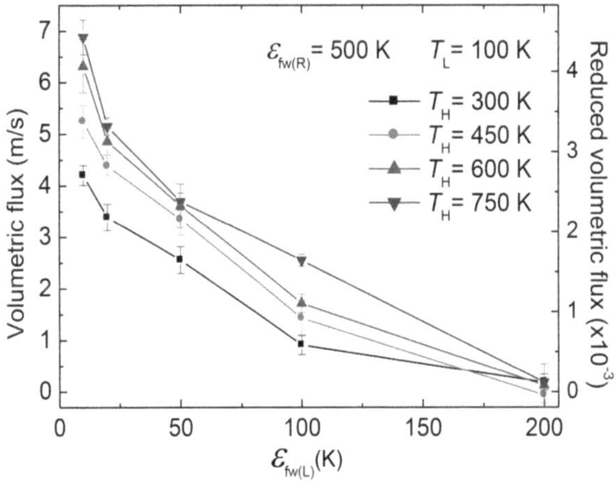

Figure 6.47 Volumetric flux versus $\varepsilon_{fw(L)}$ under different temperature gradients in 4 nm channels. The reduced flux is the flux scaled by $\sqrt{eV/m}$ with m being the mass of a fluid molecule. The fluid–wall binding energy for the right half of the channel $\varepsilon_{fw(R)} = 500$ K and $T_L = 100$ K. (Reprinted with permission from Liu and Li, 2010, Copyright (2010) by the American Physical Society.)

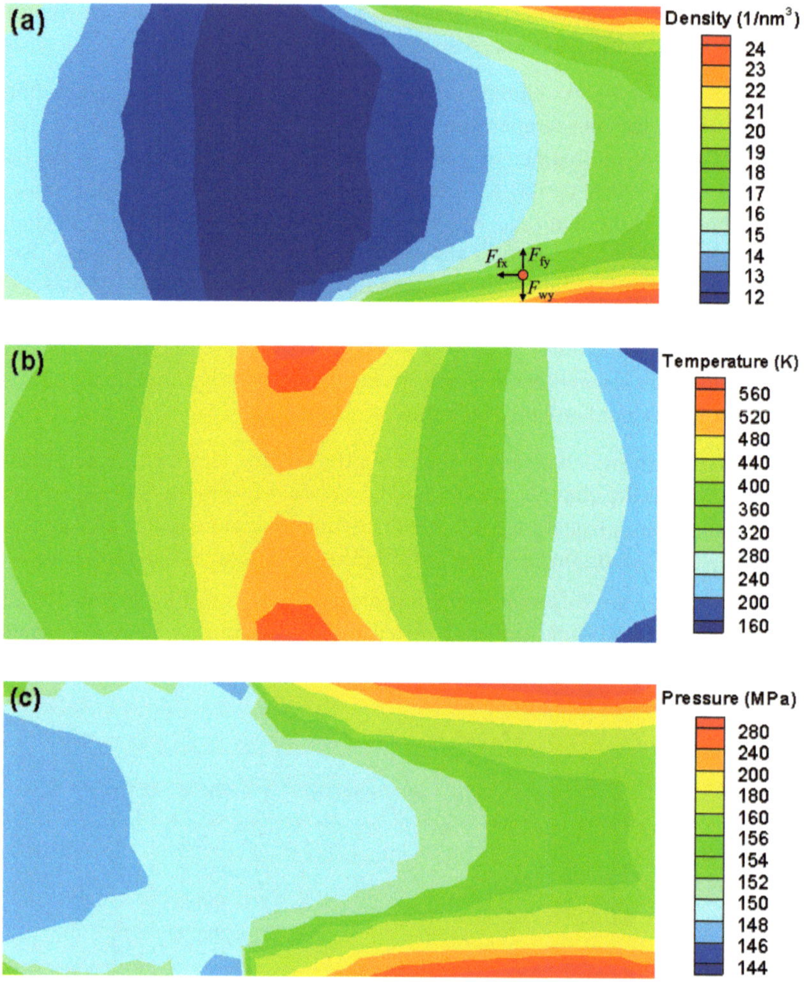

Figure 6.48 Contour plots of (a) density, (b) temperature, and (c) pressure in a 4-nm channel with $T_H = 600$ and $T_L = 100$ K. The fluid–wall binding energy $\varepsilon_{fw(L)} = 10$ and $\varepsilon_{fw(R)} = 500$ K. (Reprinted with permission from Liu and Li, 2010, Copyright (2010) by the American Physical Society.)

distribution is consistent with the temperature gradient applied to the channel walls.

In the left half (low surface energy) of the channel, the binding energy $\varepsilon_{fw(L)}$ is weak and fluid adsorption is not favored. In this part, the interaction between fluid molecules and the channel walls are mainly in the form of elastic collisions. In this case, the role of temperature is to affect the kinetic energy and pressure of the fluid. In the high temperature area (close to the center of the channel), the kinetic energy and the pressure of the fluid are higher than those at the low temperature region, as shown in Figure 6.48c. Thus, the fluid in the low surface energy part moves from the middle (high temperature) to the left end (low temperature) of the channel and the flow can be viewed as a pressure driven flow. It should be noted that the high temperature near the middle of the channel can reduce the fluid density and pressure. However, the density variation is insignificant in this part (Figure 6.48a) and the pressure variation is mainly caused by the kinetic component, as shown in Figure 6.49.

In the right part (high surface energy part), the fluid–wall binding energy $\varepsilon_{fw(R)}$ is strong and fluid adsorption on the channel walls is observed, especially in the low temperature area, as seen in Figure 6.48a. This is because the kinetic energy of the fluid molecules at a low temperature is insufficient to overcome the wall attraction. Therefore, the strong $\varepsilon_{fw(R)}$ and the temperature gradient generate a large density gradient in this part, as depicted in Figure 6.48a. As density

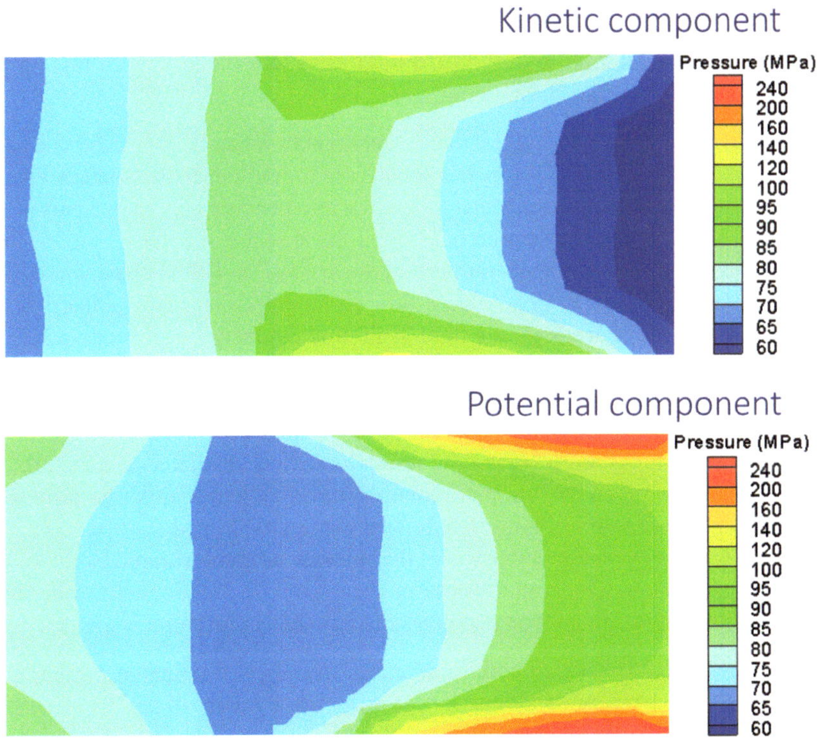

Figure 6.49 Kinetic and potential components of the pressure for the case in Figure 6.48c.

is correlated to pressure, the pressure distribution shows a similar pattern to the density variation. This is why the pressure at low temperature area is higher than that near the center of the channel. Microscopically, a fluid molecule in this part is under asymmetric forces from its neighboring molecules due to the density variation. The total force acting on a fluid molecule by its surrounding fluid molecules can be decomposed into F_{fx} and F_{fy} in the x- and y-directions, as illustrated in Figure 6.48a. As the force from wall atoms is mainly in the y-direction, the component F_{fx} caused by surrounding fluid molecules cannot be balanced, which will cause the fluid molecule to move to the left (Figure 6.48a). This is also consistent with the pressure distribution in this part. It should be noted that in this part the potential component of the pressure due to the large density variation dominates over the kinetic component (Figure 6.49).

The average velocity field for $\varepsilon_{fw(L)} = 10\,K$ and $T_H = 600\,K$ is plotted in Figure 6.50a. It is clear that fluid molecules move from the right to the left. The velocity distributions in the left and right parts of the channel are quite different. In the left part, the velocity in the center area is higher than that around the wall, which is the feature of pressure driven flows. In the right part, however, the highest velocity takes place next to the wall, except that the velocity near the wall at low temperature is almost zero due to fluid adsorption. The flow in this part is somewhat like a surface driven flow.

The results in Figure 6.47 suggest that an apparent flow can be generated in a composite nanochannel by a symmetric temperature gradient if the surface energy heterogeneity is sufficient such that fluid adsorption occurs only in one part of the channel. As discussed in Section 6.3.1, with the absence of external forces, fluid adsorption on a solid surface requires that the $\varepsilon_{fw}/k_B T \gg 1$. Therefore, for a given temperature gradient, to guarantee a flow in a composite nanochannel similar to that shown in Figure 6.46, the fluid–wall binding energy ε_{fw} for one part, say, $\varepsilon_{fw(R)}$, should be sufficiently high (usually $\varepsilon_{fw} > 100\,K$), while ε_{fw} for the other part, $\varepsilon_{fw(L)}$, needs to be small to ensure a high surface energy ratio $\varepsilon_{fw(R)}/\varepsilon_{fw(L)}$. In Figure 6.50a, $\varepsilon_{fw(L)} = 10$ and $\varepsilon_{fw(R)} = 500\,K$ with $\varepsilon_{fw(R)}/\varepsilon_{fw(L)} = 50$, which satisfy the two conditions, and a clear flow rate is observed. If the conditions are not satisfied, flows driven by symmetric temperature gradients cannot be promised. An example is shown in Figure 6.50b, where $\varepsilon_{fw(L)} = 200$ and $\varepsilon_{fw(R)} = 500\,K$. As fluid adsorption occurs in both parts of the channel, as demonstrated by the velocity field at the low temperature areas in Figure 6.50b, the distinct mechanisms required for a flow are not met and consequently the net flow rate is undetectable.

The effects of channel size and fluid density are also investigated. Figure 6.51 plots the reduced flux as a function of $\varepsilon_{fw(L)}$ in channels of different heights (Figure 6.51a and b) and for different fluid densities (Figure 6.51c and d). It is found that the flux increases as the channel size decreases. This is because the surface effects become more important in smaller channels. As the fluid

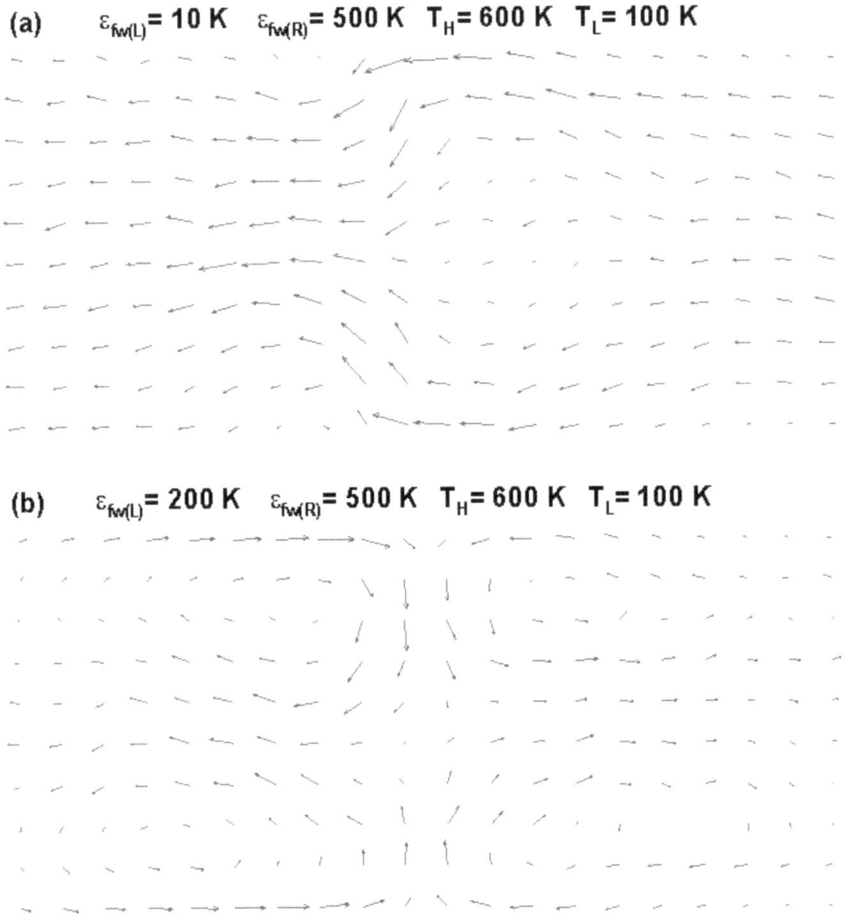

Figure 6.50 Velocity field in 4-nm channels with $T_H = 600$ and $T_L = 100$ K. (a) $\varepsilon_{fw(L)} = 10$ and $\varepsilon_{fw(R)} = 500$ K. (b) $\varepsilon_{fw(L)} = 200$ and $\varepsilon_{fw(R)} = 500$ K. (Reprinted with permission from Liu and Li, 2010, Copyright (2010) by the American Physical Society.)

density ρ is changed, however, the flux seems to assume a maximum value for $\rho/\rho_0 \sim 0.5$, as shown in Figure 6.51c and d, where ρ_0 is the density in Figure 6.47. This might be caused by the changes in the fluid–fluid intermolecular forces because the density determines the mean distance of the fluid molecules. Nevertheless, Figure 6.51 suggests a complex coupling among various parameters, including channel size, temperature gradient, surface energy, and fluid density, and there might be an optimal configuration to maximize the flow rate. This requires further studies.

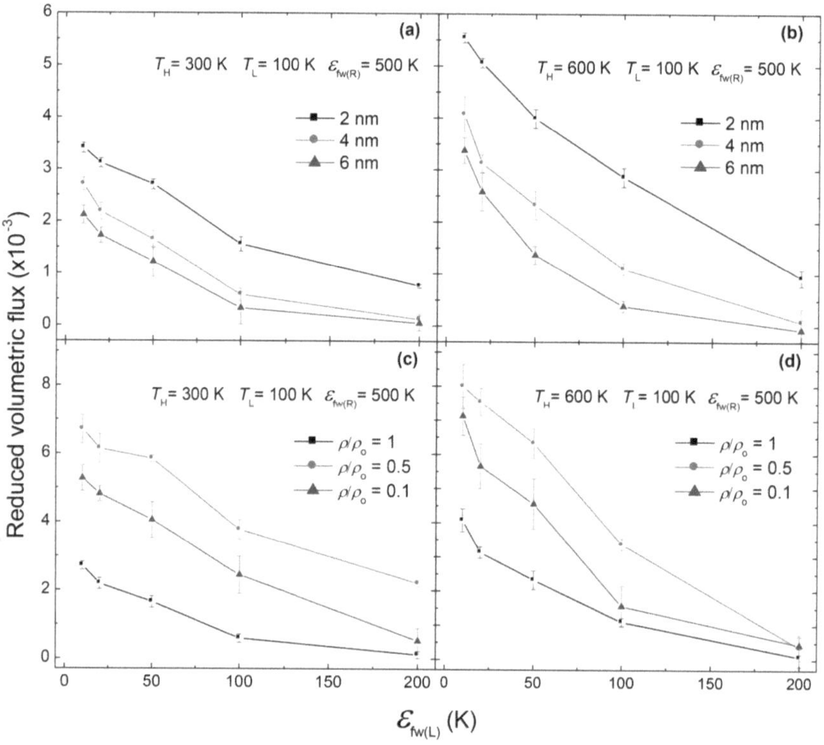

Figure 6.51 Reduced volumetric flux in 2-, 4-, and 6-nm channels [(a) and (b)] and for different fluid densities [(c) and (d)]. In (c) and (d), the channel height is 4 nm. $\rho/\rho_0 = 1$ is the case in Figure 6.47. (Reprinted with permission from Liu and Li, 2010, Copyright (2010) by the American Physical Society.)

Finally, it is worth mentioning that the temperature gradients in the MD simulations are extremely high, $\nabla T \sim 100$ K/nm, which is unpractical. This is because the numerical error in MD simulations is large. If the temperature gradient is too small, the flow flux will be masked by the numerical error and becomes unobservable. The flux under reasonable temperature gradients and the feasibility of the idea for practical applications, such as chip-level cooling, need to be investigated through experiments.

6.4 Fluidic diodes without moving parts

6.4.1 Fluidic rectifiers

A diode is a device or structure that allows flows in a specific, say, the forward, direction but blocks flows in the backward direction. A diode is a special case of a rectifier, which allows flows in both the forward and the backward directions but with different flow rates. The ratio of the flow rates (or pressure drops) in the two directions is referred to as the diodicity, D_i, which characterizes the performance of fluidic rectifiers. Once flows in one of the directions are completely blocked, a rectifier works as a diode. A semiconductor diode permits electron flows, i.e., electric current, in one direction, while blocking electron flows in the opposite direction. In electronics, electronic diodes can be used to control electric currents in desired directions, which are important in the design of various circuits (Sze and Ng, 2007). In thermal systems, if heat flows can be regulated in a unidirectional manner, thermal diodes can be designed (Li et al., 2004).

For fluid flows, unidirectional flows can be easily achieved by fluidic diodes using moving parts, such as check valves. Figure 6.52 shows a simple check valve, where flows are allowed in the forward direction (from left to right) with the flap opened (top panel of Figure 6.52). In the backward direction, however, the flap is closed and flows are prohibited (bottom panel of Figure 6.52). Various fluidic diodes can be designed using similar ideas to control flows in a desired direction through moving parts (Smith and Zappe, 2004).

Moving parts like the flap in Figure 6.52 make fluidic diodes very efficient in regulating fluid flows. They work well in macroscale fluidic systems. In small-scale fluidic systems, however, moving parts can be annoying. In addition to high fabrication cost, they usually have reliability issues and require external actuations to provide driving forces (Nabavi, 2009). Therefore, in micro- or nanofluidic fluidic systems, fluidic diodes without moving parts (fixed structures), or passive fluidic diodes, are preferred for flow regulation.

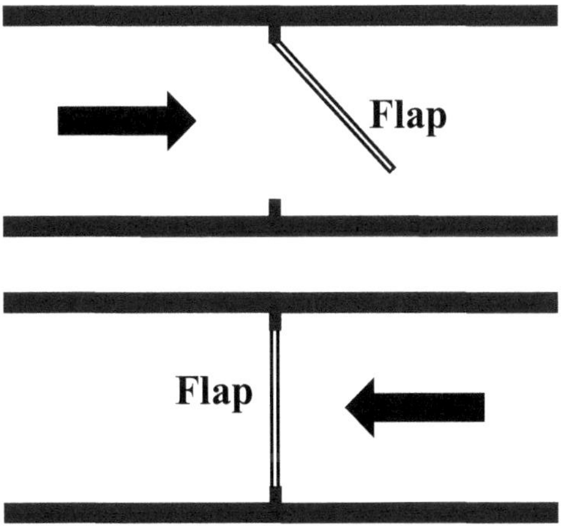

Figure 6.52 Schematic of a simple check valve. Top panel: the flap is open when a fluid flows in the forward direction (from left to right). Bottom panel: the flap is closed and flows are not allowed in the backward direction.

Flow manipulation through nanoscale passive diodes has significant applications in a variety of areas, including medicine, biology, analytical chemistry, and engineering. However, fluid transport is fundamentally different from electron transport and it is very challenging to accomplish unidirectional flows through passive fluidic diodes, especially for simple fluids. In the literature, most efforts actually are focused on the fabrication of fluidic rectifiers (Groisman and Quake, 2004; Sousa et al., 2010), which commonly employ asymmetric structures to generate direction-dependent flow resistance. The most popular structure is the Tesla design (Tesla, 1920), which is shown in Figure 6.53. It can be seen that the flow channels are asymmetric, which generate distinct flow resistances in different directions, as demonstrated by the velocity fields in the forward and backward directions in Figure 6.53.

Similar ideas using asymmetric structures have been employed to develop microscale rectifiers. Figure 6.54 shows a fluidic rectifier using a triangular-shaped microchannel to generate anisotropic flow resistances in different

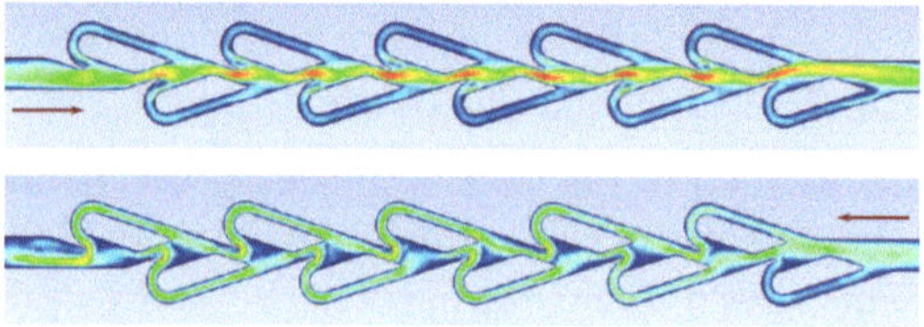

Figure 6.53 A fluidic rectifier: Tesla's design. Arrows show the flow direction. Top panel: velocity contour of a flow in the forward direction. Bottom panel: velocity contour of a flow in the backward direction.

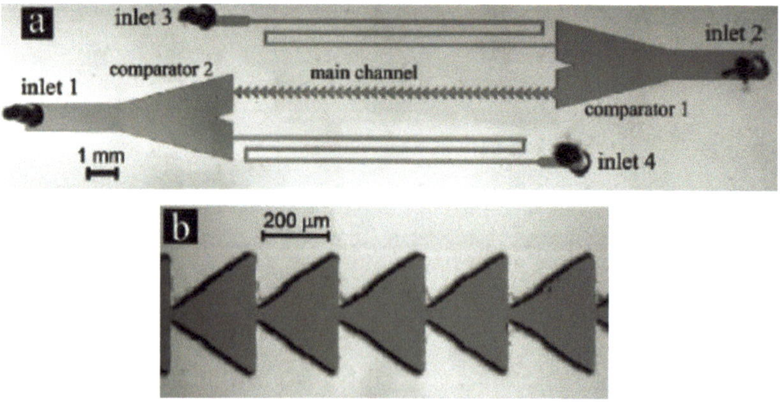

Figure 6.54 A microfluidic rectifier using a triangular-shaped channel. (a) Photograph of the rectifier. (b) A part of the asymmetric channel. (Reprinted with permission from Groisman and Quake, 2004, Copyright (2004) by the American Physical Society.)

directions (Groisman and Quake, 2004). Figure 6.55 depicts the streaklines of a non-Newtonian fluid, polymer solution, in the channel under different pressure drops. The flow patterns in the forward and backward directions indicate that the flow resistance anisotropy depends on the pressure drop or Re. At low-pressure drops, the streaklines are similar in both directions, indicating that the rectification effect is insignificant. At high-pressure drops, however, the flow patterns become quite different in different directions, which give rise to different flow rates in the forward and backward directions, as shown in Figure 6.56.

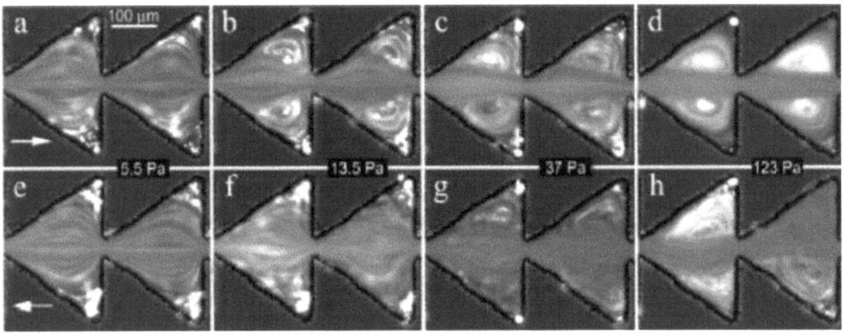

Figure 6.55 Flow patterns (streaklines) of a polymer solution in the forward and backward directions at different pressure drops. (a)–(d) Forward flows. (e)–(h) Backward flows. (Reprinted with permission from Groisman and Quake, 2004, Copyright (2004) by the American Physical Society.)

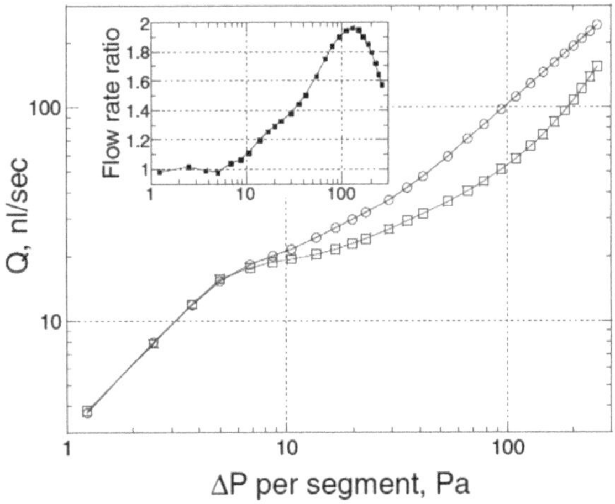

Figure 6.56 Volumetric flow rate as a function of pressure drop. Circles denote the forward flows and squares are for the backward flows. The inset shows the flow rate ratio, i.e., diodicity of the rectifier. (Reprinted with permission from Groisman and Quake, 2004, Copyright (2004) by the American Physical Society.)

To achieve flow rectification through microscale fixed structures similar to those in Figures 6.53 and 6.54, theoretically, the nonlinear inertial term in the equation of motion (Navier–Stokes equation) should be significant. This requires either non-Newtonian fluids (e.g., viscoelastic fluids) or high Re to guarantee flow anisotropy. Non-Newtonian fluids have nonlinear fluid properties (e.g., viscosity), which result in direction-dependent flow resistance in asymmetric structures. For Newtonian fluids, the Re has to be high to ensure turbulent flows for generating anisotropic flow resistances. For Newtonian fluids at low Re, flow rectification cannot be achieved in fixed structures. Nonetheless, even if the above conditions are satisfied and flow rectification is achieved, the diodicity D_i of most microrectifiers is quite small (less than 3) (Sousa et al., 2010) and a real fluidic diode ($D_i \rightarrow \infty$) has not been developed. The poor performance of microscale rectifiers is caused by the fluidity property of fluids, which makes it challenging to generate highly anisotropic flow resistances in fixed structures.

6.4.2 Nanofluidic rectifiers

At the nanoscale, the issues encountered in microfluidic rectifiers can be solved by taking advantage of molecular interactions and surface-dependent capillary behavior, as discussed in Chapter 5. In Section 5.4, it is shown that the Young–Laplace equation falls apart for hydrophobic nanochannels due to the entrance barrier pressure $p_{\Delta E}$ caused by molecular interactions between the fluid and the channel at the entrance. The entrance barrier pressure requires a pressure higher than the prediction of the Young–Laplace equation to drive a fluid into a hydrophobic nanochannel. For hydrophilic nanochannels, however, the capillary filling of a fluid does not require a pressure because it is ensured by the surface tension of the fluid, as illustrated in Figure 5.1a. The different mechanisms for fluid infiltration and capillary filling in hydrophobic and hydrophilic nanochannels suggest that a passive nanofluidic diode can be developed using nanochannels of heterogeneous surface properties.

The idea is illustrated in Figure 6.57, where a heterogeneous nanochannel is used to reach flow rectification. The left half of the channel is hydrophilic and the right

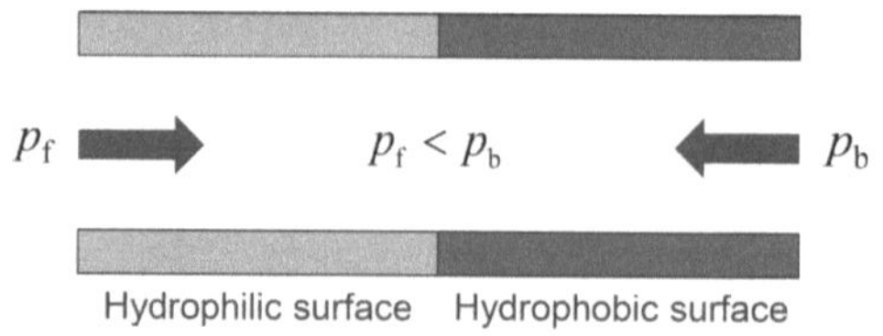

Figure 6.57 Schematic of a heterogeneous nanochannel-based fluidic diode. The pressures needed to drive flows in the forward and backward directions are different due to the difference in surface wettabilities for the left and right parts.

half is hydrophobic. For forward flows (from left to right), the upstream pressure p_f is not high with the help of surface tension during capillary filling. In the backward direction, however, a relatively high upstream pressure p_b is needed to drive a fluid into the hydrophobic part. This indicates that the heterogeneous nanochannel can work as a passive fluidic diode when the upstream pressure p_{up} is between p_b and p_f, i.e., $p_f < p_{up} < p_b$.

To demonstrate the nanofluidic diode, a fluidic system containing heterogeneous nanochannels is fabricated (Li et al., 2015). The schematic of the fluidic chip is depicted in Figure 4.21. It contains an array of 200 parallel nanochannels with heterogeneous surface energies. The left half of the nanochannels is hydrophilic and the right half is hydrophobic. The structure of a single heterogeneous nanochannel is illustrated in Figure 6.58. To form a fluidic system, two microchannels perpendicularly connecting the nanochannels are fabricated with four reservoirs at the ends of the two microchannels (Figure 4.21).

The heterogeneous nanochannels are constructed using the sacrificial machining method. The fabrication is performed on a silicon wafer and the detailed process is introduced in Chapter 4 (see Figure 4.22). The dimensions of the nanochannels are characterized by SEM. The height and width of the nanochannel are 97.8 nm and 10 μm, respectively, as shown in the top panel of Figure 4.24. To circulate a fluid through the micro- and nanochannels, a syringe pump is used to generate flows of desired flow rates. The whole experimental setup of the nanofluidic system is illustrated in Figure 6.59. The pressures of the reservoirs, P_{1-4}, are measured by pressure transducers (Figure 6.59). The pressure difference between the upstream and downstream reservoirs (e.g., $\Delta P = P_1 - P_3$ for the forward flows) is

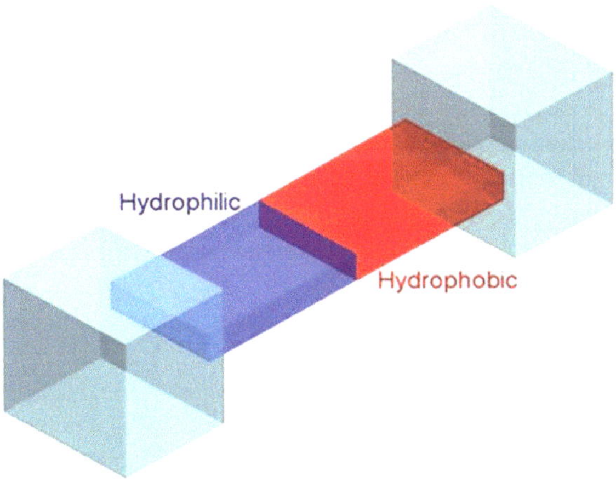

Figure 6.58 Structure of a single heterogeneous nanochannel in Figure 4.21.

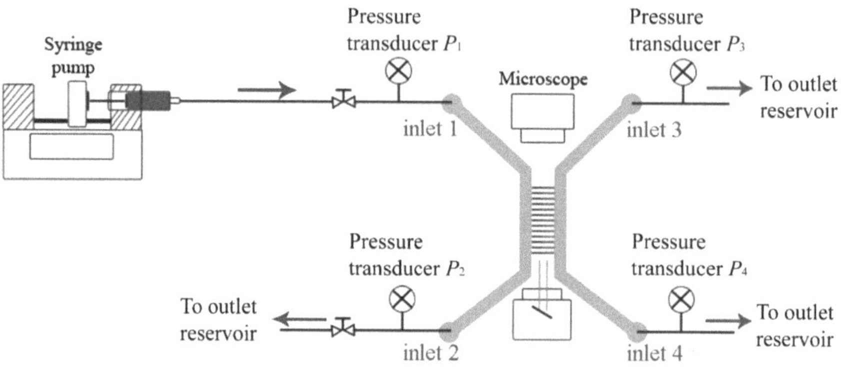

Figure 6.59 Experimental setup of the nanofluidic system. The flow rate of DI water flows is controlled by a syringe pump. Arrows show the directions of forward flows. Pressures of the reservoirs P_{1-4} are measures by pressure transducers. (Reprinted with permission from Li et al., 2015, Copyright (2015) by the American Physical Society.)

considered as the pressure drop over the nanochannels because the flow resistance in microchannels is much smaller than that in nanochannels.

In experiments, the valves connecting inlets 2 and 4 are shut off after the air is removed from the system and the downstream inlet is open to atmosphere. The forward direction is from inlet 1 to inlet 3, as shown in Figure 6.59, and the backward direction is from inlet 3 to inlet 1. DI water is employed as the fluid. The pressure drops for the forward and the backward flows are measured for various flow rates.

Figure 6.60 plots the total volumetric flow rate Q through all the nanochannels as the pressure drop ΔP is varied for both the forward and backward flows. The prediction of the classic Navier–Stokes equation Q_{NS} is also shown (solid line) in Figure 6.60. It can be seen that the flow rates in both directions are lower than Q_{NS}. This is because the flow resistance at the nanoscale is usually larger than that at the micro- and macroscale (Liu and Li, 2011). In the forward direction, water flows can be generated regardless of the pressure drop. In the backward direction, however, flows can only be maintained when the pressure drop is higher than 0.63 MPa, as shown in Figure 6.60. For $\Delta P < 0.63$ MPa, flows are blocked as an infiltration pressure is needed to drive water into the hydrophobic part of the nanochannels, as discussed in Chapter 5. The flow rate difference in the forward and backward directions in Figure 6.60 indicates that the fluidic system acts as a fluidic diode for $\Delta P < 0.63$ MPa. For $0.63 < \Delta P < 3$ MPa, it becomes a fluidic rectifier as water flows in both directions are allowed but the flow rates are different. The pressure drop ratio between the two directions, i.e., the diodicity D_i, is shown in the inset of Figure 6.60 as a function of flow rate Q. The largest diodicity is

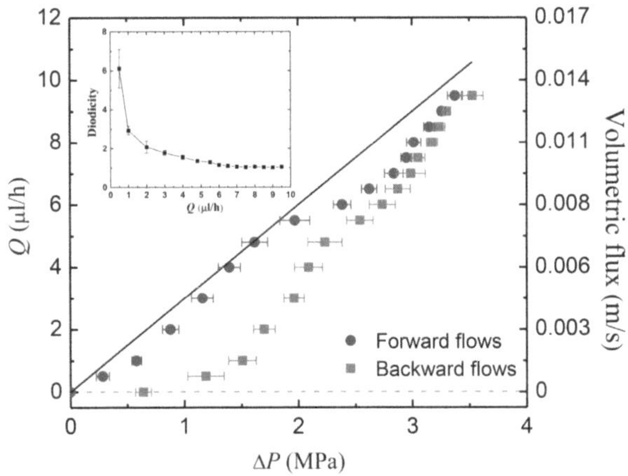

Figure 6.60 Total volumetric flow rate and flux as a function of pressure drop ΔP. The solid line is the prediction of the Navier–Stokes equation. Diodicity in the inset denotes the pressure drop ratio between the forward and backward directions. (Reprinted with permission from Li et al., 2015, Copyright (2015) by the American Physical Society.)

about 6, which is higher than previous fluidic rectifiers (Sousa et al., 2010). As the pressure drop is further increased, for $\Delta P > 3$ MPa, the flow rates in both directions tend to be the same and flow rectification vanishes. This is similar to the breakdown of electronic diodes.

Therefore, the heterogeneous nanochannel fluidic system has three modes: fluidic diode, rectifier, and breakdown, as the pressure drop is varied. The flow rectification effect is caused by the energy barrier at the entrance of the hydrophobic part, which requires a relatively high infiltration pressure for fluids to pass through in the backward direction. If the flow rate is expressed as the ratio of the pressure drop ΔP to the flow resistance R, $Q = \Delta P/R$, the flow rates in the forward and backward directions can be written as follows:

$$Q_F = \Delta P/R_{ch} \quad \text{and} \quad Q_B = \Delta P/(R_{en} + R_{ch}) \tag{6.39}$$

respectively, where R_{ch} and R_{en} are the resistances inside and at the entrance of the channels. As R_{en} is caused by the entrance energy barrier, which is only important for hydrophobic nanochannels, the upstream pressure P_{up} for backward flows has to reach a critical value P^* to drive water through the channels. If $P_{up} < P^*$, backward flows are blocked and the fluidic system works as a diode. If P_{up} is higher than P^*, flows in both directions are accepted and the anisotropic flow resistances make the fluidic system a rectifier. When P_{up} is sufficiently high such that $P_{up} \gg P^*$, the entrance energy barrier plays a minor role and the flow rectification disappears, as shown in Figure 6.60. Since the entrance barrier pressure depends on the channel size and surface hydrophobicity, the pressure drop range for the heterogeneous nanochannel system to be a fluidic diode can be adjusted by changing the channel height and surface properties.

The three flow modes are also confirmed by MD simulations. The MD system is composed of a 3.3-nm-high composite nanochannel and two reservoirs, as shown in Figure 6.61. Numerical details of the simulation, such as potentials and time step, are similar to those described in Section 6.3.3. Water molecules in the reservoirs are confined by two rigid walls, which also serve as pistons for generating flows by applying constant forces on them, as shown in Figure 6.61. Therefore, a PBC is not employed in the horizontal direction. The Berendsen thermostat is used to maintain the temperature of the system at 300 K. The water–surface binding energy ε_{ws} is set at 300 K and 50 K for the hydrophilic (left) and hydrophobic (right) parts, respectively, for which the water contact angles are 27° and 131° (Mo et al., 2015).

Figure 6.62 depicts the potential energy distribution in the channel in an x–y plane next to the bottom surface. It can be seen that the potential energy at the entrances is greatly affected by the surface wettability. The average potential energies at the hydrophilic and hydrophobic entrances in Figure 6.62 are −0.023 and −0.003 eV, respectively. As the mean kinetic and potential energies of water molecules at 300 K are around 0.1 and −0.4 eV (Nguyen et al., 2008), water molecules need assistance to transport to locations of potential energy higher than −0.3 eV. Therefore, the potential difference at the entrances requires a relatively higher pressure in the backward direction (from right to left).

The flow flux for the MD system is plotted in Figure 6.63 as a function of pressure difference ΔP between the reservoirs. It is clear that the dependence of flow flux on ΔP is similar to that in Figure 6.60. As ΔP is increased, the heterogeneous nanochannel shows three different flow modes: fluidic diode ($\Delta P < 3$ MPa),

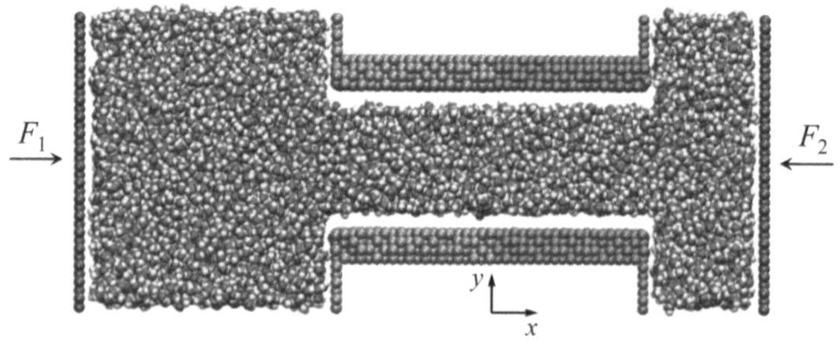

Figure 6.61 Molecular dynamics simulation system. The left (bright gray) and right (dark gray) parts of the channel are hydrophilic and hydrophobic, respectively. (Reprinted with permission from Li et al., 2015, Copyright (2015) by the American Physical Society.)

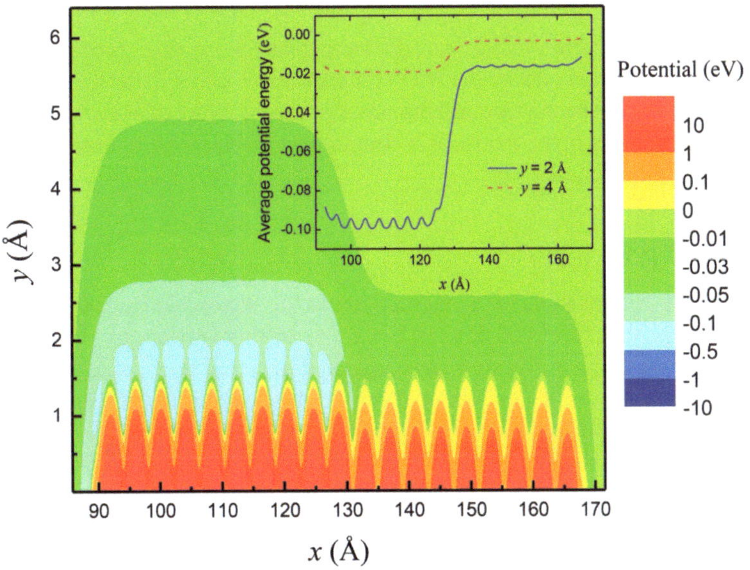

Figure 6.62 Distribution of potential energy in an x–y plane near the bottom surface ($y = 0$) of the nanochannel in Figure 6.61. The hydrophilic and hydrophobic entrances are at $x = 90$ and 168 Å, respectively. The inset shows the potential energies along $y = 2$ and 4 Å. (Reprinted with permission from Li et al., 2015, Copyright (2015) by the American Physical Society.)

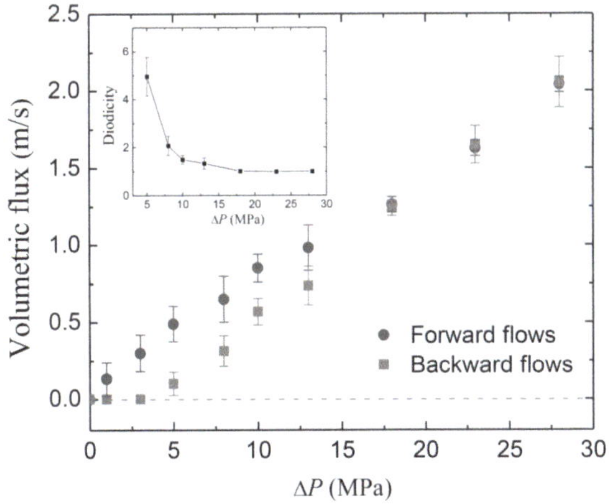

Figure 6.63 Volumetric flux as a function of pressure drop in MD simulations. Diodicity in the inset represents the flux ratio between the forward and backward directions. (Reprinted with permission from Li et al., 2015, Copyright (2015) by the American Physical Society.)

rectifier ($3 < \Delta P < 18$ MPa), and breakdown ($\Delta P > 18$ MPa). The pressure drop windows for the three modes are different from those in Figure 6.60. This is because the channel size in the MD simulation is much smaller than that in the experiments and the surface effects are significant. Nevertheless, the three different flow modes observed in MD simulations are in good agreement with the experiments in Figure 6.60.

6.4.3 Other designs

The unique role of molecular interactions at the nanoscale can be employed to design other structures for flow rectification. In this section, two novel designs by using anisotropic infiltration pressures and different flow processes are introduced. The ideas are demonstrated through MD simulations. Experimental conformation is unavailable and open for future work.

6.4.3.1 Fluidic diodes using nested nanochannels One design uses a nested nanochannel structure, which contains two slit-like nanochannels of different dimensions. Figure 6.64 illustrates such a design (Mo et al., 2016b). The outer nanochannel is longer than the inner one. The lengths of them in the x-direction are 2 and 6 nm. The heights of the two channels are 2 and 5 nm. There are two reservoirs at the ends of the outer channel. The sizes of the simulation system in the y- and z-directions are 11.5 and 4.1 nm, respectively. Similar to the system in Figure 6.61, the flows are controlled by constant forces acting on two pistons confining the reservoirs, as shown in Figure 6.64. For the channel walls, there are four layers of atoms and the outermost layers are fixed to make the system stationary. The thermal motions of the atoms in the other layers (orange spheres

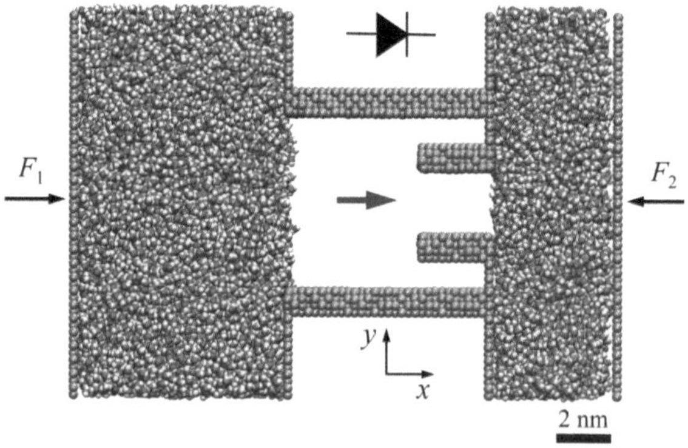

Figure 6.64 A nested nanochannel system working as a passive fluidic diode. In certain pressure drop range, flows are only accepted in the forward direction (left to right). (Reprinted with permission from Mo et al., 2016b, Copyright (2016) by the American Physical Society.)

in Figure 6.64) are considered and used to regulate the temperature of the system. Totally, there are 11,500 water molecules in the system.

Interactions among water molecules are described by the SPC/E model (see Section 3.2.2). The channel walls are modeled by the tight-binding potential and the parameters for Ag are used (see Section 3.2.4). The LJ potential is employed for water–channel interaction with $\sigma_{wc} = 2.85$ Å. The binding energy ε_{wc} is calculated based on the Lorentz–Berthelot mixing rule using the self-interaction LJ parameters for oxygen and Ag given in the literature (Chowdhuri and Chandra, 2001; Li, 2009). The surface properties of the two channels can be different and their wettabilities are tuned by changing the water–channel binding energies ε_{wc}^{in} and ε_{wc}^{out}. Periodic boundary conditions are applied in the *y*- and *z*-directions for water molecules in the reservoirs and in the *z*-direction for the channels.

First, the forward (from left to right) and backward (from right to left) flows are studied in the nested channels of same surface energy, $\varepsilon_{wc}^{in} = \varepsilon_{wc}^{out} = 130$ K, for which the water contact angle is 92° (Mo et al., 2015). The pressure of the upstream reservoir P_1 in both directions is varied from 5 to 65 MPa by changing the force acting on the corresponding reservoir wall (piston), while the pressure of the downstream reservoir P_2 is kept constant at 2 MPa. The volumetric fluxes in both directions as a function of P_1 are shown in Figure 6.65. It can be seen that water flows are allowed only when P_1 is higher than a critical value P^* in both directions. The

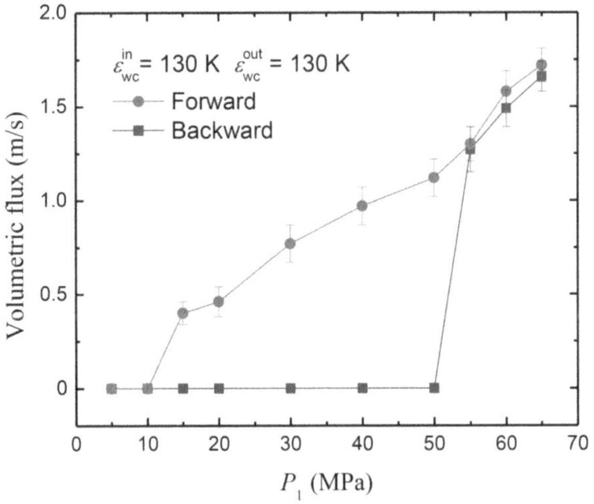

Figure 6.65 Volumetric flux as a function of the driving pressure P_1 in both directions. The heights of the inner and outer channels are 2 and 5 nm, respectively. The surface energy of both channels are the same, $\varepsilon_{wc}^{in} = \varepsilon_{wc}^{out} = 130$ K. The pressure of the downstream reservoir P_2 is controlled at 2 MPa. (Reprinted with permission from Mo et al., 2016b, Copyright (2016) by the American Physical Society.)

critical pressures are $P_F^* \sim 10$ MPa and $P_B^* \sim 50$ MPa in the forward and backward directions, respectively. If $P_1 < P^*$, water flows are blocked. Therefore, P_F^* and P_B^* can be considered as the activation pressures required for turning on water flows. When the driving pressure P_1 is between the two activation pressures, $P_F^* < P_1 < P_B^*$, flows are allowed only in the forward direction. Hence, in this pressure range, 10 MPa $< P_1 <$ 50 MPa, the nested nanochannel structure acts as a fluidic diode. If $P_1 > 55$ MPa, flows in both directions are accepted and the fluxes in the two directions tend to be the same, which corresponds to the breakdown of the fluidic diode. It should be noted that the nanostructure appears to be a flow rectifier in a very small pressure window, i.e., $50 < P_1 < 55$ MPa, which seems to be an unimportant mode compared with the function of fluidic diode.

The mechanisms for the nest nanochannels to work as a fluidic diode lie in the distinct transport processes in different directions. In the forward direction, to go through the nanostructure, water molecules should infiltrate into the outer channel first, and then the inner channel, as shown in Figure 6.66a1, a2. As the

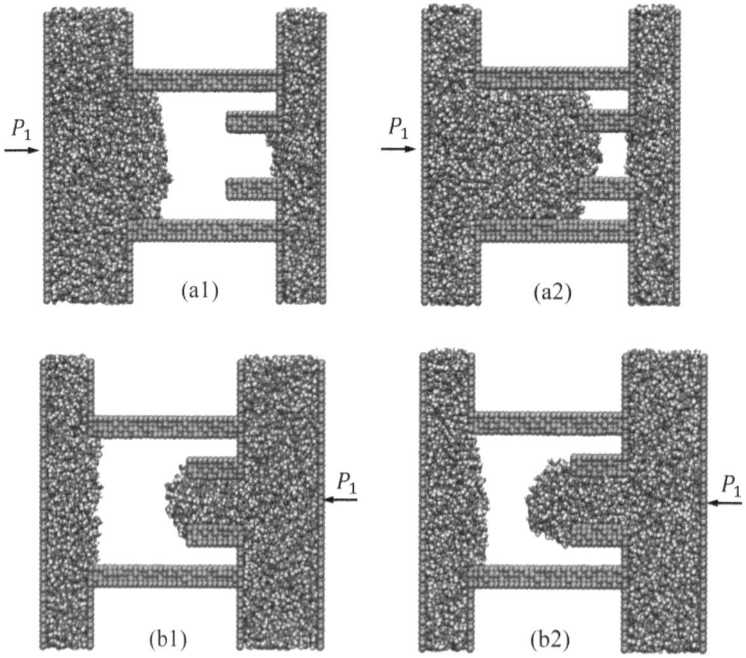

Figure 6.66 Snapshots of MD simulations demonstrating different transport processes in the forward (a1 and a2) and backward (b1 and b2) directions in the nested nanochannel structure. Both channels are slightly hydrophobic with $\varepsilon_{wc}^{in} = \varepsilon_{wc}^{out} = 130$ K. (Reprinted with permission from Mo et al., 2016b, Copyright (2016) by the American Physical Society.)

surface energies of the two channels are the same, i.e., slightly hydrophobic, an infiltration pressure is required to drive water molecules into the channels. Furthermore, the infiltration pressure for the inner channel $P_{\mathrm{inf}}^{\mathrm{i}}$ is higher than that for the outer channel $P_{\mathrm{inf}}^{\mathrm{o}}$. Using the method described by Mo et al. (2015), $P_{\mathrm{inf}}^{\mathrm{i}}$ is obtained as 10 MPa, which is the activation pressure in the forward direction, i.e., $P_{\mathrm{F}}^{*} = P_{\mathrm{inf}}^{\mathrm{i}} = 10$ MPa. In the backward direction, the flow needs to infiltrate into the inner channel first. This requires the pressure of the right reservoir P_{l} to be at least 10 MPa. However, this pressure cannot drive water to go through the whole structure. When water molecules move to the left end of the inner channel, the surface tension of the convex water surface generates a force in the positive x-direction, which will stop the water flow if P_{l} is not further increased, as depicted in Figure 6.66b1.

To overcome the resistance due to the surface tension, P_{l} needs to be raised to be higher than the highest pressure predicted by the Young–Laplace equation, $p = -2\gamma \cos\theta_{\mathrm{max}}/H$, as water goes from the inner to the outer channel (herein, θ_{max} is the largest contact angle for the water meniscus as water moves). For the case in Figure 6.65, by fitting the density distribution of water meniscus using the isopycnic line of density equal to 300 kg/m^3, as shown in Figure 6.67a (Nijmeijer et al., 1989; Mo et al., 2015), the θ_{max} is obtained as 138.3° ± 2.5°, for which the capillary pressure is 53.5 MPa if water surface tension $\gamma = 71.6$ mN/m is used. If the upstream pressure is higher than 53.5 MPa, water flows can be achieved. This highest capillary pressure is the activation pressure P_{B}^{*} in the backward direction.

Therefore, the transport processes in the forward and backward directions are different. The difference between the infiltration pressure for the inner hydrophobic channel (P_{F}^{*}) in the forward direction and the capillary pressure for the inner channel in the backward direction (P_{B}^{*}) makes the nested nanochannel structure a good fluidic diode. When the upstream pressure is much higher than P_{B}^{*}, $P_{\mathrm{l}} \gg P_{\mathrm{B}}^{*}$, the fluxes in both directions are mainly governed by P_{l} and tend to be the same. This corresponds to the breakdown of the fluidic diode, as shown in Figure 6.65.

It has to be noted that the activation pressures, P_{F}^{*} and P_{B}^{*}, are determined by the size and surface energy of the inner channel. This does not mean that the outer channel is unimportant. Actually, the role of the outer channel is also critical. The height of the outer channel has to be small and the surface has to be sufficiently hydrophobic to ensure that water in the left reservoir cannot infiltrate into the outer channel for backward flows. Otherwise, the capillary resistance due to water surface tension when water moves to the left end of the inner channel in the backward direction, as shown in Figure 6.66b2, will not play a role and the activation pressures in both directions will be the same, i.e., the infiltration pressure for the inner channel. In this case, flow rectification cannot be achieved.

By changing the surface energy and height of the inner channel, the activation pressures, P_{F}^{*} and P_{B}^{*}, which define the pressure window for the nested nanochannel

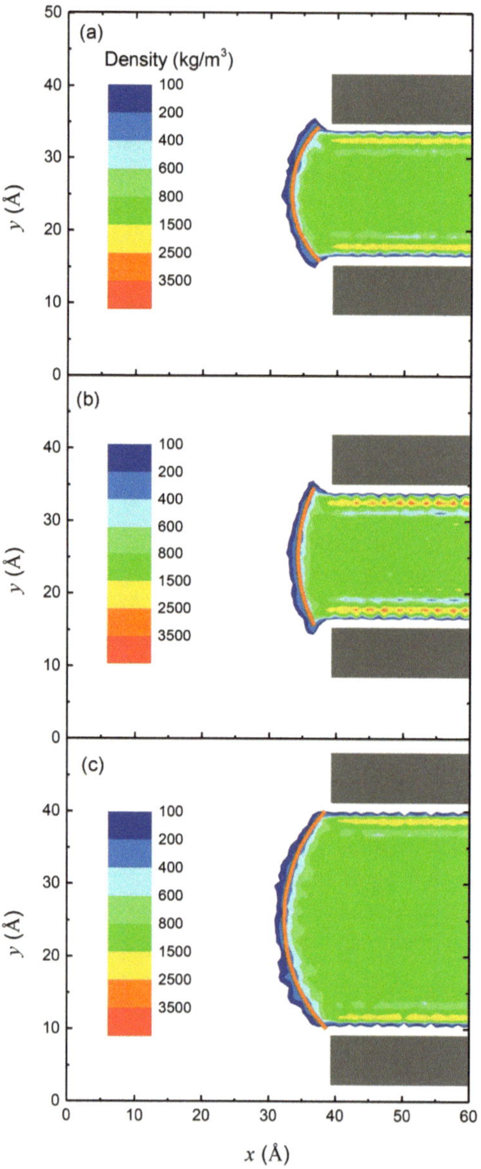

Figure 6.67 Contour of water density in the inner channel. (a) $\varepsilon_{wc}^{in} = 130$ K and $H = 2$ nm, (b) $\varepsilon_{wc}^{in} = 200$ K and $H = 2$ nm, and (c) $\varepsilon_{wc}^{in} = 130$ K, and $H = 3$ nm. The contact angle of water is determined by fitting the density contour using the isopycnic line of 300 kg/m^3, as shown by the solid lines. (Reprinted with permission from Mo et al., 2016b, Copyright (2016) by the American Physical Society.)

to work as a fluidic diode, can be adjusted. Figure 6.68 shows the flux as a function of the upstream pressure P_1 when the surface energy of the inner channel is increased with $\varepsilon_{wc}^{in} = 200$ K, for which the water contact angle is 60°. As the inner surface becomes hydrophilic, water infiltration into the inner channel turns out to be automatic and does not require extra pressure. Hence, in the forward direction, water flows are allowed when P_1 is higher than 2 MPa, which is the infiltration pressure for the outer channel (Mo et al., 2015). In the backward direction, the activation pressure P_B^* is also reduced to 35 MPa. This activation pressure is solely due to the surface tension when water flows to the left end of the inner channel, where the largest contact angle is 121.0° ± 3.2° (Figure 6.67b). The capillary pressure predicted by the Young–Laplace equation at this contact angle is 36.8 MPa, which is close to the MD result (35 MPa).

If the inner channel height is varied, the corresponding infiltration pressure is changed and the window of P_1 for the diode can be shifted. Figure 6.69 plots the flux when the inner channel height is increased to 3 nm for $\varepsilon_{wc}^{in} = \varepsilon_{wc}^{out} = 130$ K. It can be seen that P_F^* is reduced from 10 to 5 MPa due to the decrease of the infiltration pressure for the inner channel. In addition, P_B^* is lowered to 30 MPa because the increase of the channel height, according to the Young–Laplace equation, decreases the capillary pressure in the backward direction. The largest contact angle during the release of water from the inner to the outer channel is

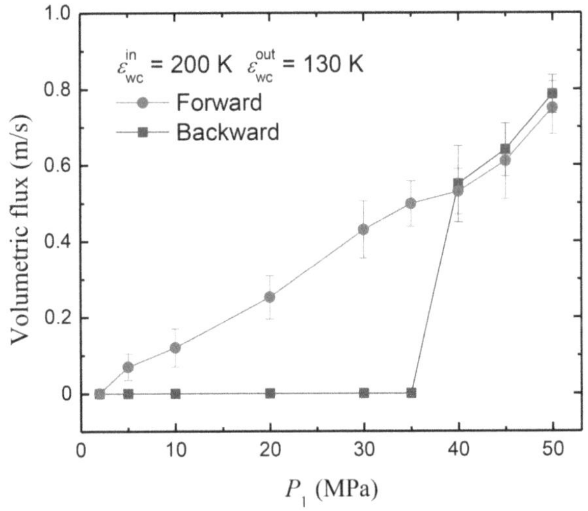

Figure 6.68 Volumetric flux as a function of the upstream pressure P_1 when the inner channel becomes hydrophilic ($\varepsilon_{wc}^{in} = 200$ K, $\varepsilon_{wc}^{out} = 130$ K). P_2 is maintained at 2 MPa. The heights of the inner and outer channels are 2 and 5 nm, respectively. (Reprinted with permission from Mo et al., 2016b, Copyright (2016) by the American Physical Society.)

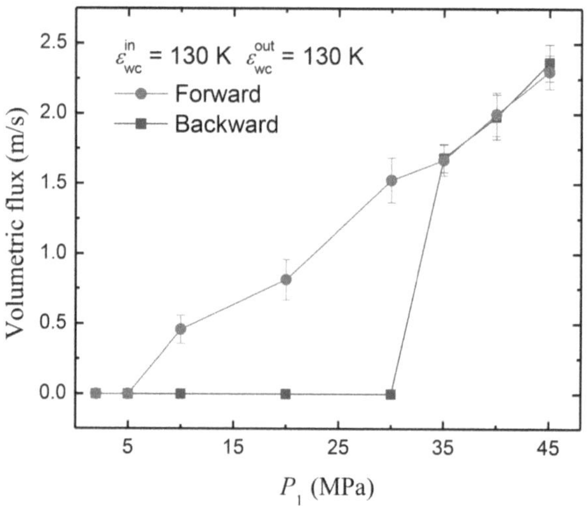

Figure 6.69 Volumetric flux as a function of the upstream pressure P_1 when the height of the inner channel is increased to 3 nm ($\varepsilon_{wc}^{in} = \varepsilon_{wc}^{out} = 130$ K). P_2 is kept at 2 MPa. (Reprinted with permission from Mo et al., 2016b, Copyright (2016) by the American Physical Society.)

136.8° ± 2.8°, as demonstrated in Figure 6.67c, for which the capillary pressure is 34.8 MPa, in good agreement with the MD result, 30 MPa, in Figure 6.69.

Compared with the heterogeneous nanochannel introduced in the previous section, the nested nanochannel structure offers more flexibility in controlling the pressure range for the system to be a fluidic diode. In terms of the performance of the diode, there are definitely optimal channel sizes and surface wettabilities, which currently are unknown and need to be further investigated. In addition to the channel height and surface property, the geometry of the nanostructure can also improve the performance of the diode. This will be discussed in the next section.

6.4.3.2 Fluidic diodes using nonuniform nanochannels Another design of passive fluidic diode takes advantage of asymmetric geometry of nonuniform nanochannels (Mo et al., 2016a). Figure 6.70 illustrates the structure, which is a converging nanochannel connecting with two reservoirs. The heights of the channel at the left and right entrances are $H_L = 3.7$ nm and $H_R = 2$ nm in the y-direction and the channel length in the x-direction is 7.1 nm. The lengths of the simulation system in the y- and z-directions are 8.6 and 4.1 nm, respectively. The reservoirs are filled with 9,800 water molecules described by the SPC/E water model. The pressures of the reservoirs are regulated by constant external forces acting on the left and right walls of the reservoirs, as shown in Figure 6.70. The material

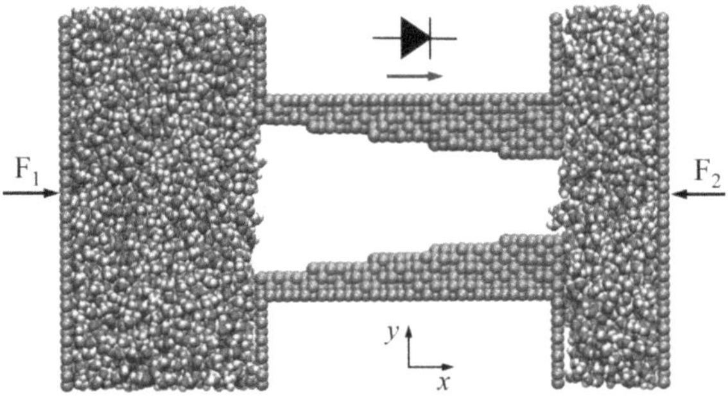

Figure 6.70 Nonuniform nanochannel system as a fluidic diode. The forward direction is from the left to the right. (Reprinted from Mo et al., 2016a, with the permission of AIP Publishing.)

of the wedge-shaped walls of the channel is Ag, which is constructed from an fcc structure with the lattice constant equal to 4.086 Å. The walls contain three to seven layers of Ag atoms and are modeled by the tight-binding potential. The water–wall interaction is calculated using the LJ potential with ε_{ww} and σ_{ww} being the LJ parameters ($\sigma_{ww} = 2.85$ Å). ε_{ww} is changed to tune the surface wettability of the channel. Other simulation details are the same as those for the nested nano-channel structure discussed previously.

First, the channel is set hydrophobic with $\varepsilon_{ww} = 50$ K, for which the water contact angle is ~130° (Mo et al., 2015). Again, the pressures of the upstream and down-stream reservoirs are denoted as P_1 and P_2 regardless of the flow direction. P_1 is varied to change the flow rate, while P_2 is kept constant at 2 MPa. As the cross-sectional area of the channel is not a constant, the flow rate, instead of the flow flux, is computed and presented. Figure 6.71 depicts the volumetric flow rates in the forward (left to right) and the backward (right to left) directions as P_1 is varied. The flow rate behavior is similar to the flux in Figure 6.65. There is an activation pressure P^* in each direction. In the forward direction, $P_F^* = 45$ MPa and in the backward direction, $P_B^* = 60$ MPa. If P_1 is lower than the activation pressure in a direction, the channel shuts off and water flows are not allowed in that direction. However, if $P_F^* < P_1 < P_B^*$, flows are only accepted in the forward direction. In this pressure range, the nonuniform nanochannel acts as a passive fluidic diode. For $P_1 > P_B^*$, water flows are allowed in both directions and the flow rates tend to be the same, which is the breakdown of the fluidic diode.

The direction-dependent activation pressures, P_F^* and P_B^*, are caused by the distinct flow transport processes in the forward and backward directions. As the channel surface is hydrophobic, the infiltration pressure required to drive water into the channel includes the entrance barrier pressure $P_{\Delta E}$ and the capillary pressure. The later, for nonuniform channels, is given by

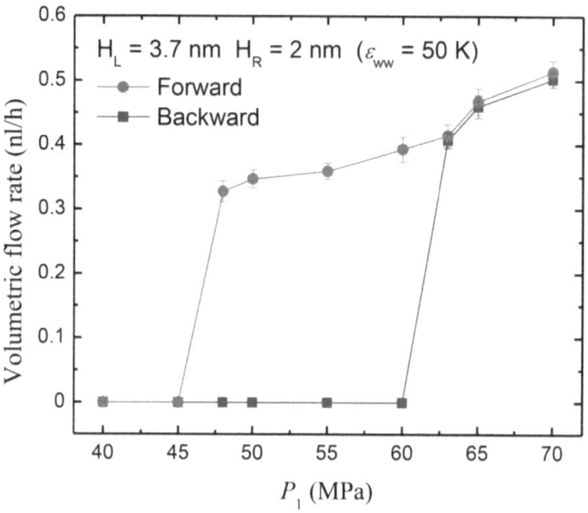

Figure 6.71 Volumetric flow rate as a function of the upstream pressure P_1 in the forward and the backward directions ($H_L = 3.7$ nm, $H_R = 2$ nm, $\varepsilon_{ww} = 50$ K). (Reprinted from Mo et al., 2016a, with the permission of AIP Publishing.)

$$p = -\frac{2\gamma \cos(\theta \pm \alpha)}{H}, \tag{6.40}$$

where θ is the contact angle of the fluid in uniform channels and α is the tilt angle of the channel (in Figure 6.70, $\alpha = 7°$). In Equation 6.40, $\theta \pm \alpha$ can be treated as the effective contact angle of a fluid in the nonuniform channel. However, it must be noted that Equation 6.40 is only valid for small α. If α is large, it may affect the contact angle θ and the effective contact angle may be different from $\theta \pm \alpha$ (Liu et al., 2009).

In the forward direction, the entrance barrier pressure is estimated to be $P_{\Delta E} \approx 10$ MPa (Mo et al., 2015) and the capillary pressure, predicted by Equation 6.40, is about 21 MPa (the entrance channel height is 3.7 nm, $\theta - \alpha = 123°$, and water surface tension $\gamma = 71.6$ mN/m). Hence, the total pressure needed to drive water into the channel from the left reservoir is about 31 MPa. Unfortunately, this pressure is not sufficient to generate a flow through the whole channel because of the converging cross-sectional area in the forward direction, which increases the capillary pressure as water advances in the channel. Therefore, P_1 has to be increased to push water forward. As water moves in the channel and reaches a location sufficiently close to the right reservoir, the molecular interactions (van der Waals and Coulombic forces) between water molecules in the channel and those the right reservoir cause water molecules to form a linking bridge, as shown in Figure 6.72. The linking bridge induces water coalescence and assists water transport in

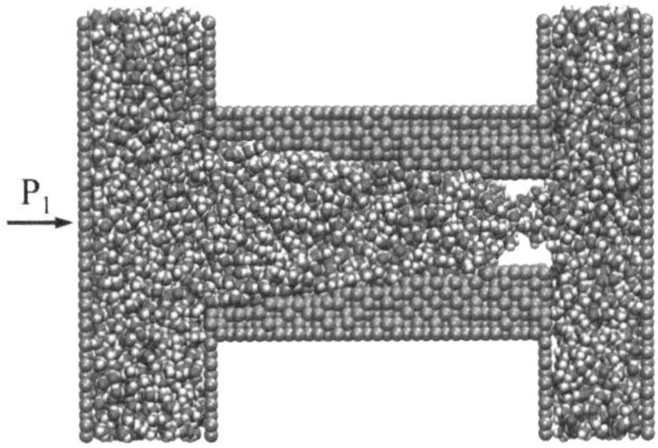

Figure 6.72 Formation of a linking bridge at the right end of the channel for a forward flow. (Reprinted from Mo et al., 2016a, with the permission of AIP Publishing.)

the forward direction (Koplik and Banavar, 1992). The distance for the water molecules in the channel and the right reservoir to attract each other to form the linking bridge is about 1 nm, which is in good agreement with other MD simulations (Zhao and Choi, 2004). Hence, the pressure required to generate a forward flow, i.e., P_F^*, is about 45 MPa, which includes the capillary pressure (~35 MPa) at the location, where the linking bridge forms, and entrance barrier pressure (~10 MPa).

In the backward direction, the entrance barrier pressure is similar to that in the forward direction, $P_{\Delta E} \approx 10$ MPa. The capillary pressure, however, is high, about 52 MPa, as predicted by Equation 6.40 with $\theta + \alpha = 137°$ and $H = 2$ nm. Thus, the total infiltration pressure is ~62 MPa, which agrees well with the numerical result shown in Figure 6.71, where the activation pressure in the backward direction, P_B^*, is 60 MPa.

As the activation pressures depend on the surface property, channel size, and title angle, the pressure window for the structure to work as a diode can by adjusted by changing these parameters. If the wall surface energy is increased, the infiltration pressure will decrease and the flow rate curves in Figure 6.71 will shift to the left. This is confirmed by a case with $\varepsilon_{ww} = 100$ K, for which the contact angle is 110° and the entrance barrier pressure is $P_{\Delta E} \approx 7.5$ MPa. The flow rate for this case is depicted in Figure 6.73, where the activation pressures, P_F^* and P_B^*, are 25 and 42 MPa, respectively, much smaller than the values in Figure 6.71. It is also found that the flow rates in Figure 6.73 are lower than those in Figure 6.71. This is because the surface friction in Figure 6.73 is higher than that in Figure 6.71 when ε_{ww} is raised from 50 K to 100 K.

The same effect can be observed when the channel size is varied. Figure 6.74 plots the flow rate for $\varepsilon_{ww} = 100$ K while the channel heights at the left and right ends are increased to $H_L = 5.7$ and $H_R = 3$ nm, respectively, for which the tilt angle $\alpha = 11°$.

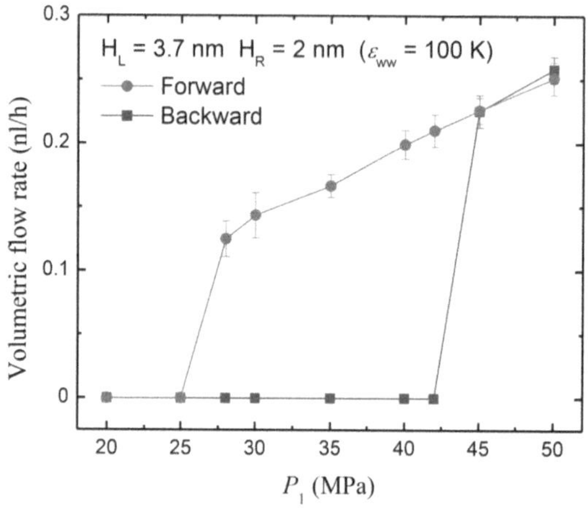

Figure 6.73 Volumetric flow rate as a function of P_l for $\varepsilon_{ww} = 100$ K ($H_L = 3.7$ nm, $H_R = 2$ nm). (Reprinted from Mo et al., 2016a, with the permission of AIP Publishing.)

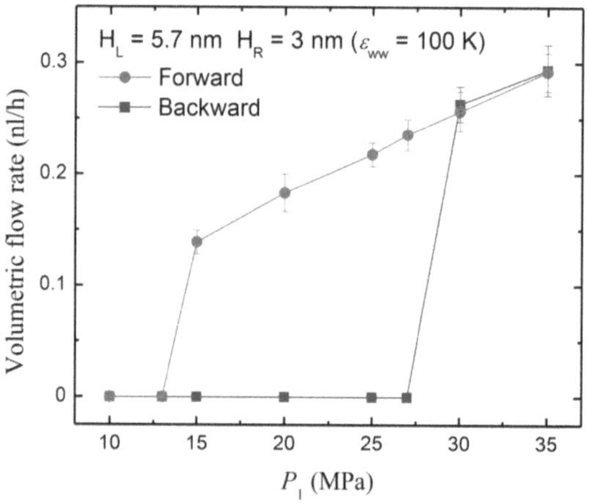

Figure 6.74 Volumetric flow rate as a function of P_l as the channel height is increased ($H_L = 5.7$ nm, $H_R = 3$ nm, $\varepsilon_{ww} = 100$ K). (Reprinted from Mo et al., 2016a, with the permission of AIP Publishing.)

It can be seen that activation pressures, P_F^* and P_B^*, are lowered to 13 and 27 MPa, respectively. This is mainly caused by the reduction in the capillary pressures.

Other than the surface energy and channel size, the tilt angle α also plays an important role. The transport process in the forward direction and Equation 6.40 indicate that a large α can improve the performance of the fluidic diode such that P_F^* can be further reduced and the pressure range, $\left(P_F^*, P_B^*\right)$, for the nonuniform nanochannel to serve as a fluidic diode can be further widened. Nevertheless, α has to be smaller than a value to make sure that the effective contact angle is larger than 90°. Otherwise, the channel will become hydrophilic and water infiltration into the channel from the left reservoir cannot be prevented. This may result in the formation of a linking bridge at the channel entrance in the backward direction and flow rectification cannot be achieved.

To understand the effects of the titled angle, the length of the channel in the x-direction is set as 5 nm and the height of right end of the channel is fixed as $H_R = 2$ nm, while the height of the left entrance is varied to change α. The critical tilt angle α^*, beyond which flow rectification disappears, is obtained under different water–wall binding energies, as shown in Figure 6.75. The values of α^* in Figure 6.75 give the upper limits of the titled angle for the design of nonuniform channel-based fluidic diodes.

For $\alpha < \alpha^*$, the activation pressures,P_F^* and P_B^*, are computed as ε_{ww} and α are varied, as depicted in Figure 6.76. It is clearly that P_F^* and P_B^* can be reduced by increasing the channel surface energy and the pressure range, $\Delta P = P_B^* - P_F^*$, can be expanded by raising the tilt angle. Therefore, the performance of the fluidic diode can be flexibly controlled by altering the tilt angle α and channel surface wettability to meet the requirements of practical applications.

When $\alpha > \alpha^*$, the automatic fluid infiltration into the channel from the left reservoir can cause the fluidic diode to break down. This can be prevented by modifying the structure of the nonuniform channel. Figure 6.77 illustrates a design,

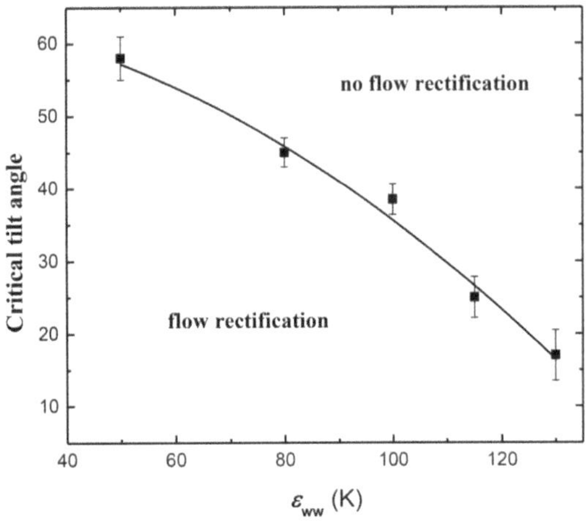

Figure 6.75 Critical tilt angle α^* as a function of water–wall binding energy ε_{ww}. $H_R = 2$ nm and the channel length is 5 nm. (Reprinted from Mo et al., 2016a, with the permission of AIP Publishing.)

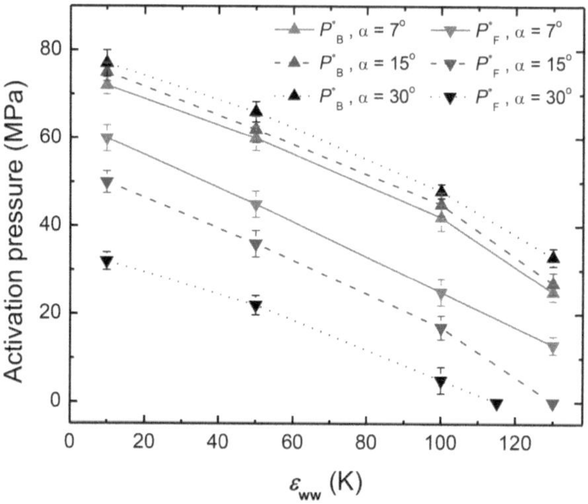

Figure 6.76 Activation pressures, P_F^* and P_B^*, as a function of ε_{ww} at different tilt angles. $H_R = 2$ nm and the channel length is 5 nm. (Reprinted from Mo et al., 2016a, with the permission of AIP Publishing.)

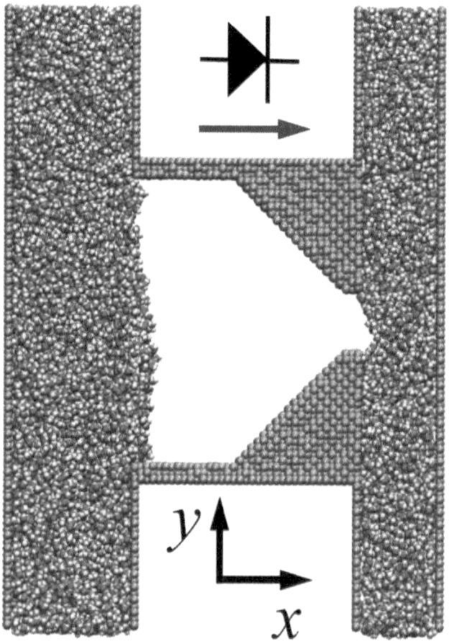

Figure 6.77 Modified nonuniform nanochannel. The left entrance part is uniform. (Adapted from Mo et al., 2016a, with the permission of AIP Publishing.)

where the original converging nanochannel is reconstructed by changing the left entrance part to be uniform. In such a structure, water flows are studied for $\varepsilon_{ww} = 100$ K with a very large title angel, $\alpha = 45°$, which is larger than the critical title angle ($\alpha^* = 40°$). The heights of the left and right entrances are $H_L = 10$ and $H_R = 2$ nm, respectively. The volumetric flow rate as a function of the driving pressure P_1 is shown in Figure 6.78. It can be seen that P_F^* is lowered to 5 MPa and P_B^* is raised to 50 MPa. The pressure range, $\left(P_F^*, P_B^*\right)$, is significantly increased, compared with those in Figures 6.73 and 6.74.

Finally, it is worth mentioning that that the channel surface has to be hydrophobic such that it requires an infiltration pressure to circulate a fluid through the channel. If the channel is hydrophilic, the fluid can flow into the channel easily even at the atmospheric pressure. In this case, the structure cannot work as a fluidic diode, as demonstrated in Figure 6.79. In Figure 6.79, the channel structure is the same as that in Figure 6.74 but $\varepsilon_{ww} = 300$ K, for which the contact angle is 27°. It is found that flow rates are roughly direction independent and flow rectification cannot be reached. This is consistent with previous experiments (Groisman and Quake, 2004).

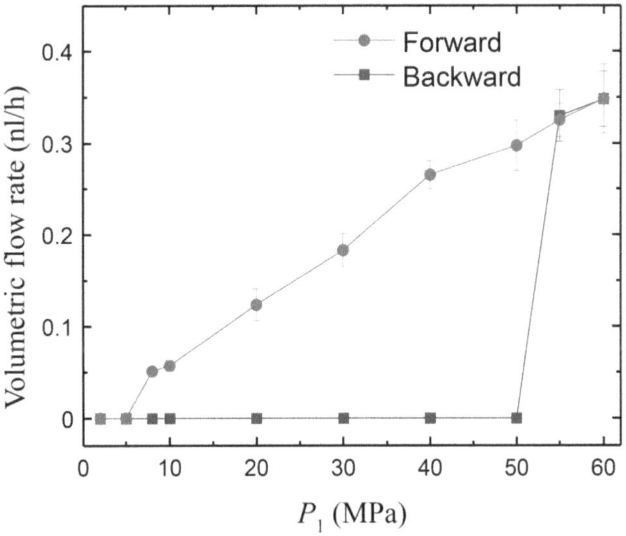

Figure 6.78 Volumetric flow rate as a function of the driving pressure P_1 for the structure in Figure 6.77. Tilt angle $\alpha = 45°$ ($H_L = 10$ nm, $H_R = 2$ nm, $\varepsilon_{ww} = 100$ K). (Adapted from Mo et al., 2016a, with the permission of AIP Publishing.)

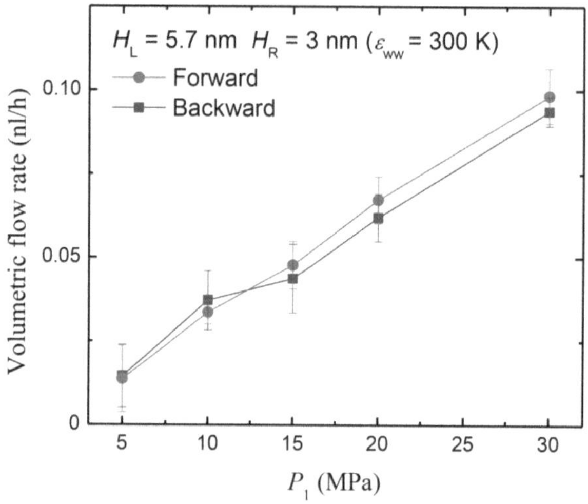

Figure 6.79 Volumetric flow rate when the channel surface is modified to hydrophilic with $\varepsilon_{ww} = 300$ K ($H_L = 5.7$ nm and $H_R = 3$ nm). (Reprinted from Mo et al., 2016a, with the permission of AIP Publishing.)

6.5 Water flows in CNTs

Due to the unique structure and special surface properties, CNTs have attracted great attention in the past decade for fluid transport. The ultra-smooth, hydrophobic surface of CNTs has been shown to promote flow slip and greatly enhance the flow rate. Therefore, as a specific nanoscale flow example, water flows in CNTs deserve a section to itself. In this section, the static structure of water in CNTs is discussed first. This is followed by a brief review of water transport in CNTs. Finally, possible explanation about the discrepancy in water flow rates in the literature is offered.

6.5.1 Water structure in CNTs

Water is a very interesting fluid. It exhibits versatile structures and behavior under different conditions. This is because water molecules are polar (oxygen atoms are more electronegative than hydrogen atoms) and interact through hydrogen bonds, which form a metastable hydrogen bond network (HBN), as illustrated in Figure 6.80. The dynamics of the HBN is quite complex and is sensitive to the variations of external parameters, such as temperature and pressure. When water is confined in a CNT, the molecular interactions between water molecules and carbon atoms also affect the structure of water.

At a given temperature and pressure, the water structure in a CNT is largely determined by the size and chirality of the CNT. Due to the small size of CNTs,

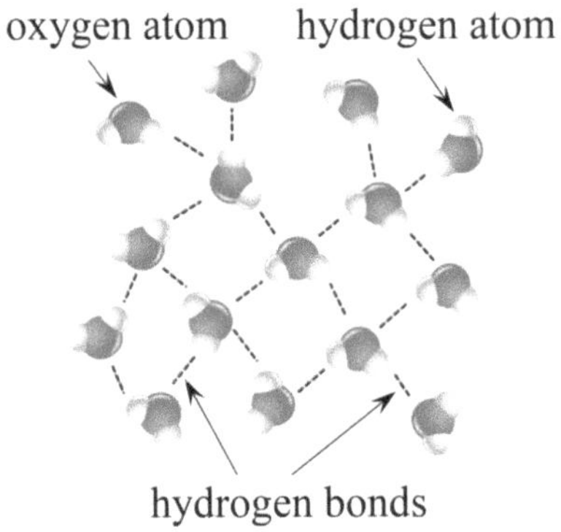

Figure 6.80 Schematic of hydrogen bonds of water forming a hydrogen bond network.

it is very difficult to directly probe the water structures in CNTs through experiments. Hence, MD simulations have been employed as the major mean for examining water structures in CNTs. As intermolecular forces change the dynamics of HBN of water, the potential function for modeling water molecules should have an effect on the water structure. This casts uncertainty to water structures and has to be kept in mind in interpreting the results in the literature.

It has been shown that for CNTs with diameters below 6.75 Å, which corresponds to (5,5) CNTs, water molecules cannot be filled into the CNTs, while in CNTs with diameters above 20 Å, i.e., (15,15) or larger CNTs, the confinement effects become insignificant and water behaves like bulk water (Noon et al., 2002). In CNTs with diameters below 6.75 Å, the water–CNT repulsive force can be extremely large because the size of a water molecule is about 3 Å (Hippler et al., 1983), which causes the rejection of water molecules. Carbon nanotubes with diameters above 8.3 Å, which is the diameter of (6,6) CNTs, are large enough to accept water molecules. In CNTs of intermediate sizes, the structure of water undergoes a transition from a single-file chain to ring-like configurations as the CNT diameter increases.

In small CNTs, water molecules cannot pass across or exchange positions with their neighbor molecules due to strong confinements. They usually form a single-file chain, as illustrated in Figure 6.81. Such water conformations have been observed

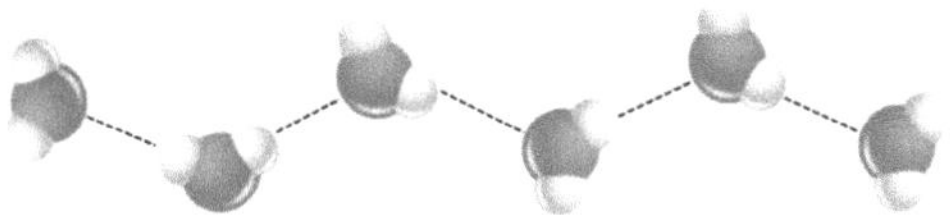

Figure 6.81 Schematic of a single-file chain formed by water molecules in small CNTs.

in (6,6) CNTs, as shown in Figures 6.82 and 6.83 (Meng, et al., 2008; Thomas and McGaughey, 2009). When the CNT diameter is relatively large such that water molecules have sufficient space to move around, ring-like structures with different conformations can be observed due to water–carbon interactions. Ring-like, helical structures of water around CNT surfaces are found in (7,7), (8,8), and (9,9) CNTs, as depicted in Figure 6.84 (Noon et al., 2002). Other ring-like structures, such as square, pentagonal, hexagonal, and even octagonal rings, are also observed in simulations (Koga et al., 2001; Meng et al., 2008; Thomas and McGaughey, 2009). Figure 6.82 depicts the structural transition of water from a single-file chains in a (6,6) CNT to ring structures in (8,8) and (9,9) CNTs (square and hexagonal rings) as the CNT diameter is increased (Meng et al., 2008). In Figure 6.83, similar structure transition is also found, where pentagonal rings are observed in both (7,7) and (8,8) CNTs and hexagonal rings are found in (9,9) CNTs (Thomas and McGaughey, 2009). Hexagonal structure in (9,9) CNTs is also found in another work, but similar ordered ring structures are not observed in (7,7) and (8,8) CNTs (Mashl et al., 2003). Different from the bulk-like water in (10,10) CNTs in Figure 6.83, an octagonal structure with a chain at the center is found in a (10,10) CNT, as demonstrated in Figure 6.85 (Kolesnikov et al., 2004). In some work, however, ordered water structures are not found in (7,7), (8,8), (9,9), and (10,10) CNTs (Wang et al., 2004).

In Figures 6.82 and 6.83, it can be seen that ordered structures only take place in (9,9) and smaller CNTs. Water structures in (10,10) or larger CNTs are not different from that of bulk water at ambient conditions. However, at high pressures,

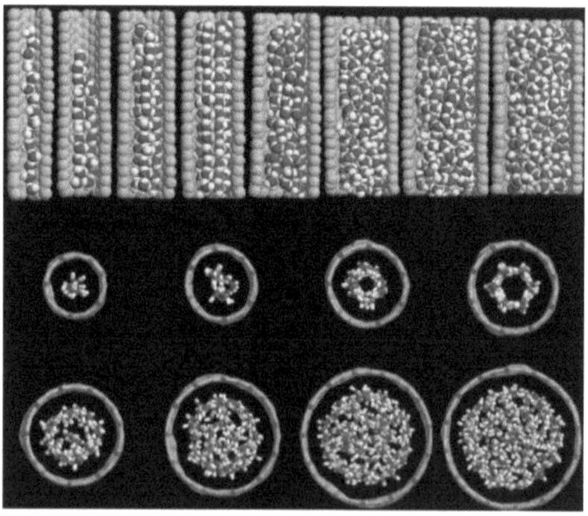

Figure 6.82 Side and top views of water structures in CNTs. Black and white spheres are oxygen and hydrogen atoms, respectively. CNT size in the top panel (from left to right): (6,6), (7,7), (8,8), (9,9), (10,10), (12,12), (14,14), and (15,15). (Reprinted from Meng, et al., 2008, with the permission of AIP Publishing.)

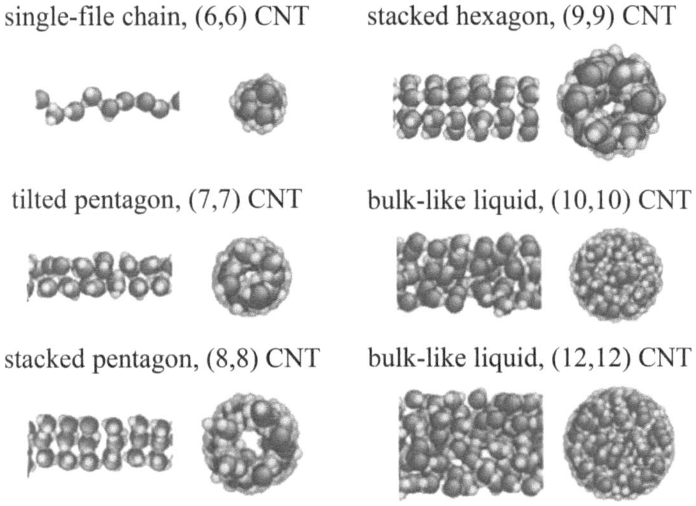

Figure 6.83 Water structures in CNTs of different sizes (CNTs are not shown for clarity). The diameters for (6,6), (7,7), (8,8), (9,9), (10,10), and (12,12) CNTs are 0.83, 0.96, 1.10, 1.25, 1.39, and 1.66 nm, respectively (Reprinted with permission from Thomas and McGaughey, 2009, Copyright (2009) by the American Physical Society.)

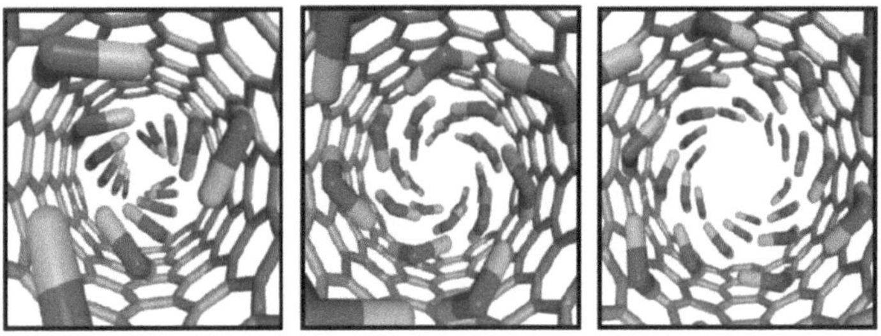

Figure 6.84 Helical water structures in (7,7) (left), (8,8) (middle), and (9,9) (right) CNTs. (Reprinted from Noon et al., 2002, Copyright (2002) with permission from Elsevier.)

ordered, ice-like ring structures are found in large CNTs, as shown in Figure 6.86, where the pressure is 50 MPa (Koga et al., 2001).

Water structures in CNTs have been extensively studied in the literature (Alexiadis and Kassinos, 2008). Although the findings in some work are consistent, considerable discrepancy exists in previous MD simulations. As stated at the beginning of this section, the potential functions for modeling water molecules and water–CNT interactions could be critical in affecting the water structure.

Figure 6.85 A shell-chain structure of water in a (10,10) CNT observed in MD simulations (Reprinted with permission from Kolesnikov et al., 2004, Copyright (2004) by the American Physical Society.)

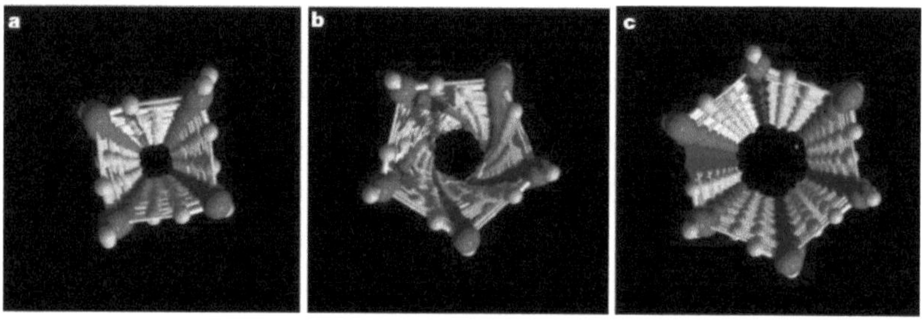

Figure 6.86 Square, pentagonal, and hexagonal ice-like structures in (14,14) (a), (15,15) (b), and (16,16) (c) CNTs at 50 MPa. (Reprinted by permission from Springer Nature, Koga et al., 2001, Copyright (2001).)

Furthermore, how the temperature of a simulation system is controlled may also change the organization of water molecules in a CNT. A comparison by Alexiadis and Kassinos (2008) shows that low values of LJ binding energy ε and collision diameter σ for C-O interactions and/or SPC/E water model tend to generate ring structures, such as pentagon and hexagonal configurations. TIP3P model favors the formation of helical structures. Nevertheless, MD simulations demonstrate the possibility that water molecules can form various conformations in nanoconfinements. These diverse structures may affect the transport of water and lead to different flow slippages and flow phenomena.

6.5.2 Slip length

The first experimental study of water flows in CNTs was conducted in 2005 (Majumder et al., 2005). The flow rates of pressure driven water flows in a 7-nm diameter CNT membrane are found to be four to five orders of magnitude higher than the predictions of the classic fluid mechanics theory. The flow rates indicate a slip length up to ~70 μm. One year later, experiments on water flows in even smaller CNTs (1.3–2.0 nm in diameter) were carried out (Holt et al., 2006). The flow enhancement, which is the ratio of the flow rate measured in experiments to that predicted by the classic theory, is between 560 and 9,600, which corresponds to a slip length ranging from 140 nm to 1.4 μm. The surprising flow enhancement observed in these experiments, together with earlier MD simulations of water transport in 0.81-nm-diameter CNTs (Hummer et al., 2001), have triggered considerable interest in the study of water transport in CNTs with a focus mainly on the slip length.

However, the subsequent studies find very dispersed slip lengths, from several nanometers to hundreds of micrometers. Through experiments, slip lengths as large as 485 μm are obtained for water flows through 10-nm-diameter CNT membranes (Du et al., 2011; Kannam et al., 2013). Although large slip lengths of 40–60 μm are reported again by the same group (Majumder et al., 2011), all the other experimental studies find much smaller slip lengths. For instance, the slip lengths are 8–53 nm for water translocation in CNTs with diameters ranging from 0.81 to 1.59 nm (Qin et al., 2011). In a recent study on the dynamics of water jets from single CNTs to a surrounding fluid, the slip length is found to depend on the CNT diameter (Secchi et al., 2016). It is about 300 nm in 30-nm-diameter CNTs and decreases quickly as the CNT diameter increases.

Other than experiments, MD simulations have also been extensively conducted to investigate the slip length. It is found that the slip lengths in 2.71-, 4.07-, 5.42-nm-diameter CNTs are 11, 13, and 15 nm, respectively, at room temperature (Kotsalis et al., 2004). The slip length is even smaller (9 nm) for water–nitrogen mixture flows in 2.71-nm-diameter CNTs. Another MD work shows that the slip length decreases monotonically from about 100 to 30 nm when the CNT diameter is increased from 1.66 to 4.99 nm and it can modeled by (Thomas et al., 2010)

$$L_s = L_{s,\infty} + C/d^3, \tag{6.41}$$

where $L_{s,\infty} = 30$ nm, which is the slip length for planar graphene, $C = 352$ nm, and d is the CNT diameter. For CNTs with diameters larger than 7 nm, the slip length approaches $L_{s,\infty}$.

In a study toward understanding the fast water transport in CNTs, a relatively large slip length of 556 nm is found in a 2.17-nm-diameter CNT due to the velocity variation in a depletion region and the water orientation at the water–CNT interface (Joseph and Aluru, 2008). In a 4-nm-diameter CNT, the slip length is obtained as 260 nm (Du et al., 2011). In a work studying the effects of curvature on the friction, it is found that the CNT curvature influences the friction and slip length greatly. The results indicate slip lengths of 500 and 120 nm for CNTs with diameters equal to 1 and 7 nm, respectively (Falk et al., 2010). However, slip lengths of only a few nanometers are found in CNTs with diameters between 1.09 and 1.62 nm (Wang et al., 2012). The dependence of water flows on the wettability of CNTs is also studied and the slip length is found to vary from 205 to 1,120 nm in a 2.72 nm diameter CNT when the contact angle of the CNT is changed from 45° to 135° (Ma et al., 2011). A theoretical work using a mathematical model, which considers a depletion layer with a reduced viscosity at the interface, predicts that the slip length is 8.57 and 35 nm for 1.4 and 40 nm diameter CNTs, respectively (Myers, 2011). By considering the effects of the external force, a recent study using both equilibrium and non-equilibrium MD simulations shows that the slip length is between 75 and 230 nm for CNTs of diameters from 1.62 to 6.5 nm (Kannam et al., 2013).

In summary, the slip lengths determined in experiments vary over five orders of magnitude in CNTs with diameters ranging from 0.81 to 10 nm. In numerical simulations, the slip lengths change from about 1 to 1,100 nm in 0.81–7 nm diameter CNTs. Table 6.4 lists CNT sizes and the corresponding slip lengths for water flows in the literature. The slip lengths are also shown in Figure 6.87. It is clear that the slip lengths for water flows in CNTs reported in the literature are quite dispersed. Even though they are affected by many parameters, the widespread results are surprising and confusing.

Table 6.4 Slip Lengths of Water Flows in CNTs Reported in the Literature

Method	CNT Diameter (nm)	L_s (nm)	References
Experiment	7	$(3.9–6.8) \times 10^4$	Majumder et al. (2005)
	1.3–2.0	140–1,400	Holt et al. (2006)
	10	4.85×10^5	Du et al. (2011)
	7	$(4.0–5.3) \times 10^4$	Majumder et al. (2011)
	0.81–1.59	8–53	Qin et al. (2011)
	30	300	Secchi et al. (2016)
Simulation	2.71, 4.07, 5.42	11, 13, 15	Kotsalis et al. (2004)
	2.17	556	Joseph and Aluru (2008)
	1.66–4.99	30–100	Thomas et al. (2010)
	1, 7	120, 500	Falk et al. (2010)
	4	260	Du et al. (2011)
	2.71	205–1,120	Ma et al. (2011)
	1.09–1.62	~10	Wang et al. (2012)
	1.62–6.5	75–230	Kannam et al. (2013)

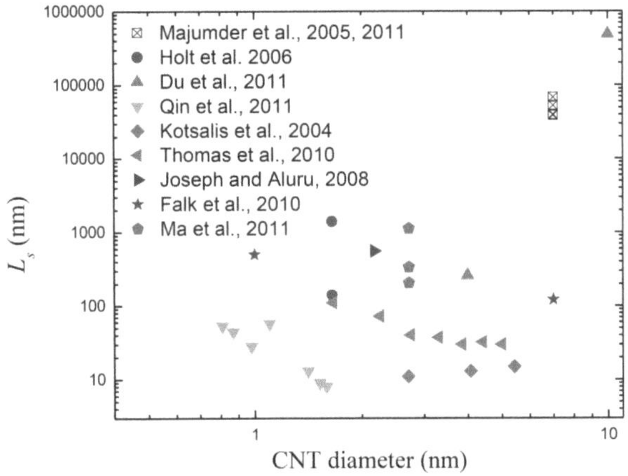

Figure 6.87 Slip lengths for water transport in CNTs reported in the literature.

Aiming at clarifying the confusing slip lengths of water flows in CNTs, delicate MD simulations, including both EMD and NEMD, are performed to carefully evaluate the slip length (Kannam et al., 2013). Through EMD simulations, where the external driving force is absent, the slip length is obtained by calculating the interfacial friction coefficient (see Equations 6.18 and 6.23). In NEMD simulations, the slip length is determined by extrapolating the velocity profiles and by using the slope of slip velocity–external force lines. Figure 6.88 depicts the slip lengths obtained in NEMD simulations. In Figure 6.88, the solid and semi-filled symbols are the slip lengths determined by fitting the streaming velocity using a quadratic equation and extrapolating the velocity using Equation 2.2. The open circles with a connected line represent the slip length calculated using

$$L_s = \frac{4s\mu}{\rho d}, \tag{6.42}$$

where s is the slope of slip velocity lines versus the external force. It can be seen that L_s using Equation 6.42 overlaps with the slip lengths obtained by extrapolating the velocity profiles. Nonetheless, in Figure 6.88, it is found that the slip length generally decreases with increasing CNT diameter and increases as the external force rises.

Figure 6.89 shows the slip length obtained from both EMD and NEMD, where the dashed line represents the slip length of water on a planar graphene surface (Kannam et al., 2012). It can be seen that the slip lengths obtained from both methods are in good agreement and they converge to the value for graphene as the CNT diameter increases. Figure 6.89 indicates that the numerical approaches are quite reliable if the water viscosity is assumed to be a constant, which is the

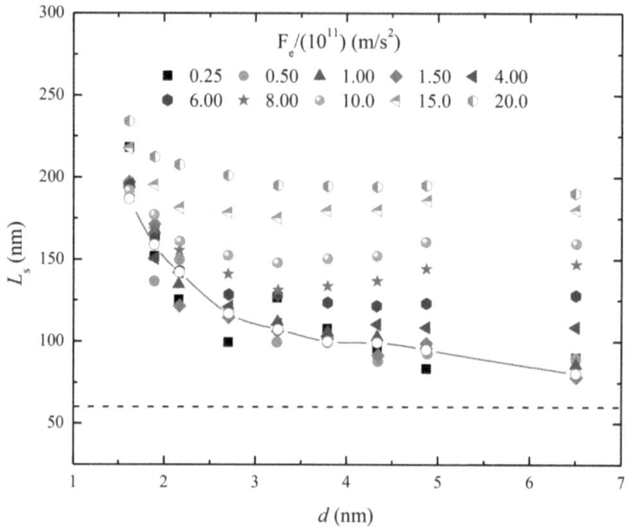

Figure 6.88 Slip length as a function of CNT diameter. Open circles with a connected line are the slip lengths predicted by Equation 6.42. The dashed line is the slip length of water on graphene. (Data are extracted from Kannam et al., 2012.)

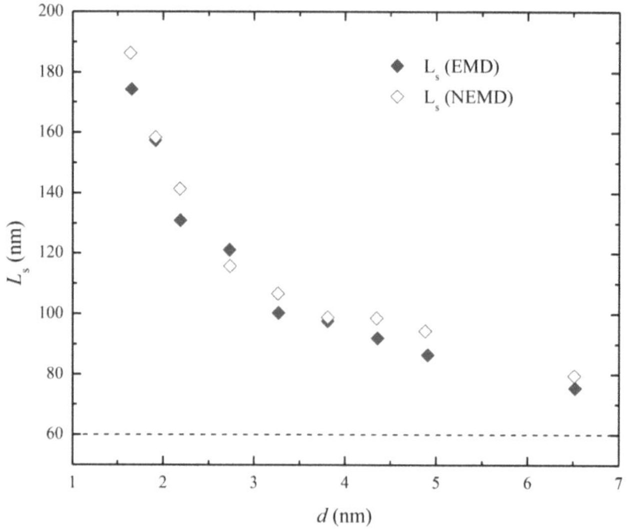

Figure 6.89 Slip length versus CNT diameter obtained using EMD and NEMD. The dashed line indicates the slip length of water on graphene. (Data are extracted from Kannam et al., 2012.)

basis for the determination of the slip length. The slip lengths in Figure 6.89 fall between 60 and 200 nm in CNTs with diameters larger than 1.6 nm. They are in good agreement with some of the previous work, but significantly smaller than microscale slip lengths found in some work (Table 6.4).

The discrepancy in the slip lengths in the literature is the consequence of several issues. For experiments, in addition to the uncertainty in flow rate measurement, which could be a critical problem, the classic theory used to determine the slip length is on the basis of parabolic velocity distribution predicted by the Navier–Stokes equation and the assumption of constant fluid viscosity. Their validity at the nanoscale is discussed in Chapter 2. In CNTs of a few nanometers in diameter, the large fluctuation of water viscosity may lead to considerable errors in the slip length. In very small CNTs, such as (6,6) CNTs, water molecules form single-file chains, as demonstrated in Section 6.5.1, for which both the velocity profile and viscosity are not well-defined. In this case, it is inappropriate to use the traditional approach to find the slip length, or a new quantity, instead of slip length, should be sought to characterize the flow rate.

In numerical simulations, first and probably the most important factor is the empirical potential functions for describing water and water–CNT interactions. Variations in the parameters of the potentials may cause significant changes in water dynamics. This is supported by experimental observation that the slip length of water flows in CNTs can be as high as 300 nm, while in boron nitride nanotubes flow slippage is not found (Secchi et al., 2016), indicating that the interactions between water molecules and the tube atoms play an important role.

Another issue in MD simulations is the thermostat used to maintain the temperature of the system. A reasonable approach is to control the velocities of carbon atoms of the CNT. However, in most of the previous simulations, the CNT is treated as a rigid tube and the carbon atoms are fixed during the simulation. Such a treatment is to save computational time because the potentials for modeling CNTs are time-consuming. If carbon atoms are fixed, the only way to control the system temperature is to scale the velocities of water molecules, which has an undesired effect on water dynamics.

Finally, it is worth mentioning that the driving force (e.g., pressure gradient) in MD simulations is usually very high and the resulting flow velocity can range from tens to hundreds of meters per second, which are comparable or even higher than the thermal velocity of molecules. This is because the numerical error of MD simulations is large and a high flow velocity is required to mask the numerical error. Such high flow velocities are unpractical and can cause large interfacial and internal frictions, which may heat up the fluid (Li, 2009) and lead to large errors when a thermostat is employed. Furthermore, the flow rate under an extremely high pressure gradient may fall into a nonlinear regime, which makes the theoretical determination of the slip length (e.g., Equation 2.2) invalid. Whether the pressure gradient and the resulting flow rate in MD simulations can be scaled down to a practical level requires further investigations.

References

Akhmatskaya, E., Todd, B.D., Daivis, P.J., Evans, D.J., Gubbins, K.E., and Pozhar, L.A. 1997. A study of viscosity inhomogeneity in porous media. *J. Chem. Phys.* 106:4684–4695.

Alexiadis, A. and Kassinos, S. 2008. Molecular simulation of water in carbon nanotubes. *Chem. Rev.* 108:5014–5034.

Bonaccurso, E., Kappl, K., and Butt, H.J. 2002. Hydrohynamic force measurements: Boundary slip of water on hydrophilic surfaces and electrokinetic effects. *Phys. Rev. Lett.* 88:076103.

Braun, O.M. and Kivshar, Y.S. 1998. Nonlinear dynamics of Frenkel-Kontorova model. *Phys. Rep.* 306:1–108.

Chowdhuri, S. and Chandra, A. 2001. Molecular dynamics simulations of aqueous NaCl and KCl solutions: Effects of ion concentration on the single-particle, pair, and collective dynamical properties of ions and water molecules. *J. Chem. Phys.* 115:3732–3741.

Cottin-Bizonne, C., Cross, B., Steinberger, A., and Charlaix, E. 2005. Boundary slip on smooth hydrophobic surfaces: Intrinsic effects and possible artifacts. *Phys. Rev. Lett.* 94:056102.

Craig, V.S.J., Neto, C., and Williams, D.R.M. 2001. Shear-dependent boundary slip in an aqueous Newtonian liquid. *Phys. Rev. Lett.* 87:054504.

Du, F., Qu, L., Xia, Z., Feng, L., and Dai, L. 2011. Membranes of vertically aligned superlong carbon nanotubes. *Langmuir* 27:8437.

Evans, D.J. and Morriss, G.P. 1990. *Statistical Mechanics of Nonequillibrium Liquids*. New York: Academic.

Falk, K., Sedlmeier, F., Joly, L., Netz, R.R., and Bocquet, L. 2010. Molecular origin of fast water transport in carbon nanotube membranes: Superlubricity versus curvature dependent friction. *Nano Lett.* 10:4067–4073.

Groisman, A. and Quake, S.R. 2004. A microfluidic rectifier: Anisotropic flow resistance at low Reynold numbers. *Phys. Rev. Lett.* 92:094501.

Hansen, J.S., Todd, B.D., and Daivis, P.J. 2011. Prediction of fluid velocity slip at solid surfaces. *Phys. Rev. E* 84:016313.

Hippler, H., Troe, J., and Wendelken, H.J. 1983. Collisional deactivation of vibrationally highly excited polyatomic molecules. II. Direct observations for excited toluene. *J. Chem. Phys.* 78:6709–6717.

Holt, J.K., Park, H.G., Wang, Y., Stadermann, M., Artyukhin, A.B., Grigoropoulos, C.P., Noy, A., and Bakajin, O. 2006. Fast mass transport through sub-2-nanometer carbon nanotubes. *Science* 312:1034–1037.

Hummer, G., Rasaiah, J.C., and Noworyta, J.P. 2001. Water conduction through the hydrophobic channel of a carbon nanotube. *Nature* 414:188–190.

Joly, L., Ybert, C., and Bocquet, L. 2006. Probing the nanohydrodynamics at liquid-solid interfaces using thermal motion. *Phys. Rev. Lett.* 96:046101.

Joseph, P. and Tabeling, P. 2005. Direct measurement of the apparent slip length. *Phys. Rev. E.* 71:035303.

Joseph, S. and Aluru, N.R. 2008. Why are carbon nanotubes fast transporters of water? *Nano Lett.* 8:452–458.

Kannam, S.K., Todd, B.D., Hansen, J.S., and Daivis, P.J. 2012. Slip length of water on graphene: Limitations of non-equilibrium molecular dynamics simulations. *J. Chem. Phys.* 136:024705.

Kannam, S.K., Todd, B.D., Hansen, J.S., and Daivis, P.J. 2013. How fast does water flow in carbon nanotubes? *J. Chem. Phys.* 138:094701.

Koga, K., Gao, G.T., Tanaka, H., and Zeng, X.C. 2001. Formation of ordered ice nanotubes inside carbon nanotubes. *Nature* 412:802–805.

Kolesnikov, A.I., Zanotti, J.M., Loong, C.K, and Thiyagarajan, P. 2004. Anomalously soft dynamics of water in a nanotube: A revelation of nanoscale confinement. *Phys. Rev. Lett.* 93:035503.

Koplik, J. and Banavar, J.R. 1992. Molecular structure of the coalescence of liquid interfaces. *Science* 257:1664.

Kotsalis, E.M., Walther, J.H., and Koumoutsakos, P. 2004. Multiphase water flow inside carbon nanotubes. *Int. J. Multiphase Flow* 30:995.

Leal, L.G. 2007. *Advanced Transport Phenomena*. New York: Cambridge University Press.

Li, B., Wang, L., and Casati, G. 2004. Thermal diode: Rectification of Heat Flux. *Phys. Rev. Lett.* 93:184301.

Li, C., Huang, B.L., Cao, L., and Li, Z.G. 2014. Molecular diffusion on surfaces in the weak friction limit. *J. Appl. Phys.* 115:214906.

Li, L., Mo, J.W., and Li, Z.G. 2014. Flow and slip transition in nanochannels. *Phys. Rev. E* 90:033003.

Li, L., Mo, J.W., and Li, Z.G. 2015. Nanofluidic diode for simple fluids without moving parts. *Phys. Rev. Lett.* 115:134503.

Li, Z.G. 2009. Surface effects on friction-induced fluid heating in nanochannel flows. *Phys. Rev. E* 79:026312.

Li, Z.G. and Drazer, G. 2008. Hydrodynamic interactions in dissipative particle dynamics. *Phys. Fluids* 20:103601.

Li, Z.G. and Hong, L. 2007. On the Knudsen transport of gases in nanochannels. *J. Chem. Phys.* 127:074706.

Li, Z.G. and Wang, H. 2005. Gas-nanoparticle scattering: A molecular view of momentum accommodation function. *Phys. Rev. Lett.* 95:014502.

Liu, C. and Li, Z.G. 2009. Flow regimes and parameter dependence in nanochannel flows. *Phys. Rev. E* 80:036302.

Liu, C. and Li, Z.G. 2010. Molecular dynamics simulation of composite nanochannels as nanopumps driven by symmetric temperature gradients. *Phys. Rev. Lett.* 105:174501.

Liu, C. and Li, Z.G. 2011. On the validity of the Navier-Stokes equations for nanoscale liquid flows: The role of channel size. *AIP Adv.* 1:032108.

Liu, C., Fan, H.B., Zhang, K., Yuen, M.M.F., and Li, Z.G. 2010. Flow dependence of interfiacial thermal resistance in nanochannels. *J. Chem. Phys.* 132:094703.

Liu, L., Zhao, J., Yin, C.Y., Culligan, P.J., and Chen, X. 2009. Mechanisms of water infiltration into conical hydrophobic nanopores. *Phys. Chem. Chem. Phys.* 11:6520.

Ma, M.D., Shen, L., Sheridan, J., Liu, J.Z., Chen, C., and Zheng, Q. 2011. Friction of water slipping in carbon nanotubes. *Phys. Rev. E* 83:036316.

Maali, A., Cohen-Bouhacina, T., and Kellay, H. 2008. Measurement of the slip length of water flow on graphite surface. *Appl. Phys. Lett.* 92:053101.

Majumder, M., Chopra, N., Andrews, R., and Hinds, B.J. 2005. Nanoscale hydrodynamics: Enhanced flow in carbon nanotubes. *Nature* 438:44.

Majumder, M., Chopra, N., and Hinds, B.J. 2011. Mass transport through carbon nanotube membranes in three different regimes: Ionic diffusion and gas and liquid flow. *ACS Nano* 5:3867–3877.

Martini, A., Hsu, H.Y., Parankar, N.A., and Lichter, S. 2008. Slip at high shear rates. *Phys. Rev. Lett.* 100:206001.

Mashl, R.J., Joseph, S., Aluru, N.R., and Jakobsson, E. 2003. Anomalously immobilized water: A new water phase induced by confinement in nanotubes. *Nano Lett.* 3:589–592.

Meng, L., Li, Q., and Shuai, Z. 2008. Effects of size constraint on water filling process in nanotube. *J. Chem. Phys.* 128:134703.
Mo, J.W., Li, L., Zhou, J.F., Xu, D.Y., Huang, B.L., and Li, Z.G. 2015. Fluid infiltration pressure for hydrophobic nanochannels. *Phys. Rev. E* 91:033022.
Mo, J.W., Li, C., Li, L., Wang, J., and Li, Z.G. 2016a. Passive nanofluidic diode for using nonuniform nanochannels. *Phys. Fluids* 28:082005.
Mo, J.W., Li, L., Wang, J., and Li, Z.G. 2016b. Passive fluidic diode for simple fluids using nested nanochannel structures. *Phys. Rev. E* 93:033101.
Myers, T.G. 2011. Why are slip lengths so large in carbon nanotubes? *Microfluid. Nanofluid.* 10:1141–1145.
Nabavi, M. 2009. Steady and unsteady flow analysis in microdiffusers and micropumps: A critical review. *Microfluid. Nanofluid.* 7:599.
Nguyen, N.T., Lam, Y.C., Ho, S.S., and Low, C.L.N. 2008. Improvement of rectification effects in diffuser/nozzle structures with viscoelastic fluids. *Biomicrofluidics* 2:034101.
Nijmeijer, M.J.P., Bruin, C., Bakker, A.F., and Van Leeuwen, J.M.J. 1989. A visual measurement of contact angles in a molecular dynamics simulations. *Physica A* 160:166–180.
Noon, W.H., Ausman, K.D., Smalley, R.E., and Ma, J. 2002. Helical ice-sheets inside carbon nanotubes in physiological condition. *Chem. Phys. Lett.* 355:445–448.
Pit, R., Hervet, H., and Leger, L. 2000. Direct experimental evidence of slip in hexadecane: Solid interfaces. *Phys. Rev. Lett.* 85:980–983.
Priezjev, N.V. 2007. Rate-dependent slip boundary conditions for simple fluids. *Phys. Rev. E* 75:051605.
Qin, X., Yuan, Q., Zhao, Y., Xie, S., and Liu, Z. 2011. Measurement of the rate of water tranlocation through carbon nanotubes. *Nano Lett.* 11:2173.
Secchi, E., Marbach, S., Nigues, A., Stein, D., Siria, A., and Bocquet, L. 2016. Massive radius-dependent flow slippage in carbon nanotubes. *Nature* 537:210–213.
Sendner, C., Horineck, D., Bocquet, L., and Netz, R.R. 2009. Interfacial water at hydrophobic and hydrophilic surfaces: Slip, viscosity, and diffusion. *Langmuir* 25:10768–10781.
Smith, P. and Zappe, R.W. 2004. *Valve Selection Handbook*. Burlington, MA: Elsevier.
Sousa, P.C., Pinho, F.T., Oliveira, M.S.N., and Alves, M.A. 2010. Efficient microfluidic rectifiers for viscoelastic fluid flow. *J. Non-Newtonian Fluid Mech.* 165:652.
Sze, S.M. and Ng, K.K. 2007. *Physics of Semiconductor Devices*. New York: Wiley.
Tesla, N. 1920. Valvular Conduit, U.S. Patent No. 1329559.
Thomas, J.A. and McGaughey, A.J.H. 2009. Water flows in carbon nanotubes: Transition to subcontinuum transport. *Phys. Rev. Lett.* 102:184502.
Thomas, J.A., McGaughey, A.J.H., Kuter-Arnebeck, O. 2010. Pressure-driven water flow through carbon nanotubes: Insights from molecular dynamics simulation. *Int. J. Therm. Sci.* 49:281–289.
Thompson, P.A. and Troian, S.M. 1997. A general boundary condition for liquid flow at solid surfaces. *Nature* 389:360–362.
Todd, B.D., Evans, D.J., and Daivis, P.J. 1995. Pressure tensor for inhomogeneous fluids. *Phys. Rev. E* 52:1627–1638.
Travis, K.P., Todd, B.D., and Evans, D.J. 1997. Departure from Navier-Stokes hydrodynamics in confined liquids. *Phys. Rev. E* 55:4288–4295.
Tretheway, D.C. and Meinhart, C.D. 2002. Apparaent fluid slip at hydrophobic microchannel walls. *Phys. Fluids* 14:L9–L12.
Vinogradova, O.I. 1995. Drainage of a thin liquid film confined between hydrophobic surfaces. *Langmuir* 11:2213–2220.
Vinogradova, O.I. and Yakubov, G.E. 2003. Dynamic effects on force measurements. 2. Lubrication and the atomic force microscope. *Langmuir* 19:1227–1234.

Wang, J., Zhu, Y., Zhou, J., and Lu, X.H. 2004. Diameter and helicity effects on static properties of water molecules confined in carbon nanotubes. *Phys. Chem. Chem. Phys.* 6:829–835.

Wang, L., Dumont, R.S., and Dickson, J.M. 2012. Nonequilibrium molecular dynamics simulation of water transport through carbon nanotube membranes at low pressure. *J. Chem. Phys.* 137:044102.

Zhao, L. and Choi, P. 2004. Molecular dynamics simulation of the coalescence of nanometer-sized water droplets in n-heptane. *J. Chem. Phys.* 120:1935.

Zhu, Y. and Granick, S. 2001. Rate-dependent slip of Newtonian liquid at smooth surfaces. *Phys. Rev. Lett.* 87:096105.

Yong, X. and Zhang, L.T. 2013. Slip in nanoscale shear flow: Mechanisms of interfacial friction. *Microfluid. Nanofluid.* 14:299–308.

Chapter 7 External flows

7.1 Introduction

In the classic fluid mechanics, the topic of external flows is focused on the drag and lift forces acting on a macroscale object immersed in a fluid. The forces are caused by a relatively motion between the fluid and the object, which generates nonuniform shear stresses and pressures on the surface of the object. The forces depend on the relative velocity, the shape of the object, and the properties of the fluid, such as density and viscosity. Lift forces are of particular importance mainly in aeronautical engineering. Drag forces, however, have wide applications in a variety of areas. For the drag force, the only case that has an analytical solution is the flow over a sphere at a low Re, for which the drag force is given by the Stokes law, $F_D = -6\pi\mu RV$ (see Equation 1.11a). For other shapes, such as cylinders, or at a high Re, theoretical expressions for the drag force are unavailable.

In a Newtonian liquid, the Stokes law works well for both macro- and microparticles. In a gas, however, the validity of the Stokes law depends on the Kundsen number, Kn, a dimensionless number defined as $\mathrm{Kn} = \lambda/R$ with λ being the mean free path of the gas. For macroparticles, $\mathrm{Kn} \ll 1$ (as a reference, λ for air at the standard condition is about 60 nm), which is considered as the continuum regime and the stick boundary condition assumed in the Stokes law is valid. For microparticles, however, Cunningham (1910) found that the stick boundary condition might be invalid and the Stokes law should be corrected by a factor, $1/(1 + A'\mathrm{Kn})$, where A' is a constant. This factor was later revised for all the values of Kn (Knudsen and Weber, 1911) and the new drag force formula is given by

$$F_D = -\frac{6\pi\mu RV}{1 + \mathrm{Kn}\left[A + B\exp\left(-E/\mathrm{Kn}\right)\right]}, \tag{7.1}$$

where A, B, and E are constants, whose values are usually obtained by fitting the oil droplet data of Millikan's experiments (1910, 1917, 1923). Some popular values of these constants are given in Table 7.1. Equation 7.1 is known as the Stokes–Cunningham formula and has been widely used for macro- and microparticles. It is also the theoretical basis for the measurement of particle size and mobility in modern instruments.

In the free molecular regime, where $\mathrm{Kn} \gg 1$, however, it is shown that the drag force depends on how gas molecules are scattered upon collisions with a small

Table 7.1 Values of the Parameters in the Cunningham Slip Correction Factor in Equation 7.1

Publications	λ (μm)	*A*	*B*	*E*
Knudsen and Weber (1911)	0.09417	0.772	0.40	1.63
Millikan (1923)	0.09417	0.864	0.29	1.25
Davies (1945)	0.066	1.257	0.40	1.10
Dahneke (1973)	0.066	1.234	0.414	0.87
Allen and Raabe (1982)	0.0673	1.155	0.471	0.596

particle (Epstein, 1924). There are two ideal models of scattering: specular and diffuse scatterings. For specular scattering, the incident angle of a gas molecule is equal to the angle of reflection, as shown in Figure 7.1. For diffuse scattering, the angle of reflection is random and the velocities of gas molecules after collisions follow the Maxwellian distribution (Figure 7.1). Using the gas kinetic theory and assuming rigid-body collisions, i.e., no interaction between a gas molecule and a particle until they physically contact, Epstein derived the drag forces on small spherical particles for both specular and diffuse scatterings (1924). The drag forces are given as follows:

$$F_{\mathrm{D}} = -\frac{8}{3}\delta\sqrt{2\pi m_{\mathrm{g}} k_{\mathrm{B}} T}\, N R^2 V \tag{7.2}$$

where δ is a factor that depends on the scattering model and $\delta = 1$ and $(8+\pi)/8$ for specular and diffuse scatterings, respectively, m_{g} is the mass of a gas molecule, and N is the number density of the gas.

The real collisions between gas molecules and a particle may not exactly follow the ideal specular and diffuse scatterings, but can be treated as a mixture of the

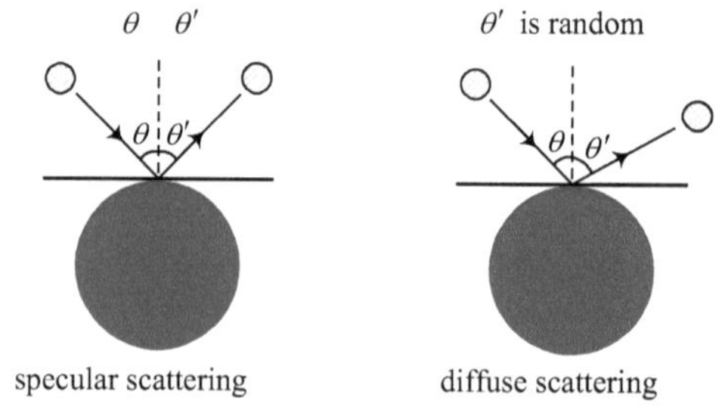

Figure 7.1 Models for gas–particle collisions. θ and θ' are incident angle and angle of reflection, respectively. Left: Specular scattering. Right: Diffuse scattering.

two cases. Therefore, practically, the drag force is evaluated as a linear combination of the forces for the two scattering scenarios,

$$F_{\mathrm{D}} = \varphi F_{\mathrm{D,diffuse}} + (1-\varphi) F_{\mathrm{D,specular}} = -\frac{8+\pi\varphi}{3}\sqrt{2\pi m_{\mathrm{g}} k_{\mathrm{B}} T} N R^2 V \tag{7.3}$$

where φ is called the momentum accommodation coefficient and $\varphi \approx 0.9$ is widely used in applications.

The drag forces in Equations 7.1–7.3 are only valid for microparticles. For nanoparticles, the molecular interactions between fluid molecules and the nanoparticles can be significant, which are not considered in the classic approaches. The importance of molecular interactions can be demonstrated by comparing these drag forces with the Chapman–Enskog theory for the transport of molecules, which gives the drag force on a molecule (Li and Wang, 2003a)

$$F_{\mathrm{D}} = -\frac{8}{3}\sqrt{2\pi m_{\mathrm{r}} k_{\mathrm{B}} T} N \sigma^2 \Omega^{(1,1)*} V \tag{7.4}$$

where m_{r} is the reduced mass of the gas molecule and particle, $m_{\mathrm{r}} = m_{\mathrm{g}} m_{\mathrm{p}} / (m_{\mathrm{g}} + m_{\mathrm{p}})$, and m_{p} is the mass of the particle, σ is the collision diameter, and $\Omega^{(1,1)*}$ is a reduced collision integral, which depends on molecular interactions. Although Equation 7.4 is for molecules, it is consistent with Epstein's solution of Equation 7.2 if it is applied to microparticles under the assumption of rigid-body collision, for which $\Omega^{(1,1)*} = 1$, $m_{\mathrm{r}} \approx m_{\mathrm{g}}$, and $\sigma \approx R$.

Figure 7.2 qualitatively shows $F_{\mathrm{D}}/(RV)$ as a function of particle radius R for various drag force expressions. It can be seen that the Stokes–Cunningham formula converges to the Stokes law for macroparticles and is close to Epstein's

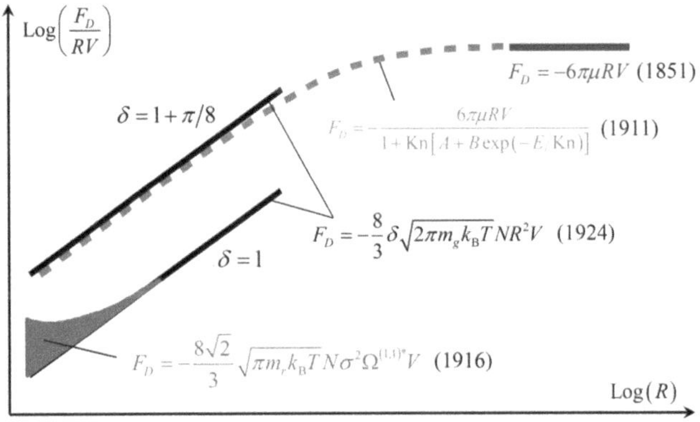

Figure 7.2 Drag forces in different regimes.

solution for diffuse scattering for small particles. However, it does not approach the Chapman–Enskog theory for the drag force on molecules or molecular scale particles (the shaded area for the Chapman–Enskog theory in Figure 7.2 arises from intermolecular interactions, which depend on the materials of the particle and gas). It is clear that there is a gap between the Stokes–Cunningham formula and the Chapman–Enskog theory. This gap takes place at the nanoscale and is caused by the negligence of molecular interactions in the classic theories and the transition from specular to diffuse scattering as the particle size is increased, as will be discussed later.

Therefore, molecular interactions play an important role for nanoparticle transport and need to be considered in theoretical treatments. In the following sections, the drag force and diffusion coefficient of nanospheres and nanocylinders in rarefied gases and liquids, i.e., free molecular and continuum regimes, will be discussed. In the free molecular regime, the gas kinetic theory is employed to derive analytical expressions for the drag force and diffusion coefficient. In the continuum regime, theoretical analysis is difficult and only MD simulations are discussed.

7.2 Drag force and diffusion coefficient of nanospheres

7.2.1 Free molecular regime

7.2.1.1 Drag force In this subsection, the transport properties of spherical nanoparticles, which hereafter are simply referred to as nanoparticles, in dilute gases with $\mathrm{Kn} \gg 1$ is discussed. This is one of the few sections dealing with gases in this book. It is because the free molecular regime is the only regime, where the transport properties of nanoparticles/nanocylinders can be analytically obtained. In the continuum regime, theoretical analysis is nontrivial and numerical simulations are the primary approach for studying the dynamics of nanoscale objects in liquids.

Equation 1.18 gives a simple relationship between the diffusion coefficient and the drag coefficient. It is valid if the drag force is linearly related to the relative velocity, i.e., $F_\mathrm{D} = -\alpha V$. For nanoparticles, Equation 1.18 should work if the relative velocity is not very large. Therefore, the diffusion coefficient is readily obtained if the drag force is determined. In the following discussion, the drag force on nanoparticles will be derived using the gas kinetic theory and the diffusion coefficient will be acquired through Equation 1.18.

Let us consider a gas of number density N in local equilibrium with the center-of-mass velocity equal to 0. The thermal velocity of the gas molecules is denoted as

v. A nanoparticle is moving in the gas at a constant velocity **V**. After a gas molecule collides with the particle, the velocities of the particle and the gas molecule are referred to as **V**′ and **v**′, respectively. If the coordinate system is attached to the particle and the origin O is set at the center of the particle, as shown in Figure 7.3a, the velocities of the gas molecules before and after collisions with the particle are $\mathbf{g} = \mathbf{v} - \mathbf{V}$ and $\mathbf{g}' = \mathbf{v}' - \mathbf{V}'$, respectively. Let the direction of vector **g** be collinear with the z-axis of the coordinate system and **i**, **j**, and **k** be the unit vectors in the x-, y-, and z-directions, respectively (Figure 7.3a). The drag force on the particle in this coordinate system can be obtained by calculating the momentum change of the gas molecules per unit time during collisions with the particle. If

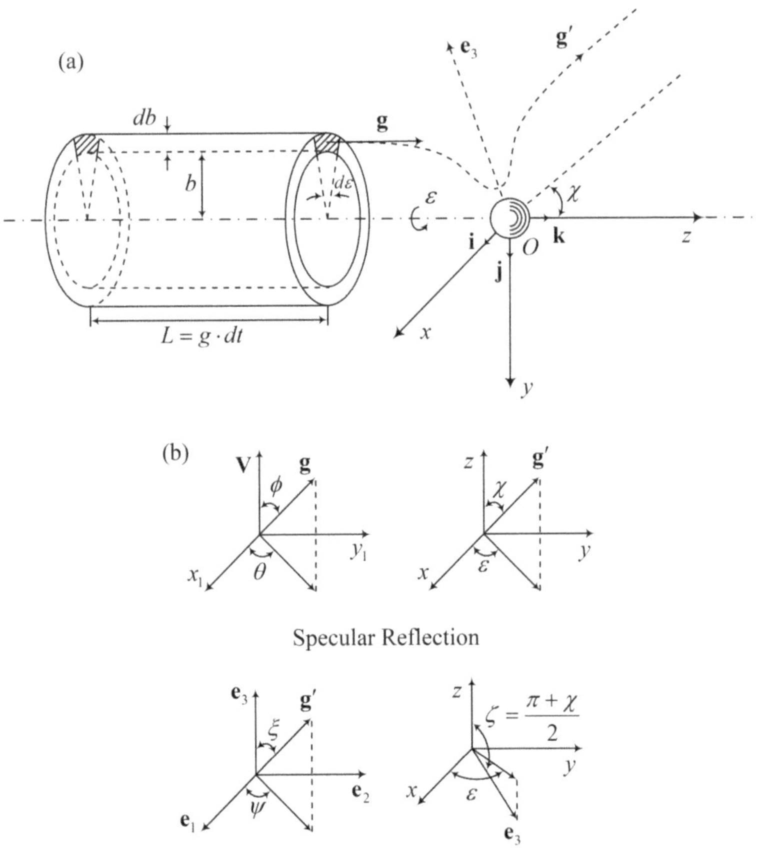

Figure 7.3 Gas–nanoparticle collision model. (a) Coordinate system. (b) Relationship among various vectors for specular and diffuse scatterings. (Reprinted with permission from Li and Wang, 2003a, Copyright (2003) by the American Physical Society.)

the momenta of the gas molecules before and after collisions are denoted as $\mathbf{p}$ and $\mathbf{p}'$, the drag force on the particle can be expressed as follows:

$$\mathbf{F} = -\frac{d\mathbf{p}}{dt} = \frac{\mathbf{p} - \mathbf{p}'}{\Delta t} \tag{7.5}$$

where Δt is the collision time.

Let us consider a differential cylindrical shell with the z-axis as its axis, as shown in Figure 7.3a. The length of the cylindrical shell is $L = g \cdot dt$, which is the distance traveled by the gas molecules in a differential time interval dt. All the molecules in this cylindrical shell are moving toward the particle and the radius of the shell is equal to the impact factor b. For a small sector of this cylindrical shell (patterned area in Figure 7.3a) characterized by a differential angle $d\varepsilon$, the number of molecules in this sector at a time t is given by

$$n = fgbdbd\varepsilon dt \tag{7.6}$$

where f is the velocity distribution function of the gas molecules (Li and Wang, 2003a),

$$f = \frac{N}{\left(2\pi k_B T/m_r\right)^{3/2}} \exp\left(-\frac{v^2}{2k_B T/m_r}\right) \tag{7.7}$$

Here the use of the reduced mass m_r is required by the coordinate system. The total momentum $\delta\mathbf{p}$ carried by the molecules in the sector before colliding with the particle is

$$\delta\mathbf{p} = m_r n\mathbf{g} \tag{7.8}$$

The momentum of the gas molecules after collisions $\delta\mathbf{p}'$ depends on how the gas molecules are reflected after collisions. For specular scattering (Figure 7.1), the incident angle and the angle of reflection are the same and the magnitude of $\mathbf{g}$ is equal to that of $\mathbf{g}'$. For diffuse scattering, the angle of reflection is random and the magnitude of $\mathbf{g}'$ follows the Maxwellian distribution. Therefore, the momentum changes of the gas molecules for the two scattering cases are different. This leads to different drag forces.

7.2.1.1.1 Specular scattering Due to molecular interactions, the angle of scattering χ depends on the impact parameter b and the velocity $\mathbf{g}$. If the molecular interactions are described by a potential function $\Phi(r)$, where r is the distance between a gas molecule and the particle, the angle of scattering is given by (Hirschfelder et al., 1954)

$$\chi(g,b) = \pi - 2b\int_{r_m}^{\infty} \frac{dr}{r^2\sqrt{1 - \frac{b^2}{r^2} - \frac{\Phi(r)}{m_r g^2/2}}} \tag{7.9}$$

where r_m is the minimum distance during the collision. The momentum of the reflected molecules is then given by

$$\delta\mathbf{p}' = m_r ng(\mathbf{k}\cos\chi + \mathbf{i}\sin\chi\cos\varepsilon + \mathbf{j}\sin\chi\sin\varepsilon) \tag{7.10}$$

According to Equation 7.5, the differential force due to the collisions between the gas molecules in the differential sector in Figure 7.3a and the particle is

$$\delta\mathbf{F}_s = \frac{\delta\mathbf{p} - \delta\mathbf{p}'}{\Delta t} \tag{7.11}$$

where the subscript "s" denotes specular scattering. To obtain the total drag force, Equation 7.11 needs to be integrated over ε, b, and $\mathbf{g}$. This causes the second and third terms in the bracket of Equation 7.10 to vanish. Hence, Equation 7.11 can be rewritten as follows:

$$d\mathbf{F}_s = \frac{\delta\mathbf{p} - \delta\mathbf{p}'}{dt} = m_r n\mathbf{g}(1 - \cos\chi) \tag{7.12}$$

Integrating Equation 7.12, the drag force is obtained as

$$\mathbf{F}_s = \frac{m_r N}{(2\pi k_B T/m_r)^{3/2}} \int_v g\mathbf{g}\exp\left(-\frac{v^2}{2k_B T/m_r}\right) Q_s(g)\, d\mathbf{v} \tag{7.13}$$

where $Q_s(g)$ is the collision cross-section for specular scattering, which is defined by

$$Q_s(g) = 2\pi\int_0^{\infty} (1 - \cos\chi)\, b\, db \tag{7.14}$$

As V is the velocity of the particle, it is assumed to be much smaller than the thermal velocity v of the gas molecules, i.e., $V \ll v$, which leads to $d\mathbf{v} \approx d\mathbf{g}$. Let ϕ and θ be the colatitude and azimuthal angles of $\mathbf{g}$ in a reference frame, where $\mathbf{V}$ is collinear with the z-axis, as shown in Figure 7.3b. Then, Equation 7.13 can be rewritten as follows:

$$\mathbf{F}_s = \frac{m_r N}{\left(2\pi k_B T/m_r\right)^{3/2}} \int_{\mathbf{g}} g\mathbf{g} \exp\left(-\frac{g^2+V^2+2gV\cos\phi}{2k_B T/m_r}\right) Q_s(g)d\mathbf{g}$$

$$= \frac{2\pi m_r N\mathbf{V}}{\left(2\pi k_B T/m_r\right)^{3/2} V} \int_0^\infty g^4 \exp\left(-\frac{g^2+V^2+2gV\cos\phi}{2k_B T/m_r}\right) Q_s(g)dg \int_0^\pi \cos\phi \sin\phi\, d\phi$$

(7.15)

Given that $g \gg V$ and $m_p \gg m_g$, it is obtained that $k_B T \sim m_g g^2 \sim m_r g^2 \gg m_r gV$, which leads to $gV/\left(k_B T/m_r\right) \ll 1$. Then, the exponential term in Equation 7.15 can be approximated as

$$\exp\left(-\frac{g^2+V^2+2gV\cos\phi}{2k_B T/m_r}\right) \approx \exp\left(-\frac{g^2}{2k_B T/m_r}\right)\left(1-\frac{gV\cos\phi}{k_B T / m_r}\right) \quad (7.16)$$

Substituting Equation 7.16 into Equation 7.15, the drag force becomes

$$\mathbf{F}_s = -\frac{4\pi m_r^2 N\mathbf{V}}{3k_B T\left(2\pi k_B T/m_r\right)^{3/2}} \int_0^\infty g^5 \exp\left(-\frac{g^2}{2k_B T/m_r}\right) Q_s(g)dg \quad (7.17)$$

Let $\gamma = g/\sqrt{2k_B T/m_r}$, Equation 7.17 is simplified as

$$\mathbf{F}_s = -\frac{8}{3\pi}\sqrt{2\pi m_r k_B T} N\mathbf{V} \int_0^\infty \gamma^5 \exp\left(-\gamma^2\right) Q_s(g)d\gamma \quad (7.18)$$

With the definition of a collision integral,

$$\Omega_s^{(1,q)} = \int_0^\infty \exp\left(-\gamma^2\right)\gamma^{2q+3} Q_s(g)d\gamma \quad (7.19)$$

the drag force in Equation 7.18 is given by

$$\mathbf{F}_s - \frac{8}{3\pi}\sqrt{2\pi m_r k_B T} N\Omega_s^{(1,1)}\mathbf{V} \quad (7.20)$$

The collision integral $\Omega_s^{(1,1)}$ in Equation 7.20 contains the molecular interactions between the gas molecules and the particle. Following the Chapman–Enskog approach (Hirschfelder et al., 1954), a reduced collision integral, which is defined as

$$\Omega_s^{(1,q)*} \frac{2\Omega_s^{(1,q)}}{(q+1)!\pi R^2} \tag{7.21}$$

is used and the drag force in Equation 7.20 becomes

$$\mathbf{F}_s = -\frac{8}{3}\sqrt{2\pi m_r k_B T} N R^2 \Omega_s^{(1,1)*} \mathbf{V} \tag{7.22}$$

It should be noted that if $\pi\sigma^2$ is used to scale $\Omega_s^{(1,1)}$ in Equation 7.21 when the particle size approaches that of a molecule, Equation 7.22 is reduced to the drag force on a molecule, i.e., Equation 7.4.

For rigid-body collisions, it can be shown that $\Omega_s^{(1,1)*} = 1$, which is valid for micro- and macroscale particles, and Equation 7.22 is simplified to the specular case of Epstein's solution, Equation 7.2 with $\delta = 1$.

7.2.1.1.2 Diffuse scattering For diffuse scattering, the velocity of the reflected molecules follows the Maxwellian distribution

$$f' = cgf \exp\left(-\frac{g'^2}{2k_B T/m_r}\right) \tag{7.23}$$

where c is a constant and gf is the incident flux. By mass conservation, i.e., $gf = \int_{\mathbf{g}'} \mathbf{g}' \cdot \mathbf{e}_3 f' d\mathbf{g}'$, the constant c is calculated as $c = (m_r/k_B T)^2/(2\pi)$. Let the polar angles between $\mathbf{g}'$ and $\mathbf{e}_3$ be ξ and ψ, where $\mathbf{e}_3$ is a unit vector normal to the surface of the particle, as shown in Figure 7.3b. Then, the momentum of the reflected molecules can be written as follows:

$$\begin{aligned} \delta\mathbf{p}' &= bdbd\varepsilon dt \int_{\mathbf{g}'} m_r g' \mathbf{g}' \cos\xi f' d\mathbf{g}' \\ &= \frac{m_r n \mathbf{e}_3}{(k_B T/m_r)^2} \int_0^\infty g'^4 \exp\left(-\frac{g'^2}{2k_B T/m_r}\right) dg' \int_0^{\pi/2} \cos^2\xi \sin\xi \, d\xi \end{aligned} \tag{7.24}$$

Integrating Equation 7.24, it is obtained that

$$\delta\mathbf{p}' = \frac{\sqrt{2\pi m_r k_B T}}{2} n\mathbf{e}_3 = \frac{\sqrt{2\pi m_r k_B T}}{2} gfbdbd\varepsilon dt\mathbf{e}_3 \tag{7.25}$$

Decomposing the unit vector $\mathbf{e}_3$ into the x-, y-, and z-directions, $\delta\mathbf{p}'$ can be rewritten as

$$\delta\mathbf{p}' = \frac{\sqrt{2m_r\pi k_B T}}{2}\left[\mathbf{g}\cos\zeta + \mathbf{i}g\sin\zeta\cos\varepsilon + \mathbf{j}g\sin\zeta\sin\varepsilon\right] fbdbd\varepsilon dt \quad (7.26)$$

where $\zeta = (\chi + \pi)/2$, as shown in Figure 7.3b. With Equations 7.8 and 7.26, the drag force for diffuse scattering is obtained as

$$\mathbf{F}_d = m_r \int_{\mathbf{v}} g\mathbf{g}f d\mathbf{v} \int_0^\infty 2\pi\left(1 + \frac{1}{g}\sqrt{\frac{\pi k_B T}{2m_r}}\sin\frac{\chi}{2}\right) b\,db \quad (7.27)$$

It is noted that the integral $\int_0^\infty \sin(\chi/2)\,b\,db$ in Equation 7.27 is divergent due to the infinite upper limit. Physically, when the impact parameter b becomes sufficiently large, gas molecules will not collide with the particle and do not contribute to the drag force. Hence, a critical impact parameter b_0 needs to be found to be the upper integration limit in Equation 7.27. For gas–nanoparticle collisions, there are three types of scatterings when the impact parameter is changed, as illustrated in Figure 7.4 (Hirschfelder et al., 1954). It is seen that the definition of diffuse scattering is only valid when $b < b_0$ (Figure 7.4a). For $b = b_0$, a molecule will orbit about the particle, which is termed orbiting scattering (Figure 7.4b). When $b > b_0$, grazing collisions take place, as shown in Figure 7.4c, for which the collision scenario of specular scattering applies. Considering these collision scenarios, the drag force in Equation 7.27 is revised as (Li and Wang, 2003a)

$$\mathbf{F}_d = 2\pi m_r \int_{\mathbf{v}} g\mathbf{g}f d\mathbf{v}\left[\int_0^{b_0}\left(1 + \frac{1}{g}\sqrt{\frac{\pi k_B T}{2m_r}}\sin\frac{\chi}{2}\right) b\,db + \int_{b_0}^\infty (1-\cos\chi)bdb\right] \quad (7.28)$$

With the definition of the diffuse-scattering collision cross-section

$$Q_d(g) = 2\pi\left[\int_0^{b_0}\left(1 + \frac{1}{g}\sqrt{\frac{\pi k_B T}{2m_r}}\sin\frac{\chi}{2}\right) b\,db + \int_{b_0}^\infty (1-\cos\chi)bdb\right] \quad (7.29)$$

Equation 7.28 becomes

$$\mathbf{F}_d = -\frac{8}{3}\sqrt{2\pi m_r k_B T}NR^2\Omega_d^{(1,1)*}\mathbf{V} \quad (7.30)$$

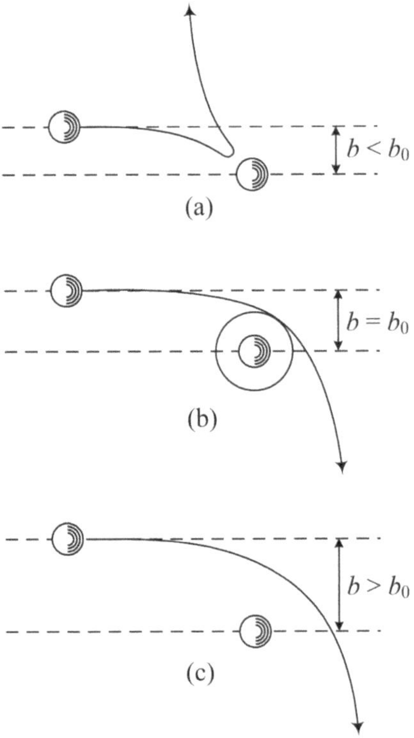

Figure 7.4 Gas–particle collision scenarios. (a) Collision with scattering angle $-\infty < \chi \leq \pi$ for $b < b_0$. (b) Orbiting scattering with $\chi \to -\infty$ at $b = b_0$. (c) Grazing scattering with $\chi < 0$ for $b > b_0$. (Reprinted with permission from Li and Wang, 2003a, Copyright (2003) by the American Physical Society.)

where $\Omega_{\mathrm{d}}^{(1,1)*}$ is the reduced collision integral for diffuse scattering defined through

$$\Omega_{\mathrm{d}}^{(1,q)*} = \frac{2\int_0^{\infty} \exp\left(-\gamma^2\right)\gamma^{2q+3} Q_{\mathrm{d}}\left(g\right) d\gamma}{\left(q+1\right)!\pi R^2} \tag{7.31}$$

For rigid-body collisions, it can be shown that $\Omega_{\mathrm{d}}^{(1,1)*} = 1 + \pi/8$, and Equation 7.30 is simplified to the Epstein solution for diffuse scattering, Equation 7.2 with $\delta = 1 + \pi/8$.

7.2.1.1.3 Parametrization The Chapman–Enskog theory assumes specular scattering for molecular collisions, which is a reasonable assumption and has been widely accepted. For microparticles, however, diffuse scattering appears to dominate over specular scattering, as indicated in Equation 7.3. Following the method in Equation 7.3, the drag force for nanoparticles can be written as follows:

$$\mathbf{F}_{\mathrm{D}} = -\frac{8}{3}\sqrt{2\pi m_{\mathrm{r}} k_{\mathrm{B}} T}\, N R^{2} \Omega_{\mathrm{avg}}^{(1,1)*} \mathbf{V} \tag{7.32}$$

where

$$\Omega_{\mathrm{avg}}^{(1,1)*} = \varphi \Omega_{\mathrm{d}}^{(1,1)*} + (1-\varphi)\Omega_{\mathrm{s}}^{(1,1)*} \tag{7.33}$$

In Equation 7.33, φ ($0 \le \varphi \le 1$) is termed the momentum accommodation function. For microparticles, it is treated as a constant, $\varphi \approx 0.9$, in practical applications. For nanoparticles, however, its value depends on the particle size, as will be shown later. The parametrization in Equation 7.33 accounts for the limiting case of diffuse scattering with $\varphi = 1$ and specular scattering with $\varphi = 0$. The collision integrals $\Omega_{\mathrm{s}}^{(1,1)*}$ and $\Omega_{\mathrm{d}}^{(1,1)*}$ can be calculated if the gas–particle interaction potential $\Phi(r)$ is given. In the next section, $\Omega_{\mathrm{s}}^{(1,1)*}$ and $\Omega_{\mathrm{d}}^{(1,1)*}$ will be computed for a representative potential function. The switch function φ will also be proposed on the basis of available experimental data.

7.2.1.1.4 Collision integrals To evaluate the collision integrals, $\Omega_{\mathrm{s}}^{(1,1)*}$ and $\Omega_{\mathrm{d}}^{(1,1)*}$, the potential function $\Phi(r)$ in Equation 7.9 is needed. The most popular potential is the LJ 12–6 potential function (see Chapter 3), which, however, is usually used for the van der Waals interactions of molecular species. For the interaction between a gas molecule and a nanoparticle, a potential constructed by integrating the LJ potential between the gas molecule and differential parts of the nanoparticle is developed by Rudyak and Krasnolutski (2001). This potential consists of a repulsive part $\Phi_9(r)$ and an attractive term $\Phi_3(r)$,

$$\Phi(r) = \Phi_9(r) - \Phi_3(r) \tag{7.34}$$

Here

$$\Phi_i(r) = C_i\left\{\left[(r-R)^{-i} - (r+R)^{-i}\right] - a_i\left[(r-R)^{-i+1} - (r+R)^{-i+1}\right]\right\}, \tag{7.35}$$

where $a_9 = 9/(8r)$, $a_3 = 3/(2r)$, $C_9 = 4\pi\varepsilon\sigma^{12}/(45\mathcal{V})$, and $C_3 = 2\pi\varepsilon\sigma^{6}/(3\mathcal{V})$ with $\mathcal{V}$ being the effective volume per one constituent atom of the particle. ε and σ in C_3 and C_9 are the LJ parameters, i.e., the binding energy and collision diameter, which, for gas–particle interactions, are estimated using the Loretz–Berthelot combining rules (see Chapter 3),

$$\sigma = (\sigma_g + \sigma_p)/2, \varepsilon = \sqrt{\varepsilon_g \varepsilon_p} \tag{7.36}$$

where the subscripts g and p denote gas molecules and the constituent atoms of the particle.

To be consistent with previous theories, the values of the reduced collision integrals $\Omega_s^{(1,1)*}$ and $\Omega_d^{(1,1)*}$ are computed by nondimensionalizing the potential function and relevant quantities. Thus, the potential in Equation 7.34 is given by

$$\begin{aligned}\Phi'(r') = \frac{2}{15}\sigma'^9 &\left\{\frac{1}{(r'-1)^9} - \frac{1}{(r'+1)^9} - \frac{9}{8r'}\left[\frac{1}{(r'-1)^8} - \frac{1}{(r'+1)^8}\right]\right\} \\ -\sigma'^3 &\left\{\frac{1}{(r'-1)^3} - \frac{1}{(r'+1)^3} - \frac{3}{2r'}\left[\frac{1}{(r'-1)^2} - \frac{1}{(r'+1)^2}\right]\right\},\end{aligned} \tag{7.37}$$

where $\Phi'(r') = \Phi(r')/\varepsilon'$ with $\varepsilon' = 2\pi\varepsilon\sigma^3/(3\mathcal{V})$, $r' = r/R$, and $\sigma' = \sigma/R$. The reduced potential $\Phi'(r')$ is a function of r' and σ'. If a reduced temperature $T^* = k_B T/\varepsilon'$ is used, the collision integrals become functions that depend on dimensionless quantities σ' and T^* only.

Figures 7.5 and 7.6 plot the values of $\Omega_s^{(1,1)*}$ and $\Omega_d^{(1,1)*}$, respectively, as functions of T^* and σ'. It is seen that $\Omega_s^{(1,1)*}$ and $\Omega_d^{(1,1)*}$ converge to their rigid-body limits, 1 and $1+\pi/8$, respectively, at high temperatures (or small binding energy ε) and/or large particle size (or small σ). However, for very small particles at a low temperature and/or a strong gas–particle binding energy ε, $\Omega_s^{(1,1)*}$ and $\Omega_d^{(1,1)*}$ greatly deviate from their rigid-body limits, indicating that the molecular interactions between the gas molecules and the particle are important. In this case, the Stokes–Cunningham formula, Equation 7.1, and Epstein's solution, Equation 7.2, are inaccurate. The values of $\Omega_s^{(1,1)*}$ and $\Omega_d^{(1,1)*}$ are also tabulated in Tables A.1 and A.2 as functions of T^* and σ' for $0.1 \le T^* \le 100$ and $0.01 \le \sigma' \le 0.6$, which cover most practical cases. For instance, $\sigma' = 0.6$ approximately corresponds to particles of 0.5 nm in radius (σ typically is around 3 Å). For practical applications, the values of $\Omega_d^{(1,1)*}$ and $\Omega_d^{(1,1)*}$ can also be evaluated using the following two equations:

$$\Omega_s^{(1,1)*} = 1 + \left[0.351 + \frac{1.34}{T^{*1/4}} + \frac{0.57}{T^{*1/2}}\right]\sigma' + \left[1.008 - \frac{2.744}{T^{*1/4}} + \frac{1.454}{T^{*1/2}}\right]\sigma'^2 \tag{7.38}$$

and

$$\Omega_d^{(1,1)*} = 1 + \frac{\pi}{8} + \left[1.248 + \frac{1.644}{T^{*1/4}} + \frac{1.45}{T^{*1/2}}\right]\sigma' + \left[2.374 - \frac{5.46}{T^{*1/4}} + \frac{1.695}{T^{*1/2}}\right]\sigma'^2 \tag{7.39}$$

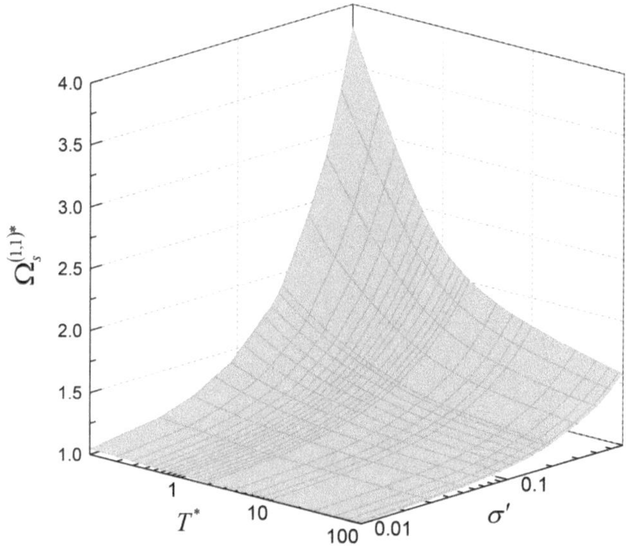

Figure 7.5 Reduced collision integral for specular scattering $\Omega_s^{(1,1)*}$ as a function of T^* and σ'. (Data are the same as those in Table A.1.)

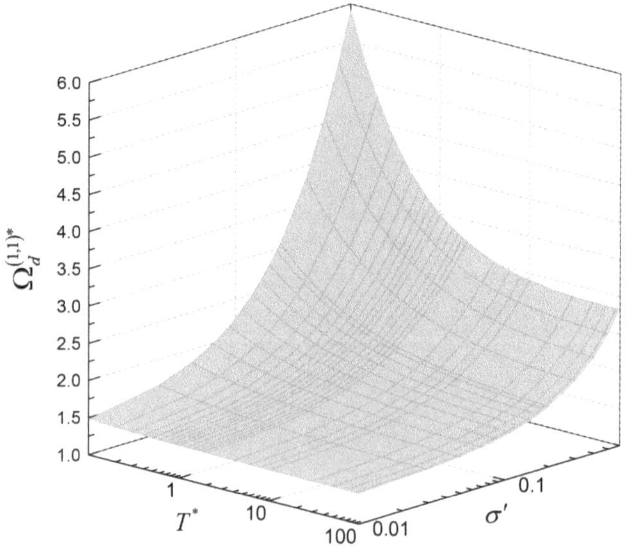

Figure 7.6 Reduced collision integral for diffuse scattering $\Omega_d^{(1,1)*}$ as a function of T^* and σ'. (Data are from Equation 7.39.)

These equations are obtained by fitting the data in Tables A.1 and A.2 using the method of least squares. The maximum fitting errors are 4.14% and 4.93% for $\Omega_s^{(1,1)*}$ and $\Omega_d^{(1,1)*}$, respectively, which occur at small T^* and large σ'.

It is noted that the numerical calculation of $\Omega_s^{(1,1)*}$ and $\Omega_s^{(1,1)*}$ uses intrinsic functions of a Fortran complier for performing integrations, especially the calculation of the angle of scattering χ. Different Fortran compliers may use different numerical libraries, which may make the values of $\Omega_s^{(1,1)*}$ and $\Omega_d^{(1,1)*}$ different, especially for the diffuse-scattering case because of large numerical errors when $\chi \to \infty$ at $b \approx b_0$ The values in Tables A.1 and A.2 are different from those in Li and Wang (2003b). The difference is well below 5% for large T^* and small σ'. Large discrepancy takes place only for a couple of values at very small T^* and large σ' ($T^* < 0.2$ and $\sigma' > 0.5$).

Nevertheless, it is worth mentioning that the values of $\Omega_s^{(1,1)*}$ and $\Omega_d^{(1,1)*}$ in Figures 7.5 and 7.6 and those in Tables A.1 and A.2 are based on the potential in Equation 7.34, whose accuracy for describing gas–particle interactions is unknown. If new potentials are developed in the future, $\Omega_s^{(1,1)*}$ and $\Omega_d^{(1,1)*}$ can be readily calculated using the above approach.

7.2.1.1.5 Momentum accommodation function Now, to evaluate the drag force using Equation 7.32, the only quantity that needs specification is the momentum accommodation function φ. Previous experiments and theoretical analysis show that φ should approach 0.9 for micro- and macroscale particles and assume 0 for particles at the molecular scale (Li and Wang, 2003a). Therefore, φ experiences a transition as the particle size is varied from the molecular scale to the microscale. As there is no a priori formula for φ, an empirical function is proposed for φ, which takes the form of

$$\varphi = \frac{1 + 0.9\mathrm{Kn}\left[1 - 1/\left(1 + R/\beta\right)^{\alpha}\right]}{1 + \mathrm{Kn}}, \tag{7.40}$$

where α and β are two parameters. β roughly presents the critical particle size, where the transition of φ takes place, and α determines how fast the transition is. It is clear that φ in Equation 7.40 reduces to unity for $\mathrm{Kn} \ll 1$ and approaches 0 for $\mathrm{Kn} \gg 1$. In Equation 7.40, the constant, 0.9, ensures that the gas–particle collisions assume the combination of 90% diffuse and 10% specular scatterings for large particles, in agreement with Millikan's observation (Millikan, 1923).

By fitting experimental data for Ag and Cu_2O nanoparticles (Scheibel and Porstendörfer, 1983; Kuga et al., 2001; Rudyak et al., 2002) using Equation 7.32 with φ given in Equation 7.40, the values of α and β are obtained as $\alpha = 15$ and $\beta = 2.5$ nm (Li and Wang, 2003b), and Equation 7.40 can be rewritten as follows:

$$\varphi = \frac{1+0.9\text{Kn}\left[1-1/\left(1+R/2.5\right)^{15}\right]}{1+\text{Kn}} \tag{7.41}$$

where the unit of the particle radius R is nanometer. Figure 7.7 plots φ as a function of particle radius. It can be seen that the transition from specular to diffuse scattering occurs at $R \sim 2.5$ nm and the transition is quite fast. Diffuse scattering appears to be popular for particles of radius larger than 4 nm, while specular collisions dominate for particles with radii below 2 nm.

7.2.1.2 Diffusion coefficient As discussed in Section 1.6, Chapter 1, the diffusion coefficient of a particle D_p can be related to the drag coefficient through Equation 1.18, which, for convenience, is reproduced here

$$D_p = \frac{k_B T}{\alpha} \tag{7.42}$$

It must be noted that this relationship is valid if the drag force is a linear function of velocity, i.e., the drag force can be written as $F_D = -\alpha V$ (Equation 1.11b). For nanoparticles, Equation 7.32 gives the drag force, which satisfies the linear relationship. Hence, the diffusion coefficient of nanoparticles can be expressed as follows:

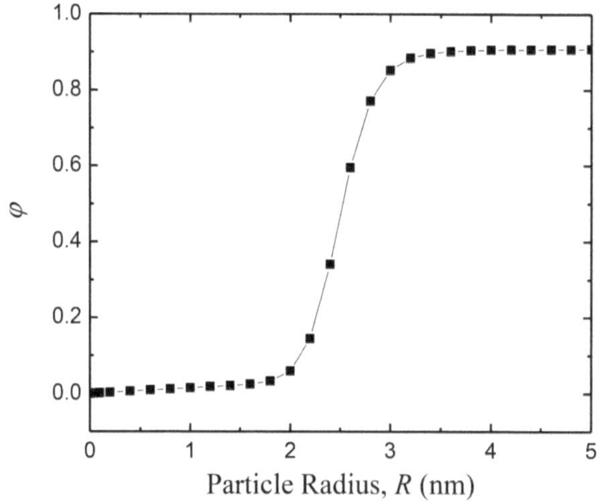

Figure 7.7 Dependence of momentum accommodation function on particle radius, given by Equation 7.41. Mean free path $\lambda = 65$ nm is used for the calculation of Kn.

$$D_{\mathrm{p}} = \frac{3}{8}\sqrt{\frac{k_{\mathrm{B}}T}{2\pi m_{\mathrm{r}}}}\frac{1}{NR^{2}\Omega_{\mathrm{avg}}^{(1,1)*}} \tag{7.43}$$

In Equation 7.43, if the particle radius and the collision integral $\Omega_{\mathrm{avg}}^{(1,1)*}$ are replaced by σ and $\Omega^{(1,1)*}$, the diffusion coefficient is reduced to the Chapman–Enskog theory for molecular diffusion, which can be obtained by combining Equations 7.4 and 7.42.

7.2.2 Continuum regime

7.2.2.1 Drag force The drag force on nanoparticles moving in a liquid is quite different from that in the free molecular regime. In the free molecular regime, the drag force is the consequence of a large number of collisions between gas molecules and a nanoparticle. Therefore, the K is used to characterize the collision dynamics. In this section, the "continuum regime" is referred to as the cases of nanoparticles moving in a liquid, where nanoparticles interact with many surrounding fluid molecules simultaneously. In this situation, the fluid–particle interactions are usually multi-body collisions and the fluid–fluid interactions around the particle also play an important role. In this regime, the drag force is mainly caused by viscous effects and molecular interactions, and the collision models employed in the free molecular regime cannot precisely catch the mechanism of drag force generation.

The study of the drag force on nanoparticles in a liquid is challenging. Experimental measurement is difficult because the size of nanoparticles is small and the diffusion of nanoparticles can cause large errors in particle tracking. For MD simulations, there are also some issues. The most critical one is that particles of a few nanometers in diameter are not spherical, as will be shown in the next section. To consider the shape effect on the drag force, nanoparticles cannot be fixed and the rotation of the particles has to be taken into account. This can make the drag force calculation time-consuming. In this case, it is easier to study the diffusion of nanoparticles than the drag force. This is why much attention has been paid to nanoparticle diffusion instead of drag force calculation in the literature.

7.2.2.2 Diffusion coefficient For large particles, the diffusion coefficient is given by the Stokes–Einstein relation,

$$D_{\mathrm{p}} = \frac{k_{\mathrm{B}}T}{6\pi\mu R} \tag{7.44}$$

As this relation is based on the Stokes law, which does not consider the molecular interactions, its applicability for nanoparticles in liquids is questionable.

The validity of the Stokes–Einstein relation has been an important scientific problem and attracted great attention in the past decades. Both experimental and numerical studies have been carried out to examine the validity of the Stokes–Einstein relation. The techniques of X-ray photon correlation spectroscopy and dark-field microscopy have shown that Equation 7.44 works well for nanoparticles of diameters larger than 140 nm (Lurio et al., 2000; Koenig et al., 2009). For smaller particles, however, particle tracking through experiments becomes challenging and the validation of the Stokes–Einstein relation mainly relies on MD simulations.

Actually, the importance of molecular interactions in nanoparticle diffusion can be qualitatively revealed by looking into the limiting cases: macroscale particle transport and molecular diffusion. For large particles, the Stokes–Einstein relation is valid and the diffusion coefficient depends on the particle size and fluid viscosity at a given temperature. For molecular diffusion, however, molecule size and fluid viscosity are insufficient to determine the diffusion coefficient. For instance, the sizes of Kr and Ar molecules are very close (the size of a Kr molecule is about 5% larger than that of an Ar molecule). According to the Stokes–Einstein relation, their diffusion coefficients should be similar in the same fluid. However, experiments show that the diffusion coefficient of Kr is about three times larger than that of Ar in olive oil (Osburn et al., 1969). This suggests that other than the fluid viscosity and particle size, molecular interactions play a critical role when the particle size is sufficiently small.

The Stokes–Einstein relation for large particles and experimental evidence of molecular diffusion indicate that there is a critical particle size, below which microscopic parameters become important and the Stokes–Einstein relation breaks down, as illustrated in Figure 7.8. Previous experiments verifying the validity of the Stokes–Einstein relation for 140-nm or larger particles (Lurio

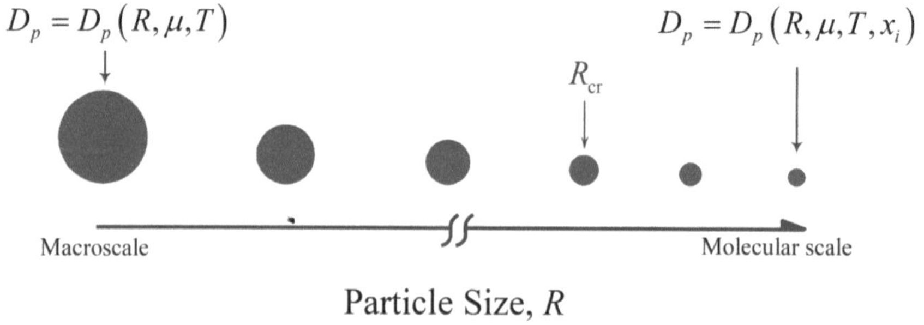

Figure 7.8 Schematic of parameter dependence of particle diffusion coefficient. x_i in $D_p = D_p(R, \mu, T, x_i)$ represents microscale parameters, which are not considered in the Stoke-Einstein relation. R_{cr} denotes the critical particle radius, below which the Stokes–Einstein relation falls apart.

et al., 2000; Koenig et al., 2009) suggest that this critical particle size lies at the nanoscale.

Due to the technical difficulty, it is hard to find the critical particle size through experiments. Molecular dynamics simulations offer a convenient way to study the diffusion coefficient of small particles. Figure 7.9 shows a cubic MD simulation system, where a solid particle is diffusing in a liquid (Li, 2009). The liquid is modeled as liquid Ar by the LJ potential with parameters $\sigma = 3.395$ Å and $\varepsilon = 116.8$ K. The particle is generated by keeping all the atoms of an fcc structure in a sphere of radius R (the lattice constant of the fcc structure is equal to 4.08 Å). The thermal motion of the atoms in the particle is considered using the tight-binding potential function (see Chapter 3). The interactions between the liquid molecules and the particle atoms are also described by the LJ potential. To study the effect of particle material on the diffusion coefficient, the liquid–particle binding energy ε_{LP} is artificially varied (σ_{LP} is set as a constant equal to 3.395 Å). The mass of the particle atoms is chosen to be the same as that of Ar molecules. The temperature of the system is 94.4 K to ensure that Ar is in liquid state. As PBCs are applied in all the directions, the side length of the simulation systems is set to be five to ten times the particle diameter to reduce the effect of particle–particle interactions on the diffusivity.

To compare with the Stokes–Einstein relation, the viscosity of the liquid is computed by the Green–Kubo relation, i.e., Equation 3.40, using a similar MD system

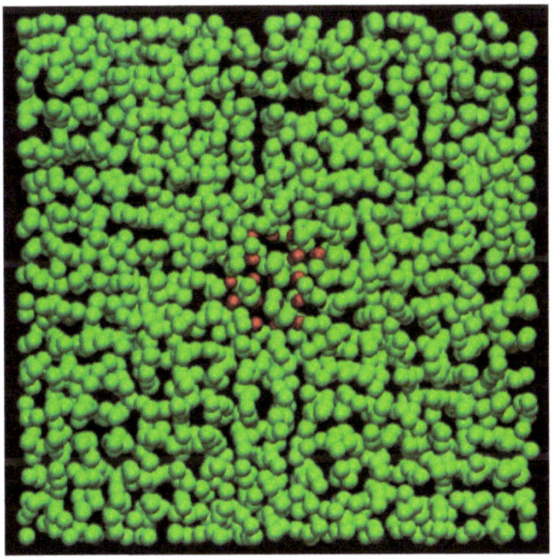

Figure 7.9 An MD simulation system for studying particle diffusion. Green and red spheres are liquid and particle atoms, respectively.

without the particle. It is obtained as $\mu = 2.13 \pm 0.12 \times 10^{-4}$ Pa · s at $T = 94.4$ K, which is in good agreement with experiments and other MD simulations.

Nanoparticles may not be spherical if their sizes are too small. Figure 7.10 shows the molecular shapes of some nanoparticles. It can be seen that a nanoparticle containing few atoms is just a cluster and cannot be treated as a sphere (e.g., the last particle in Figure 7.10). For very small particles/clusters, the hydrodynamic radius R_H is usually used to characterize their sizes. The hydrodynamic radius R_H is defined as follows:

$$1/R_H = \left\langle \sum_{i \neq j} 1/r_{ij} \right\rangle \Big/ N_p^{\,2}, \tag{7.45}$$

where r_{ij} is the distance between a pair of particle atoms and N_p is the total number of atoms in the particle. In Figure 7.10, R_H for various particles/clusters is also given, which is smaller than the radius used to generate the particles.

The MSDs of particles of various sizes are calculated and the diffusion coefficients of the particles are obtained using Equation 1.9. Figure 7.11 plots the particle diffusion coefficient as a function of $1/R_H$. It can be seen that D_p increases and becomes sensitive to the liquid–particle interactions as the particle size decreases. For relatively large particles (small $1/R_H$ in Figure 7.11), liquid–particle interactions tend to be unimportant and the diffusion coefficients approach the predictions of the Stokes–Einstein relation regardless of the values of the liquid–particle binding energy ε_{LP}. As numerical simulations for large particles are

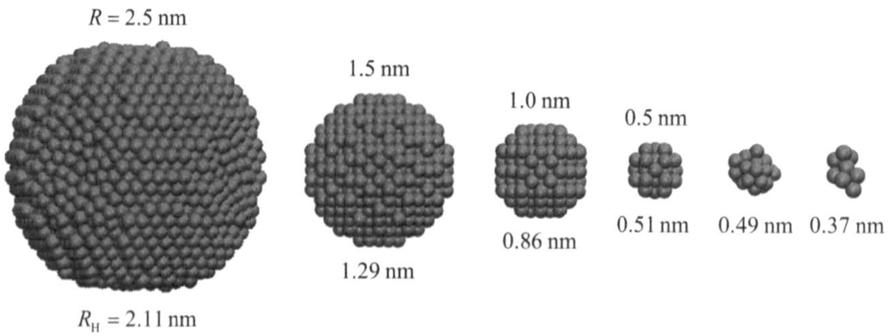

Figure 7.10 Shapes and sizes of particles containing different numbers of atoms. The sizes above the particles are the radii used to construct the particles and those below the particles are the hydrodynamic radii.

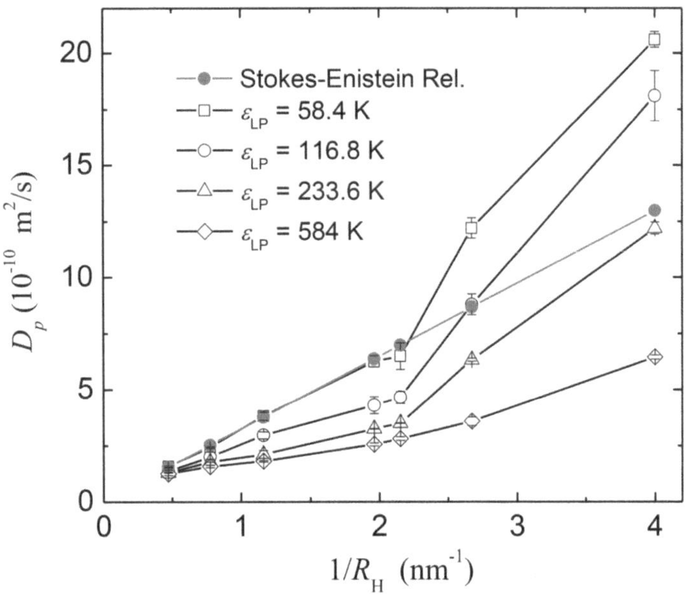

Figure 7.11 Diffusion coefficient of nanoparticles as a function of $1/R_H$ for different liquid–particle binding energies ε_{LP}. (Reprinted with permission from Li, 2009, Copyright (2009) by the American Physical Society.)

time-consuming, the largest particle size studied is $R_H = 2.11$ nm. The critical particle size has not been found. Nevertheless, the results in Figure 7.11 suggest that the critical particle size is of a few nanometers (Li, 2009). For particles of tens of nanometers in diameter, the Stokes law and the Stokes–Einstein relation are expected to work well.

In Figure 7.11, it is found that the liquid–particle interaction plays an important role for small nanoparticles. One reason is because the liquid molecules around a small nanoparticle can greatly affect the size of the particle. For a large ε_{LP} value, a layer of fluid molecules are adsorbed by the particle, as shown in Figure 7.12, which increases the particle size and reduces the particle diffusivity. However, such a change in particle size cannot fully explain the discrepancy between simulations and the Stokes–Einstein relation (Li, 2009). Another reason could be the fluid property change around the particle arising from the van der Waals interactions (Zhang et al., 2016). At strong fluid–particle interactions, the density and viscosity of the fluid will increase, which leads to a large stress and drag force on the particle. As a consequence, the diffusion coefficient decreases. Nevertheless, quantitative studies of these effects are needed to provide a clear picture about nanoparticle diffusion in liquids.

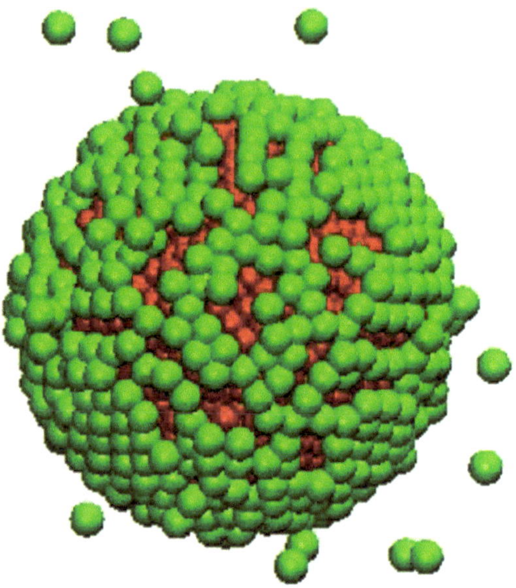

Figure 7.12 Adsorption of liquid molecules on the surface of a nanoparticle at $\varepsilon_{LP} = 584$ K. (Reprinted with permission from Li, 2009, Copyright (2009) by the American Physical Society.)

7.3 Drag force and diffusion coefficient of nanocylinders

7.3.1 Free molecular regime

7.3.1.1 Drag force Transport properties of nanocylinders are important in many applications, including the synthesis and separation of CNTs, CNT-based sensing, and the dynamics of long-chain molecules. In the free molecular regime, $\mathrm{Kn} = \lambda/R \gg 1$, where R is the radius of a cylinder, the drag force can be obtained using the theoretical approach in Section 7.2.1.1 (Li, 2005).

Let us consider a nanocylinder of length L and diameter D moving in a stationary gas with a velocity $\mathbf{V}$. The thermal velocity of gas molecules is denoted by $\mathbf{v}$. The velocities of the cylinder and gas molecules after collisions are referred to as $\mathbf{V}'$ and $\mathbf{v}'$. For convenience, the coordinate system is attached to the cylinder with the origin O locating at the mass center of the cylinder. The axis of the cylinder is the y-axis and the gas velocity, which is equal to $-\mathbf{V}$ in this system, is collinear with the z-axis, as shown in Figure 7.13a (Li, 2005). The velocity of the gas molecules is then given by $\mathbf{g} = \mathbf{v} - \mathbf{V}$ before collisions and $\mathbf{g}' = \mathbf{v}' - \mathbf{V}'$ after collisions. Another coordinate system, ($\mathbf{e}_1$, $\mathbf{e}_2$, $\mathbf{e}_3$), is constructed, where $\mathbf{g}$ is parallel to $\mathbf{e}_3$ and $\mathbf{e}_1$ is collinear with the axis of the cylinder, as shown in Figure 7.13b. The polar angles between $\mathbf{g}$ and $\mathbf{V}$ are referred to as ϕ, θ, and ψ, as demonstrated in Figure 7.13c. The effective length of the cylinder, L_e, in ($\mathbf{e}_1$, $\mathbf{e}_2$, $\mathbf{e}_3$) coordinates is then given by $L_e = L\sin\psi = L\sqrt{\cos^2\phi + \sin^2\phi\cos^2\theta}$ (Figure 7.13b).

Similar to the derivations for the drag force on nanoparticles, here, gas molecules in a rectangular differential element are considered to calculate the momentum change. As shown in Figure 7.13b, the differential element is of impact parameter b, height db, width dL, and length $L'' = gdt$. All the gas molecules in this element have velocity $\mathbf{g}$ and will collide with the cylinder in a time interval dt. The number of gas molecules in the element is given as follows:

$$n = gfdbdLdt \tag{7.46}$$

where f is the velocity distribution function of the gas molecules, given by Equation 7.7. The total momentum $\delta\mathbf{p}$ of these gas molecules is $\delta\mathbf{p} = m_r n\mathbf{g}$ (Equation 7.8), where m_r is the reduced mass of a gas molecule and the cylinder.

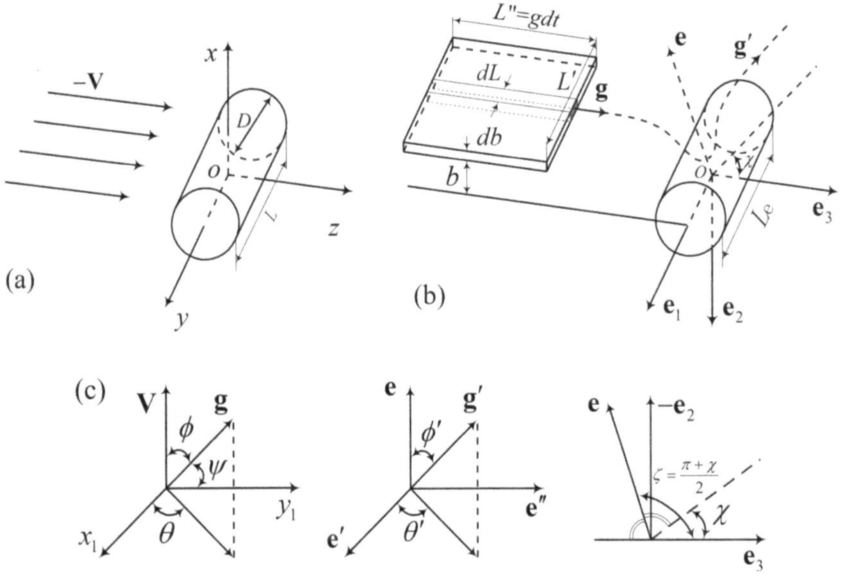

Figure 7.13 Gas-nanocylinder collision model. (a) and (b) Coordinate systems. (c) Relationships among various vectors. (Li, 2005, reprinted by courtesy of the author.)

7.3.1.1.1 Specular scattering For specular scattering, the scattering angle χ is the same as that given in Equation 7.9. Given that $g = g'$ and $\mathbf{g}' = \mathbf{g}\cos\chi - g\sin\chi \cdot \mathbf{e}_2$, the momentum of gas molecules after collisions reads

$$\delta\mathbf{p}' = m_r n(\mathbf{g}\cos\chi - g\sin\chi \cdot \mathbf{e}_2) \tag{7.47}$$

According to Equation 7.11, the differential drag force due to the collisions between the gas molecules in the element with the cylinder is given by

$$d\mathbf{F}_{s,\perp} = \frac{\delta\mathbf{p} - \delta\mathbf{p}'}{dt} = m_r n\left[\mathbf{g}(1-\cos\chi) + g\sin\chi \cdot \mathbf{e}_2\right] \tag{7.48}$$

where the subscript "⊥" denotes that the flow direction is perpendicular to the cylinder. Integrating Equation 7.48 (term $g\sin\chi \cdot \mathbf{e}_2$ vanishes upon integration over $-\infty < b < \infty$) yields

$$\mathbf{F}_{s,\perp} = \frac{m_r N}{\left(2\pi m k_B T/m_r\right)^{3/2}} \int_{\mathbf{v}} g\mathbf{g}\exp\left(-\frac{v^2}{2k_B T/m_r}\right) d\mathbf{v} \int_{-\infty}^{\infty} (1-\cos\chi)\, db \int_0^{L'} dL \tag{7.49}$$

where L' is the length in $\mathbf{e}_1$ direction, beyond which gas molecules do not collide with the cylinder. It is noted that L' should be slightly larger than L_e due to the molecular interactions between the gas molecules and the cylinder and the ratio, L'/L_e, should approach unity when the aspect ratio of the cylinder $L/D \gg 1$.

As the drift velocity $\mathbf{V}$ is much smaller than the thermal velocity of the gas molecules $\mathbf{v}$, it is reasonable to assume that $d\mathbf{v} \approx d\mathbf{g}$, which leads to

$$\mathbf{F}_{s,\perp} = \frac{m_r N}{\left(2\pi k_B T/m_r\right)^{3/2}} \int_{g} g\mathbf{g}\exp\left(-\frac{g^2 + V^2 + 2gV\cos\phi}{2k_B T/m_r}\right) \times\sqrt{\cos^2\phi + \sin^2\phi\cos^2\theta}\, d\mathbf{g} \int_{-\infty}^{\infty} L(1-\cos\chi)\, db \tag{7.50}$$

If the collision cross-section $Q_s(g)$ is defined as

$$Q_s(g) = L\int_{-\infty}^{\infty} (1-\cos\chi)\, db \tag{7.51}$$

and by using Equation 7.16, the drag force is obtained as

$$\mathbf{F}_{s,\perp} = -\frac{3}{4}\sqrt{2\pi m_r k_B T}\, NLD\Omega_s^{(1,1)*}\mathbf{V} \tag{7.52}$$

where $\Omega_s^{(1,1)*}$ is the reduced collision integral for specular scattering given by

$$\Omega_s^{(1,1)*} = \frac{\int_0^\infty \gamma^5 \exp(-\gamma^2) Q_s(g)\, d\gamma}{LD} \tag{7.53}$$

where $\gamma = g/\sqrt{2k_B T/m_r}$.

For rigid-body collisions, it can be shown that $\Omega_s^{(1,1)*} = 4/3$ and the drag force is reduced to

$$\mathbf{F}_{s,\perp}^{\text{rigid}} = -\sqrt{2\pi m_r k_B T}\, NLD\mathbf{V}. \tag{7.54}$$

If $\mathbf{V}$ is parallel with the cylinder axis, $L_e = L\sin\phi$ and the drag force is obtained as

$$\mathbf{F}_{s,\parallel} = -\frac{1}{2}\sqrt{2\pi m_r k_B T}\, NLD\Omega_s^{(1,1)*}\mathbf{V} \tag{7.55}$$

$F_{s,\perp}$ and $F_{s,\parallel}$, with $F_{s,\perp} = 1.5F_{s,\parallel}$, set the bounds for any arbitrary orientation of the cylinder.

It is noted that a more comprehensive analysis about the drag force on nanocylinders has been performed recently (Liu et al., 2016), where the definition of $\Omega_s^{(1,1)*}$ is different from Equation 7.53, but the drag force is the same as Equation 7.52.

7.3.1.1.2 Diffuse scattering For diffuse scattering, the velocity distribution of the reflected molecules is given by Equation 7.23. In this case, the derivation of the drag force can be simplified by recognizing that the mean angle of reflection is in the **e** direction (Figure 7.13b) because the net momentum transfer is zero in the direction perpendicular to **e**. Given the angles in Figure 7.13c, the momentum of the reflected molecules in the element can be written as follows:

$$\delta\mathbf{p}' = dbdLdt\int_{\mathbf{g}'} m_r g\mathbf{g}\cos\phi' f' d\mathbf{g}' = \frac{\sqrt{2\pi m_r k_B T}}{2} gf\mathbf{e}dbdLdt \tag{7.56}$$

If **V** is perpendicular to the cylinder axis, the differential drag force is given by

$$d\mathbf{F}_{\mathrm{d},\perp} = \left[m_{\mathrm{r}} g\mathbf{g} - \frac{\sqrt{2\pi m_{\mathrm{r}} k_{\mathrm{B}} T}}{2} g\left(\cos\zeta \cdot \mathbf{e}_3 - \sin\zeta \cdot \mathbf{e}_2\right) \right] f db dL \tag{7.57}$$

Considering that $\sin\zeta \cdot \mathbf{e}_2$ is equal to zero upon the integration over b, the drag force is obtained as

$$\begin{aligned} \mathbf{F}_{\mathrm{d},\perp} &= \frac{m_{\mathrm{r}} N}{\left(2\pi k_{\mathrm{B}} T/m_{\mathrm{r}}\right)^{3/2}} \int_{\mathbf{g}} g\mathbf{g} \exp\left(-\frac{g^2 + V^2 + 2gV\cos\phi}{2k_{\mathrm{B}}T/m_{\mathrm{r}}} \right) d\mathbf{g} \\ &\times \int_{-\infty}^{\infty} \left(1 + \frac{1}{g}\sqrt{\frac{\pi k_{\mathrm{B}} T}{2m_{\mathrm{r}}}} \sin\frac{\chi}{2} \right) db \int_0^L dL \end{aligned} \tag{7.58}$$

Similar to gas–nanoparticle collisions, there is also a critical impact parameter b_0, beyond which gas molecules do not physically collide with the cylinder (Figure 7.4c). Then, the collision cross-section for diffuse scattering, $Q_{\mathrm{d}}(g)$, can be defined as follows:

$$Q_{\mathrm{d}}(g) = L\left[\int_{-b_0}^{b_0} \left(1 + \frac{1}{g}\sqrt{\frac{\pi k_{\mathrm{B}} T}{2m_{\mathrm{r}}}} \sin\frac{\chi}{2} \right) db + \int_{b_0}^{\infty} (1 - \cos\chi) db + \int_{-\infty}^{-b_0} (1 - \cos\chi) db \right] \tag{7.59}$$

and the drag force is determined as

$$\mathbf{F}_{\mathrm{d},\perp} = -\frac{3}{4}\sqrt{2\pi m_{\mathrm{r}} k_{\mathrm{B}} T} NLD\Omega_{\mathrm{d}}^{(1,1)*} \mathbf{V} \tag{7.60}$$

where $\Omega_{\mathrm{d}}^{(1,1)*}$, the reduced collision integral for diffuse scattering, is given by

$$\Omega_{\mathrm{d}}^{(1,1)*} = \frac{\int_0^{\infty} \gamma^5 \exp(-\gamma^2) Q_{\mathrm{d}}(g) d\gamma}{LD} \tag{7.61}$$

For rigid-body collisions, it can be shown that $\Omega_{\mathrm{d}}^{(1,1)*} = \left(1+3\pi^2/64\right)LD$ and the corresponding drag force reads

$$\mathbf{F}_{\mathrm{d},\perp}^{\mathrm{rigid}} = -\frac{192+9\pi^2}{256}\sqrt{2\pi m_{\mathrm{r}} k_{\mathrm{B}} T}\, NLD\mathbf{V} \tag{7.62}$$

If $\mathbf{V}$ is parallel to the cylinder axis, the drag force is derived as

$$\mathbf{F}_{\mathrm{d},\parallel} = -\frac{1}{2}\sqrt{2\pi m_{\mathrm{r}} k_{\mathrm{B}} T}\, NLD\Omega_{\mathrm{d}}^{(1,1)*}\mathbf{V} \tag{7.63}$$

and the solution for rigid-body collisions is given by

$$\mathbf{F}_{\mathrm{d},\parallel}^{\mathrm{rigid}} = -\frac{64+3\pi^2}{128}\sqrt{2\pi m_{\mathrm{r}} k_{\mathrm{B}} T}\, NLD\mathbf{V} \tag{7.64}$$

Equations 7.62 and 7.64 are slightly different from Liu's results (2016) because the momentum transfer in the plane perpendicular to $\mathbf{V}$ is considered in Liu's work (Liu et al., 2016). For $\mathbf{F}_{\mathrm{d},\perp}^{\mathrm{rigid}}$, the difference is less than 5%, while for $\mathbf{F}_{\mathrm{d},\parallel}^{\mathrm{rigid}}$, Equation 7.64 is about 45% larger than that in Liu's work.

7.3.1.1.3 Drag force on single-walled CNTs To calculate the drag force on nanocylinders, the reduced collision integrals, $\Omega_{\mathrm{s}}^{(1,1)*}$ and $\Omega_{\mathrm{d}}^{(1,1)*}$, need to be determined. As $\Omega_{\mathrm{s}}^{(1,1)*}$ and $\Omega_{\mathrm{d}}^{(1,1)*}$depend on molecular interactions, a gas-nanocylinder interaction potential $\Phi(r)$ is required to obtain the drag force.

In the literature, the only available gas–cylinder interaction potential is for the interactions between gas molecules and single-walled CNTs (SWCNTs). The potential is obtained by summing the LJ potentials between a gas molecule and carbon atoms of a CNT through MD simulations (Wong et al., 2012),

$$\Phi_i(r) = \sum_{j=1}^{N_{\mathrm{O}}} 4\varepsilon\left[\left(\sigma/r_{ij}\right)^{12} - \left(\sigma/r_{ij}\right)^{6}\right] \tag{7.65}$$

where the subscripts "i" and "j" denote gas molecule i and carbon atom j, r_{ij} is the distance between gas molecule i and carbon atom j, r in $\Phi_i(r)$ is the distance between gas molecule i and the mass center of the CNT, and N_{o} is the total number of carbon atoms in the CNT. The interaction model is demonstrated in Figure 7.14.

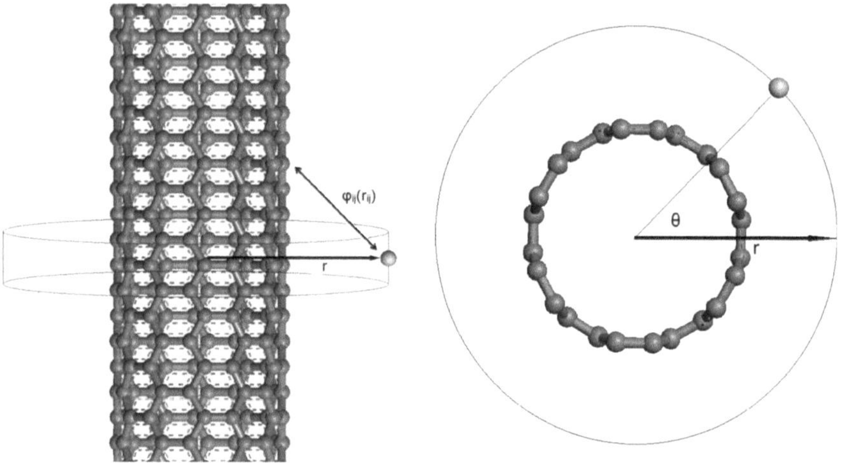

Figure 7.14 Interaction model for the determination of gas–CNT interaction potential. Left panel: Side view. Right panel: Top view. (Reprinted with permission from Wong et al., 2012, Copyright © American Scientific Publishers.)

By considering $M = 1000$ random locations on a portion of a cylindrical surface with radius r, as shown in the left panel of Figure 7.14, the potential is determined as

$$\Phi(r) = \frac{1}{M}\sum_{i=1}^{M}\Phi_i(r) \tag{7.66}$$

By changing the CNT diameter D, a universal potential in a dimensionless format is obtained as follows:

$$\Phi^*\left(r^*\right) = a_1 r^{*n_1} - a_2 r^{*n_2} \tag{7.67}$$

where $a_1 = 0.635$, $a_2 = 1.63$, $n_1 = -10.1$, and $n_2 = -4.1$ are empirical constants. In Equation 7.66, $r^* = (r - R)/\sigma_{gc}$ with R being the radius of the CNT and $\Phi^*(r) = \Phi(r)/\varepsilon_{\text{gas-CNT}}$. Herein, ε and σ are the LJ parameters, i.e., the binding energy and collision diameter. However, σ_{gc} stands for the interaction between a gas molecule and a carbon atom, while $\varepsilon_{\text{gas-CNT}}$ denotes the interaction between a gas molecule and the CNT (all carbon atoms are included). An empirical expression for $\varepsilon_{\text{gas-CNT}}$ have been proposed, which is given by (Wong et al., 2012)

$$\varepsilon_{\text{gas-CNT}} = a_0\left(\frac{R}{\sigma_{gc}}\right)^m n_C\sigma_{gc}\varepsilon_{gc}R \tag{7.68}$$

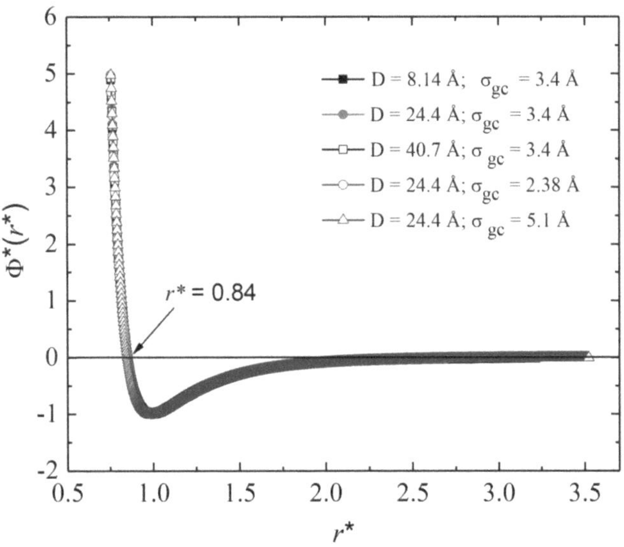

Figure 7.15 Reduced potential $\Phi^*(r^*)$ as a function of r*. (Adapted with permission from Wong et al., 2012, Copyright © American Scientific Publishers.)

where $a_0 = 5.81$ and $m = -0.86$ are constants, and $n_C = 0.381$ Å^{-2} is the number of carbon atoms per unit surface area of CNTs. Equation 7.68 is valid for all the values of ε_{gc} and a wide ranges of D (4 – 40 Å) and σ_{gc} (2.38 – 5.1 Å). With Equation 7.68, the gas–CNT potential $\Phi(r^*)$ can be obtained through $\Phi(r^*) = \varepsilon_{\text{gas-CNT}} \cdot \Phi^*(r^*)$ if the LJ parameters, ε_{gc} and σ_{gc}, and the CNT diameter are given. Figure 7.15 shows $\Phi^*(r^*)$ for different CNT sizes and σ_{gc} values. It can be seen that all the potentials collapse on a master curve, which is given by Equation 7.67.

Using the above potential, the reduced collision integral $\Omega_s^{(1,1)*}$ and $\Omega_d^{(1,1)*}$ for gas–SWCNT interactions are calculated as functions of the reduced temperature $T^* = k_B T/\varepsilon_{\text{gas-CNT}}$ and reduced collision diameter $\sigma^* = \sigma_{gc}/R$. Figure 7.16a and b plots $\Omega_s^{(1,1)*}$ and $\Omega_d^{(1,1)*}$, respectively, at different values of T^* and σ^*. It is seen that they converge to the rigid-body collision limits, $\Omega_{s,\text{rigid}}^{(1,1)*} = 4/3$ and $\Omega_{d,\text{rigid}}^{(1,1)*} = 1 + 3\pi^2/64$ for specular and diffuse scatterings, as $T^* \to \infty$ and $\sigma^* \to 0$. For $0.1 < T^* < 100$ and $0.1 < \sigma^* < 2$, which cover a wide range of CNT sizes and gas–CNT interaction strengths, the values of $\Omega_s^{(1,1)*}$ and $\Omega_d^{(1,1)*}$ can be approximately evaluated by (Wong et al., 2012)

$$\Omega_s^{(1,1)*} = \frac{4}{3} + \left(0.422 + \frac{0.852}{T^{*1/4}} + \frac{0.169}{T^{*1/2}}\right)\sigma^* + \left(0.0935 - \frac{0.288}{T^{*1/4}} + \frac{0.150}{T^{*1/2}}\right)\sigma^{*2} \quad (7.69)$$

$$\Omega_{\mathrm{d}}^{(1,1)*} = 1 + \frac{3\pi^2}{64} + \left(0.867 + \frac{0.661}{T^{*1/4}} + \frac{0.412}{T^{*1/2}}\right)\sigma^{*} + \left(0.133 - \frac{0.406}{T^{*1/4}} + \frac{0.156}{T^{*1/2}}\right)\sigma^{*2} \quad (7.70)$$

The largest fitting error of these two expressions to the data in Figure 7.16 is about 5% for small T^* and large σ^*. For most values of σ^* and T^*, the errors are below 1%.

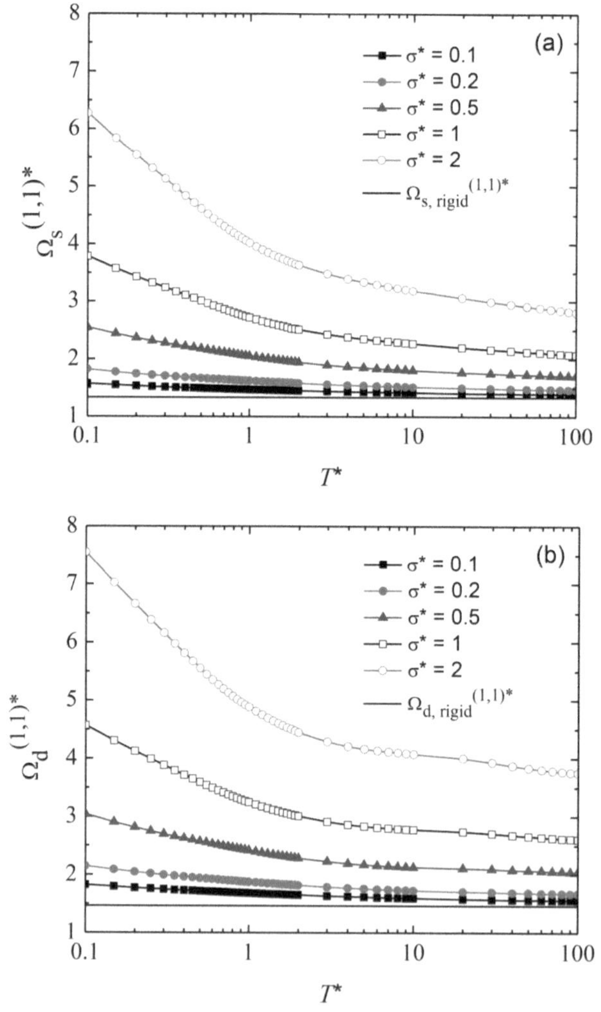

Figure 7.16 Reduced collision integrals for gas–SWCNT interactions. (a) $\Omega_{\mathrm{s}}^{(1,1)*}$ for specular scattering. (b) $\Omega_{\mathrm{d}}^{(1,1)*}$ for diffuse scattering. (Reprinted with permission from Wong et al., 2012, Copyright © American Scientific Publishers.)

Molecular dynamics simulations have been conducted to directly calculate the drag force on SWCNTs (Wong et al., 2012). In the simulations, an SWCNT of $D = 2.034\,\text{nm}$ and $L = 10\,\text{nm}$ is placed and fixed at the center of the simulation system, which is of 40, 20, and 10 nm in the x-, y-, and z-directions, respectively (the axis of the CNT is in the z-direction). The system is filled with gas molecules with the number density equal to $0.065/\text{nm}^3$, which corresponds to a mean free path of $\lambda = 28.8$ nm (Kn = 14.16). The gas is moving in the positive x-direction under an external force on each gas molecule. The LJ parameters for the gas molecules are $\varepsilon_{gg} = 0.0098$ eV and $\sigma_{gg} = 3.47$ Å, and those for gas–carbon atom interactions are set as $\varepsilon_{gc} = 0.0053$ eV and $\sigma_{gc} = 3.485$ Å. The mean force on the CNT is computed as the drag force. Figure 7.17 shows the drag force as the flow velocity is varied at two different values of T^*. It can be seen that the MD results agree well with Equations 7.52 and 7.60 as they fall between the theoretical predictions for specular and diffuse scatterings.

7.3.1.1.4 Parametrization Similar to the treatment for nanoparticles, the drag force on nanocylinders should include both specular and diffuse scatterings. If φ is used to denote the momentum accommodation function, the drag forces for **V** being perpendicular and parallel to the cylinder axis are written as follows:

$$\mathbf{F}_{\perp} = -\frac{3}{4}\sqrt{2\pi m_r k_B T}\,NLD\Omega_{avg}^{(1,1)*}\mathbf{V} \text{ and } \mathbf{F}_{\parallel} = -\frac{1}{2}\sqrt{2\pi m_r k_B T}\,NLD\Omega_{avg}^{(1,1)*}\mathbf{V} \tag{7.71}$$

where $\Omega_{avg}^{(1,1)*} = \varphi\Omega_d^{(1,1)*} + (1-\varphi)\Omega_s^{(1,1)*}$. For nanoparticles, φ is well understood and is shown dependent on particle size. For nanocylinders, unfortunately, φ has not been studied. Figure 7.17 shows that φ assumes a value between 0 and 1. A comparison between theory and experiments of 2-nm-diameter CNTs shows that $\varphi = 0$ describes gas–CNT collisions well (Liu et al., 2016). However, how it depends on the size of nanocylinders is unknown.

7.3.1.2 Diffusion coefficient For spherical particles, the diffusion coefficient can be determined using Equation 1.18 if the drag force is known. For nanocylinders, however, it is not straightforward because the drag force depends on the orientation of the cylinder. In the previous section, the drag forces on nanocylinders for flows perpendicular and parallel to the cylinder are given. If the orientation of a nanocylinder is considered, the average drag force on a freely rotating cylinder can be determined by equally accounting for the contributions in all the directions (Liu et al., 2016), which yields

$$\mathbf{F}_{avg} = -\frac{2}{3}\sqrt{2\pi m_r k_B T}\,NLD\Omega_{avg}^{(1,1)*}\mathbf{V} \tag{7.72}$$

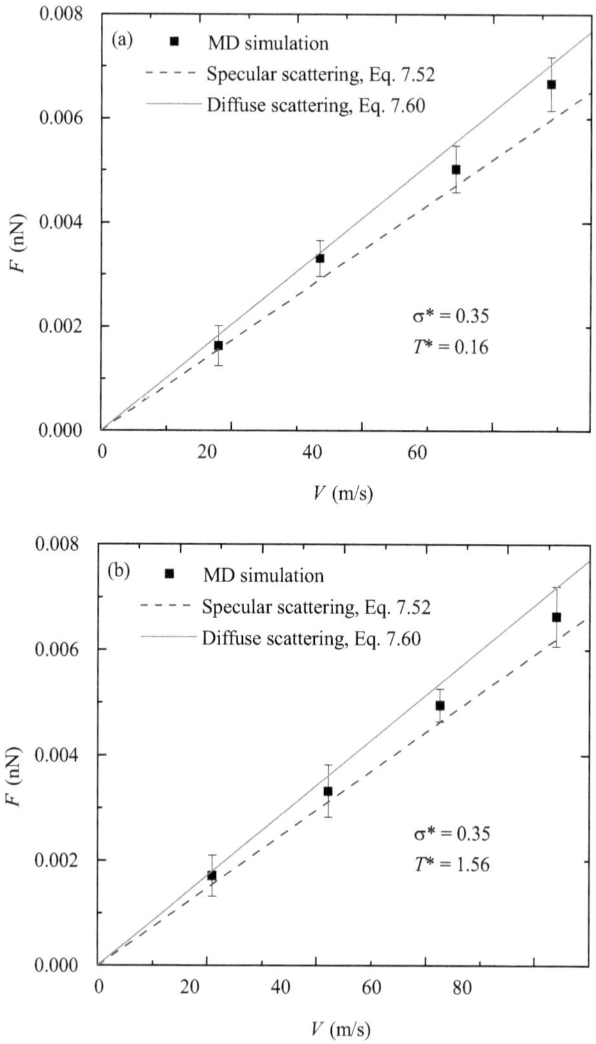

Figure 7.17 Drag forces on an SWCNT of $D = 2.034$ nm and $L = 10$ nm obtained through MD simulations (symbols) and predicted by Equations 7.52 and 7.60. (a) $\sigma^* = 0.35$ and $T^* = 0.16$ and (b) $\sigma^* = 0.35$ and $T^* = 1.56$. (Reprinted with permission from Wong et al., 2012, Copyright @ American Scientific Publishers.)

Thus, the diffusion coefficient of a nanocylinder D_c can be written as follows:

$$D_c = \frac{3}{2}\sqrt{\frac{k_B T}{2\pi m_r}}\frac{1}{NLR\Omega_{avg}^{(1,1)*}} \tag{7.73}$$

7.3.2 Continuum regime

Analytical solutions for the drag force on macroscale cylinders in a liquid are unavailable. For nanocylinders, theoretical analysis is even more complex. Therefore, the major approach for studying the transport properties of nanocylinders in liquids is MD simulations. Compared with nanoparticles, the dynamics of nanocylinders in the continuum regime is less examined in the literature. Most of the previous work focuses on the drag force on CNTs and the diffusion coefficient has not been extensively investigated. Hence, in the section, only the drag force on CNTs in liquids is discussed.

Molecular dynamics simulations have been conducted to calculate the drag force on CNTs in liquid Ar flows (Tang and Advani, 2006). Figure 7.18 sketches the simulation system, where an SWCNT is fixed in the simulation cell with a liquid Ar flow moving perpendicularly over the CNT. The length, width, and height of the simulation domain are $L = 2.13\,\text{nm}$, $W = 30.56\,\text{nm}$, and $H = 25.47\,\text{nm}$, which are sufficiently large to ensure that the influence of the simulation domain size on the drag force is insignificant. The temperature, density, and viscosity of the liquid Ar are 95 K, $1342\,\text{kg/m}^3$, and 2.084×10^{-4} kg/(m·s), respectively. Initially, a uniform velocity V_0 is assigned to all the Ar molecules in the y-direction (Figure 7.18). Then, the flow velocity is maintained by controlling the velocity of Ar molecules in the region of 0.01 W at the inlet to be equal to V_0 every 50 time steps. Liquid Ar is described by the LJ potential, while the CNT is treated as a rigid body.

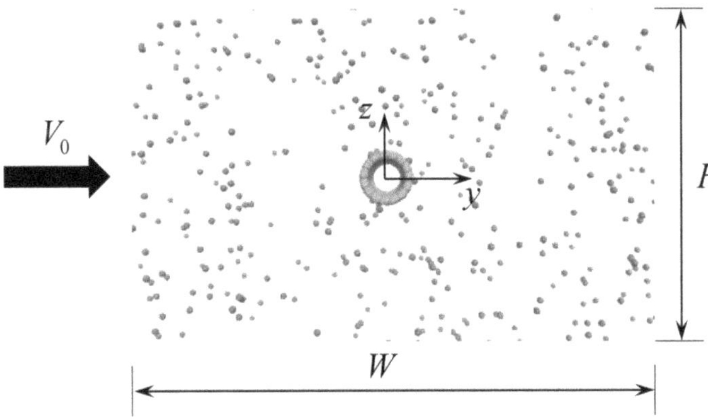

Figure 7.18 Simulation system for computing the drag force on CNTs in liquid Ar flows (for clarity, only part of the Ar molecules are shown).

The drag force F_c and drag coefficient C_d are computed for (6,0) and (12,0) CNTs at different flow velocities. C_d is defined as follows:

$$C_d = \frac{2F_c}{\rho U_0^2 d} \tag{7.74}$$

where d is the diameter of the CNT. The MD results are also compared with empirical drag forces for macroscale cylinders obtained from experimental data, which is given by (Huner and Hussey, 1977)

$$F_c^{cm} = 4\pi\mu U_0 \alpha \left\{1 - 0.87\alpha^2 + 0.514\left[1 - \exp(-\mathrm{Re})\right]\alpha^3\right\} \tag{7.75}$$

where $\alpha = \left[-0.077 - \ln\left(\mathrm{Re}/8\right)\right]^{-1}$ and this Equation is only valid for Re < 3.0.

Table 7.2 summarizes the drag forces and drag coefficients from MD simulations and those predicted by Equation 7.75. It is seen that the drag forces from MD simulations are larger than those given by Equation 7.75, especially at low flow velocities ($U_0 = 1$ corresponds to 157.9 m/s). As flow velocity increases, the difference between the drag force obtained from MD simulations and that from Equation 7.75 becomes small. This is due to the flow slip at the CNT surface at high flow velocities, which tends to reduce the drag force (Tang and Advani, 2006). Nevertheless, it can be seen that molecular interactions is critical for nanocylinders.

Further simulations show that the size of the CNT is more important than the structure of the CNT. For the same outer diameter, the drag forces on SWCNTs and multi-walled CNTs are similar. This indicates that the drag forces in Table 7.2 may apply to nanowires of similar diameters.

7.4 Thermophoresis of nanoparticles

7.4.1 Introduction

Thermophoresis is about the motion of particles suspended in a fluid caused by a temperature gradient. It is of great importance in quite a few areas, including aerosol science, combustion, and nanofluids (fluids dispersed with nanoparticles). The force acting on a particle induced by a temperature gradient is called the thermophoretic force. As the particle moves, a drag force is developed on it. When the drag force becomes equal to the thermophoretic force, the particle moves with a constant velocity, which is termed the thermophoretic velocity. The mechanism responsible for thermophoresis depends on the flow regime.

Table 7.2 Drag Forces and Drag Coefficients Per Unit Length for (6,0) and (12,0) CNTs Obtained from MD Simulations and Predicted by Equation 7.75 (Tang and Advani, 2006)

Inlet Velocity u_o (MD unit)	**Drag Force**		**Drag Coefficient**		C_d^{MD}/C_d^{em}
	F_c^{MD}	F_c^{em}	C_d^{MD}	C_d^{em}	
(6,0) CNT					
0.25	3.32	1.66	94.95	47.50	2.00
0.5	6.46	3.91	46.21	27.94	1.65
0.75	10.04	6.51	31.94	20.72	1.54
1.0	13.37	9.42	23.92	16.86	1.42
1.25	16.31	12.59	18.67	14.42	1.30
1.5	18.85	15.99	14.99	12.72	1.18
1.75	21.70	19.61	12.67	11.46	1.11
2.0	23.67	23.43	10.59	10.48	1.01
(12,0) CNT					
0.1	1.50	0.63	135.40	56.99	2.38
0.25	3.73	1.95	53.90	28.12	1.92
0.5	7.70	4.70	27.78	16.96	1.64
0.75	11.77	7.98	18.87	12.79	1.48
1.0	15.86	11.68	14.30	10.54	1.36
1.25	19.69	15.77	11.37	9.10	1.25
1.5	22.63	20.18	9.07	8.09	1.12
1.75	26.02	24.90	7.67	7.34	1.05
2.0	29.17	29.92	6.58	6.75	0.98

In the free molecular regime, e.g., a small particle in a rarefied gas, the thermophoretic force is mainly the consequence of a large number of molecular collisions between gas molecules and the particle. At the high temperature side of the particle, gas molecules carry relatively high kinetic energies and the gas–particle collision frequency is higher than that at the low temperature side. These usually lead to a strong momentum transfer between gas molecules and the particle at the high temperature side and consequently the thermophoretic force causes the particle to move from high to low temperature. However, the gas density at the high temperature side is lower than that at the low temperature side, which tends to reduce the momentum transfer. Therefore, under certain situations, the thermophoretic force can drive the particle to travel from low to high temperature area, which is referred to as negative thermophoresis. For nanoparticles, as discussed in Section 7.2, molecular interactions are important and the thermophoretic force and velocity also depend on the molecular nature of the gas and particle.

In liquids, i.e., the continuum regime, the thermophoretic force is largely generated by viscous effects, which are affected by the temperature gradient because the fluid viscosity depends on temperature. Similar to the free molecular regime, the thermophoretic force in liquids can be in or against the direction of the temperature gradient, especially for nanoparticles, for which the particle material plays a role in thermophoresis.

7.4.2 Free molecular regime

In the free molecular regime (Kn $\gg$ 1), for relatively large particles (e.g., microparticles), the gas–particle collisions can be treated as rigid-body collisions and the thermophoretic force is derived by Waldmann (1959), which is given by

$$F_{\mathrm{T}} = -\frac{8}{15}\sqrt{\frac{2\pi m_{\mathrm{g}}}{k_{\mathrm{B}}T}}\kappa R^2|\nabla T| \tag{7.76}$$

where κ is the thermal conductivity of the gas and ∇T is the temperature gradient. The corresponding thermophoretic velocity reads

$$V_{\mathrm{T}} = -\frac{\kappa|\nabla T|}{5\left(1+\pi\varphi/8\right)Nk_{\mathrm{B}}T} \tag{7.77}$$

where φ bears the same meaning as that for the drag force, i.e., the momentum accommodation coefficient. As Equation 7.77 is derived using the drag force under rigid-body gas–particle collisions, $\varphi = 0.9$ is usually used in practical applications.

For a nanoparticle, the analyses in Section 7.2 show that the molecular interactions between gas molecules and the particle can be significant and Equation 7.76 may not work well. The thermophoretic force on nanoparticles with the consideration of molecular interactions can be obtained by employing the same theoretical framework introduced in Section 7.2. For a nanoparticle suspended in a stationary gas with a temperature gradient, ∇T, let us use $\mathbf{v}$ and $\mathbf{V}$ to denote the velocities of the gas molecules and nanoparticle. Upon collisions between the gas molecules and particle, the velocities of the gas molecules and particle are referred to as $\mathbf{v}'$ and $\mathbf{V}'$, respectively.

Using the collision model and coordinate systems shown in Figure 7.3, the velocities of the gas molecules before and after collisions with the particle are given by $\mathbf{g} = \mathbf{v} - \mathbf{V}$ and $\mathbf{g}' = \mathbf{v}' - \mathbf{V}'$, respectively. As the idea of calculating the momentum change of gas molecules is the same as that for the drag force, Equations 7.5, 7.6, and 7.8 are valid for the thermophoretic force. The only difference herein is the velocity distribution function, which, with a temperature gradient, reads

$$f = f^0\left(1 - f^{\mathrm{T}}\right) \tag{7.78}$$

where f^0 is the velocity distribution given in Equation 7.7 and f^{T} accounts for the contribution of ∇T (Li and Wang, 2004),

$$f^{\mathrm{T}} = \frac{2m_{\mathrm{r}}\kappa}{5N(k_{\mathrm{B}}T)^2}\left(\frac{m_{\mathrm{r}}v^2}{2k_{\mathrm{B}}T} - \frac{5}{2}\right)\frac{\mathbf{V}\cdot\mathbf{v}}{V}|\nabla T| \tag{7.79}$$

7.4.2.1 Specular scattering For specular scattering, with the angle of scattering χ (Figure 7.3) given in Equation 7.9, the momentum $\delta\mathbf{p}'$ of the reflected molecules, which are in the differential cylindrical shell in Figure 7.3 before collisions, can be evaluated using Equation 7.10. Integrating Equation 7.11 using the velocity distribution in Equation 7.78, the force due to the momentum change of gas molecules, **F**, is obtained as

$$\mathbf{F} = m_{\mathrm{r}}\int_{\mathbf{v}} g\mathbf{g}\left(f^0 - f^0 f^{\mathrm{T}}\right)Q_{\mathrm{s}}(g)\,d\mathbf{v} \tag{7.80}$$

where $Q_{\mathrm{s}}(g)$, the collision cross-section, is given by Equation 7.14. It is clear that Equation 7.80 contains two forces:

$$\mathbf{F}_{\mathrm{D,s}} = m_{\mathrm{r}}\int_{\mathbf{v}} g\mathbf{g}f^0 Q_{\mathrm{s}}(g)\,d\mathbf{v} \quad \text{and} \quad \mathbf{F}_{\mathrm{T,s}} = -m_{\mathrm{r}}\int_{\mathbf{v}} g\mathbf{g}f^0 f^{\mathrm{T}} Q_{\mathrm{s}}(g)\,d\mathbf{v} \tag{7.81}$$

The first force, $\mathbf{F}_{\mathrm{D,s}}$, is nothing but the drag force derived in Section 7.2, where it is denoted as $\mathbf{F}_{\mathrm{s}}$, and the second force, $\mathbf{F}_{\mathrm{T,s}}$, is the thermophoretic force. Substituting Equations 7.7 and 7.79 into $\mathbf{F}_{\mathrm{T,s}}$, the thermophoretic force is obtained as follows:

$$\mathbf{F}_{\mathrm{T,s}} = -\frac{8}{5\pi^{3/2}}\left(\frac{m_{\mathrm{r}}}{2k_{\mathrm{B}}T}\right)^{7/2}\kappa\nabla T\mathbf{u}\int_{\mathbf{v}} g\mathbf{g}e^{-\frac{v^2}{2k_{\mathrm{B}}T/m_{\mathrm{r}}}}\left(\frac{m_{\mathrm{r}}v^2}{2k_{\mathrm{B}}T} - \frac{5}{2}\right)\mathbf{v}Q_{\mathrm{s}}(g)\,d\mathbf{v} \tag{7.82}$$

where $\mathbf{u} = \mathbf{V}/V$ is a unit vector. Using Equation 7.16, $\gamma = g/\sqrt{2k_{\mathrm{B}}T/m_{\mathrm{r}}}$, and

$$\left[\frac{m_{\mathrm{r}}v^2}{2k_{\mathrm{B}}T} - \frac{5}{2}\right]\mathbf{u}\cdot\mathbf{v} \approx \left(\gamma^2 - \frac{5}{2}\right)\left(\gamma\sqrt{2k_{\mathrm{B}}T/m_{\mathrm{r}}}\cos\phi + V\right) + 2\gamma^2 V\cos^2\phi \tag{7.83}$$

Equation 7.82 can be written as follows:

$$\mathbf{F}_{\mathrm{T,s}} = -\frac{16}{5\sqrt{\pi}}\left(\frac{m_{\mathrm{r}}}{2k_{\mathrm{B}}T}\right)\kappa\nabla T\int_0^{\infty}\int_0^{\pi}\gamma^4 e^{-\gamma^2}$$

$$\times\left[\left(\gamma^2-\frac{5}{2}\right)\left(\gamma\sqrt{2k_{\mathrm{B}}T/m_{\mathrm{r}}}\cos\phi+V\right)+2\gamma^2 V\cos^2\phi\right]Q_{\mathrm{s}}(g)\sin\phi\cos\phi d\phi d\gamma \tag{7.84}$$

With the integration of the inner integral, it is obtained that

$$\mathbf{F}_{\mathrm{T,s}} = \frac{8}{3}\sqrt{\frac{2m_{\mathrm{r}}}{\pi k_{\mathrm{B}}T}}\kappa\nabla T\left[\int_0^{\infty}\gamma^5 e^{-\gamma^2}Q_{\mathrm{s}}(g)d\gamma-\frac{2}{5}\int_0^{\infty}\gamma^7 e^{-\gamma^2}Q_{\mathrm{s}}(g)d\gamma\right] \tag{7.85}$$

Using the reduced collision integral defined in Equation 7.21, the thermophoretic force can be rewritten as follows:

$$\mathbf{F}_{\mathrm{T,s}} = \frac{8}{3}\sqrt{\frac{2\pi m_{\mathrm{r}}}{k_{\mathrm{B}}T}}R^2\kappa\nabla T\left(\Omega_{\mathrm{s}}^{(1,1)*}-\frac{6}{5}\Omega_{\mathrm{s}}^{(1,2)*}\right) \tag{7.86}$$

For large particles ($m_p \gg m_g$), the assumption of rigid-body collision is valid. It can be shown that $\Omega_{\mathrm{s}}^{(1,1)*}=1$ and $\Omega_{\mathrm{s}}^{(1,2)*}=1$, and Equation 7.86 is reduced to the Waldmann solution (see Equation 7.76). Therefore, Equation 7.86 is a general expression that applies to both nano- and microparticles.

It is noted that in most cases, $\Omega_{\mathrm{s}}^{(1,1)*}-6\Omega_{\mathrm{s}}^{(1,2)*}/5<0$, which indicates that the thermophoretic force is in the opposite direction of the temperature gradient, i.e., it points from high to low temperature. However, in certain situations, $\Omega_{\mathrm{s}}^{(1,1)*}-6\Omega_{\mathrm{s}}^{(1,2)*}/5$ could be positive, which corresponds to negative thermophoresis. This will be discussed later.

7.4.2.2 Diffuse scattering For diffuse scattering, the velocity distribution of the reflected molecules follows the Maxwellian given by

$$f' = \frac{1}{2\pi}\left(\frac{m_{\mathrm{r}}}{k_{\mathrm{B}}T}\right)^2 g f^0\left(1-f^{\mathrm{T}}\right)e^{-\frac{g'^2}{2k_{\mathrm{B}}T/m_{\mathrm{r}}}} \tag{7.87}$$

With the momentum in Equation 7.26, Equation 7.11 leads to a force

$$\mathbf{F} = m_{\mathrm{r}}\int_{\mathrm{v}} g\mathbf{g}f^0\left(1-f^{\mathrm{T}}\right)Q_{\mathrm{d}}(g)d\mathbf{v} \tag{7.88}$$

where $Q_d(g)$ is the same as that given in Equation 7.29. Again, this force has two parts: the drag force and the thermophoretic force,

$$\mathbf{F} = \mathbf{F}_{D,d} + \mathbf{F}_{T,d} = m_r \int_{\mathbf{v}} g\mathbf{g}f^0 Q_d(g)d\mathbf{v} - m_r \int_{\mathbf{v}} g\mathbf{g}f^0 f^T Q_d(g)d\mathbf{v} \qquad (7.89)$$

where $\mathbf{F}_{D,d}$ is obtained as Equation 7.30 and the thermophoretic force is derived as

$$\mathbf{F}_{T,d} = \frac{8}{3}\sqrt{\frac{2\pi m_r}{k_B T}} R^2 \kappa \nabla T \left(\Omega_d^{(1,1)*} - \frac{6}{5}\Omega_d^{(1,2)*} \right) \qquad (7.90)$$

where $\Omega_d^{(1,1)*}$ and $\Omega_d^{(1,2)*}$ are the reduced collision integrals for diffuse scattering defined in Equation 7.31.

For rigid-body collisions, it can be shown that $\Omega_d^{(1,1)*} = 1 + \pi/8$ and $\Omega_d^{(1,2)*} = 1 + 5\pi/48$, and Equation 7.90 is simplified to be the Waldmann solution (see Equation 7.76). Therefore, for relatively large particles, especially microparticles, the collision model, either specular or diffuse scattering, does not affect the thermophoretic force.

7.4.2.3 Parametrization To consider the contributions of both specular and diffuse scatterings, a generalized thermophoretic force may be expressed by the averaged collision integral, which is a linear combination of $\Omega_d^{(1,q)*}$ and $\Omega_d^{(1,q)*}$

$$\Omega_{avg}^{(1,q)*} = \varphi \Omega_d^{(1,q)*} + (1 - \varphi)\Omega_s^{(1,q)*}, \quad q = 1,2 \qquad (7.91)$$

where φ is the momentum accommodation function introduced in Section 7.2.1 and Equation 7.41 can be used to calculate φ. Based on Equation 7.91, the thermophoretic force can be written as

$$\mathbf{F}_T = \frac{8}{3}\sqrt{\frac{2\pi m_r}{k_B T}} R^2 \kappa \nabla T \left(\Omega_{avg}^{(1,1)*} - \frac{6}{5}\Omega_{avg}^{(1,2)*} \right) \qquad (7.92)$$

7.4.2.4 Collision integrals The collision integrals $\Omega_{s/d}^{(1,1)*}$ and $\Omega_{s/d}^{(1,2)*}$ depend on the gas–particle interaction potential. If the potential in Equation 7.34 is employed, $\Omega_{s/d}^{(1,2)*}$ can be evaluated ($\Omega_{s/d}^{(1,1)*}$ is given in Section 7.2.1). Figures 7.19 and 7.20 plot $\Omega_s^{(1,2)*}$ and $\Omega_d^{(1,2)*}$, respectively, as a function of T^* and σ'. They are similar to $\Omega_s^{(1,1)*}$ and $\Omega_d^{(1,1)*}$. At high temperatures and large particle size (small σ'),

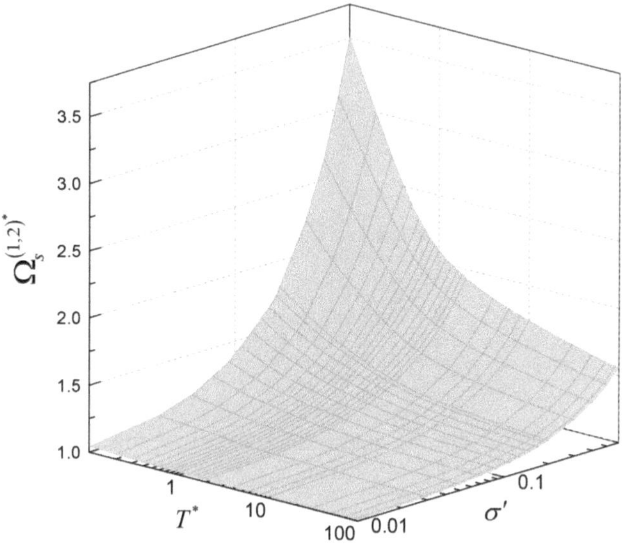

Figure 7.19 Reduced collision integral $\Omega_s^{(1,2)*}$ for specular scattering as a function of T^* and σ'. (Data are the same as those in Table B.1.)

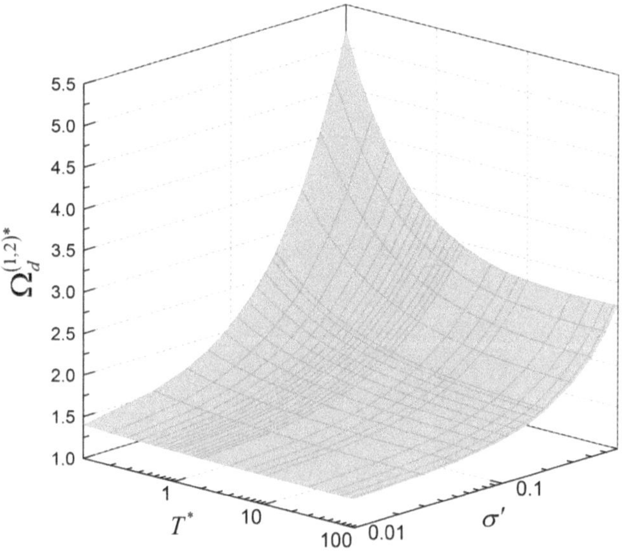

Figure 7.20 Reduced collision integral $\Omega_d^{(1,2)*}$ for diffuse scattering as a function of T^* and σ'. (Data are from Equation 7.94.)

they approach the rigid-body collision limits, 1 and $1+5\pi/48$ for specular and diffuse scatterings. Molecular interactions become important at low temperatures and/or for small particles. The values of $\Omega_{\mathrm{s}}^{(1,2)*}$ and $\Omega_{\mathrm{d}}^{(1,2)*}$ are also given in Tables B.1 and B.2 for $0.1 \le T^* \le 100$ and $0.01 \le \sigma' \le 0.6$. The values in Tables B.1 and B.2 can be well predicted using the following equations:

$$\Omega_{\mathrm{s}}^{(1,2)*} = 1+\left[0.397+\frac{1.11}{T^{*1/4}}+\frac{0.539}{T^{*1/2}}\right]\sigma'+\left[0.986-\frac{2.553}{T^{*1/4}}+\frac{1.264}{T^{*1/2}}\right]\sigma'^2 \quad (7.93)$$

$$\Omega_{\mathrm{d}}^{(1,2)*} = 1+5\pi/48+\left[1.406+\frac{1.899}{T^{*1/4}}+\frac{1.474}{T^{*1/2}}\right]\sigma'+\left[2.128-\frac{4.5}{T^{*1/4}}+\frac{1.209}{T^{*1/2}}\right]\sigma'^2 \quad (7.94)$$

7.4.2.5 Thermophoretic velocity Based on Equations 7.32 and 7.92, the terminal velocity of a nanoparticle under a thermophoretic and a drag force is obtained as follows:

$$V_{\mathrm{T}} = \left(1-\frac{6}{5}\frac{\Omega_{\mathrm{avg}}^{(1,2)*}}{\Omega_{\mathrm{avg}}^{(1,1)*}}\right)\frac{\kappa|\nabla T|}{Nk_{\mathrm{B}}T} \quad (7.95)$$

For the case of specular or diffuse scattering, the thermophoretic velocity can be evaluated using Equation 7.95 by replacing $\Omega_{\mathrm{avg}}^{(1,2)*}/\Omega_{\mathrm{avg}}^{(1,1)*}$ by $\Omega_{s}^{(1,2)*}/\Omega_{s}^{(1,1)*}$ or $\Omega_{\mathrm{d}}^{(1,2)*}/\Omega_{\mathrm{d}}^{(1,1)*}$. It is not surprising that Equation 7.77 is just a special case of Equation 7.95 for rigid-body collisions.

The ratio of the thermophoretic velocity given by Equation 7.95 to that predicted by Equation 7.77 for rigid-body collisions is obtained for Ag, CuO_2, and soot (carbon) particles, as depicted in Figure 7.21. Soot is assumed to be composed of carbon only and 1.8 g/cm^3 is used as the density. The thermophoretic velocity ratios for the three types of particles follow similar fashion. The variation of the ratios are caused by the momentum accommodation function (at $R = 2$–3 nm) and the gas–particle intermolecular interactions. For very small particles, at relatively low temperatures, the deviation of Waldmann's thermophoretic velocity from Equation 7.95 is significant, indicating that gas–nanoparticle interactions play an essential role. For $R > 10$ nm, the ratios approach unity and the assumption of rigid-body collisions tends to work well.

7.4.2.6 Negative thermophoresis The Waldmann solution given by Equation 7.76 suggests that microparticles only move from high to low temperature, which is referred to as positive thermophoresis. For nanoparticles, however, Equations 7.86 and 7.90 show the possibility of negative thermphoresis, i.e., particles move

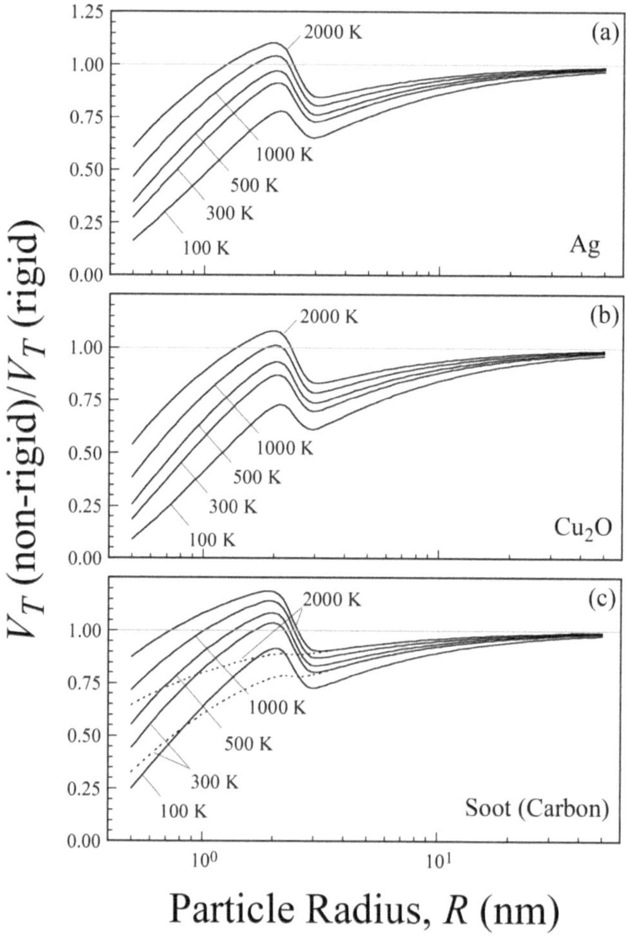

Figure 7.21 Thermophoretic velocity ratio between the cases of nonrigid-body (Equation 7.95) and rigid-body (Equation 7.77) collisions. Solid lines are the ratios using the φ values given by Equation 7.41 for nonrigid-body collisions and $\varphi = 0.9$ for rigid-body collisions (Equation 7.77). Dashed lines are the ratios using Equation 7.41 for both Equations 7.95 and 7.77. The LJ parameters for Ag, CuO_2, and Soot (carbon) are $(\sigma, \varepsilon) = (2.54\ \text{Å}, 3995.4\ \text{K})$, $(4.124\ \text{Å}, 2909.1\ \text{K})$, and $(3.5\ \text{Å}, 33.3\ \text{K})$, respectively. (Reprinted with permission from Li and Wang, 2004, Copyright (2004) by the American Physical Society Adopted from.)

from low to high temperature, if $\Omega_{s/d}^{(1,1)*} - 6\Omega_{s/d}^{(1,2)*}/5 < 0$. Actually, the Chapman–Enskog theory for molecular diffusion under a temperature gradient indicates that molecules can transport form low to high temperature (negative thermal diffusion) (Wang and Li, 2012). If a molecule is viewed as a small particle, the Waldmann solution and the Chapman–Enskog theory reveal that there is a critical particle size, below which negative thermophoresis occurs under proper conditions.

To have negative thermophoresis, the condition, $\Omega_{s/d}^{(1,1)*} - 6\Omega_{s/d}^{(1,2)*}/5 < 0$, is required. As $\Omega_{s/d}^{(1,1)*}$ and $\Omega_{s/d}^{(1,2)*}$ are complex functions of $\Phi(r)$ and temperature T, it is unclear what this criterion means. To fundamentally understand negative thermophoresis, dimensionless quantities, $T^* = k_B T/\varepsilon$, $g^{*2} = m_r g^2/(2\varepsilon)$, and $Q^* = Q/(\pi R^2)$, are introduced. Using these dimensionless groups, the reduced collision integrals can be expressed as follows (for simplicity, hereafter $\Omega^{(1,1)*}$ and $\Omega^{(1,2)*}$ are used to replace $\Omega_{s/d}^{(1,1)*}$ and $\Omega_{s/d}^{(1,2)*}$, respectively):

$$\Omega^{(1,1)*} = \frac{1}{T^{*3}} \int_0^\infty g^{*5} e^{-g^{*2}/T^*} Q^*(g^*) dg^* \text{ and } \Omega^{(1,2)*} = \frac{1}{3T^{*4}} \int_0^\infty g^{*7} e^{-g^{*2}/T^*} Q^*(g^*) dg^* \tag{7.96}$$

Based on these two equations, it is easy to obtain that

$$\frac{\Omega^{(1,2)*}}{\Omega^{(1,1)*}} = 1 + \frac{1}{3} \frac{d \ln \Omega^{(1,1)*}}{d \ln T^*} \tag{7.97}$$

which leads to

$$\left(\frac{6}{5} \Omega^{(1,2)*} - \Omega^{(1,1)*} \right) = \Omega^{(1,1)*} \left[\frac{1}{5} + \frac{2}{5} \frac{d \ln \Omega^{(1,1)*}}{d \ln T^*} \right] \tag{7.98}$$

As $\Omega^{(1,1)*} > 0$, Equation 7.98 suggests that negative thermophoresis requires

$$\Omega^{(1,1)*} = a\left(T^*\right)^{-s}, \quad \text{with} \quad s > \frac{1}{2} \tag{7.99}$$

where a is a constant. If $\ln \Omega^{(1,1)*}$ is plotted as a function of $\ln T^*$, the exponent, $-s$, in Equation 7.99 is nothing but the slope of the $\ln \Omega^{(1,1)*}$ curve.

Equation 7.99 offers an alternative criterion for negative thermophoresis, which has a clearer physical picture compared with $\left(6\Omega^{(1,2)*}/5 - \Omega^{(1,1)*}\right) < 0$. Fundamentally, $\Omega^{(1,1)*}$ is a function of the angle of scattering χ, which governs the momentum change of gas molecules during collisions with a nanoparticle. For rigid-body interactions, $\Omega^{(1,1)*}$ is a constant and $s = 0$, therefore negative thermophoresis will not take place. However, if the gas–particle collisions are nonrigid, $\Omega^{(1,1)*}$ varies when the temperature T is changed, as shown in Figures 7.5 and 7.6. For an appropriate $\Phi(r)$, $s > 0.5$ in Equation 7.99 may be satisfied in certain ranges of T^*.

Figure 7.22 illustrates the collision scenarios between a gas molecule and a small particle at different mean velocities $\bar{g}$, which are governed by the temperature of the gas. The top panel shows two trajectories of a gas molecule with different angles of scattering χ at different temperatures. A low temperature corresponds to a small $\bar{g}$ (dashed line in Figure 7.22) and a large χ, which is associated with a large momentum change. This indicates that the force acting on the cold side of the particle is larger than that on the hot side. However, such a situation depends on $\Phi(r)$ and it is only valid in a certain range of temperatures. If the temperature is very high or low, as depicted in the middle and bottom panels of Figure 7.22, the collisions become close to rigid-body interactions and negative thermophoresis will not occur. It must be noted that the velocity of gas molecules follows the Maxwellian distribution. At a given temperature gradient, whether negative thermophoresis will take place or not depends on the overall momentum change of the gas molecules. If the case in the top panel of Figure 7.22 dominates the contribution to the gas momentum change, negative thermophoresis will occur.

Nevertheless, Equation 7.99 and Figure 7.22 suggest that under a proper $\Phi(r)$ it is possible for a nanoparticle to experience negative thermophoresis when the temperature is varied. As a quick check, the collision integrals are evaluated by using the LJ potential. Figure 7.23 plots $\left(6\Omega^{(1,2)*}/5 - \Omega^{(1,1)*}\right)$ and s as a function of T^*. It is clear that $\left(6\Omega^{(1,2)*}/5 - \Omega^{(1,1)*}\right)$ is positive and $s > 0.5$ for $0.45 < T^* < 0.95$, where negative thermophoresis is expected to take place.

For gas–particle collisions, the LJ potential is not a proper potential. However, if the gas–particle interaction is considered as the overall effect of the LJ interactions between the gas molecules and each constituent atom of the particle, the gas–particle interaction potential should show similar characteristics to the LJ potential. The potential given by Equation 7.34 is an example of such a treatment. Then, gas–particle interactions can be described by a general LJ type potential (Wang and Li, 2012)

$$\Phi(r) = \varepsilon\left[c_1\left(\sigma/r\right)^{\alpha} - c_2\left(\sigma/r\right)^{\beta}\right] \tag{7.100}$$

where c_1, c_2, α, and β are parameters. If $c_1\varepsilon$ and σ are used to nondimensionalize $\Phi(r)$ and r [note that $k_B T$ is also scaled by $c_1\varepsilon$, such that $T^* = k_B T/(c_1\varepsilon)$], the dimensionless form of the potential is given by

$$\Phi^*(r^*) = r^{*-\alpha} - cr^{*-\beta} \tag{7.101}$$

where $\Phi^* = \Phi/(c_1\varepsilon)$, $r^* = r/\sigma$, and $c = c_2/c_1$. This equation is denoted as the LJ $\alpha-\beta$ potential.

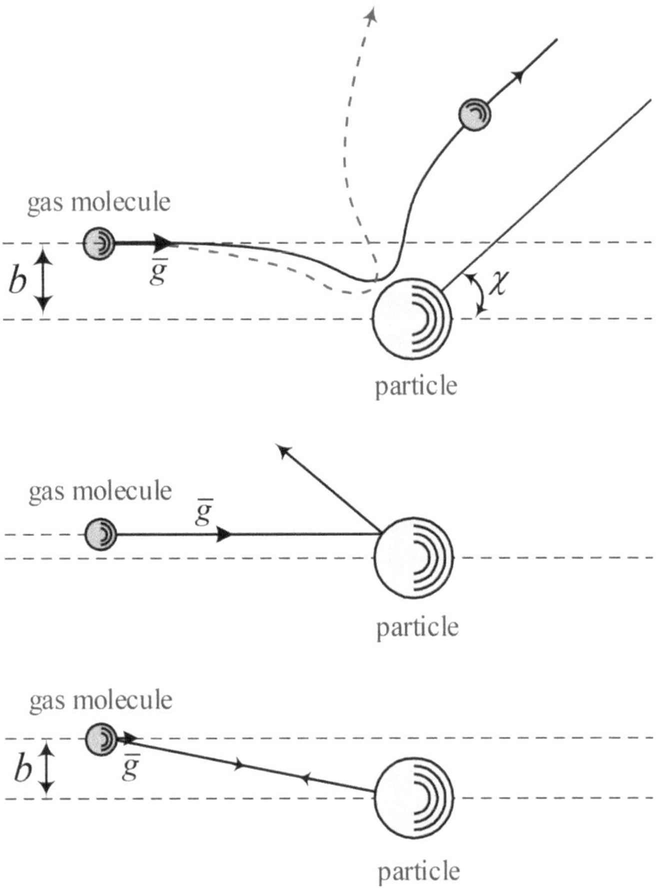

Figure 7.22 Temperature effect on the collision between a gas molecule and a nanoparticle ($\bar{g}$ is the mean velocity of the gas molecule, which is a measure of temperature). Top: Two possible trajectories of a gas molecule at different temperatures. Middle and bottom: Rigid-body-like collisions at very high (middle) and low (bottom) temperatures. (Reprinted with permission from Wang and Li, 2012, Copyright (2012) by the American Physical Society.)

Figure 7.24a shows the exponent s in Equation 7.99 using the LJ $\alpha - \beta$ potential with $c = 1$. It is seen that negative thermophoresis only occurs at proper α and β values. For a given α, β should be sufficiently small to have negative thermophoresis. This is understandable because the second term in Equation 7.101 describes attractive forces. It becomes dominant when $r^* \gg 1$ (usually $\alpha > \beta$). As β increases, the attractive force weakens and the gas–particle interaction tends to approach the rigid-body interaction, for which negative thermophoresis does not occur. Therefore,

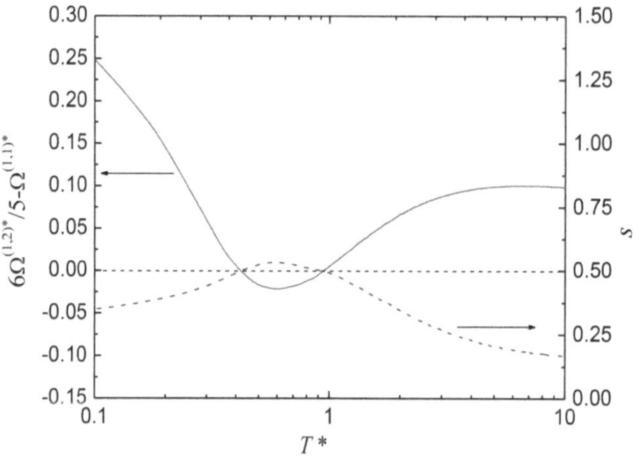

Figure 7.23 The values of $\left(6\Omega^{(1,2)*}/5-\Omega^{(1,1)*}\right)$ and s as a function of T^* based on the LJ potential. (Reprinted with permission from Wang and Li, 2012, Copyright (2012) by the American Physical Society.)

there is a critical β value, β_{cr}, below which negative thermophoresis becomes possible. It is easy to find that β_{cr} is independent of c if the potential $\Phi(r)$ is reduced to a c-independent format $\Phi^*(r^*) = r^{*-\alpha} - r^{*-\beta}$, which can be achieved by using $\varepsilon c_1 c^{\alpha/(\alpha-\beta)}$ and $\sigma c^{1/(\beta-\alpha)}$ to scale $\Phi(r)$ and r, respectively. Figure 7.24b plots β_{cr} as α is varied. By fitting the data in Figure 7.24b, a necessary condition for the occurrence of negative thermophoresis is obtained as follows:

$$\beta < 4.8 + 18.8\exp\left(-\alpha/5.05\right), \alpha \in (9,24) \tag{7.102}$$

If the condition in Equation 7.102 is satisfied, negative thermophoresis can be expected in a temperature range ($T^*_{\min}$, $T^*_{\max}$). Figure 7.25 shows $T^*_{\min}$ and $T^*_{\max}$ for two pairs of α and β values as c is varied, where $T^*_{\min}$ and $T^*_{\max}$ are plotted in the logarithmic scale. The linear dependence in Figure 7.25 and extensive calculations suggest an expression for both $T^*_{\min}$ and $T^*_{\max}$

$$T^*_{\min(\max)} = T^0_{\min(\max)} c^{\frac{\alpha}{\alpha-\beta}} \tag{7.103}$$

where $T^0_{\min}$ and $T^0_{\max}$ are coefficients depending on α and β. Under the condition of Equation 7.102, $T^0_{\min}$ and $T^0_{\max}$ are calculated for various α and β values. Figure 7.26

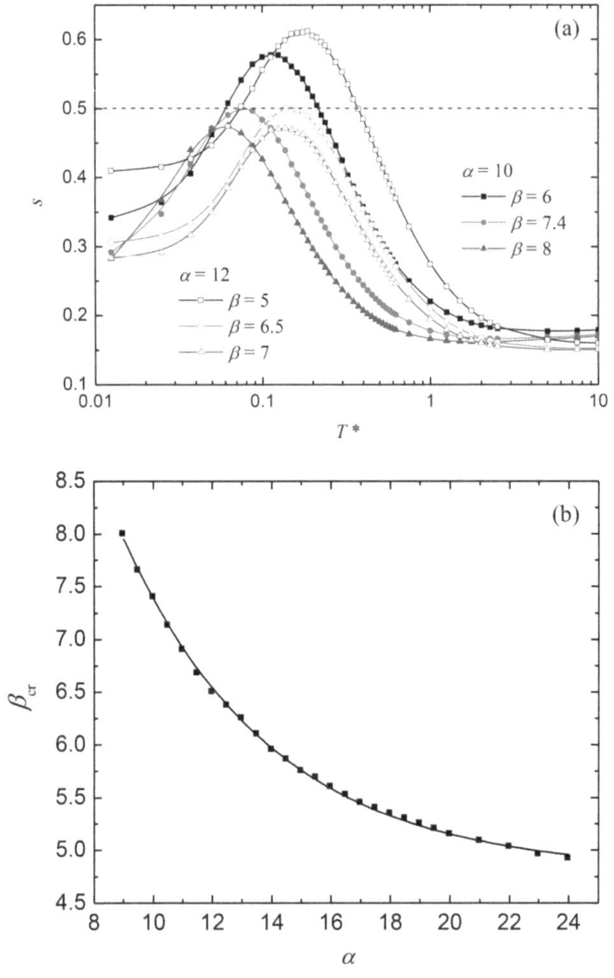

Figure 7.24 Condition for negative thermophoresis. (a) s as a function of T^* for various α and β values. (b) Critical value of β, β_{cr}, in a range of α values (solid line is an exponential fit to the data). (Reprinted with permission from Wang and Li, 2012, Copyright (2012) by the American Physical Society.)

presents T^0_{min} and T^0_{max} as a function of β for different α values. For a given gas–particle interaction, whether negative thermophoresis will take place or not can be predicted using Equations 7.102 and 7.103 with the help of Figure 7.26.

7.4.2.7 Binary gas mixtures Thermophoresis in gas mixtures is quite common in applications, such as the dynamics of soot particles in flames. In a gas mixture,

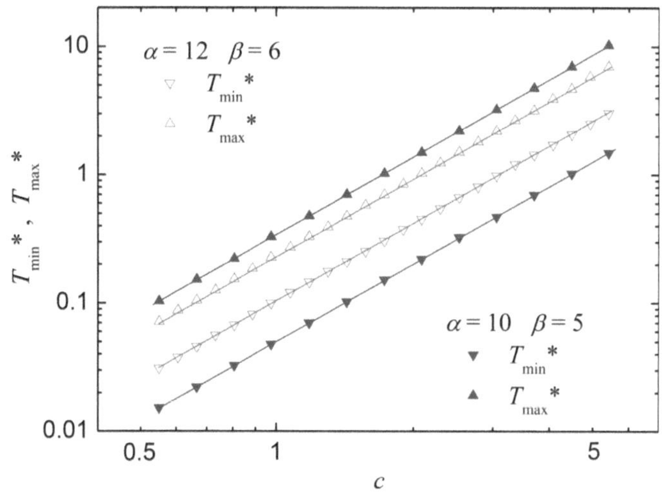

Figure 7.25 T^*_{min} and T^*_{max} for the occurrence of negative thermophoresis as a function of c. Solid lines are linear fits to the data and their slopes are the same, equal to $\alpha/(\alpha-\beta)=2$. (Reprinted with permission from Wang and Li, 2012, Copyright (2012) by the American Physical Society).

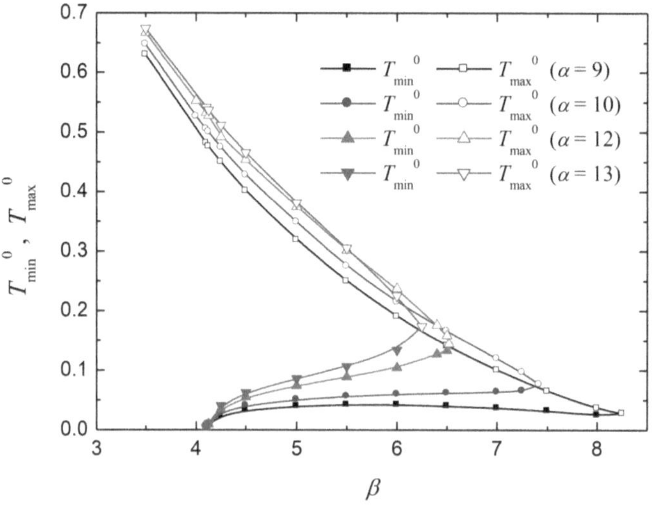

Figure 7.26 T^0_{min} and T^0_{max} in Equation 7.103 as a function of β for different α values. Solid lines are guide to the eye. (Reprinted with permission from Wang and Li, 2012, Copyright (2012) by the American Physical Society).

the velocity distribution of the gas molecules is affected by both the temperature gradient and the distribution of gas concentration. Therefore, the theoretical analysis is nontrivial. In this section, the thermophoretic force on spherical particles in monatomic binary gas mixtures in the free molecular regime is discussed.

For a binary mixture of monatomic gases, the velocity distributions of the two gases, which are denoted as, f_1 and f_2, are governed by the Boltzmann equation (Wang and Li, 2011):

$$\frac{\partial f_i}{\partial t} + \mathbf{v}_i \cdot \frac{\partial f_i}{\partial \mathbf{r}} + \mathbf{F}_i \cdot \frac{\partial f_i}{\partial \mathbf{v}_i} = \frac{\partial_e f_i}{\partial t}, \quad i = 1,2 \tag{7.104}$$

where $\mathbf{v}_i$ and $\mathbf{F}_i$ are the molecular velocity of and the external force acting on species i, $\partial_e f_i / \partial t$ is the change of f_i due to molecular collisions, by which f_1 and f_2 are coupled with each other.

The equilibrium solution of Equation 7.104 gives the first-order approximations of f_1 and f_2, $f_1^{(0)}$ and $f_2^{(0)}$, which are given by

$$f_i^{(0)} = N_i \left(\frac{m_i}{2\pi k_B T} \right)^{3/2} \exp\left(-\frac{m_i v_i^2}{2k_B T} \right), \quad i = 1,2 \tag{7.105}$$

Under a temperature gradient, the velocity distribution is varied and is given by the second-order approximation of f_i, $f_i^{(1)}$, which reads

$$f_i^{(1)} = f_i^{(0)} \Phi_i^{(1)} (\nabla T, \nabla N_i), \quad i = 1,2 \tag{7.106}$$

where $\Phi_i^{(1)}$ is a function of ∇T and ∇N_i given by

$$\Phi_i^{(1)} (\nabla T, \nabla N_i) = -\mathbf{A}_i \cdot \nabla \ln T - \mathbf{D}_i \cdot \mathbf{d}_{ij}, \quad i,j = 1,2; i \neq j \tag{7.107}$$

where $\mathbf{A}_i$ and $\mathbf{D}_i$ are vectors and the vector $\mathbf{d}_{ij}$ is given by

$$\mathbf{d}_{ij} = \frac{N_i}{N} \nabla \ln p_i - \frac{\rho_i \rho_j}{p\rho} (\mathbf{F}_i - \mathbf{F}_j) - \frac{\rho_i}{p\rho} \nabla p \tag{7.108}$$

where $p = Nk_B T$ and $p_i = N_i k_B T$ are the total and partial pressures, ρ and $\rho_i = m_i N_i$ are the total and partial mass densities. If the binary mixture is at rest, i.e., $\mathbf{F}_i = \mathbf{0}$, Equation 7.108 becomes

$$\mathbf{d}_{ij} = \nabla x_i \tag{7.109}$$

where, $x_i = N_i/N$ is the concentration. In the steady state, x_i is related to the temperature gradient through

$$\nabla x_i = -k_T \nabla \ln T \tag{7.110}$$

where $k_{\mathrm{T}} \propto D_i^{\mathrm{T}}/D_{ij}$ is the thermal diffusion ratio (D_i^{T} is the thermal diffusion coefficient of species i and D_{ij} is the binary diffusion coefficient). Using Equations 7.109 and 7.110, $\Phi_i^{(1)}$ can be rewritten as

$$\Phi_i^{(1)} = -\tilde{\mathbf{A}}_i \cdot \nabla \ln T \tag{7.111}$$

where $\tilde{\mathbf{A}}_i = \mathbf{A}_i - k_{\mathrm{T}}\mathbf{D}_i$ and can be expanded as a convergent Sonine polynomial. If the first term in the polynomial is used, $\tilde{\mathbf{A}}_i$ is given by

$$\tilde{\mathbf{A}}_i = a_i\left(\frac{5}{2} - \zeta_i^2\right)\zeta_i \tag{7.112}$$

where a_i is a coefficient depending on temperature, m_i, and the intermolecular interaction potential, $\zeta_i = \mathbf{v}_i\sqrt{m_i/2k_{\mathrm{B}}T}$, and $\zeta_i = |\zeta_i|$.

For a rarefied binary gas mixture at rest, using the theoretical analysis in Sections 7.4.2.1 and 7.4.2.2, the thermophoretic force on a particle mainly due to gas species i, $\mathbf{F}_{\mathrm{T},i}$, is calculated as follows:

$$\mathbf{F}_{\mathrm{T},i} = m_{ip}\int_{\mathbf{v}_i} g_i \mathbf{g}_i f_i^{(0)} \Phi_i^{(1)} Q(g_i)\, d\mathbf{v}_i \tag{7.113}$$

where $m_{ip} = m_i m_p/(m_i + m_p)$ is the reduced mass for gas species i and the particle. Substituting Equations 7.105 and 7.111 into Equation 7.113, $\mathbf{F}_{\mathrm{T},i}$ is obtained as (Wang and Li, 2011)

$$\mathbf{F}_{\mathrm{T},i} = -\frac{20\sqrt{\pi}k_{\mathrm{B}}R^2 a_i N_i}{3}\left(\frac{m_i}{m_{ip}}\right)^2\left(\Omega_{ip}^{(1,1)*} - \frac{6}{5}\Omega_{ip}^{(1,2)*}\right)\nabla T, \quad i = 1,2, \tag{7.114}$$

where the reduced collision integral $\Omega_{ip}^{(l,q)*} = \int_0^\infty e^{-\gamma_i^2}\gamma_i^{2q+3}Q^l(g_i)\,d\gamma_i\, /$ $\left\{\left[(q+1)!/2\right]\left[1+\left(1+(-1)^l\right)/2(1+l)\right]\pi R^2\right\}$ with $\gamma_i = g_i\sqrt{m_{ip}/2k_{\mathrm{B}}T}$. The subscript

"ip" in $\Omega_{ip}^{(l,q)}$ means that the interaction potential between gas species i and the particle should be used in calculating the angle of scattering χ. The coefficient a_i in Equation 7.114 is given by

$$a_i = \frac{\alpha_i a_{jj} + \alpha_j a_{ij}}{\det|\mathbf{A}|}, \quad i,j = 1,2;\ i \neq j \tag{7.115}$$

where $\alpha_i = -\dfrac{15N_i}{4N^2}\sqrt{\dfrac{2k_\mathrm{B}T}{m_i}}$ and matrix $A = \begin{bmatrix} a_{ii} & a_{ij} \\ a_{ji} & a_{jj} \end{bmatrix}$. The elements of A are given by

$$a_{ii} = \frac{4N_i^2}{N^2}\sqrt{\frac{k_\mathrm{B}T}{\pi m_{ii}}}\Omega_{ii}^{(2,2)} + 8M_j\frac{N_iN_j}{N^2}\sqrt{\frac{k_\mathrm{B}T}{2\pi m_{ij}}}\Pi_{ii} \tag{7.116}$$

$$a_{jj} = \frac{4N_j^2}{N^2}\sqrt{\frac{k_\mathrm{B}T}{\pi m_{jj}}}\Omega_{jj}^{(2,2)} + 8M_i\frac{N_iN_j}{N^2}\sqrt{\frac{k_\mathrm{B}T}{2\pi m_{ij}}}\Pi_{jj} \tag{7.117}$$

$$a_{ij} = a_{ji} = -8\left(M_iM_j\right)^{-3/2}\frac{N_iN_j}{N^2}\sqrt{\frac{k_\mathrm{B}T}{2\pi m_{ij}}}\Pi_{ij} \tag{7.118}$$

where $M_i = m_i/\left(m_i + m_j\right)$, $m_{ij} = m_im_j/\left(m_i + m_j\right)$, and Π_{ii}, Π_{jj}, and Π_{ij} are the combinations of collision integrals,

$$\Pi_{ii} = \frac{5}{4}\left(6M_i^2 + 5M_j^2\right)\Omega_{ij}^{(1,1)} - M_j^2\left(5\Omega_{ij}^{(1,2)} - \Omega_{ij}^{(1,3)}\right) + 2M_iM_j\Omega_{ij}^{(2,2)} \tag{7.119}$$

$$\Pi_{jj} = \frac{5}{4}\left(6M_j^2 + 5M_i^2\right)\Omega_{ij}^{(1,1)} - M_i^2\left(5\Omega_{ij}^{(1,2)} - \Omega_{ij}^{(1,3)}\right) + 2M_iM_j\Omega_{ij}^{(2,2)} \tag{7.120}$$

$$\Pi_{ij} = \frac{55}{4}\Omega_{ij}^{(1,1)} - \left(5\Omega_{ij}^{(1,2)} - \Omega_{ij}^{(1,3)}\right) - 2\Omega_{ij}^{(2,2)} \tag{7.121}$$

where $\Omega_{ij}^{(l,q)}$ is the collision integral for gas-gas interactions. If the LJ potential is used for gas-gas interactions, $\Omega_{ij}^{(l,q)}$ can be found in standard handbooks about gas kinetic theory. They can also be found in Appendix C.

Based on Equation 7.114, the total thermophoretic force on the particle is then obtained as follows:

$$\mathbf{F}_{\mathrm{T}} = -\frac{20\sqrt{\pi}}{3} R^2 k_{\mathrm{B}} \nabla T \left[\begin{array}{l} a_1 N_1 \left(\frac{m_1}{m_{1p}} \right)^2 \left(\Omega_{1p}^{(1,1)*} - \frac{6}{5} \Omega_{1p}^{(1,2)*} \right) + \\ a_2 N_2 \left(\frac{m_2}{m_{2p}} \right)^2 \left(\Omega_{2p}^{(1,1)*} - \frac{6}{5} \Omega_{2p}^{(1,2)*} \right) \end{array} \right] \quad (7.122)$$

It is noted that a_1 and a_2 in this equation depend on the molecular interactions between the two gas species. Therefore, $\mathbf{F}_{\mathrm{T}}$ is not a simple linear combination of the thermophoretic forces caused by the two gases. If the two terms in the square bracket in Equation 7.122 are denoted by $\Omega_{\mathrm{T},1}^*$ and $\Omega_{\mathrm{T},2}^*$, $\mathbf{F}_{\mathrm{T}}$ can be rewritten as follows:

$$\mathbf{F}_{\mathrm{T}} = -\frac{20\sqrt{\pi}}{3} R^2 k_{\mathrm{B}} \left(\Omega_{\mathrm{T},1}^* + \Omega_{\mathrm{T},2}^* \right) \nabla T \quad (7.123)$$

The unit of $\Omega_{\mathrm{T},1}^*$ and $\Omega_{\mathrm{T},2}^*$ is the same as $1/R^2$, while $\Omega_{ip}^{(l,q)*}$ in Equation 7.122 is dimensionless.

Equations 7.122 and 7.123 are general expressions for the thermophoretic force. The value of $\mathbf{F}_{\mathrm{T}}$ is affected by the gas scattering scenario. Again, if specular and diffuse scatterings are considered as the limiting cases, Equation 7.123 can be written as follows:

$$\mathbf{F}_{\mathrm{T,s/d}} = -\frac{20\sqrt{\pi}}{3} R^2 k_{\mathrm{B}} \Omega_{\mathrm{T,s/d}}^* \nabla T \quad (7.124)$$

where $\Omega_{\mathrm{T}}^* = \Omega_{\mathrm{T},1}^* + \Omega_{\mathrm{T},2}^*$, which can be called binary thermophoretic collision integral and the subscript "s/d" denotes that the specular or diffuse collision cross-section should be used in calculating the collision integrals. In Equation 7.124, if the concentration of one gas component is zero, say, $x_2 = 0$ ($x_1 = 1$), then $a_1 N_1 = -15\sqrt{2\pi} \big/ \left(16 \Omega_{11}^{(2,2)} \right)$ and $\mathbf{F}_{\mathrm{T,s/d}}$ is reduced to

$$\mathbf{F}_{\mathrm{T,s/d}} = \frac{25\sqrt{2\pi}}{4\Omega_{11}^{(2,2)}} R^2 k_{\mathrm{B}} \nabla T \left(\Omega_{1p}^{(1,1)*} - \frac{6}{5} \Omega_{1p}^{(1,2)*} \right)_{\mathrm{s/d}} \quad (7.125)$$

by assuming that $m_p \gg m_1$. Equation 7.125 is the same as Equations 7.86 and 7.90 considering that the thermal conductivity of a gas $\kappa = 75 k_{\mathrm{B}}^2 T \big/ \left(32\sqrt{k_{\mathrm{B}} T m_1 / \pi}\, \Omega_{11}^{(2,2)} \right)$ (Chapman and Cowling, 1970).

For microparticles, if the rigid-body collision is assumed, it can be shown that all the collision integrals in Equation 7.122 are equal to 1, i.e., $\Omega_{1p}^{(1,1)*} = \Omega_{1p}^{(1,2)*} = \Omega_{2p}^{(1,1)*} = \Omega_{2p}^{(1,2)*} = 1$. In this case, with $m_p \gg m_i$, the thermophoretic force is given by

$$\mathbf{F}_\mathrm{T} = \frac{4\sqrt{\pi}}{3} R^2 k_\mathrm{B} \nabla T (a_1 N_1 + a_2 N_2) \tag{7.126}$$

This expression reproduces the Waldmann solution, Equation 7.76, for simple gases by setting either N_1 or N_2 to be zero.

For nanoparticles, the collision integrals $\Omega_{ip}^{(l,q)*}$ are not unity due to gas–particle intermolecular interactions. If the Rudyak–Krasnolutski potential, Equation 7.34, is employed, $\Omega_{ip}^{(l,q)*}$ can be evaluated. Figure 7.27 shows the quantity $\Omega_T^* R^2$ as a function of the mean temperature of the gas mixture T for the thermophoresis of Ag nanoparticles in dilute He-Ar mixtures for specular (Figure 7.27a) and diffuse (Figure 7.27b) scatterings at different He concentrations x_1. It is seen that F_T increases monotonically but nonlinearly with increasing T at a given x_1. Figure 7.28 depicts $\Omega_T^* R^2$and $\Omega_{T,i}^* R^2$ when the He concentration x_1 is varied at $T = 300$ K for specular (Figure 7.28a) and diffuse (Figure 7.28b) scatterings. It is found that $F_{\mathrm{T},1}$ (He) increases while $F_{\mathrm{T},2}$ (Ar) decreases nonlinearly as x_1 is increased from 0 to 1. $F_{\mathrm{T},1}$ appears to be more sensitive to x_1 than $F_{\mathrm{T},2}$ does. This causes the total thermophoretic force F_T to increase with increasing x_1, as shown in Figure 7.27. The nonlinear dependence of $\Omega_T^* R^2$ on x_1 leads to a maximum value of $\Omega_T^* R^2$ at $x_1 \sim 0.9$, where F_T is larger than that in either pure He or pure Ar, as shown in Figure 7.28. This indicates that the thermophoretic force in binary mixtures is not a simple linear superposition of the forces caused by the two gas components.

7.4.3 Continuum regime

The theoretical analysis of nanoparticle thermophoresis in liquids is not an easy topic. In the literature, most attempts are focused on the gas kinetic analysis for small Kn, Kn $\ll$ 1, which, with certain modification, is applied to the thermophoresis of nanoparticles in liquids. Using the gas kinetic theory, Epstein performed the first study of particle dynamics caused by a temperature gradient in a gas with small Kn (Epstein, 1929). By allowing discontinuities in the velocity and temperature at the gas–particle interface, the thermophoretic force and velocity are derived as follows:

$$F_\mathrm{T} = -9\pi\mu\nu R \frac{\kappa_\mathrm{g}}{2\kappa_\mathrm{g} + \kappa_\mathrm{p}} \frac{|\nabla T|}{T_0} \quad \text{and} \quad V_\mathrm{T} = -\frac{3\nu}{2} \frac{\kappa_\mathrm{g}}{2\kappa_\mathrm{g} + \kappa_\mathrm{p}} \frac{|\nabla T|}{T_0} \tag{7.127}$$

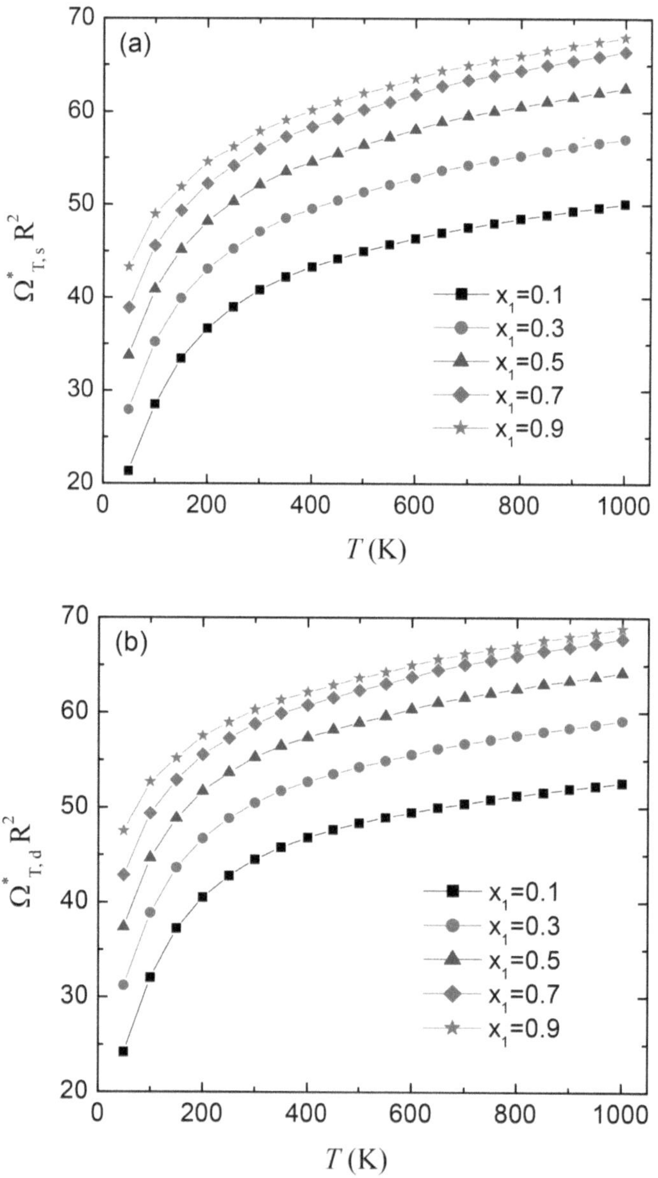

Figure 7.27 $\Omega_T^* R^2$ at different He concentrations x_1 as a function of the mean temperature of the gas mixture T for specular (a) and diffuse (b) scatterings. (Reprinted with permission from Wang and Li, 2011, Copyright (2011) by the American Physical Society.)

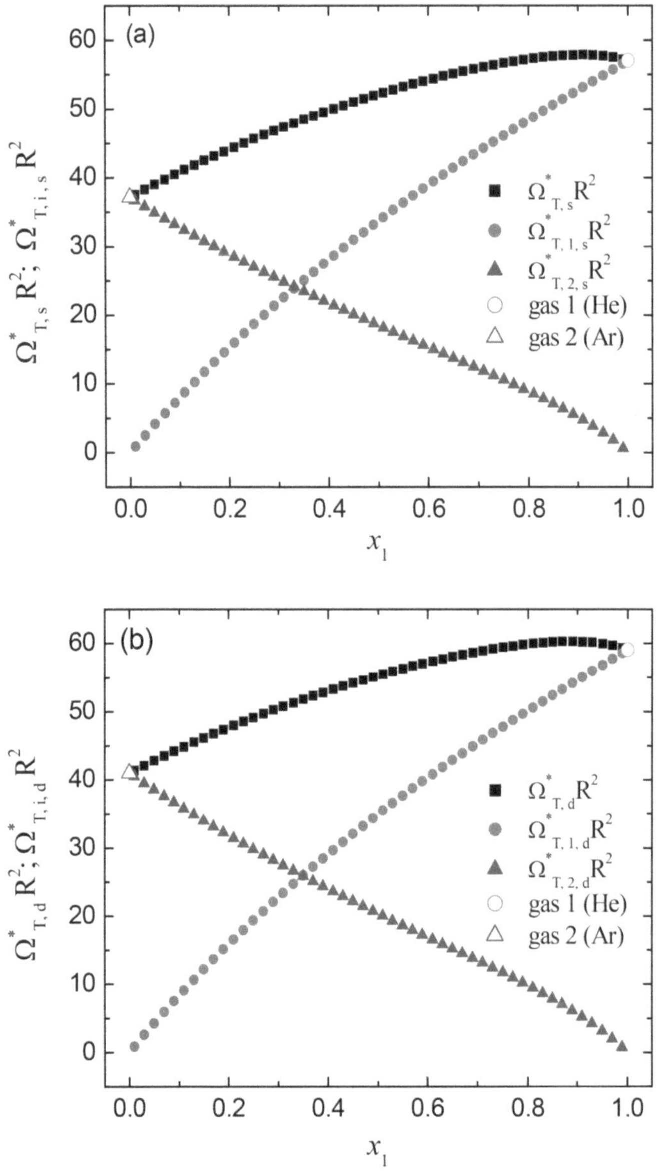

Figure 7.28 $\Omega_T^* R^2$ and $\Omega_{T,i}^* R^2$ as a function of the He concentrations x_1 for specular (a) and diffuse (b) scatterings at $T = 300$ K. Open triangle and circle represent the results in pure Ar and pure He, respectively. (Reprinted with permission from Wang and Li, 2011, Copyright (2011) by the American Physical Society).

where $\nu = \mu/\rho$ is the kinematic viscosity of the gas, κ_g and κ_p are the thermal conductivities of the gas and particle, respectively, and T_0 is the mean gas temperature in the vicinity of the particle. Equation 7.127 works well for particles having a thermal conductivity comparable to that of the gas, $\kappa_g/\kappa_p \sim 1$. For particles with a high thermal conductivity, Epstein's solution underestimates the thermophoretic force.

Epstein's analysis was later improved with the consideration of velocity slip (Brock, 1962) and the revised thermophoretic force and velocity are given by

$$F_T = -12\pi\mu\nu R \frac{C_s\left(\kappa_g + \kappa_p C_t \mathrm{Kn}\right)}{(1+3C_m\mathrm{Kn})\left(2\kappa_g + \kappa_p + 2\kappa_p C_t \mathrm{Kn}\right)} \frac{|\nabla T|}{T_0} \tag{7.128}$$

and

$$V_T = -\frac{2\nu C_s\left(\kappa_g + \kappa_p C_t \mathrm{Kn}\right)}{(1+3C_m\mathrm{Kn})\left(2\kappa_g + \kappa_p + 2\kappa_p C_t \mathrm{Kn}\right)} \frac{|\nabla T|}{T_0} \tag{7.129}$$

where $C_s = 0.75$, $C_t = 2.18$, and $C_m = 1.14$ are empirical parameters. These two equations are only valid for small Kn. Based on these expressions, a thermophoretic force and velocity for the entire Kn range is proposed by using Millikan's drag force formula (Talbot et al., 1980):

$$F_T = -12\pi\mu\nu R \frac{C_s\left(\kappa_g + \kappa_p C_t \mathrm{Kn}\right)\left\{1 + \left[1.2 + 0.41\exp(-0.88/\mathrm{Kn})\right]\mathrm{Kn}\right\}}{(1+3C_m\mathrm{Kn})\left(2\kappa_g + \kappa_p + 2\kappa_p C_t \mathrm{Kn}\right)} \frac{|\nabla T|}{T_0} \tag{7.130}$$

$$V_T = -\frac{2\nu C_s\left(\kappa_g + \kappa_p C_t \mathrm{Kn}\right)\left\{1 + \left[1.2 + 0.41\exp(-0.88/\mathrm{Kn})\right]\mathrm{Kn}\right\}}{(1+3C_m\mathrm{Kn})\left(2\kappa_g + \kappa_p + 2\kappa_p C_t \mathrm{Kn}\right)} \frac{|\nabla T|}{T_0} \tag{7.131}$$

The best agreement between Equations 7.130 and 7.131 and experimental data suggests that $C_s = 1.17$, $C_t = 2.18$, and $C_m = 1.14$.

Although the above formulas are derived for small Kn, which is also considered as the continuum regime, they are supposed to work well for gases. In liquids, particle thermophoresis is fundamentally different from that in gases. In gases, the motion of particles is mainly caused by molecular collisions between gas molecules and the particles. In liquids, however, viscous effects play a dominant role. Furthermore, the dependence of viscosity on temperature for liquids is completely different from that for gases. The viscosity of gases increases while the viscosity of liquids usually decreases with increasing temperature. These reasons make Equations 7.127–7.131 inaccurate for particle thermophoresis in liquids.

Due to the difficulty of theoretical analysis for the thermophoretic phenomenon in liquids, practically, the thermophoretic force/velocity formulas for gases are modified to describe the motion of particles in liquids. For instance, Epstein's thermophoretic force in Equation 7.127 is revised to describe the thermophoresis of latex particles in water by a correction factor

$$F_T = -1.56\pi\mu\nu R \frac{\kappa_g}{2\kappa_g + \kappa_p} \frac{|\nabla T|}{T_0} \tag{7.132}$$

For nanoparticles, however, none of the above equations is suggested because they do not consider molecular interactions. This is manifested by other analysis that metal nanoparticles experience much larger thermophoretic forces than those predicted by Equation 7.132. The topic of nanoparticle thermophoresis in liquids remains a challenging scientific problem. For practical applications and on the basis of previous theoretical analyses, it would be reasonable to propose a general thermophoretic force law as

$$F_T = -K_{th}\mu\nu R \frac{|\nabla T|}{T_0} \tag{7.133}$$

where K_{th} is a coefficient depending on the physical properties of the liquid and nanoparticle. The values of K_{th} for different particles and liquids require extensive investigations.

References

Allen, M.D. and Raabe, O.G. 1982. Re-evaluation of Millikan's oil drop data for the motion of small particles in air. *J. Aerosol Sci.* 13:537–547.

Brock, J.R. 1962. On the theory of thermal forces acting on aerosol particles. *J. Colloid Sci.* 17:768–780.

Chapman, S. and Cowling, T.G. 1970. *The Mathematical Theory of Nonuniform Gases.* Cambridge: Cambridge University Press.

Cunningham, E. 1910. On the velocity of steady fall of spherical particles through fluid medium. *Proc. R. Soc. London, Ser. A* 83:357–365.

Dahneke, B.E. 1973. Slip correction factors for nonspherical bodies—I Introduction and continuum flow. *J. Aerosol Sci.* 4:139–145.

Davies, C.N. 1945. Definitive equations for the fluid resistance of spheres. *Proc. Phys. Soc.* 57:259–270.

Epstein, P.S. 1924. On the resistance experienced by spheres in their motion through gases. *Phys. Rev.* 23:710–733.

Epstein, P.S. 1929. Zur theorie des radiometers. *Z. Phys.* 54:537–563.

Hirschfelder, J.O., Curtiss, C.F., and Bird, R.B. 1954. *Molecular Theory of Gases and Liquids*. New York: Wiley.

Huner, B. and Hussey, R.G. 1977. Cylinder drag at low Reynolds number. *Phys. Fluids* 20:1211.

Knudsen, M. and Weber, S. 1911. Resistance to motion of small spheres. *Ann. Phys. (Leipzig)* 36:981–984.

Koenig Jr., G.M., Ong, R., Cortes, A.D., Moreno-Razo, J.A., de Pablo, J.J., and Abbott, N.L. 2009. Single nanoparticle tracking reveals influence of chemical functionality of nanoparticles on local ordering of liquid crystals and nanoparticle diffusion coefficients. *Nano Lett.* 9:2794–2801.

Kuga, Y., Okauchi, K., Takeda, D., Ohira, Y., and Ando, K. 2001. Classification performance of a low pressure differential mobility analyzer for nanometer-sized particles. *J. Nanoparticle Res.* 3:175–183.

Li, Z.G. 2005. Gas kinetic theory and molecular dynamics simulation of nanomaterial transport in dilute gases. Ph.D. Thesis, University of Delaware.

Li, Z.G. 2009. Critical particle size where the Stokes-Einstein relation breaks down. *Phys. Rev. E* 80:061204.

Li, Z.G. and Wang, H. 2003a. Drag force, diffusion coefficient, and electric mobility of small particles. I. Theory applicable to the free-molecule regime. *Phys. Rev. E* 68:061206.

Li, Z.G. and Wang, H. 2003b. Drag force, diffusion coefficient, and electric mobility of small particles. II. Application. *Phys. Rev. E* 68:061207.

Li, Z.G. and Wang, H. 2004. Thermophoretic force and velocity of nanoparticles in the free molecule regime. *Phys. Rev. E* 70:021205.

Liu, C.R., Li, Z.G., and Wang, H. 2016. Drag force and transport property of a small cylinder in free molecule flow: A gas-kinetic theory analysis. *Phys. Rev. E* 94:023102.

Lurio, L.B., Lumma, D., Sandy, A.R., Borthwick, M.A., Falus, P., and Mochrie, S.G., Pelletier, J.F., Sutton, M., Regan, L., Malik, A., and Stephenson, G.B. 2000. Absence of scaling for the intermediate scattering function of a hard-sphere suspension: Static and dynamic X-ray scattering from concentrated polystyrene latex spheres. *Phys. Rev. Lett.* 84:785–788.

Millikan, R.A. 1910. The isolation of an ion, a precision measurement of its charge, and the correction of Stokes's law. *Science* 32:436–448.

Millikan, R.A. 1917. A new determination of e, N, and related constants. *Philos. Mag.* 34:1–31.

Millikan, R.A. 1923. The general law of fall of a small spherical body through a gas, and its bearing upon the nature of molecular reflection from surfaces. *Phys. Rev.* 22:1–23.

Osburn, J.O., Stitzell, J.A., and Peterson, R.E. 1969. Diffusion of argon, krypton, and xenon in olive oil. *J. Appl. Physiol.* 27:624–629.

Rudyak, V.Ya. and Krasnolutski, S.L. 2001. Kinetic description of nanoparticle diffusion in rarefied gas. *Dokl. Phys.* 46:897–899.

Rudyak, V.Ya., Krasnolutski, S.L., Nasibulin, A.G., and Kauppinen, E.I. 2002. Methods of measuring the diffusion coefficient and sizes of nanoparticles in a rarefied gas. *Dokl. Phys.* 47:758–761.

Scheibel, H.G. and Porstendörfer, J. 1983. Generationof monodispersed Ag- and NaCl-aerosols with particle diameters between 2 and 300 nm. *J. Aerosol Sci.* 14:113–126.

Talbot, L., Cheng, R.K., Schefer, R.W., and Willis, D.R. 1980. Thermophoresis of particles in a heated boundary layer. *J. Fluid Mech.* 101:737–758.

Tang, W.Z. and Advani, S.G. 2006. Drag on a nanotube in uniform liquid argon flow. *J. Chem. Phys.* 125:174706.

Waldmann, L. 1959. Uber die Kraft eines inhomogenen Gases auf kleine suspendierte Kugeln. *Z. Naturforsch. A* 14a:589–599.

Wang, J. and Li, Z.G. 2011. Thermophoretic force on micro- and nanoparticles in dilute binary gas mixtures. *Phys. Rev. E* 84:021201.

Wang, J. and Li, Z.G. 2012. Negative thermophoresis of nanoparticles in the free molecular regime. *Phys. Rev. E* 86:011201.

Wong, R.Y.M., Liu, C., Wang, J., Chao, C.Y.H., and Li, Z.G. 2012. Evaluation of the drag force on single-walled carbon nanotubes in rarefied gases. *J. Nanosci. Nanotechnol.* 3:2311–2319.

Zhang, X., Tran, S., and Gray-Weale, A. 2016. Hydrodynamic drag on diffusing nanoparticles for size determination. *J. Phys. Chem. C* 120:21888–21896.

Chapter 8 Nanoscale electrokinetic phenomena

8.1 Classification of electrokinetic effects

Electrokinetics is concerned with the dynamics of charged species (e.g., ions) and the properties associated with the motion of charged species, such as fluid flows, ionic currents, and potentials generated by moving charges. Generally, electrokinetic phenomena include electroosmosis, electrophoresis, and streaming current/potential.

Electroosmosis is about the motion of a fluid caused by the movement of charged particles under an external electric field. For an electrolyte next to a charged surface, an electric double layer (EDL) develops. Once an electric field is applied through the EDL, the electrostatic forces on the counterions in the EDL drive ions and cause the fluid to move, generating an electroosmotic flow. In micro- and nanofluidics, electroosmotic flows have attracted great attention due to the potential applications in various areas, including biology, medicine, and engineering.

Electrophoresis is referred to as the motion of charged particles or molecules dispersed in a fluid under an electric field.

Streaming current/potential is induced by driving a charge-containing fluid (e.g., electrolyte) through a charged channel by a non-electrostatic force, say, a pressure gradient. If the upstream and downstream of the flow is connected, the motion of the charged particles generates a current, which is termed the streaming current. However, if the flow is not a part of a closed "circuit," an electric field is generated over the channel due to the motion of charges and the potential difference between the ends of the channel is denoted as the streaming potential.

8.2 Nanoscale electrokinetic effects

Electrokinetic effects at the nanoscale possess unique features that are significantly different from those at the micro- and macroscale (Schoch et al., 2008). As introduced in Chapter 1, an EDL develops next to a charged surface in an electrolyte. The ion concentration and mobility in the EDL are different from their bulk values. The Debye length, λ_D, which characterizes the thickness of EDL, depends on the bulk concentration of the charged species, n_0, and can vary roughly from a few angstroms (at $n_0 \sim 1$ M) to hundreds of nanometers (at $n_0 \sim 10^{-6}$ M). In a micro- or macroscale system, the Debye length can be neglected as compared with the characteristic length of the system. In regions away from a charged surface, the electrolyte solution is neutral and can be treated as a bulk solution. Therefore, the EDL barely affects the ion and fluid dynamics, as demonstrated in the top panel of Figure 8.1. In a nanofluidic system, however, the Debye length can be comparable to or even larger than the dimension of a flow channel. This results in the overlapping of EDLs and makes the flow domain counterion-dominant, as shown in the bottom panel of Figure 8.1. In such a case, the transport of counterions becomes predominant, while the passage of co-ions (ions having the same charge sign to the surface charge) is not favored.

Due to the overlapping of EDLs and the changes of fluid properties, ion transport at the nanoscale shows versatile phenomena that cannot be observed at large scales.

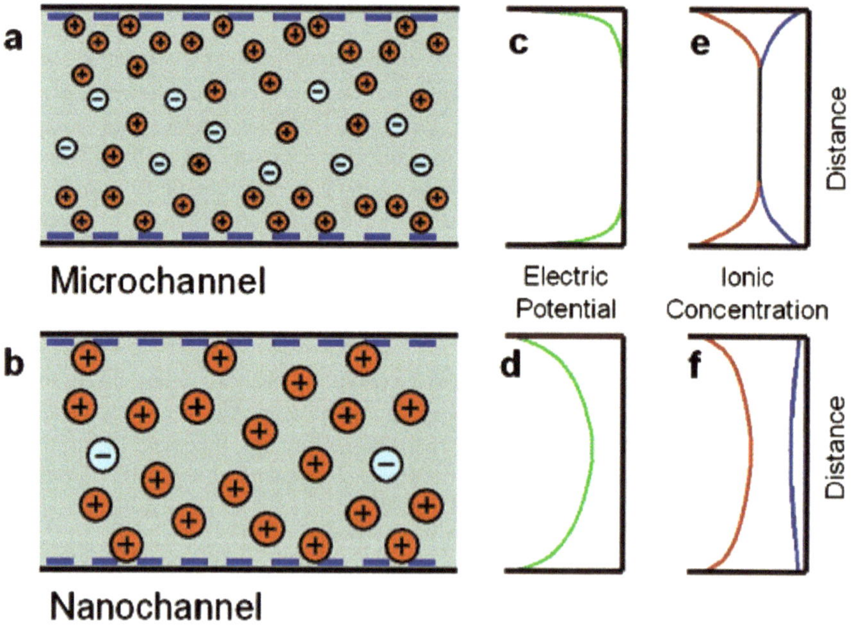

Figure 8.1 Electric potential and ion density distributions in a charged microchannel (top panel) and a charged nanochannel with EDL overlapping (bottom panel). (a) and (b) Schematic of ion distribution. (c) and (d) Illustration of the electric potential. (e) and (f) Sketch of the ionic concentration distribution (red and blue lines are for counterions and co-ions, respectively). (Reprinted with permission from Karnik et al., 2005, Copyright (2005) American Chemical Society.)

For instance, surface-charge governed ion transport has been widely observed in nanochannels (Stein et al., 2004), where the ionic conductance can assume roughly a constant value at a low solution concentration rather than showing the classic, linear dependence on the strength of the external electric field. Moreover, nanopores with asymmetric geometries or surface charges can be used to rectify ion transport such that the passage of ions is preferred in a particular direction and hindered in the reverse direction, exhibiting similar transport behavior to the fluidic diodes discussed in Chapter 6.

Although the classic electrostatic and electrokinetic theories, such as the Poisson–Boltzmann equation and the Poisson–Nernst–Planck equation, have been shown to work well for nanoscale ion transport, they do not consider the surface effects and the overlapping of EDLs. Such intrinsic limitations can cause the classic theories to break down in describing ion dynamics in nanochannels, especially in nanoconfinements with characteristic dimensions <2 nm, as will be shown later. This is why many surprising results and ion transport phenomena at the nanoscale have been found in experiments and numerical investigations, including fast ion transport in CNTs, ionic current rectification, stochastic ionic current fluctuations, and ion sieving. These new ion transport phenomena at the nanoscale provide unique opportunities in a variety of applications, such as biomolecule separation, seawater desalination, sensing, energy conversion and harvesting, as well as the design of novel functional nanofluidic devices.

8.3 Theoretical framework for electrostatics and electrokinetics

In Chapter 1, basic concepts of electrostatics and electrokinetics are briefly introduced. Although the classic theories have limitations, they play critical roles in describing fluid and ion transport at the nanoscale. In many cases, they are capable of providing important and quite accurate information about the electric potential, charge density distribution, and velocity field. Therefore, in this section, the classic theories of electrostatics and electrokinetics will be discussed in detail (some results in Chapter 1 might be reproduced).

8.3.1 The Poisson–Boltzmann equation

The Poisson equation, given by Equation 1.49, describes the electric potential for a given charge density ρ_e and electric permittivity $\acute{\varepsilon}$ of the medium. For an electrolyte, $\rho_e = \sum_i z_i e n_i$ (Equation 1.51), which depends on the valence and number density of the ions. If it is assumed that (i) ions are considered as point charges, i.e., the size of ions is neglected, (ii) the solvent is treated as a continuous medium with the permeability, $\acute{\varepsilon}$, being a constant, and (iii) only electrostatic forces are considered for ion adsorption, it can be shown that the ion density follows the Boltzmann distribution given by Equation 1.53. Substituting Equations 1.51 and 1.53 into the Poisson equation, it is obtained that (if the notation in Figure 8.2 is used)

$$\frac{d^2\phi}{dy^2} = -\frac{e}{\dot{\varepsilon}}\sum_i n_{i,0} z_i \exp\left(-z_i e\phi/k_B T\right) \tag{8.1}$$

Equation 8.1 is the Poisson–Boltzmann equation, which is simplified to Equation 1.54 for a symmetric electrolyte. For certain cases, with proper boundary conditions, the analytical solution of the Poisson–Boltzmann equation can be obtained. However, in many situations, it needs to be numerically solved.

8.3.2 The Debye–Hückel approximation

If the surface potential is low, the exponential term in Equation 8.1 can be approximated as

$$\exp\left(-z_i e\phi/k_B T\right) \approx 1 - z_i e\phi/k_B T \tag{8.2}$$

Substituting Equation 8.2 into 8.1 and using $\sum_i n_{i,0} z_i = 0$, Equation 8.1 can be rewritten as

$$\frac{d^2\phi}{dy^2} = \kappa^2\phi(y) \tag{8.3}$$

where $\kappa = \sqrt{e^2 \sum_i n_{i,0} z_i^{\,2} / (\dot{\varepsilon} k_B T)}$ is the Debye–Hückel parameter (in Chapter 1, κ is given for symmetric electrolytes). Integrating Equation 8.3 after multiplying both sides by $2d\phi/dy$, it is obtained that

$$\int_\infty^y d\left(\frac{d\phi}{dy}\right)^2 = 2\kappa^2 \int_0^{\phi(y)} \phi\, d\phi \tag{8.4}$$

where it is assumed that $\phi = 0$ when $y \to \infty$ (far away from the surface, Figure 8.2). If the boundary condition, $d\phi/dy = 0$ for $y \to \infty$, is used, Equation 8.4 can be solved and it gives

$$\frac{d\phi}{dy} = -\kappa\phi \tag{8.5}$$

The solution of Equation 8.5 reads

$$\phi(y) = \phi_s \exp(-\kappa y) \tag{8.6}$$

where ϕ_s is the potential at the surface, $y = 0$ (Figure 8.2). It is noted that Equation 8.6 is the same as Equation 1.57 if ϕ_s is replaced by the zeta potential ζ.

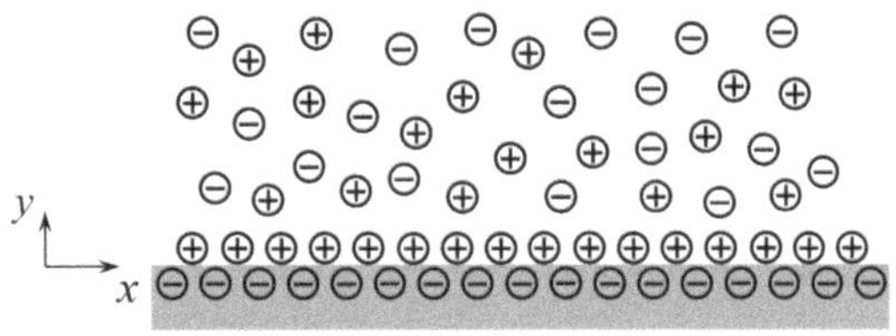

Figure 8.2 Coordinate system and schematic of ion distribution next to a charged surface.

The above linearization process is called the Debye–Hückel approximation and Equation 8.6 provides an estimation for the potential distribution near a surface. It should be noted that Equation 8.6 works well only for the cases with very low surface potential. At the nanoscale with significant EDL overlapping, the Debye–Hückel approximation may break down.

8.3.3 The Gouy–Chapman model

For symmetric electrolytes, co-ions and counterions are equal in valence. For simplicity, let us consider an electrolyte containing monovalent ions only such that $|z_i| = 1$ (e.g., NaCl solutions). The Poisson–Boltzmann equation, Equation 8.1, is then simplified as

$$\frac{d^2\phi}{dy^2} = \frac{k_B T \kappa^2}{e} \sinh\left(\frac{e\phi}{k_B T}\right) \tag{8.7}$$

If $\phi^* = e\phi/(k_B T)$ is defined, Equation 8.7 can be rewritten as

$$\frac{d^2\phi^*}{dy^2} = \kappa^2 \sinh\left(\phi^*\right) \tag{8.8}$$

Integrating this equation once with boundary conditions: $d\phi^*/dy = 0$ and $\phi^* = 0$ for $y \to +\infty$, one obtains

$$\frac{d\phi^*}{dy} = \kappa\sqrt{2\cosh\phi^* - 2} \tag{8.9}$$

Integrating Equation 8.9, it is obtained that

$$\tanh\left(\frac{\phi^*}{4}\right) = \tanh\left(\frac{\phi_s^*}{4}\right)\exp(-\kappa y) \tag{8.10}$$

where $\phi_s^* = e\phi_s/(k_B T)$. This equation is called the Gouy–Chapman equation. For small values of ϕ^* and ϕ_s^*, $\tanh(\phi^*/4) \approx \phi^*/4$ and $\tanh(\phi_s^*/4) \approx \phi_s^*/4$, and Equation 8.10 is reduced to the Debye–Hückel approximation, Equation 8.6.

The analytical solution of the Gouy–Chapman equation takes the form of

$$\phi(y)=\phi^*(y)\frac{k_BT}{e}=-\frac{2k_BT}{e}\ln\left(\frac{1+\hat{\gamma}e^{-\kappa y}}{1-\hat{\gamma}e^{-\kappa y}}\right) \tag{8.11}$$

where the coefficient $\hat{\gamma}=-\tanh(\phi_s^*/4)>0$ is determined by the potential at the surface. With Equation 8.11 and the Boltzmann distribution equation (Equation 1.53), the number density distributions for counterions and co-ions, n_+ and n_-, from the Gouy–Chapman model are given as

$$n_\pm(y)=n_0\left(\frac{1\pm\hat{\gamma}e^{-\kappa y}}{1\mp\hat{\gamma}e^{-\kappa y}}\right)^2 \tag{8.12}$$

where n_0 is the bulk density.

For the case with a constant surface-charge density n_s, the coefficient $\hat{\gamma}$ can be determined from the boundary condition. From Equation 8.11, the electric field at the surface can be calculated as

$$E\big|_{y=0}=-\frac{d\phi}{dy}\bigg|_{y=0}=\frac{4\kappa k_BT}{e}\frac{\hat{\gamma}}{1-\hat{\gamma}^2} \tag{8.13}$$

On the other hand, the electric field at the surface can also be obtained by Gauss' law,

$$E\big|_{y=0}=\frac{|n_s|}{\dot{\varepsilon}} \tag{8.14}$$

A combination of Equations 8.13 and 8.14 gives

$$\hat{\gamma}^2+2\kappa\lambda_{GC}\hat{\gamma}-1=0 \tag{8.15}$$

where $\lambda_{GC}=2k_BT\dot{\varepsilon}/(|n_s|e)$ is the Gouy–Chapman length. The positive root of Equation 8.15 gives $\hat{\gamma}$ as

$$\hat{\gamma}=-\kappa\lambda_{GC}+\sqrt{(\kappa\lambda_{GC})^2+1} \tag{8.16}$$

The ion density distributions obtained using the Gouy–Chapman model are depicted in Figure 8.3(a), which clearly shows an accumulation of excess counterions (n_+) close to the surface ($y = 0$). In contrast, the Debye–Hückel approximation provides symmetric ion density distributions, as shown in Figure 8.3(b).

8.3.4 Electric potential between two interacting electric double layers

In the analysis of micro- and macroscale electroosmotic flows, the thickness of the EDL plays a minor role because it is much smaller than the channel height. Therefore, the electric potential and net charge density can be assumed to be zero

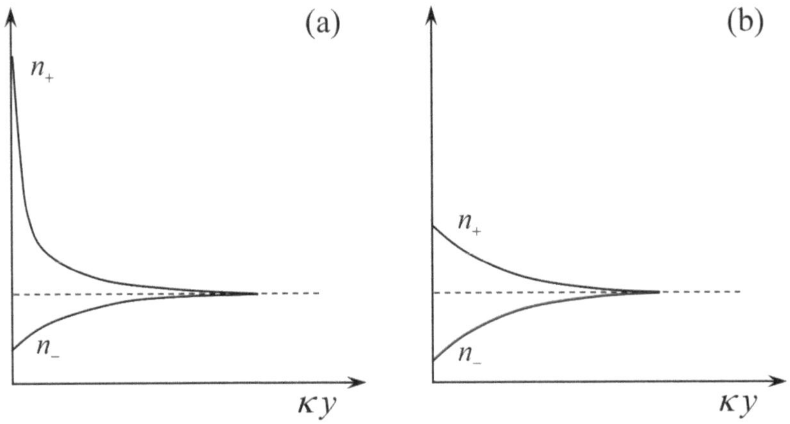

Figure 8.3 Schematic of ion density distributions predicted by the Gouy–Chapman model (Equation 8.12) (a) and the Debye–Hückel approximation (b).

in the middle part of the channel. However, in nanochannels with characteristic dimensions comparable to the Debye length, EDL overlapping is a common phenomenon. In this situation, the electric potential and net charge density in the channel are nonuniform and the results for micro- and macrochannels cannot be used. To describe electrokinetic phenomena in nanochannels, it is essential to accurately predict the electric potential under the situation of interacting EDLs.

The electric potential between two overlapping EDLs is first considered in the area of colloid science (Verwey and Overbeek, 1948). It is obtained by solving the Poisson–Boltzmann equation under the Gouy–Chapman model, Equation 8.8. For an electrolyte between two parallel, charged plates, as sketched in Figure 8.4, if Equation 8.8 is integrated once with boundary conditions: $d\phi^*/dy\big|_{y=h} = 0$ and $\phi^* = \phi^*_{\mathrm{m}}$ at the mid plane ($y = h$), it yields

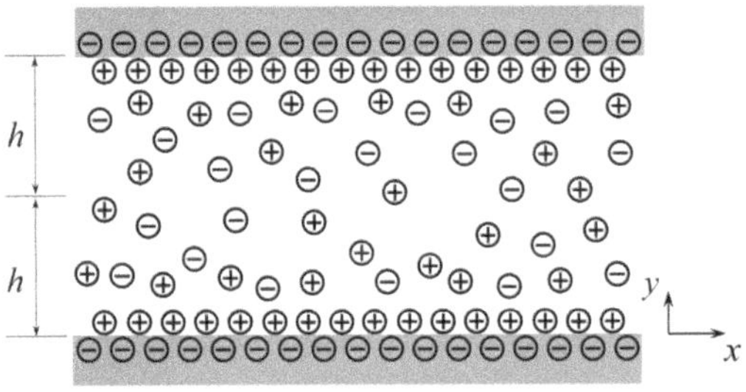

Figure 8.4 Notation and illustration of an electrokinetic system between two parallel, charged surfaces.

$$\frac{d\phi^*}{dy} = -\kappa\left(2\cosh\phi^* - 2\cosh\phi_m^*\right)^{1/2} \tag{8.17}$$

Integrating Equation 8.17 gives

$$\kappa y = -\int_{\phi_s^*}^{\phi^*} \frac{d\phi^*}{\sqrt{2\cosh\phi^* - 2\cosh\phi_m^*}} \tag{8.18}$$

which is in the form of the elliptic integral. With the introduction of two new substitutions: $m = \exp\left(-\phi_m^*\right)$ and $m\sin^2\theta = \exp\left(-\phi^*\right)$, Equation 8.18 can be rewritten as

$$\kappa y = 2\exp\left(-\phi_m^*/2\right)\int_{\theta(\phi_s^*)}^{\theta(\phi^*)} \frac{d\theta}{\left(1 - m^2\sin^2\theta\right)^{1/2}} \tag{8.19}$$

where $\theta\left(\phi_s^*\right) = \arcsin\left[\exp\left(-\left(\phi_s^* - \phi_m^*\right)/2\right)\right]$ and $\theta\left(\phi^*\right) = \arcsin\left[\exp\left(-\left(\phi^* - \phi_m^*\right)/2\right)\right]$. The integral in Equation 8.19 is the standard elliptic integral of the first kind

$$F\left(b\middle|a^2\right) = \int_0^b \frac{d\theta}{\left(1 - a^2\sin^2\theta\right)^{1/2}} \tag{8.20}$$

With Equation 8.20, the electric potential at an arbitrary position y can be obtained from the following equation:

$$\kappa y = 2\exp\left(-\phi_m^*/2\right)\left[F\left(\theta\left(\phi^*\right)\middle|m^2\right) - F\left(\theta\left(\phi_s^*\right)\middle|m^2\right)\right] \tag{8.21}$$

The inverse relation of this equation gives the electric potential, which can be written in the form of the Jacobi elliptical function, $\mathrm{CD}\left(u\middle|a^2\right)$,

$$\phi^* = \phi_m^* + 2\ln\left[\mathrm{CD}\left(\frac{\kappa(y-h)}{2}\exp\left(-\phi_m^*/2\right)\middle|\exp\left(2\phi_m^*\right)\right)\right] \tag{8.22}$$

Under specific conditions, approximate electric potential distributions of simple forms can be determined (Verwey and Overbeek, 1948; Hunter, 1981).

8.3.4.1 The Debye–Hückel approximation If the electric potential in the flow domain is small, the Debye–Hückel approximation can be employed. Multiplying both sides of Equation 8.3 by $2d\phi/dy$ and then performing integration yield

$$\left(\frac{d\phi}{dy}\right)^2 = \kappa^2\phi^2 + C \tag{8.23}$$

If the potential gradient at the mid plane vanishes, i.e., $d\phi/dy\big|_{y=h} = 0$, which is a boundary condition, the integration constant is determined as $C = -\kappa^2\phi_m^2$. Then, Equation 8.23 can be written as

$$\frac{d\phi}{dy} = \kappa\sqrt{\phi^2 - \phi_{\rm m}^2} \tag{8.24}$$

and the electric potential is given by

$$\phi(y) = \phi_{\rm s}\frac{\cosh\left[\kappa(h-y)\right]}{\cosh(\kappa h)} \tag{8.25}$$

8.3.4.2 Weak EDL overlap approximation If the interaction or overlap of EDLs is relatively weak, the electric potential between two plates can be assumed to be additive such that

$$\phi(y) = \phi_1(y) + \phi_2(2h-y) \tag{8.26}$$

where ϕ_1 and ϕ_2 are the electric potentials generated by the bottom and top plates, respectively.

Using the Gouy–Chapman equation, Equation 8.10, it is obtained that

$$\phi^*(y) = 4\left\{\operatorname{arctanh}\left(\tanh\left(\frac{\phi_{\rm s}^*}{4}\right)\exp^{-\kappa y}\right) + \operatorname{arctanh}\left(\tanh\left(\frac{\phi_{\rm s}^*}{4}\right)\exp^{-\kappa(2h-y)}\right)\right\} \tag{8.27}$$

and the mid plane potential $\phi_{\rm m}$ can be estimated through

$$\phi_{\rm m}^* = 8\tanh\left(\frac{\phi_{\rm s}^*}{4}\right)\exp(-\kappa h) \tag{8.28}$$

8.3.4.3 Strong EDL overlap For the case with strong EDL overlap (or high surface potential), Equation 8.18, with $\phi_{\rm m}^* >> 1$ and $\phi^* > \phi_{\rm m}^*$, can be revised as (Verwey and Overbeek, 1948; Hunter, 1981)

$$\kappa y = -\int_{\phi_{\rm s}^*}^{\phi^*} \frac{d\phi^*}{\left(e^{\phi^*} - e^{\phi_{\rm m}^*}\right)^{1/2}} = -e^{-\phi_{\rm m}^*/2}\int_{\phi_{\rm s}^*}^{\phi^*} \frac{e^{-\phi^*/2}d\phi^*}{\left(e^{-\phi_{\rm m}^*} - e^{-\phi^*}\right)^{1/2}} \tag{8.29}$$

With the relationship, $\exp\left(-\phi^*/2\right) = \exp\left(-\phi_{\rm m}^*/2\right)\sin\theta$, Equation 8.29 can be written as

$$\kappa y = 2\exp\left(-\phi_{\rm m}^*/2\right)\left\{\arcsin\left\{\exp\left[-\left(\phi^* - \phi_{\rm m}^*\right)/2\right]\right\} - \arcsin\left\{\exp\left[-\left(\phi_{\rm s}^* - \phi_{\rm m}^*\right)/2\right]\right\}\right\} \tag{8.30}$$

At the mid plane $y = h$, Equation 8.30 gives

$$\kappa h = 2\exp\left(-\phi_{\rm m}^*/2\right)\arccos\left\{\exp\left[-\left(\phi_{\rm s}^* - \phi_{\rm m}^*\right)/2\right]\right\} \tag{8.31}$$

If the surface potential, ϕ_s^*, is given, the electric potential distribution is determined as

$$\phi^*(y) = \phi_m^* - 2\ln\cos\left\{\arccos\exp\left[-\left(\phi_s^* - \phi_m^*\right)/2\right] - \frac{\kappa y}{2}\exp\left(\phi_m^*/2\right)\right\} \quad (8.32)$$

8.3.5 Governing equations for electrokinetics

To describe an electrokinetic problem, the Nernst–Planck equation that accounts for ion transport, the Poisson equation, which relates the electric potential with the local ion density, and the Naiver–Stokes equation, are needed. These equations for an electrokinetic system are given by

$$\mathbf{J}_i = -D_i\left(\nabla n_i + n_i \frac{z_i e}{k_B T}\nabla\phi\right) + n_i\mathbf{V} \quad (8.33)$$

$$\nabla^2\phi = -\frac{\rho_e}{\varepsilon_0\varepsilon_r} \quad (8.34)$$

$$\rho\left(\frac{\partial\mathbf{V}}{\partial t} + \mathbf{V}\cdot\nabla\mathbf{V}\right) = -\nabla p + F_E + \mu\nabla^2\mathbf{V} \quad (8.35)$$

where $\mathbf{J}_i$ is the flux of species i, $\mathbf{V}$ is the velocity, and F_E is the external force due to the electric field. F_E is given by

$$F_E = \rho_e E = \varepsilon_0\varepsilon_r\left(\nabla^2\phi\right)\nabla\phi \quad (8.36)$$

where $E = -\nabla\phi$ is the external electric field strength.

It must be noted that the Naiver–Stokes equation is usually solved with the stick boundary condition. However, at the nanoscale, as discussed in Chapter 6, flow slippage may take place at a fluid–solid interface and the no-slip boundary condition is invalid. If the Naiver–Stokes equation is solved using the stick boundary condition for an electrokinetic problem, the ion transport properties (e.g., ionic conductance) may be significantly underestimated (Liu et al., 2010). Therefore, a slip boundary condition for nanoscale electrokinetic systems may need to be used.

8.3.6 Electroosmotic flows between two parallel plates

For an electroosmotic flow solely driven by an external electric field, it has been shown in Chapter 1 that the electroosmotic flow velocity is given by Equation 1.60, which, if the surface potential is denoted as ϕ_s, reads

$$u_{eo} = \frac{\varepsilon_0\varepsilon_r E}{\mu}\left[\phi(y) - \phi_s\right] \quad (8.37)$$

For a macroscale system, the distance between the two plates is much larger than the Debye length ($\kappa h >> 1$) and the thickness of the EDL can be neglected. The electric potential $\phi(y)$ can be assumed to be zero in the flow field and Equation 8.37 is reduced to the Helmholtz–Smoluchowski equation,

$$u_{eo} = -\frac{\varepsilon_0 \varepsilon_r \phi_s}{\mu} E \tag{8.38}$$

For steady electroosmotic flows between two parallel, infinite plates (let us use Figure 8.4 for notation), with the consideration of a pressure gradient, dp/dx, the Naiver–Stokes equation, Equation 8.35, can be written as

$$-E_x \varepsilon_0 \varepsilon_r \frac{d^2\phi}{dy^2} + \mu \frac{d^2 u}{dy^2} - \frac{dp}{dx} = 0 \tag{8.39}$$

It can be shown that the mean velocity $\bar{u}$ is given by (Hunter, 1981)

$$\bar{u} = -\frac{h^2}{3\mu}\frac{dp}{dx} - \frac{\varepsilon_0 \varepsilon_r E_x \phi_s}{\mu}\left[1 - H\left(\phi_s^*, \kappa h\right)\right] \tag{8.40}$$

The first term on the right-hand side is the mean velocity due to the pressure gradient. The second term is the Smoluchowski relation with a correction function H, which is defined by (Burgreen and Nakache, 1964)

$$H\left(\phi_s^*, \kappa h\right) = \frac{\bar{\phi}(y)}{\phi_s^*} \tag{8.41}$$

where $\bar{\phi}(y) = \frac{1}{h}\int_0^h \phi(y) dy$ is the average potential.

An accurate description of the electric potential is in the elliptic form given by Equation 8.22, for which it is difficult to obtain the analytical expression for the correction function. In the Debye–Hückel limit, i.e., the electric potential is small in the flow field, Equation 8.25 applies and the correction function, H, is determined as

$$H = \frac{\tanh \kappa h}{\kappa h} \tag{8.42}$$

and the mean flow velocity reads

$$\bar{u} = -\frac{h^2}{3\mu}\frac{dp}{dx} - \frac{\varepsilon_0 \varepsilon_r E_x \phi_s}{\mu}\left(1 - \frac{\tanh \kappa h}{\kappa h}\right) \tag{8.43}$$

8.3.7 Electrokinetic flows in cylinders

Rice and Whitehead (1965) first formulized the electrolyte transport in a cylindrical capillary. Let us consider a long, uniform, circular tube with radius R filled with a 1:1 electrolyte of bulk concentration n_0. In cylindrical polar coordinates, the Poisson equation is given by

$$\frac{1}{r}\frac{d}{dr}\left(r\frac{d\phi}{dr}\right)=-\frac{\rho_c(r)}{\varepsilon_0\varepsilon_r} \tag{8.44}$$

where $\rho_c(r)$ is the net charge density at r, which is assumed to follow the Boltzmann distribution. In a cylinder, ρ_c is given by

$$\rho_c(r)=-2en_0\sinh\frac{e\phi}{k_BT} \tag{8.45}$$

In the Debye–Hückel limit, $\sinh\frac{e\phi}{k_BT}\approx\frac{e\phi}{k_BT}$, $\rho_c(r)$ can be written as

$$\rho_c(r)=-\varepsilon_0\varepsilon_r\kappa^2\phi \tag{8.46}$$

A combination of Equations 8.44 and 8.46 leads to the Poisson–Boltzmann equation in cylindrical polar coordinates,

$$\frac{1}{r}\frac{d}{dr}\left(r\frac{d\phi}{dr}\right)=\kappa^2\phi \tag{8.47}$$

The solution of Equation 8.47 gives the electric potential

$$\phi=BI_0(\kappa r) \tag{8.48}$$

where B is a constant and I_0 is the zeroth-order-modified Bessel function of the first kind. With the boundary condition at the wall of the tube, $\phi=\phi_s$, Equation 8.48 can be written as

$$\phi=\phi_s\frac{I_0(\kappa r)}{I_0(\kappa R)} \tag{8.49}$$

With this electric potential, the net charge distribution $\rho_c(r)$ can be obtained from Equation 8.46,

$$\rho_c(r)=-\varepsilon_0\varepsilon_r\kappa^2\phi_s\frac{I_0(\kappa r)}{I_0(\kappa R)} \tag{8.50}$$

If an external electric field E_z is applied in the axial direction of the tube, say, the z-direction, an electroosmotic flow will be generated and the flow velocity, $V_z=V_z(r)$, can be derived from the Naiver–Stokes equation, which, in cylindrical polar coordinates, is given by

$$\frac{d^2V_z}{dr^2}+\frac{1}{r}\frac{dV_z}{dr}=\frac{1}{\mu}\frac{dp}{dz}-\frac{E_z\rho_c(r)}{\mu} \tag{8.51}$$

Substituting Equation 8.50 into Equation 8.51 and using boundary conditions, $dV_z/dr\big|_{r=0}=0$ and $V_z(R)=0$ yield

$$V_z(r)=\frac{1}{4\mu}\frac{dp}{dz}\left(R^2-r^2\right)-\frac{\varepsilon_0\varepsilon_r\phi_s E_z}{\mu}\left[1-\frac{I_0(\kappa r)}{I_0(\kappa R)}\right] \tag{8.52}$$

where the first term is the solution of Poiseuille's flows and the second term is the electrokinetic contribution. In the absence of a pressure gradient, the velocity profile becomes

$$V_z(r)=-\frac{\varepsilon_0\varepsilon_r\phi_s E_z}{\mu}\left[1-\frac{I_0(\kappa r)}{I_0(\kappa R)}\right] \tag{8.53}$$

If $\kappa R>>1$, i.e., large tube radius, $I_0(\kappa r)/I_0(\kappa R)\approx 0$. In this case, Equation 8.53 is reduced to the classic solution,

$$V_z(r)=-\frac{\varepsilon_0\varepsilon_r\phi_s E_z}{\mu} \tag{8.54}$$

For $\kappa R<<1$, with the series expansion of I_0, the solution can be approximated as

$$V_z(r)=-\frac{\varepsilon_0\varepsilon_r\kappa^2\phi_s E_z}{4\mu}\left(R^2-r^2\right) \tag{8.55}$$

which, similar to the Poiseuille flow, exhibits a parabolic distribution.

The volumetric flow rate, Q, can be obtained by integrating Equation 8.53,

$$Q=2\pi\int_0^R V_z(r)r\,dr=-\frac{\varepsilon_0\varepsilon_r\phi_s E_z\pi R^2}{\mu}\left[1-\frac{2I_1(\kappa R)}{\kappa R I_0(\kappa R)}\right] \tag{8.56}$$

where I_1 is the first-order modified Bessel function of the first kind.

The ionic current, I, generated by the motion of charges contains two parts, the electroosmotic current, I_{eo}, and electrophoretic current, I_{ep}, i.e., $I=I_{eo}+I_{ep}$. I_{eo} can be obtained through

$$I_{eo}=2\pi\int_0^R V_z(r)\rho_c(r)r\,dr=-\frac{\left(\varepsilon_0\varepsilon_r\phi_s\right)^2\kappa^2 E_z\pi R^2}{\mu}\left[1-\frac{2I_1(\kappa R)}{\kappa R I_0(\kappa R)}-\frac{I_1^{\,2}(\kappa R)}{I_0^{\,2}(\kappa R)}\right] \tag{8.57}$$

where Equations 8.53 and 8.50 for $V_z(r)$ and $\rho_c(r)$ are used. The determination of I_{ep} is relatively complex. For large or uncharged tubes, I_{ep} can be given by the bulk conductivity,

$$I_{ep} = 2Zen_0E_z\pi R^2 \tag{8.58}$$

where Z is the mobility of cations and/or anions, as will be introduced in the next section. For small or charged tubes, however, the exact expression of I_{ep} requires the consideration of the ion concentration distribution in the tube given by Equation 8.45. The total ionic current is obtained through (Hunter, 1981)

$$I = I_{eo} + I_{ep} = \int_A \left[V_z(r)\rho_c(r) + 2Zen_0E_z\cosh\phi^*(r)\right]dA \tag{8.59}$$

8.3.8 Ionic conductance and ion mobility in nanochannels

As ions migrate under an external electric field, E, in a channel, the mobility of ions, $Z_\pm$, is an important variable. It is defined as

$$Z_\pm = \frac{v_\pm}{E} \tag{8.60}$$

where $v_\pm$ is the velocity of cations/anions. The ion mobility determines the ionic conductivity, σ_e, and the ionic current I. For a 1:1 electrolyte (e.g., NaCl solutions), the ionic conductivity is given by

$$\sigma_e = n_0e(Z_+ + Z_-) \tag{8.61}$$

and the ionic current I is given by

$$I = \sigma_eEA \tag{8.62}$$

where A is the cross-sectional area of the channel. The ionic conductance due to the motion of ions, G, is defined as the ratio of the ionic current to the voltage, V, i.e.,

$$G = \frac{I}{V} \tag{8.63}$$

As $V = EL$ with L being the length of the channel, G can rewritten as

$$G = \sigma_e\frac{A}{L} \tag{8.64}$$

As σ_e depends on the bulk ion concentration and mobilities, the conductance given by Equation 8.64 is commonly termed bulk conductance.

In nanochannels, a surface conductance G_s is induced by the counterions in the EDL. The surface conductance G_s of a slit channel (negatively charged) can be estimated as (Schoch and Renaud, 2005)

$$G_s = 2|n_s|Z_+\frac{w}{L} \tag{8.65}$$

where n_s is the surface-charge density and w is the width of the channel. For cylindrical nanochannels with a diameter d, G_s is given by

$$G_s = |n_s|Z_+\frac{\pi d}{L} \tag{8.66}$$

Taking slit nanochannels for example, the ratio of the surface conductance to the bulk conductance is $|n_s|/(n_0 eh)$, where h is the channel height. Therefore, in nanochannels (h is small), the surface conductance can be significant and the total ionic conductance is given by

$$G = n_0 e(Z_+ + Z_-)\frac{A}{L} + 2|n_s|Z_+\frac{w}{L} \tag{8.67}$$

for slit nanochannels and

$$G = n_0 e(Z_+ + Z_-)\frac{\pi d^2}{4L} + |n_s|Z_+\frac{\pi d}{L} \tag{8.68}$$

in cylindrical channels.

8.4 Nanoconfinements for electrokinetic transport

Electrokinetic transport at the nanoscale strongly depends on the geometry and surface conditions of a nanoconfinement. In the literature, a variety of nanoconfinements have been employed, including lab-fabricated nanochannels, nanotubes, two-dimensional (2D) nanopores, and lamellar membranes, as shown in Figure 8.5. In this section, these nanoconfinements will be briefly introduced before the detailed discussion of nanoscale electrokinetic transport.

8.4.1 Slit nanochannels

Lab-fabricated nanochannels are typically slit nanochannels with a rectangular cross-section but a very high width-to-height ratio. The bulk machining method introduced in Chapter 4 is capable of fabricating such channels. The height of most slit nanochannels ranges from several nanometers to about 100 nm and the length is typically of a few micrometers. To date, the smallest channels are 2 nm in height (Duan and Majumdar, 2010). Glass, silicon, PDMS, and PMMA are the most popular materials for nanochannel fabrication and the channel surfaces are usually negatively charged.

8.4.2 Nanotubes

Nanotubes, including, CNTs, boron nitride nanotubes (BNNTs), silicon carbide nanotubes (SiCNTs), and silicon nitride nanotubes (SiNNTs), are a group of

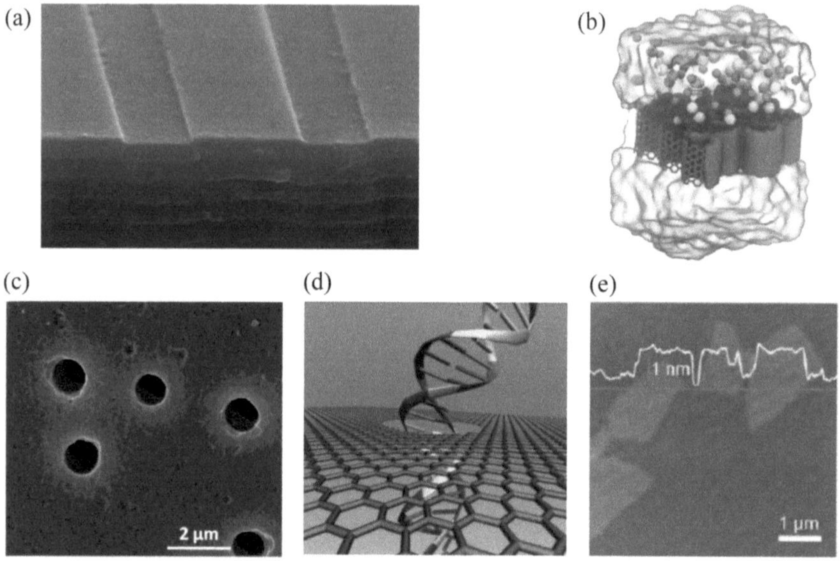

Figure 8.5 Representative nanoconfinements for ion transport. (a) Lab-fabricated slit nanochannels (Reprinted from Li, 2016, by courtesy of the author). (b) Carbon nanotubes (CNTs) (Reprinted with permission from Thomas and Corry, 2016). (c) Solid-state nanopores (Reproduced from Zhai et al., 2014, with permission of The Royal Society of Chemistry). (d) Graphene nanopores (Reprinted by permission from Springer Nature, Siwy and Davenport, 2010, Copyright (2010)). (e) Graphene oxide membranes.

nanomaterials, which can serve as one-dimensional nanochannels for electrokinetic analysis.

These nanotubes share some common properties, such as small tube diameter and smooth surface. However, flow dynamics in these tubes could be different, depending on the size and surface properties. For example, the slip length in CNTs can be much larger than that in BNNTs if the tube radius is smaller than 30 nm (Secchi et al., 2016a). Nevertheless, the surface charge of these tubes can be varied through proper doping, which offers great flexibility in studying nanoscale electrokinetic phenomena. Ion transport in these nanotubes has been shown processing great potential for energy conversion/harvesting and seawater desalination.

8.4.3 Solid-state nanopores

Solid-state nanopores can be fabricated on a substrate (e.g., SiN_x, SiO_2, or a polymer film) using focused ion beams or electron beams and the typical pore diameter ranges from a few nanometers to tens of nanometers. Recent advances in nanofabrication techniques can make the pore diameter down to 1 nm (Dekker, 2007). Solid-state nanopores usually have good mechanical and chemical stability. The pore diameter and length can be well controlled. The surface properties can also be easily tuned. These features make solid-state nanopores popular structures for investigating fluid and electrokinetic transport. A very important phenomenon observed for ion transport in solid-state nanopores is the ionic current rectification, as will be discussed later in this chapter. Solid-state nanopores have also been widely used for studying DNA translocation, which may be employed for the next-generation DNA sequencing. They can also be used for molecular separation and detection.

8.4.4 Two-dimensional nanomaterial-based nanopores

Graphene and molybdenum disulfide (MoS_2) nanosheets are representative 2D nanomaterials. Graphene is a single, atomic layer carbon material with carbon atoms positioned on hexagonal lattice sites. A single-layer MoS_2 nanosheet has three atomic layers (two S layers and one Mo layer), as will be shown later. Nanopores in a graphene or MoS_2 nanosheet can be generated by removing some atoms, as shown in Figure 8.6 (Schneider et al., 2013). 2D materials can be viewed as membranes and nanopores in such membranes have attracted great attention due to their exceptional properties in fluid and molecular transport (Wang et al., 2017). For instance, nanoporous graphene is an ideal structure for seawater desalination, molecular sieving, and DNA sequencing.

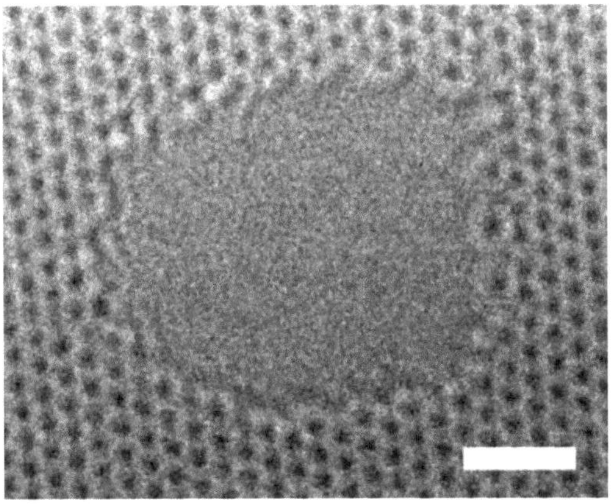

Figure 8.6 High-resolution episcopic microscopy image of a 3-nm pore in a graphene sheet. (Reprinted by permission from Springer Nature, Schneider et al., 2013, Copyright (2013).)

Pristine graphene nanosheets are hydrophobic. The hydrophobic property of nanopores in a graphene sheet may not promote the translocation of long-chain molecules. In contrast, MoS_2 is usually hydrophilic and nanoporous MoS_2 is suitable for DNA sequencing. In addition, MoS_2 nanopores are intrinsically charged due to the electron distribution between Mo and S. This endows MoS_2 unique properties in ion sieving and water desalination. The mechanical strength of MoS_2 is another advantage that favors the applications of MoS_2 nanosheets in various areas. Nonetheless, as a monolayer MoS_2 sheet has three atomic layers, the pore size and pore properties cannot be easily controlled because they depend on how Mo and S atoms are removed. The uncertainties in pore size and pore properties may affect the transport of ions, molecules, and fluids.

Nanopores in a 2D material can be generated using many techniques, including focused electron beam, UV zone etching, ion irradiation, chemical or plasma etching, ion bombardment, and chemical oxidative etching (Wang et al., 2017). These techniques have both advantages and disadvantages. Some of them (e.g., UV zone etching) can treat a large area, while the pore size may be rather dispersed. Electron beam irradiation can create nanopores with tunable and well-controlled pore sizes, but is limited to small areas. A combination of ion bombardment and oxygen plasma etching can generate nanopores with diameters of 0.4 ± 0.24 nm and pore densities of about 10^{12} pores/cm^2 (O'Hern et al., 2014; Jang et al., 2017).

8.4.5 Two-dimensional material-based lamellar membranes

Different from nanopores in graphene and MoS_2 nanosheets, which can be viewed as zero-dimensional, 2D material-based lamellar membranes, which are made of 2D laminates, contain 2D nanochannels. Typical examples are graphene oxide (GO) and MoS_2 lamellar membranes. The interlayer spacing between two laminates forms a well-defined 2D nanochannel, which allows the transport of molecules with sizes smaller than the interlayer spacing while rejecting the passage of large species. Such lamellar membranes can be fabricated through vacuum filtration, layer-by-layer deposition, and drop casting. Compared with nanoporous graphene, the preparation of GO membranes is cost-effective and scalable. The interlayer spacing of GO membranes can be controlled by changing fabrication parameters, such as chemical oxidation and sonication conditions. Therefore, 2D material-based lamellar membranes are promising materials in applications, where a high flux is desired.

8.5 Surface-charge-governed ion transport

The first experiment conducted to investigate ion transport in slit nanochannels was in 2004 (Stein et al., 2004). The channel height ranges from 70 to 1015 nm and

KCl solutions are used as the electrolytes. At low ion concentrations, the experimental results show significant enhancements for the conductance of the electrolyte compared to the bulk conductance. As shown in Figure 8.7a, the conductance G experiences two regimes as the KCl concentration is increased. At high ion concentrations, the conductance increases linearly with increasing ion concentration, which is in good agreement with the classic theory (e.g., Equation 8.64), as illustrated in Figure 8.7b. At low ion concentrations, however, the conductance becomes saturated at a value that is independent of the channel height and ion concentration. Compared with the bulk conductance, the enhancement at very low ion concentrations can reach several orders of magnitude, as shown in Figure 8.7b.

The linear dependence of the conductance on the ion concentration at high ion concentrations in Figure 8.7a is not surprising because the EDLs are thin and the effects of surface charges play a minor role. The ion concentration in the whole channel can be assumed to be uniform. As the ion concentration is increased, the number of ions moving in the channel increases, leading to a linear relationship between the ionic current and ion concentration. As the conductance is proportional to the ionic current, G shows linear dependence on the ion concentration (Equation 8.64). At low ion concentrations, especially when the ion concentration is lower than the surface-charge density, however, the surface charges become dominant and EDLs play an important role. As the surface charges attract counterions (potassium ions) and the electroneutrality of the channel requires that the counterion density in the channel should be higher than the bulk ion concentration.

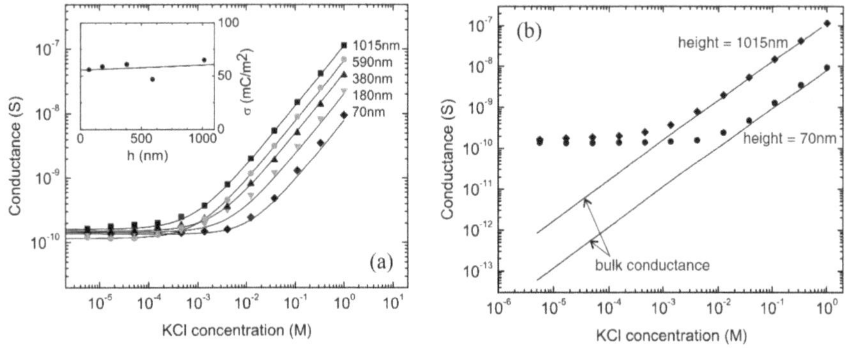

Figure 8.7 Conductance as a function of ion concentration. (a) Conductance for different channel heights, h, which are 1015, 590, 380, 180, and 70 nm. The solid lines are the fits of the data to the electrokinetic model. σ in the inset is the surface-charge density obtained by fitting the conductance data. (b) Conductance for h = 1015 and 70 nm. Lines are the conductance predicted using the conductivity of the bulk solution. (Reprinted with permission from Stein et al., 2004, Copyright (2004) by the American Physical Society.)

The motion of the excess counterions in the EDLs enhances the ionic current and the conductance. If the ion concentration is very low, the conductance is mainly caused by the counterions in the channel, whose concentration is equal to the surface-charge density. This is why the conductance in Figure 8.7a remains a constant at low ion concentrations.

Actually, the two regimes in Figure 8.7a can be easily explained using Equation 8.67, where the conductance is contributed by two parts: bulk concentration ($G \propto n_0 e(Z_+ + Z_-)$) and surface charges ($G_s \propto |n_s| Z_+$). At high ion concentrations, the surface-charge contribution can be neglected and the conductance is governed by the bulk concentration, i.e., $G \propto n_0 e$. At low ion concentrations, the bulk concentration contribution is weak and the conductance is mainly determined by the surface-charge density, $G \propto |n_s| Z_+$, which is a constant if n_s is fixed.

More rigorously, the ion transport problem can be analytically treated with electrokinetic theories. Let us consider a slit nanochannel with symmetric surface charges and a monovalent electrolyte, the electrostatic potential is described by the Poisson–Boltzmann equation, Equation 8.8. If the y-axis is perpendicular to the channel surfaces and the x-axis goes though the midplane ($y = 0$) of the channel, as shown in Figure 8.8, the solution of Poisson–Boltzmann equation is given by

$$\phi_0(y) = \phi_0(0) + 2\ln\left(\mathrm{CD}\left(\frac{\kappa y}{2} e^{-\phi_0(0)/2} \middle| e^{2\phi_0(0)}\right)\right) \tag{8.69}$$

If an electric field E is applied along the channel in the x-direction, the total electrostatic potential is $\phi = \phi_0 + Ex$. Then, the Naiver–Stokes equation for the electroosmotic flow, Equation 8.35, reads

$$\mu \frac{d^2 u}{dy^2} - E\varepsilon_r \varepsilon_0 \frac{d^2 \phi}{dy^2} = 0 \tag{8.70}$$

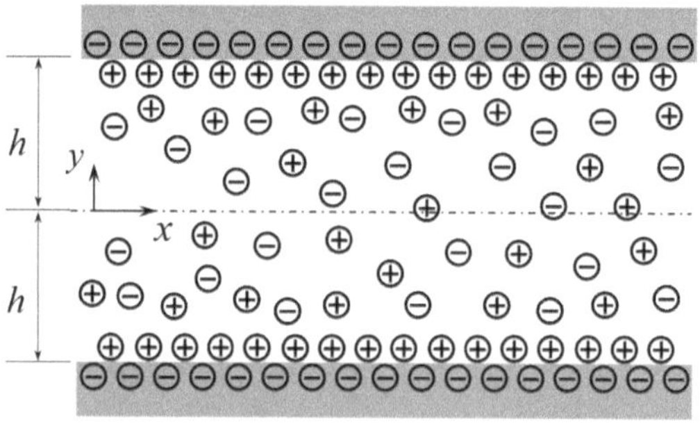

Figure 8.8 Coordinate system for the analysis of ionic conductance.

The ionic current I can be expressed as follows:

$$I = ew\int_{-h/2}^{h/2}\{[u + Z_{+}E]n_{+} - [u - Z_{-}E]n_{-}\}dy \tag{8.71}$$

where w is the width of the channel, n_{+} and n_{-} are the concentrations of cations and anions, respectively. If the ionic conductance is calculated using Equation 8.63, it can be shown that (Levine et al., 1975)

$$G = \frac{I}{EL} = \frac{2n_0eZhw(1+H)}{L} \tag{8.72}$$

where H is a correction factor to the bulk conductivity given by

$$H = \cosh\left(\frac{e\phi_0(0)}{k_BT}\right) - 1 + \frac{E_e}{4n_0k_BTh}\left(1 + \frac{2\varepsilon_r\varepsilon_0k_BT}{e\mu Z}\right) \tag{8.73}$$

where $E_e = \varepsilon_0\varepsilon_r\int_0^{h/2}\left[d\phi_0(y)/dy\right]^2 dy$ is the electrostatic field energy per unit area. The surface-charge density can be viewed as the boundary condition for Equations 8.69–8.73. It is related to the potential at the walls,

$$n_s = \mp\varepsilon_0\varepsilon_r\left.\frac{d\phi_0(y)}{dy}\right|_{y=\pm h/2} \tag{8.74}$$

Equations 8.69–8.74 are a set of transcendental equations, which can only be solved numerically. For the two limiting cases, i.e., high and low ion concentrations, the conductance can be approximately estimated. At a high ion concentration, the Debye length is small and $\kappa h >> 1$, for which $|n_s| << en_0h$ and $H \approx 0$, then the conductance is given by

$$G = \frac{2n_0eZhw}{L} \tag{8.75}$$

For the dilute limit, $\kappa h << 1$ and $|n_s| >> en_0h$, the conductance is approximated as

$$G \approx \frac{2|n_s|Zw}{L}\left(1 + \frac{2\varepsilon_r\varepsilon_0k_BT}{e\mu Z}\right) \tag{8.76}$$

which is independent of h and n_0. The two terms on the right-hand side of Equation 8.76 represent the surface conductance and the contribution of electroosmosis. Equations 8.75 and 8.76 agree well with the dependence of G on the ion concentration shown in Figure 8.7.

8.6 Ionic current rectification

Analogous to semiconductor diodes, which regulate electron transport in a preferred direction, and the nanofluidic diodes introduced in Chapter 6 that can achieve unidirectional passage of a fluid, diodes capable of rectifying ion transport can be developed by taking advantage of distinct ion transport mechanisms at the nanoscale. A diode that allows ions to transport in a specific direction (say, forward direction) while blocks the migration of ions in the opposite (backward) direction is called ionic diode.* An ionic diode is a special mode of an ionic rectifier, which allows the transport of ions in both the forward and backward directions but with different ion fluxes. As the transport of ions leads to ionic current, an ionic rectifier can also act as an ionic current rectifier. A typical ionic current–voltage (*I*–*V*) curve for an ionic diode/rectifier is illustrated in Figure 8.9. The performance of an ionic rectifier is evaluated by its diodicity, which is the ratio of ionic currents in the forward and backward directions.

Ionic current rectification can be achieved using different transport mechanisms, which usually requires asymmetric properties for a flow system, such as geometric asymmetry of the channel, surface-charge asymmetry, and ion concentration asymmetry. In the following sections, representative experimental results about ionic current rectification in nanochannels are discussed.

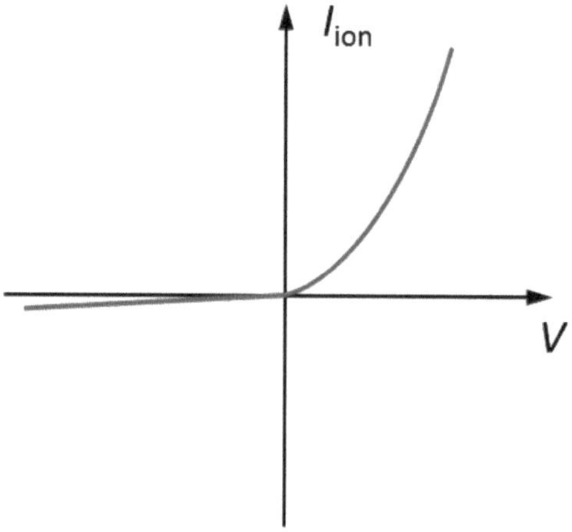

Figure 8.9 A representative *I*–*V* curve for an ionic rectifier.

* In the literature, a diode rectifying ion transport is called nanofluidic diode, which is quite confusing as it has been used for flow rectification (see Chapter 6). Therefore, such a diode, herein, is termed ionic diode.

8.6.1 Geometric asymmetry

A popular structure with geometric asymmetry for ionic rectification is a cone-shaped channel. To achieve ionic current rectification, the narrow end of a conical channel is usually at the nanoscale. Once the surface is charged, which can be realized by the fabrication process or chemical modifications, the small size of the narrow end may only allow counterions to pass through the channel from the narrow to the wide end, while the wide end does not affect the motion of ions. Therefore, the ion selectivity caused by a conical geometry generates different ion transport in different directions, leading to ionic rectification. For example, let us consider a conical channel with negatively charged surface in an electrolyte, as shown in Figure 8.10. If an electric potential is applied by two electrodes such that the electric field directs from the narrow end to the wide end of the channel (top panel of Figure 8.10), both cations and anions can go through the channel easily because the narrow end does not screen cations. However, if the electric field is switched, the motion of anions from the narrow end to the wide end is blocked because of the small opening of the narrow end, as shown in the bottom panel of Figure 8.10. Several types of conical nanopores have been shown exhibiting ion current rectification, including glass nanopipettes, silicon nitride nanopores, conical nanopores in polymer films, and conical gold nanotubes.

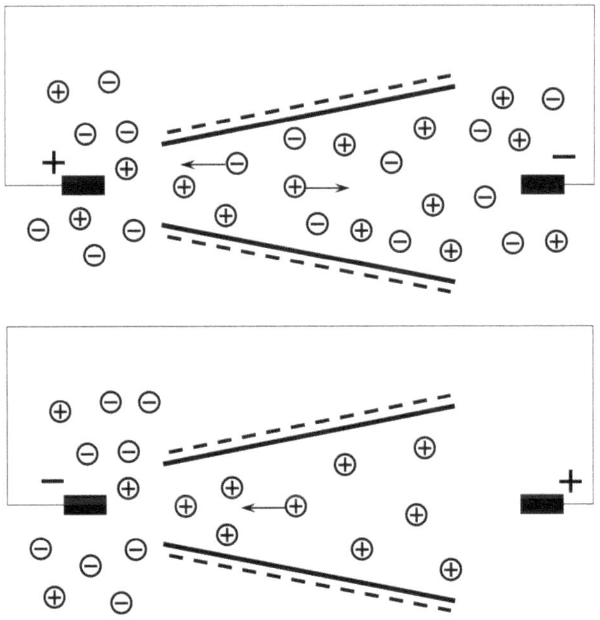

Figure 8.10 Schematic of ionic current rectification in a conical nanopore.

Ionic current rectification in conical nanopores is first demonstrated in quartz nanopipettes (Wei et al., 1997). Ionic current is measured for a KCl electrolyte after an electric potential is applied. If the tip (narrow end) radius is as small as 20 nm, the current–voltage relationship exhibits different fashions as the KCl concentration is varied, as depicted in Figure 8.11. At relatively high ion concentration, the EDL next to the nanopipette surface is thin and the tip of the nanopipette cannot effectively screen co-ions. Therefore, the ion transport is direction independent, as shown in Figure 8.11a. As the ion concentration is lowed, the thickness of the EDL increases and the nanopipette tip tends to block co-ions, leading to ionic current rectification (Figure 8.11b). At very low ion concentration, the EDL thickness can be larger than the tip radius. In this case, co-ions are largely blocked at the tip end and the ionic current in one direction is greatly suppressed, as illustrated in Figure 8.11c.

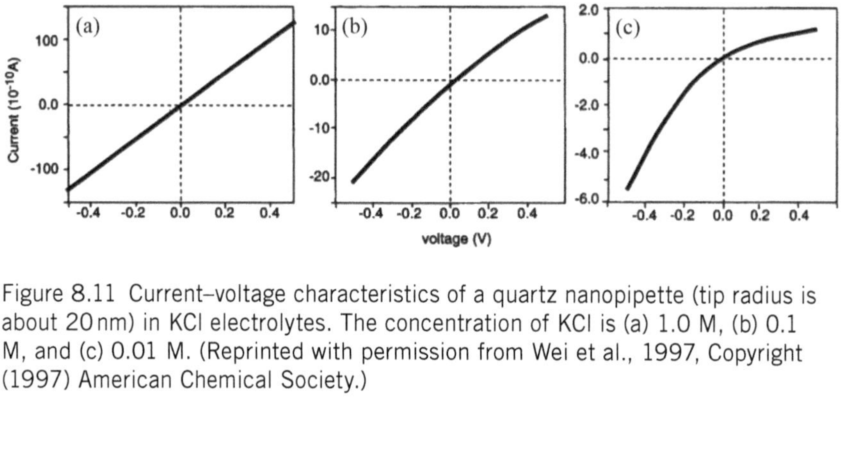

Figure 8.11 Current–voltage characteristics of a quartz nanopipette (tip radius is about 20 nm) in KCl electrolytes. The concentration of KCl is (a) 1.0 M, (b) 0.1 M, and (c) 0.01 M. (Reprinted with permission from Wei et al., 1997, Copyright (1997) American Chemical Society.)

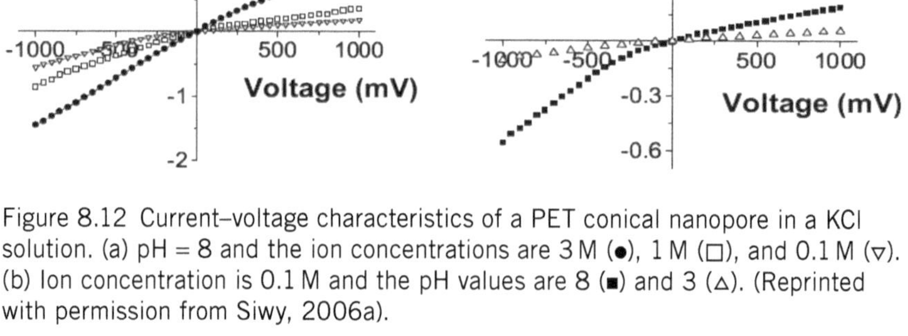

Figure 8.12 Current–voltage characteristics of a PET conical nanopore in a KCl solution. (a) pH = 8 and the ion concentrations are 3 M (●), 1 M (□), and 0.1 M (▽). (b) Ion concentration is 0.1 M and the pH values are 8 (■) and 3 (△). (Reprinted with permission from Siwy, 2006a).

Conical nanopores fabricated in polymers, such as polyethylene terephthalate (PET), can also achieve ionic current rectification. PET nanopores are widely used in studying the rectification of ionic current due to simple and reliable fabrication. During a fabrication process, the nanopore surface can be modified with functional groups, such as carboxyl groups. The typical density of functional groups is about $1\,nm^{-2}$. Such a modification causes the nanopore surface to be negatively charged when immersed in a basic or neutral electrolyte. PET nanopores show better rectification performance compared with glass nanopipettes. Even at very high ion concentration, e.g., 1 M, PET nanopores display strong asymmetric ionic currents (Siwy, 2006a), as shown in Figure 8.12a. Similar to glass nanopipettes, ionic current rectification can only be observed in charged nanopores. Figure 8.12b shows the *I–V* curves at different pH values, 8 and 3, for the electrolyte when the ion concentration is set at 0.1 M. At pH 8, the electrolyte is basic and the surface is negatively charged, which leads to strong ionic rectification. However, at pH 3, the net surface charge is zero and ionic rectification is not observed.

The surface of a polymer-based conical nanopore can be modified to endow the structure with more flexibility for ionic current rectification. The ionic rectification performance of PET conical nanopores deposited with Au has been investigated (Siwy et al., 2004). Compared with uncoated polymer nanopores, the surface properties of Au-modified nanopores can be tuned easily by using different electrolytes or through thiol chemisorption. Figure 8.13 depicts the *I–V* curves of an Au-modified PET nanopore in a 0.1 M KCl and a 0.1 M KF solution (Siwy et al., 2004). It can be seen that ionic rectification depends on the electrolyte. In a KCl solution, the Au surface can adsorb Cl^- ions such that the nanopore

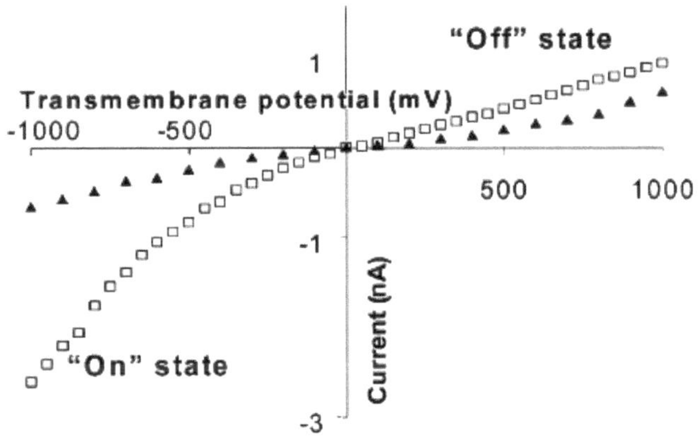

Figure 8.13 *I–V* curves for an Au-modified PET nanopore in a 0.1 M KCl (□) and a 0.1 M KF (▲) solution. (Reprinted with permission from Siwy et al., 2004, Copyright (2004) American Chemical Society.)

surface is negatively charged. As the diameter of the small opening is as small as 10 nm, ionic current rectification can be easily achieved with great performance. However, in KF solutions, the Au nanopore does not absorb F^- ions such that the surface is not charged and the pore cannot rectify ion transport. The results in Figure 8.13 indicate that surface charge is necessary for a conical nanopore to rectify ion transport.

To charge the surface of the Au-coated PET nanopore to reach ionic current rectification, the nanopore surface can be further modified. Figure 8.14a shows the current–voltage curves of an Au-coated nanopore modified with 2-mercaptopropionic acid, which causes the surface to be negatively charged at pH 6.6. If the surface is modified with mercaptoethylammonium cation, the surface can be positively charged in a KF solution and ionic current rectification can also be reached, as shown in Figure 8.14b.

Other asymmetric geometries have also been designed, including bullet-, funnel-, and step-wise-shaped channels. These shapes are more complex than conical structures, but offer more opportunities for optimizing the performance of ionic rectification. For instance, a funnel-shaped nanochannel contains two parts: a

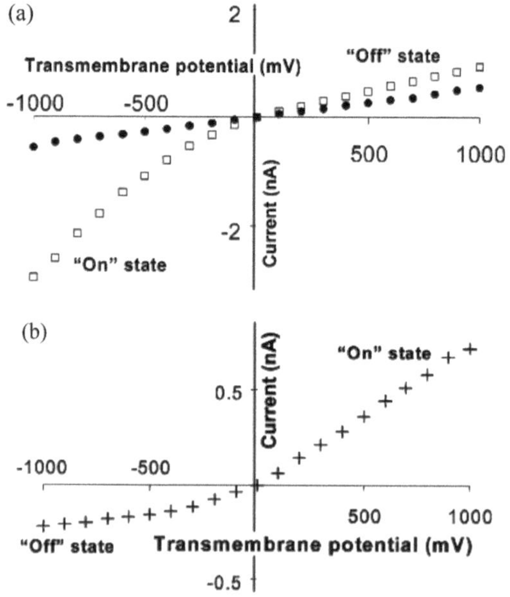

Figure 8.14 Ionic current rectification in an Au-coated PET nanopore in a 0.1 M KF solution. (a) The pore is modified with 2-mercaptopropionic acid at pH 6.6 (□) and pH 3.5 (•). (b) The pore is modified with mercaptoethylammonium cation and the surface is positively charged (pH 6.6). (Reprinted with permission from Siwy et al., 2004, Copyright (2004) American Chemical Society.)

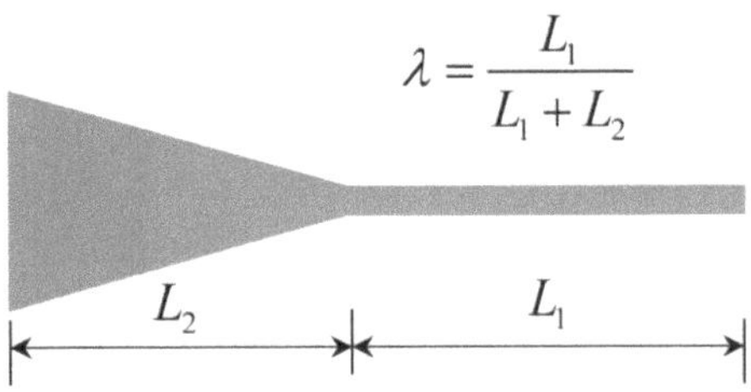

Figure 8.15 Schematic of a funnel-shaped structure for ionic current rectification.

conical segment and a cylindrical segment, as sketched in Figure 8.15. If the surface of both parts are charged, the diodicity of ionic rectification can be changed by varying the ratio of the length of the cylindrical part to the total length, λ, and the taper angle of the conical part. Experiments show that at a taper angle of about 1.4°, the maximum diodicity can reach 13 at $\lambda = 0.52$ (Xiao et al., 2016). The rectification performance can be further enhanced by changing the surface-charge density, which will be discussed in Section 8.6.4.

8.6.2 Surface-charge asymmetry

A nanochannel with asymmetric surface charges can also work as an ionic diode. The idea is first validated by theoretical modeling (Daiguji et al., 2005). Generally, there are two ways to realize charge asymmetry. One approach is to use heterogeneously charged surface, i.e., half of the surface is positively charged and the other half is negatively charged, as shown in Figure 8.16. Such an ionic diode is called the bipolar diode. When a forward voltage bias is applied, as illustrated in Figure 8.16a, counterions on both sides, i.e., anions on the left and cations on the right sides, migrate toward the center of the

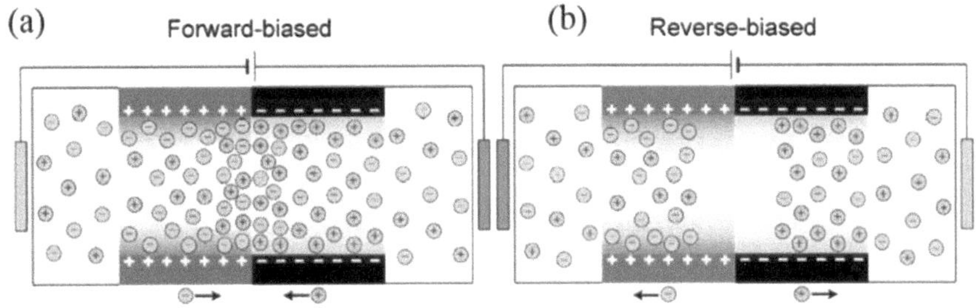

Figure 8.16 Schematic of a bipolar diode under a forward bias (a) and a reverse bias (b). (Reprinted with permission from Daiguji et al., 2005, Copyright (2005) American Chemical Society).

channel, forming an ion enriched zone in the channel and resulting in a high ionic current, which can be viewed as an open state for ion transport. However, if a reverse bias is applied, counterions on both sides will move away from the channel to the reservoirs. This results in an ion depletion zone in the channel and the ionic current is low, which corresponds to the close state. Figure 8.17 shows the theoretical results of the current–potential characteristics for the design in Figure 8.16. It is clear that the ionic currents are different in different directions and they are affected by the surface-charge density and the bulk solution concentration.

Ionic rectification using heterogeneous surface charges has been demonstrated through experiments. Heterogeneous charges can be accomplished by taking advantage of the fact that different materials may exhibit different charge properties when they are in contact with an aqueous solution, depending on the band gap of the materials and the pH value of the solution. For example, silicon oxide (SiO_2) is usually negatively charged while aluminum oxide (Al_2O_3) is positively charged in an electrolyte. A heterogeneous nanochannel made of SiO_2 and Al_2O_3 can help achieve heterogeneous surface charges, as shown in Figure 8.18 (Cheng and Guo, 2009). The surface-charge densities of SiO_2 and Al_2O_3 in an 1 mM KCl solution are −4.24 and 2.1 mC/m², respectively. The *I–V* curves for SiO_2–Al_2O_3 heterogeneous nanochannels with channel height <20 nm are shown in Figure 8.19. It can be seen that ionic rectification can be achieved in solutions with ion concentrations <0.1 M. The highest diodicity is over 300 at 1 mM.

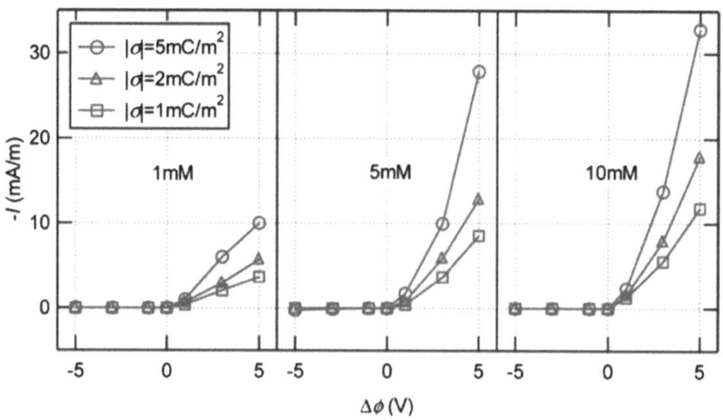

Figure 8.17 Current–potential curves for the design in Figure 8.16 at different surface-charge densities, σ, and bulk ion concentrations, 1, 5, and 10 mM. (Reprinted with permission from Daiguji et al., 2005, Copyright (2005) American Chemical Society).

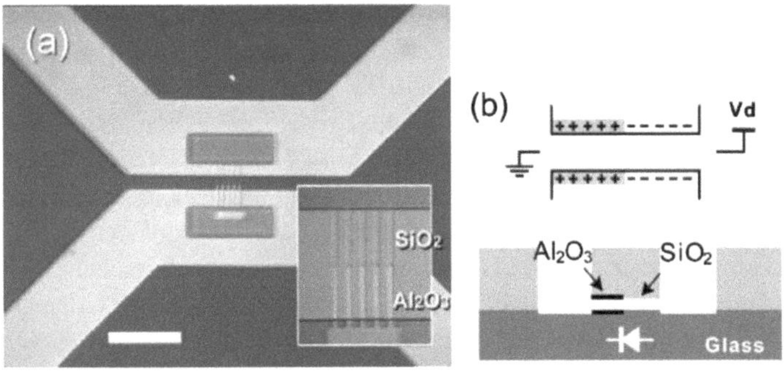

Figure 8.18 A bipolar diode using heterogeneous nanochannels (SiO_2 and Al_2O_3). (a) Microscope image of the diode. The inset shows the composition of the channels: the upper half of the channels is SiO_2 and the lower half is Al_2O_3. (b) Schematic of the surface-charge distribution in an electrolyte. (Reprinted with permission from Cheng and Guo, 2009, Copyright (2009) American Chemical Society.)

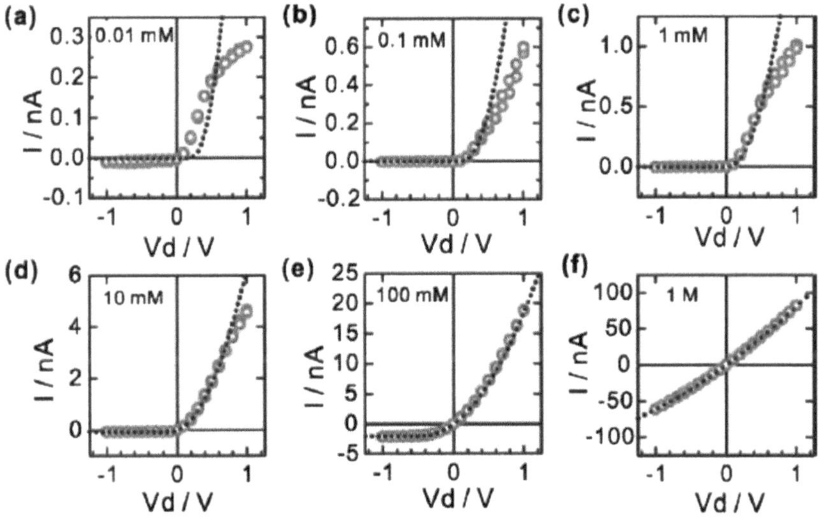

Figure 8.19 Current–voltage characteristics for the heterogeneous nanochannels in Figure 8.18 in KCl solutions at different bulk KCl concentrations: (a) 0.01 mM, (b) 0.1 mM, (c) 1 mM, (d) 10 mM, (e) 100 mM, and (f) 1 M. Circles are experimental data and dashed lines are theoretical predictions of a simplified one-dimensional transport model. (Reprinted with permission from Cheng and Guo, 2009, Copyright (2009) American Chemical Society.)

Another scenario of surface-charge heterogeneity is to charge half of a channel, either positively or negatively, while keeping the other half neutral, as demonstrated in Figure 8.20a. Ionic diodes using this type of charge pattern are called unipolar diodes. A representative method reaching such surface-charge distributions is to modify the channel surface with biomaterials. Figure 8.20b shows 30-nm high nanochannels, half of which is coated with a layer of biotin and the other half is patterned with avidin (Karnik et al., 2007). When in contact with an electrolyte, the biotin side is neutral while the avidin side becomes positively charged, resulting in asymmetric surface charges. Figure 8.21 depicts the *I–V* curves for this unipolar diode in a 10-mM KCl solution. The diodicity depends on the surface-charge density. A good performance is associated with a high degree of surface-charge heterogeneity. It must also be noted that the diodicity depends on the bulk ion concentration. In a solution of high bulk ion concentration (e.g., 1 M), ionic rectification is barely observed.

The surface-charge asymmetry created using either intrinsic material properties or through chemical modifications cannot be changed once a device is fabricated. To modulate the surface-charge distribution, an electronic gate can be employed, which can alter the surface charge by varying the gate voltage. A nanofluidic field-effect

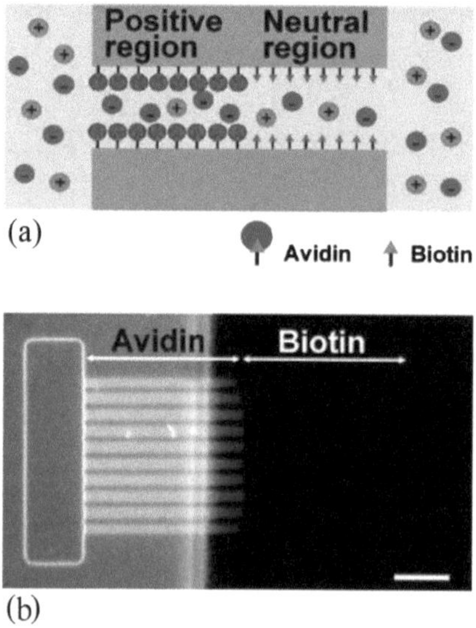

Figure 8.20 Unipolar ionic diode. (a) Schematic of a nanochannel with half of the surface patterned with avidin and the other half patterned with biotin moieties. (b) Epifluorescence image of a fabricated unipolar diode. The left half of the channels is modified with fluorescently labeled avidin. (Reprinted with permission from Karnik et al., 2007, Copyright (2007) American Chemical Society.)

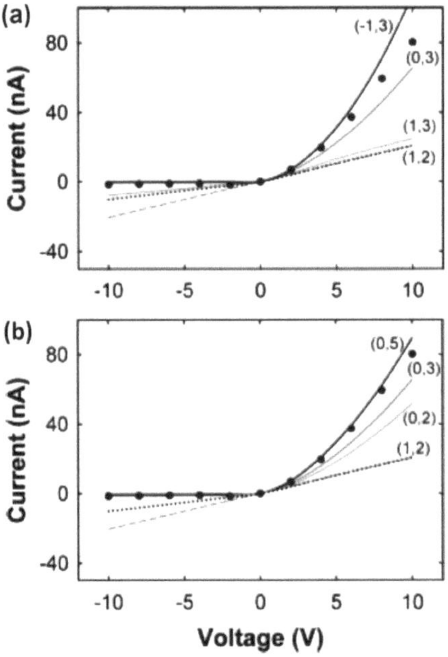

Figure 8.21 Current–voltage curves for the unipolar diode in Figure 8.20 in a 10-mM KCl electrolyte. Solid circles are experimental data. The slope of the dashed lines is equal to the conductance measured under a voltage bias range from −50 to 50 mV. Solid lines are theoretical predictions for a channel with 30 nm, 40 μm, and 120 μm in height, width, and length, respectively. The numbers in the brackets are the surface-charge densities (mC/m^2) for the biotin and avidin parts. (a) Surface-charge densities are 1, 0, and −1 mC/m^2 for the biotin part and the charge density for the avidin part is 3 mC/m^2. (b) Zero charge density for the biotin part and the charge densities for the avidin part are 2, 3, and 5 mC/m^2. (Reprinted with permission from Karnik et al., 2007, Copyright (2007) American Chemical Society).

reconfigurable diode (FERD) is a good example. Figure 8.22a–d depicts a nanofluidic FERD, where a gate terminal is integrated on top of an array of nanochannels (Guan et al., 2011). Nanochannels are fabricated in a SiO_2 substrate and the surface-charge properties of the channels can be modulated by applying a voltage to the gate. If the gate is placed close to one end of the nanochannels (Figures 8.22c and 8.23a), the device exhibits the ability of rectifying ion transport and the performance of ion rectification depends on the voltage of the gate, as indicated by the *I*–*V* curves in Figure 8.23b. If the gate voltage $V_G = 0$ V, the ionic rectification is not observed (the middle panel of Figure 8.23b). This is reasonable because the nanofluidic system in this case is symmetric. At a negative gate voltage, the current under a positive bias is higher than that under a negative bias. At a positive gate voltage, the ionic

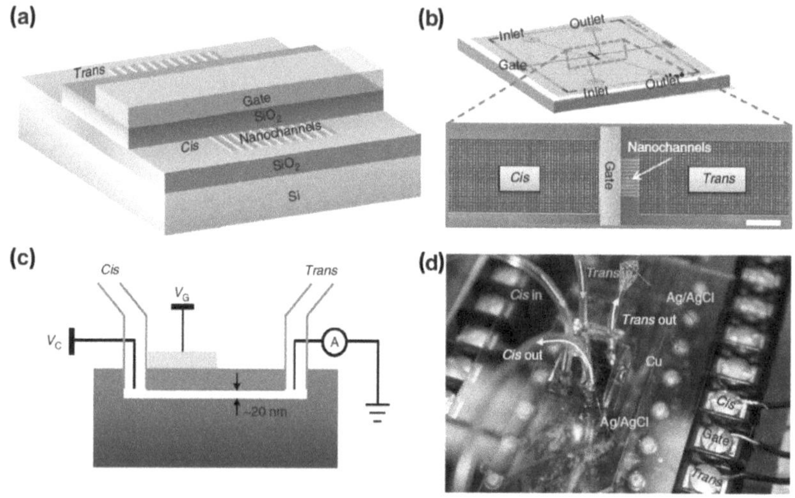

Figure 8.22 A nanofluidic FRED. (a) Materials and 3D layout of the nanofluidic FRED. (b) Planar layout of the FRED. The scale bar is 200 μm. (c) Configuration of electrical and fluidic connections (*Cis* and *Trans* represent the reservoirs close to and far away from the gate electrode). (d) Photograph of a physical device of the FRED. (Reprinted by permission from Springer Nature, Guan et al., 2011, Copyright (2011)).

current under a negative bias is higher than that under a positive bias. Therefore, the ionic rectification properties can be easily tuned by changing the gate voltage. This is the consequence of the tunable surface-charge distribution controlled by the gate voltage. If the gate is placed in the middle of the channels, the gate voltage can change the ionic conductance but does not generate asymmetric surface charges, and ionic rectification is not expected, as shown in Figure 8.23c and d.

Although the single-gate nanofluidic diode in Figure 8.23a can modulate the ionic rectification properties, the rectification performance (e.g., diodicity) is not very impressive. The diodicity for the case in Figure 8.23a with $V_G = -1.4$ V is about 1.2, which is rather low. To enhance the rectification effect, the gate configuration can be modified. Figure 8.23e shows a dual-gate design, where two gates are fabricated close to the ends of the nanochannels. Figure 8.23f shows the corresponding current–voltage characteristics. It is clear that the performance of the rectification effect is significantly enhanced. The diodicity at $V_G = 1$ V is about 80, which is much higher than that in Figure 8.23b.

8.6.3 Bath solution concentration asymmetry

Another strategy to achieve ionic rectification is to use the ion concentration asymmetry of the bath solutions in a nanofluidic system, i.e., the reservoirs at the

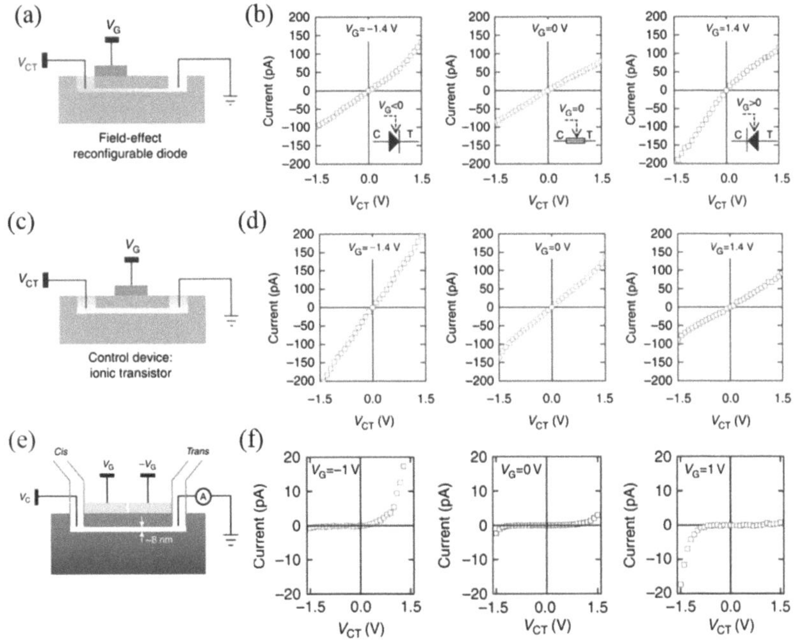

Figure 8.23 Nanofluidic FRED with different gate configurations and the corresponding *I–V* curves. (a) Schematic of a single-gate FERD device. (b) *I–V* characteristics for the FERD in (a). The electrolyte is a 100-μM KCl solution and the nanochannel width, length, and height are $W = 2\,\mu m \times 11$, $L = 100\,\mu m$, $H = 20\,nm$, respectively. (c) Schematic of a control device with the gate at the center of the nanochannels. (d) *I–V* characteristics for the control device ($W = 3\,\mu m \times 11$, $L = 100\,\mu m$, $H = 20\,nm$) using a 100-μM KCl solution. (e) Schematic of a dual split-gate FERD device. (f) *I–V* characteristics for the FERD device in (e). The nanochannel dimensions are $W = 1\,\mu m \times 11$, $L = 116\,\mu m$, $H = 8\,nm$ and the electrolyte is 1-mM KCl. (Reprinted by permission from Springer Nature, Guan et al., 2011, Copyright (2011).)

ends of nanochannels have different ion concentrations. This idea takes advantage of the dependence of EDLs on ion concentration. Due to the small size of nanochannels and the ion concentration difference between the reservoirs, it is possible to have significant EDL overlap at one end of the nanochannels, while the EDL thickness at the other end of the channels is small as compared with the channel height and EDL overlapping is unimportant, as illustrated in Figure 8.24. Nanofluidic ionic diodes using this idea have been experimentally demonstrated (Cheng and Guo, 2007). In the experiments, nanochannels with channel heights of 4 and 20 nm are fabricated. The bath concentration at one side is set at 0.1 mM and that at the other side is varied from 0.1 mM to 1 M. The current–voltage curves are shown in Figure 8.25, where the highest diodicity is ~3.5.

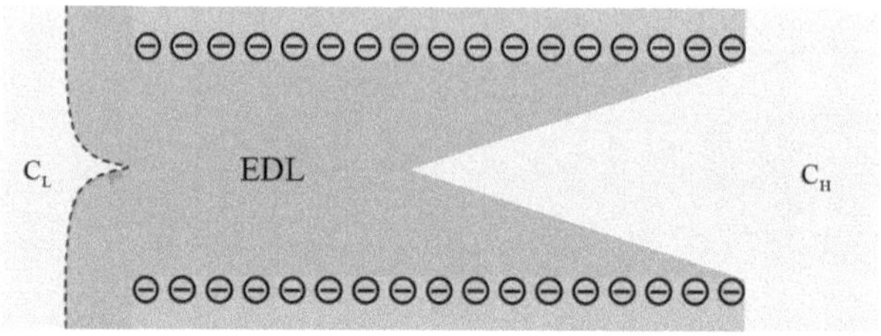

Figure 8.24 Schematic of a nanofluidic ionic diode based on the ion concentration difference in the reservoirs at the ends of nanochannels. C_L and C_H denote low and high ion concentrations. The EDLs overlap at the low concentration side.

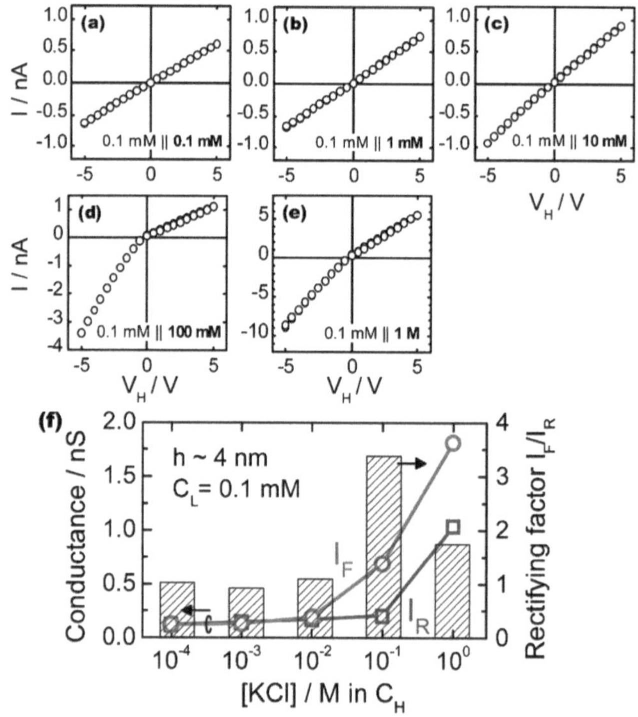

Figure 8.25 *I–V* characteristics and performance of a bath ion concentration difference-based ionic diode using KCl solutions. The channel height, width, and length are 4 nm, 2.5 μm, and 60 μm, respectively. (a)–(e) *I–V* curves at different bath concentrations (the low bath concentration is fixed at 0.1 mM and the high bath concentration is changed from 0.1 nM to 1 M). (f) Depicts the conductance in the forward and backward directions and the diodicity (rectifying factor). (Reprinted with permission from Cheng and Guo, 2007, Copyright (2007) American Chemical Society.)

8.6.4 Hybrid asymmetry

The asymmetric properties discussed in Sections 8.6.1–8.6.3 can be combined to enhance the rectification performance. Figure 8.26a shows a conical nanochannel with heterogeneous surface charges (Vlassiouk and Siwy, 2007). It is a hybrid nanofluidic system combining geometric asymmetry and surface-charge asymmetry. The radii at the tip and base ends are 2.5 nm and 500 nm, respectively. Through chemical treatments, the left half of the conical channel is positively charged and the right half is negatively charged. This hybrid system can significantly enhance the ionic rectification. The current–voltage curve for a 0.1 M KCl solution with pH = 5.5 shown in Figure 8.26b clearly demonstrates that the transport of ions is blocked once a negative voltage bias is applied. The diodicity is determined as 217, which is about five times larger than that purely caused by geometric asymmetry.

Surface-charge heterogeneity can also be applied to funnel-shaped nanochannels. Figure 8.27a shows a funnel-shaped nanochannel, where the cylindrical part is negatively charged and the conical part is positively charged by modifying the surface with different functional groups (Xiao et al., 2016). The *I*–*V* curve in Figure 8.27b shows that ions are only allowed to move from the cylindrical to the conical part if a negative voltage bias is applied to the cylindrical end. If the cylindrical and conical parts are positively and negatively charged (Figure 8.27c), ions can only transport in the reverse direction under the same voltage bias, as shown in Figure 8.27d. The diodicities in Figure 8.27b and d are 553 and 528, which are much higher than uniformly charged funnel-shaped nanochannels, as discussed in Section 8.6.1.

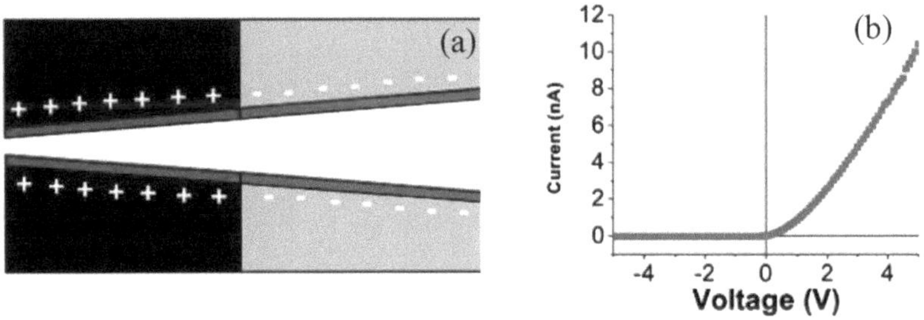

Figure 8.26 Ionic diode using geometric and surface-charge asymmetry. (a) Schematic of the nanochannel with heterogeneous surface charges. The radii of the channel at the tip and base ends are 2.5 and 500 nm, respectively. (b) *I*–*V* characteristics for a 0.1-mM KCl solution with pH = 5.5. (Reprinted with permission from Vlassiouk and Siwy, 2007, Copyright (2007) American Chemical Society.)

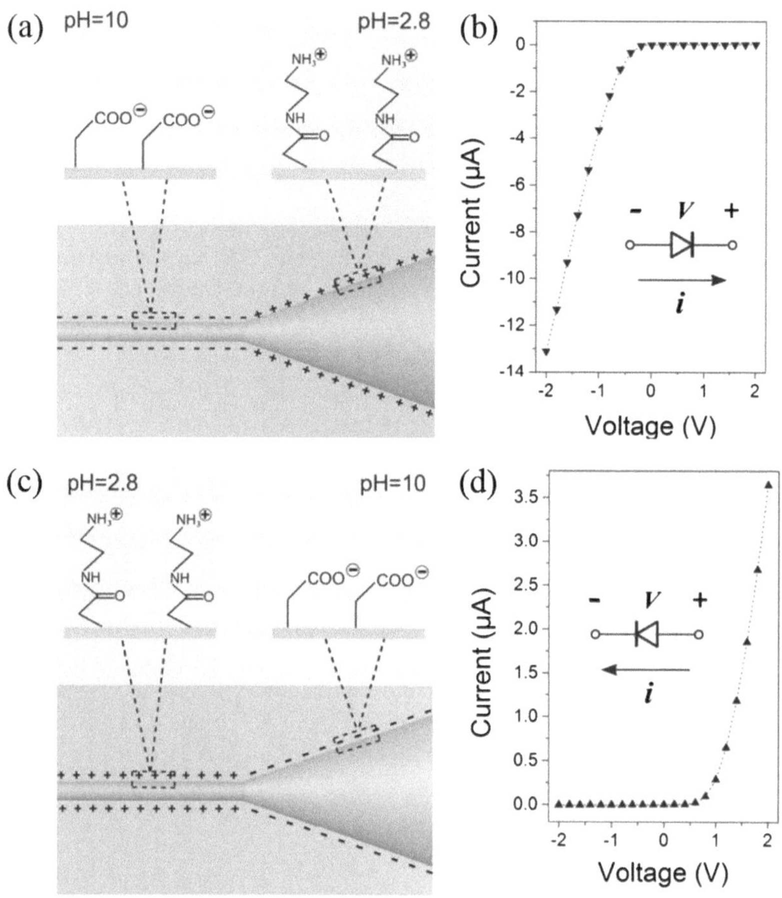

Figure 8.27 Ionic rectification using funnel-shaped nanopores with asymmetric surface charges. (a)–(b) Schematic and *I*–*V* characteristics for a funnel-shaped nanopore with a negatively charged cylindrical part and a positively charged conical part. (c)–(d) Schematic and *I*–*V* characteristics for a funnel-shaped nanopore with a positively charged cylindrical part and a negatively charged conical part. (Reprinted with permission from Xiao et al. 2016.)

8.7 Bioinspired nanochannels

Similar to many other subjects, nanofluidics also gains inspirations from nature. One example is the Aquaporins in cell membranes. Aquaporins are also called water channels because they selectively facilitate the transport of water while blocking the passage of ions. The water permeability of Aquaporins can be three orders of magnitude larger than the prediction of the classic theories for regular channels of the same size. The transport mechanisms of Aquaporins can help design new flow channels for selectively transporting specific species. Another example is the potassium ion channels in cell membranes, which only allow the transport of K^+ while rejecting the passage of other ions (e.g., Li^+ and Na^+). In terms of water permeability and ion selectivity, these biological nanopores far exceed the performance of artificial nanochannels and open new perspectives for the design of novel nanofluidic devices with potential applications in a variety of areas, including seawater desalination, molecular separation/filtration, and energy conversion.

Another important feature of biological channels is their extraordinary ability in modulating the ion transport using stimuli-responsive gates, which can open or close channels under specific external stimuli, such as transmembrane voltage, ligand binding, membrane tension, light, and temperature. Inspired by the gating properties of biological ion nanochannels, biomimetic nanochannels with similar stimuli-responsive functions have been designed by functionalizing nanochannels with responsive molecules through bio- or chemical interactions. The stimuli cover a wide range of entities, including pH value, ions, temperature, and light, and offer a variety of functions for the applications in flow control, molecular separation, biodetection, and drug delivery. Among various nanostructures, solid-state nanopores are widely employed as biomimetic, stimuli-responsive nanochannels due to their excellent mechanical and chemical stability, easy control of channel size, and tunable surface properties. In particular, conical nanochannels are good candidates for designing biomimetic nanochannels because their structure is similar to that of biological ion channels. In the following subsections, some experimental and numerical work on ion transport in bioinspired nanochannels will be introduced.

8.7.1 Biomimetic ion channels

In biological ion channels, the transport of K^+ and Na^+ plays a critical role in maintaining the physiological functions of living cells. Although K^+ and Na^+ have the same charge valence and similar hydration diameters, biological ion channels can distinguish and selectively transport these two ions. Similarly, a potassium ion channel exhibits great selectivity for K^+ transport and the flow rate ratio of K^+ to Na^+ can reach up to 1,000, which exceeds the performance of artificial channels. The superb transport properties of biological ion channels have attracted great interest to develop biomimetic nanochannels for ion transport. However, experimental investigations are quite challenging and recent progress in biomimetic nanochannels is mainly in numerical simulations.

Inspired by the carbonyl groups and the structure of K^+-selective KcsA channels (K^+ channels from *Streptomyces lividans*), CNT-based biomimetic nanochannels have been proposed. Figure 8.28a and b shows the side and top views of a (9,9)

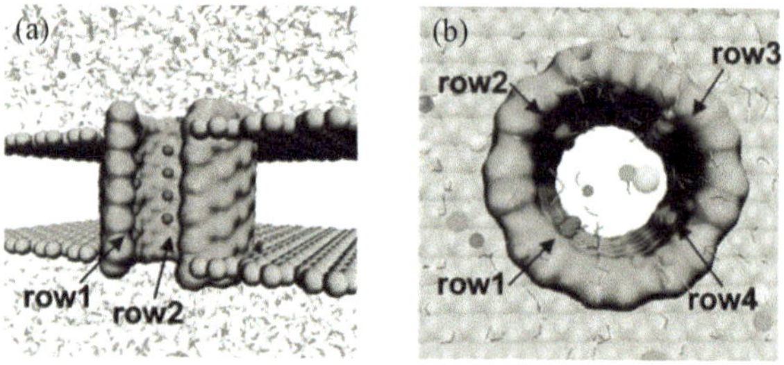

Figure 8.28 A (9,9) CNT patterned with four rows of oxygen atoms as carbonyl groups. (a) Side view. (b) Top view. (Reprinted with permission from Gong et al., 2010, Copyright (2010) American Chemical Society.)

CNT with oxygen atoms (four oxygen atoms in a row) patterned on the inner surface to form carbonyl groups (Gong et al., 2010). Three types of carbonyl oxygen patterns are designed. For type 1, four arrays of carbonyl oxygen atoms are symmetrically attached to the surface of the CNT. The two facing arrays have the same positions while adjacent arrays have a 1.2 Å displacement in the axial direction. For type 2, the four arrays have the same axial positions. In type 3 channels, there are only two arrays of oxygen atoms having the same axial position and facing each other. The density distributions of K^+ and Na^+ in the CNTs obtained through MD simulations show that type 1 channel preferentially conducts K^+ over Na^+ with a K^+ to Na^+ passage rate ratio equal to 4.6 (Figure 8.29a). In contrast, type 2 channel does not show selectivity for the two ions (Figure 8.29b). Type 3 channel, however, selectively conducts Na^+ with a passage rate 2.2 times higher than that of K^+ (Figure 8.29c). The dramatic difference in ion transport for the three types of channels are caused by the hydration number of ions in the CNTs (hydration number is the number of water molecules moving with an ion as a unit). In a bulk solution, the average coordination numbers for Na^+ and K^+ are about 6 and 7,

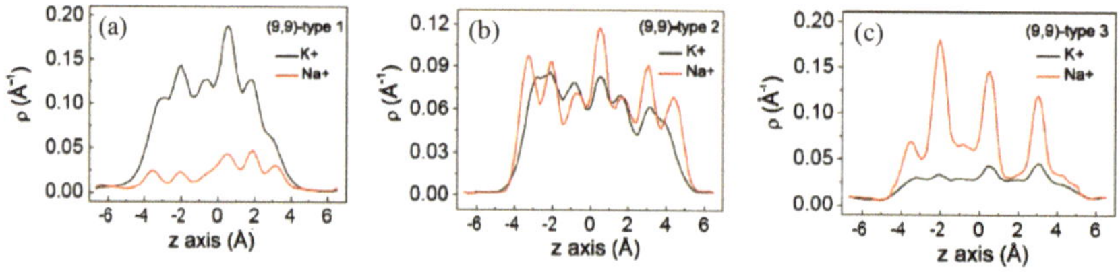

Figure 8.29 Density distributions of K^+ and Na^+ in patterned CNTs. (a) Type 1 CNT. (b) Type 2 CNT. (c) Type 3 CNT. (Reprinted with permission from Gong et al., 2010, Copyright (2010) American Chemical Society.)

respectively, as depicted in Figure 8.30a. For type 1 channel, Na^+ needs to get rid of about two water molecules from its hydration shell to enter the CNT, while the coordination number of K^+ in the CNT remains roughly the same as that in a bulk solution (Figure 8.30b). Thus, type 1 channel favors the transport of K^+. In type 2 channel, the coordination numbers for K^+ and Na^+ are close to their bulk values (Figure 8.30c), indicating that the CNT does not preferentially constrain the motion of K^+ and Na^+. In type 3 channel, K^+ has to lose 2–3 water molecules from its hydration shell to go through the CNT, while the coordination number of Na^+ is close to the bulk value (Figure 8.30d). This leads to the selective transport of Na^+.

Other than CNTs, graphene nanopores have also been designed as biomimetic ion channels. Figure 8.31 depicts three graphene nanopores functionalized with carbonyl or carboxylate groups to mimic K^+-selective KcsA channels and Na^+-selective NavAb channels (voltage-gated sodium channels from *Arcobacter butzleri*) (He et al., 2013). The nanopore with four carbonyl groups (4CO) at the rim mimics a KcsA channel. It has a diameter of ~0.65 nm. The nanopore containing four negatively charged carboxylate groups (4COO) is used to imitate a NavAb

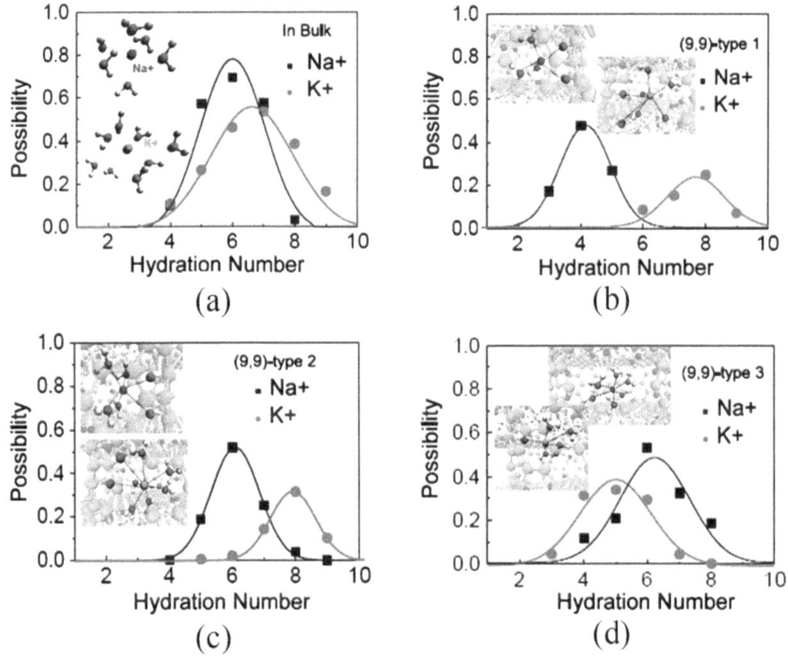

Figure 8.30 Possibility distribution of the hydration number of Na^+ and K^+. (a) In a bulk solution. (b) In type 1 CNT. (c) In type 2 CNT. (d) In type 3 CNT. (Reprinted with permission from Gong et al., 2010, Copyright (2010) American Chemical Society.)

channel, whose dimeter is ~0.79 nm. The nanopore modified with three carboxylate groups (3COO) has a relatively small diameter, ~0.43 nm.

The free energy barriers for K^+ and Na^+ to pass through the 4CO-nanopore are about 7.0 and 10.0 kJ/mol, respectively, suggesting that K^+ can go through the pore more easily than Na^+. The difference in the free energy barriers is caused by the hydration states (or hydration numbers) of K^+ and Na^+ in the nanopore. To pass through the pore, K^+ and Na^+ need to lose some water molecules in its hydration shell and coordinate with the functional groups at the pore rim. The energy penalty due to the partial dehydration is partly compensated by the coordination with the functional groups. In the 4CO nanopore, K^+ (4.5 water molecules and 2 carbonyl groups in the first hydration shell) coordinates with one more carbonyl group than Na^+ (4.5 water molecules and 1 carbonyl group in the first hydration shell), which lowers the energy barrier for K^+. This is why the 4CO nanopore conducts K^+ more favorably than Na^+.

For the 4COO nanopore, the energy barriers for K^+ and Na^+ show direction dependent due to the orientation of the carboxylate groups. Nevertheless, the energy barrier for K^+ is lower than that for Na^+, suggesting that Na^+ binds more strongly with the carboxylate groups than K^+ and the pore is in favor of the transport of K^+, as indicated by the selectivity ratio curve in Figure 8.31. This contradicts with the

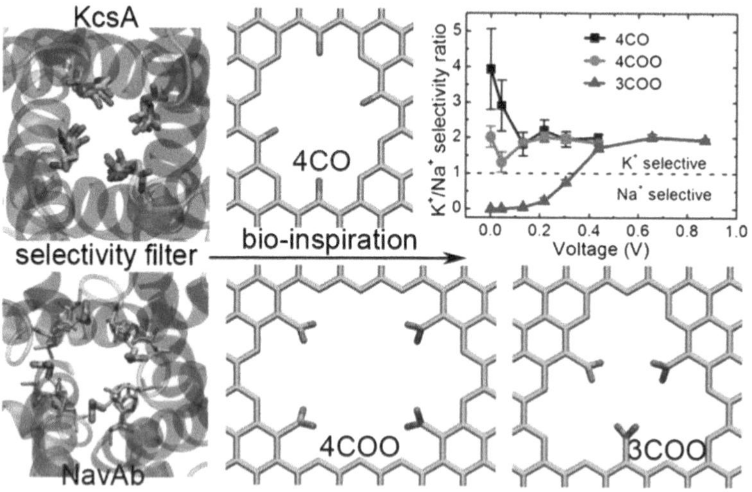

Figure 8.31 Graphene-based ion nanopores. Left: Illustration of KcsA channel (top) and NavAb channel (bottom). Middle and lower right: 4CO-, 4COO-, and 3COO-nanopores. Upper right: K^+/Na^+ selectivity ratio as a function of voltage applied across the pore. (Reprinted with permission from He et al., 2013, Copyright (2013) American Chemical Society Adopted from.)

design of the pore, which mimics the NavAb channel for Na^+ transport. The reason for the unexpected transport of K^+ is the consequence of the large pore size. For the 3COO nanopore, whose diameter is relatively small, the pass rate of Na^+ is higher than that of K^+ at relatively low electric fields (Figure 8.31). At a low electric field, the nanopore is occupied by Na^+ due to the stronger binding between Na^+ and the carboxylate groups. The bound Na^+ can be replaced by other Na^+ rather than K^+ through the knock-on mechanism, which is commonly observed in biological ion channels. In this way, the 3COO nanopore preferentially conducts Na^+. At a relatively high electric field, however, the orientation of the carboxylate groups is affected and their binding with Na^+ is weakened. In this case, the bound Na^+ can be replaced by K^+ and the nanopore conducts more K^+ than Na^+ due to the relatively low binding energy between K^+ and the carboxylate groups (Figure 8.31). Tunable selectivity for Na^+ and K^+ transport by varying external electric fields has also been observed in other graphene nanopores that model the structure of KcsA channels (Kang et al., 2014).

In addition to the selective transport of Na^+ and K^+, bioinspired ion channels for discriminating Mg^{2+} and Li^+ have also been developed (Ruan et al., 2016; Zhu et al., 2017). The separation of Mg^{2+} and Li^+ is of critical importance for improving the performance of lithium-ion batteries, but is quite challenging using regular nanochannels because Mg^{2+} and Li^+ have similar ionic radii. Inspired by biological CorA Mg^{2+} channel (Figure 8.32a), which allows a high Mg^{2+} flux while rejecting monovalent cations, graphene nanopores with different diameters

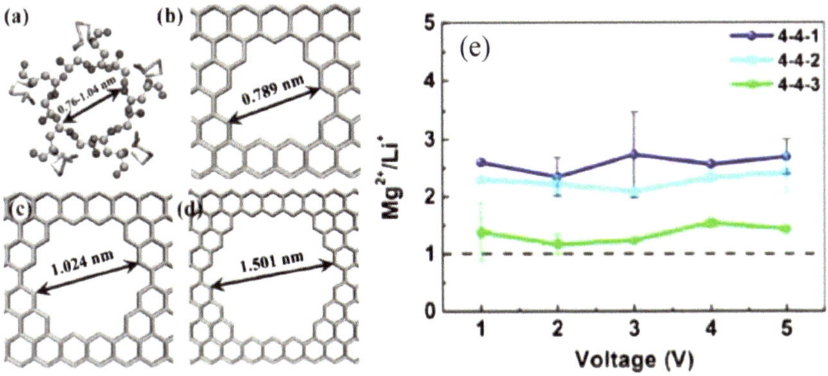

Figure 8.32 Graphene nanopores for selective transport of Mg^{2+}. (a) Schematic of biological CorA Mg^{2+} channel. (b)–(d) Pores with different diameters. (e) Pass rate ratio of Mg^{2+} to Li^+ (4-4-1, 4-4-2, and 4-4-3 are referred to as 0.789-, 1.024-, and 1.501-nm-diameter pores, respectively). (Reprinted with permission from Ruan et al., 2016, Copyright (2016) American Chemical Society.)

(0.789, 1.024, and 1.501 nm) are designed without further functionalization, as shown in Figure 8.32b–d (Ruan et al., 2016). Simulations show that 0.789- and 1.024-nm-diameter graphene nanopores exhibit favorable transport of Mg^{2+} over Li^{+}. For the 1.501-nm-diameter pore, although selective transport of Mg^{2+} is observed, the Mg^{2+}/Li^{+} pass rate ratio is not significant, as shown in Figure 8.32e.

The selectivity of Mg^{2+} can be enhanced by functionalizing the graphene pores. Figure 8.33a illustrates the structure of a *Thermotoga maritime* CorA (TmCorA) Mg^{2+} channel, which contains carbonyl and negatively charged carboxylate groups, and favorably transport Mg^{2+}. To mimic this structure, graphene nanopores functionalized with five CO and COO groups are designed, as depicted in the bottom panel of Figure 8.33b (the top panel of Figure 8.33b shows the MD simulation system) (Zhu et al., 2017). Molecular dynamics simulations demonstrate that both nanopores exhibit distinct selectivities for Mg^{2+} and Li^{+}, as depicted in Figure 8.34. The Mg^{2+}/Li^{+} selectivity ratio is insensitive to the electric field but depends on the functional groups. It is higher for the COO pore than that of the CO pore. The maximum Mg^{2+}/Li^{+} selectivity ratio for the CO and COO pores are 3.56 and 4.73, respectively, which are higher than unfunctionalized graphene pores in Figure 8.32.

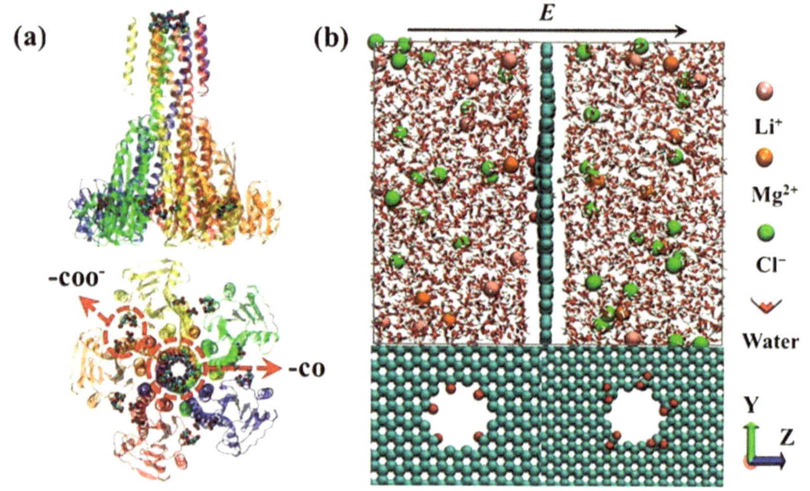

Figure 8.33 Nanopores for conducting Mg^{2+}. (a) Structure of a TmCorA Mg^{2+} channel. (b) Top: Snapshot of an MD simulation system with a functionalized graphene nanopore at the center. Bottom: Graphene nanopores modified with CO and COO groups. (Reprinted with permission from Zhu et al., 2017, Copyright (2017) American Chemical Society.)

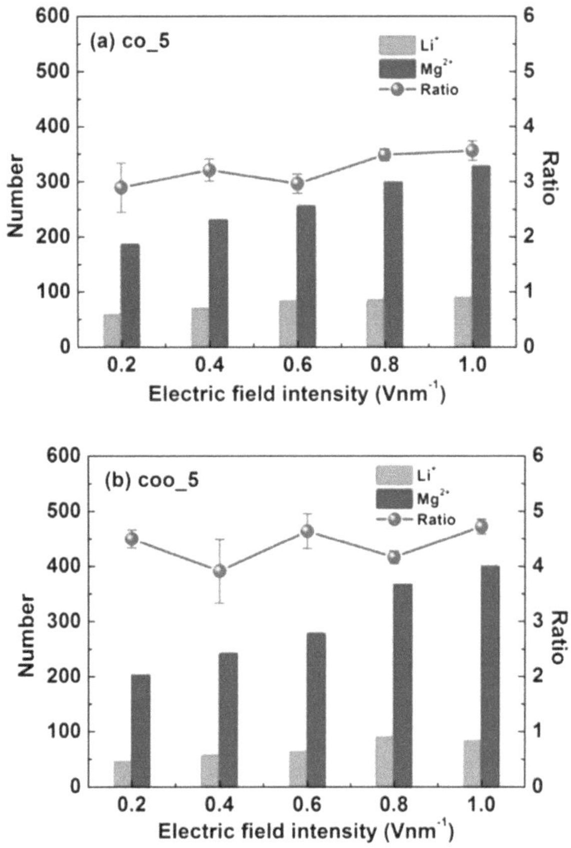

Figure 8.34 Number of Mg^{2+} and Li^{+} going through graphene nanopores and the flux ratio of Mg^{2+} to Li^{+}. (a) Graphene nanopore modified with five CO groups. (b) Graphene nanopore modified with five COO groups. (Reprinted with permission from Zhu et al., 2017, Copyright (2017) American Chemical Society.)

8.7.2 Biomimetic ionic gates

Allowing or screening the transport of ions can also be achieved by modifying channels with ion-responsive molecules. Figure 8.35a and b illustrates conical polyimide (PI) nanochannels modified with crown ether molecules to realize selective transport of ions, which are inspired by biological Na^{+} and K^{+} channels (Figure 8.35c) (Liu et al., 2015a). Crown ethers have a ring-shape structure and can bind with cations having a size similar to that of the cavity of crown ethers. To discriminate Na^{+} and K^{+}, whose radii are about 0.98 and 1.38 Å, respectively, two types of crown ethers, 4′-aminobenzo-15-crown-5 (4-AB15C5) and

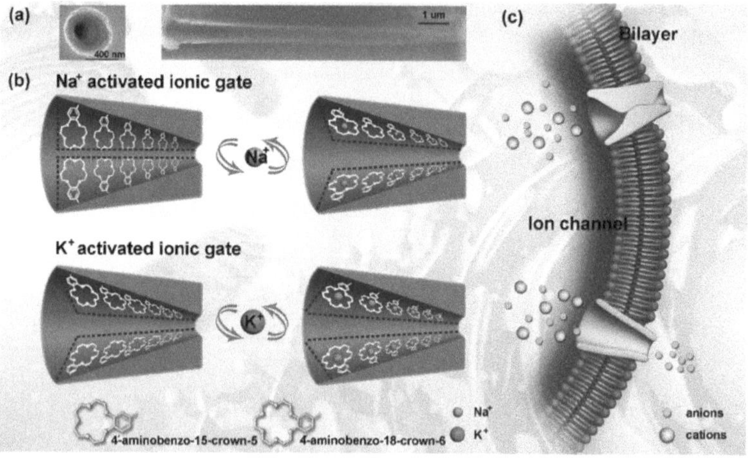

Figure 8.35 Biomimetic ionic gates. (a) SEM image of a conical polyimide nanochannel. (b) Schematic of a Na^+-activated and a K^+-activated nanochannel. (c) Schematic of biological Na^+ and K^+ channels. (Reprinted with permission from Liu et al., 2015a, Copyright (2015) American Chemical Society.)

4′-aminobenzo-18-crown-6 (4-AB18C6), are used to modify PI nanochannels. The radius of 4-AB15C5 cavity is about 0.86–1.1 Å, which fits Na^+ well. The cavity radius of 4-AB18C6 is in the range of 1.3–1.6 Å, similar to the size of K^+. Therefore, these two crown ether molecules can selectively bind with Na^+ or K^+ and form stable complexes. If they are immobilized on a PI channel surface, the channel can be converted to a Na^+-activated or K^+-activated ionic gate.

For the Na^+-activated channel (channel surface modified with 4AB15C5), before the channel is activated by Na^+, the channel exhibits very low ionic conductance due to the low surface charge and small effective radius of the channel, as shown in Figure 8.36a. After Na^+ is introduced to activate the channel, Na^+ binds with the 4-AB15C5 molecules on the channel surface and significantly increases the surface charge, leading to a high ionic conductance (Figure 8.36a). The mechanism is similar to that of uniformly charged conical nanochannels introduced in Section 8.6.1. For the K^+-activated channel (channel surface is modified with 4-AB18C6), before the channel is activated by K^+, the nanochannel surface remains negatively charged and shows ionic rectification behavior (Figure 8.36b). Once the channel is activated by K^+, the formation of K^+-4-AB18C6 complexes changes the surface charge to be positive. This enhances the transport of anions and reverses the current rectification, as shown in Figure 8.36b. Both the Na^+-activated and K^+-activated nanochannels show good stability and responsive switchability, as indicated in Figure 8.37.

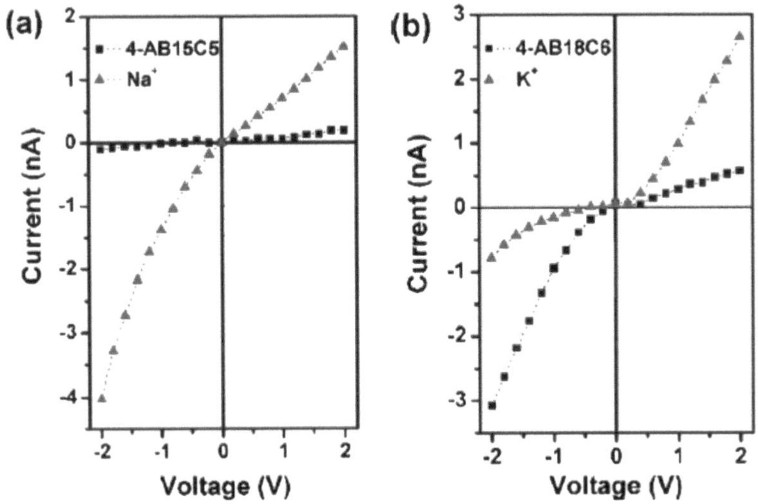

Figure 8.36 Current–voltage curves of conical, crown ether-modified, PI nanochannels. (a) Channel surface is modified with 4-AB15C5. Squares and triangles are the results before and after the activation of the channel with 10 μM Na^+. (b) 4-AB18C6-modified channel. Squares and triangles are the results before and after the activation of the channel with 10 μM K^+. (Reprinted with permission from Liu et al., 2015a, Copyright (2015) American Chemical Society.)

8.7.3 pH-gated nanochannels

Ion transport in nanochannels can also be regulated by the pH of the solution. The idea is to modify the channel surface with certain materials or molecules, which experience conformational change or charge variation when the pH of the environment is altered. Polymers and DNA molecules are popular materials that are sensitive to pH changes.

Figure 8.38a illustrates a pH-gated nanochannel modified with polymers (Yameen et al., 2009a). Cylindrical nanopores with diameters of ~15 nm and lengths of ~12 μm are prepared with PET films. The nanopores are then modified with pH-responsive polyvinylpyridine (PVP), which takes a swollen, charged, hydrophilic state at acid conditions while assumes a collapsed, neutral, hydrophobic state at high pH values, as depicted in Figure 8.38b and c. Figure 8.39 plots the ionic current as a function of the pH value. As the pH increases from 2 to 5, the ionic current decreases continuously. When the pH is higher than 5, the ionic current remains almost constant. The pH dependence of the ionic current is caused by the protonation/deprotonation of the pyridine groups in the polymer brush, which has an acid dissociation constant, pKa, of 5.2. Under pH 5.2, the number of deprotonated pyridine groups increases

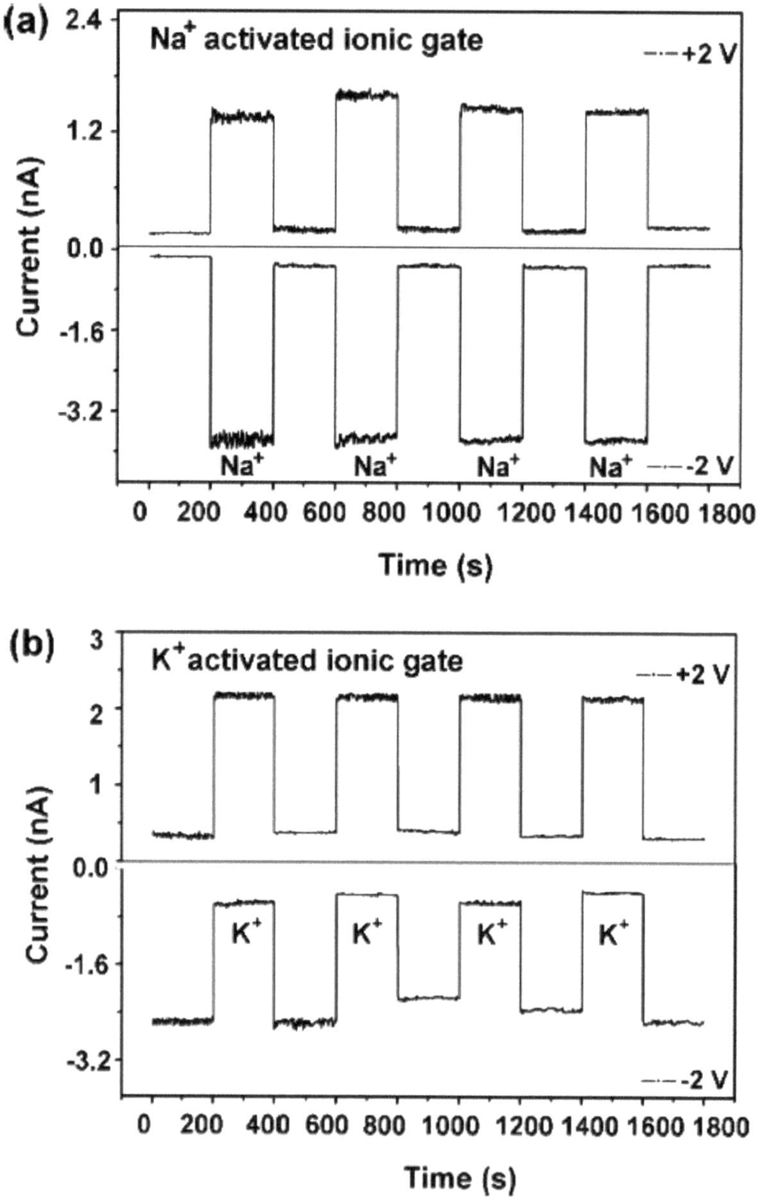

Figure 8.37 Variation of the current under an alternating voltage. (a) Na^+-activated channel. (b) K^+-activated channel. (Reprinted with permission from Liu et al., 2015a, Copyright (2015) American Chemical Society.)

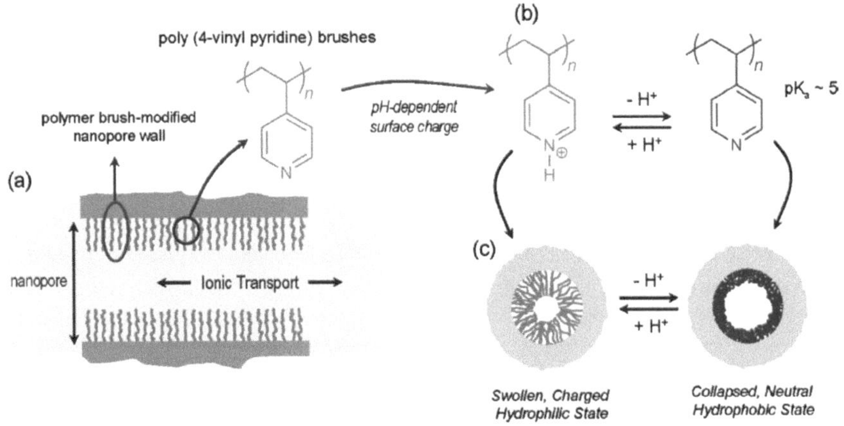

Figure 8.38 A pH-gated ionic nanochannel. (a) Schematic of a nanochannel modified by poly (4-vinyl pyridine) brushes. (b) Structure change between pyridine and pyridinium under pH changes. (c) Conformational changes of the polymer brush at different pH values. (Reprinted with permission from Yameen et al., 2009a, Copyright (2009) American Chemical Society.)

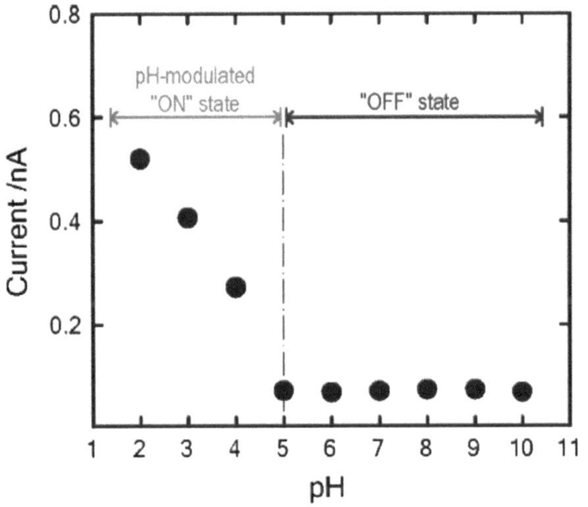

Figure 8.39 Ionic current as a function of pH at an applied voltage of 2V. (Reprinted with permission from Yameen et al., 2009a, Copyright (2009) American Chemical Society.)

with increasing pH. This reduces the surface charge and causes the ion current to decrease. Over pH 5.2, the pyridine groups are fully deprotonated and the surface becomes neutral, leading to a stable, low ionic current, which is the close state of the channel.

Another pH-gated nanopore is shown in Figure 8.40, where the surface of a conical PET nanopore is immobilized with single stranded DNA (Buchsbaum et al., 2014). The single stranded DNA contains adenosine (A) and cytidine (C) bases, which can be protonated in an acid environment. In a basic environment, without the protonation of the DNA bases, the DNA strands are negatively charged and the nanopore shows ionic rectification, as commonly observed in conical nanochannels (see Section 8.6.1). In this case, the nanopore is in an open state, as shown in Figure 8.40a and b. In an acid environment, however, A and C bases are protonated and the DNA strands become positively charged. This attracts the negatively charged backbones of the DNA strands and the DNA strands form a relatively stable network, leading to the closure of the nanopore, as depicted in Figure 8.40c and d. In a DNA-modified conical nanopore with the tip diameter

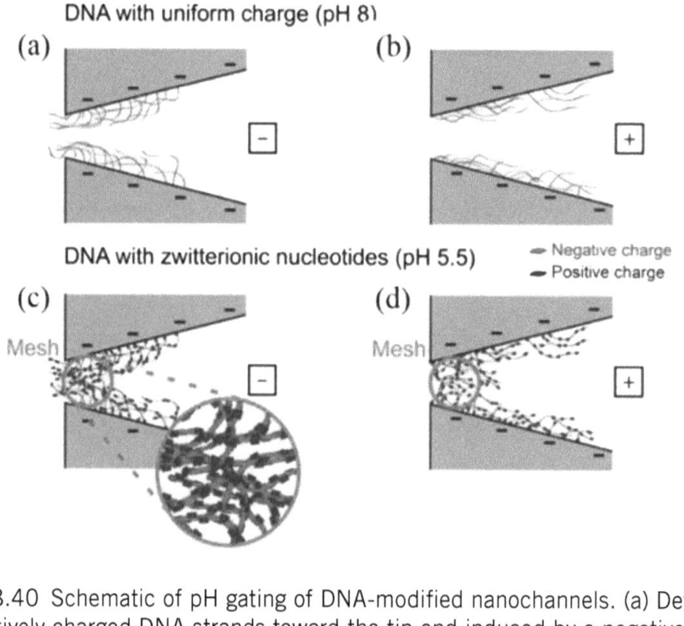

Figure 8.40 Schematic of pH gating of DNA-modified nanochannels. (a) Deflection of negatively charged DNA strands toward the tip end induced by a negative potential at pH 8 (channel is partially open). (b) Deflection of DNA strands toward to the base end caused by a positive potential at pH 8 (open state). (c) Formation of electrostatic DNA mesh at pH 5.5 (close state). (d) Deflection of DNA strands toward the base end at pH 5.5 (partial open). (Reprinted with permission from Buchsbaum et al., 2014, Copyright (2014) American Chemical Society.)

of 8 nm, the pore conductance dramatically drops by ~60 folds as the pH changes from 8 to 5.5. Therefore, DNA-modified PET nanopores are very sensitive to pH values and can work as effectiveness pH-regulated ionic gates.

8.7.4 Temperature-responsive nanochannels

Thermosensation, the ability to perceive temperature, is a basic but crucial function of a living body. It is realized by activating thermo-sensitive ion channels in the membranes of neurons belonging to the transient receptor potential superfamily. Inspired by biological, thermo-sensitive ion channels, artificial functional nanochannels that can work as temperature-gated ion channels have been developed by incorporating a nanopore with temperature-responsive materials, such as copolymers and hydrogels. One example is PI conical nanopores with surface modified by poly(*N*-isopropylacrylamide) (PNIPAM) (Yameen et al., 2009b). Compared with unmodified PI nanopores (see Section 8.6.1), which are negatively charged and show rectified ion transport, PNIPAM-modified nanopores are neutral and exhibit symmetric *I–V* curves at a relatively high temperature. At a temperature below the lower critical solubility temperature (LCST), PNIPAM displays a swollen state, as shown in Figure 8.41a, which reduces the effective diameter of the nanopore and hinders the transport of ions (Figure 8.41b). However, when the temperature is above LCST, PNIPAM changes to a collapsed state, enlarges the effective diameter of the nanopore, and enhances the ionic current (Figure 8.41a and b). With the conformational transition of PNIPAM from the swollen state (23°C) to the collapsed state (40°C), the effective diameter of the

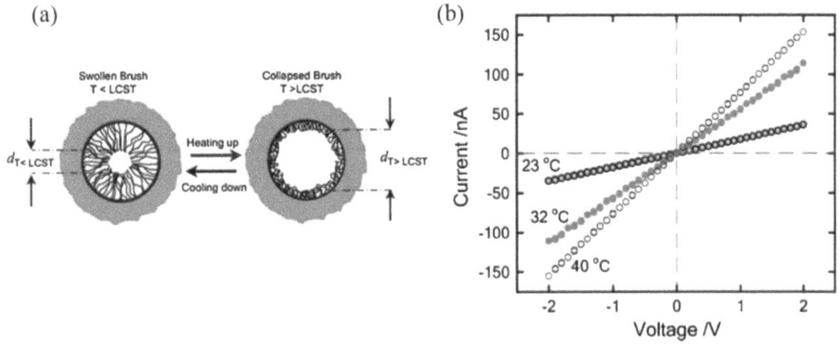

Figure 8.41 Temperature-gated ion nanochannel. (a) Schematic of conformational changes of PNIPAM brushes at different temperatures. (b) *I–V* curves for a single PI conical nanopore (tip diameter $d = 8\,\text{nm}$, base diameter $D = 1.45\,\mu\text{m}$) modified with PNIPAM brushes in a 1 M KCl electrolyte at different temperatures. (Reprinted with permission from Yameen et al., 2009b.)

nanopore is changed by a factor of 3. Accordingly, the ionic conductance varies from a low conductance state to a high conductance state with an enhancement ratio of about 4.5.

8.7.5 Light-gated ion nanochannels

The working principle of *Channelrhodopsins*, which are a type of proteins functioning as light-gated ion channels, has motivated the development of light-gated ion channels. Such ion channels usually exhibit reversible switching of ion conductance under the irradiation of UV and visible lights. Compared with other stimuli-responsive methods, light gating has attracted great attention because it possesses promising potentials for precise spatial and temporal control. Furthermore, it can be treated as a noninvasive method as lights do not affect the chemical environment of most systems. The light gating function is usually enabled by photoswitch molecules, such as azobenzenes, spiropyrans, malachite green leuconitriles, and retinal. Under the radiation of UV and visible light, photoswitches undergo a transition between at least two distinct thermodynamically stable states. For instance, spiropyrans assume a low-polarity conformation (close state) under darkness or the irradiation of visible light, while changing to a high-polarity configuration (open state) upon the irritation of UV light. Photoswitches govern light gating through the transition between these stable states, which affect the transport of ions. In the past decades, various engineering light-gated ion channels have been developed by modifying the channels with photoswitches, which, under the radiation of UV or visible light, experience structure, surface charge, or wettability changes and consequently alter the channel size or surface property to achieve ion transport regulation.

Spiropyran is a widely used photoswitch. It can switch between a neutral and a charged state upon light irradiation, and the charge polarity can be further tuned by changing the solution pH. If a nanochannel is modified with spiropyran, the photogated polarity transition of spiropyran causes the surface charge of the nanochannel to be photoregulated and controls the transport of ions or molecules through the channel. Such photoregulated transport mechanism has been demonstrated in controlling the transport of a positively charged species, $Fe(bpy)_3^{2+}$, through a spiropyran modified glass nanopore (Wang et al., 2006). As shown in Figure 8.42, spiropyran is transformed from the neutral form to either zwitterionic (ME) or pronated merocyanine (MEH^+) upon exposure to UV light. This process is reversed under visible light. In a conical glass nanochannel modified with spiropyran, under visible light, the transport of $Fe(bpy)_3^{2+}$ is allowed, as indicated by the ionic current in Figure 8.43 (blue curve). If the channel is exposed to UV light, however, the flux of $Fe(bpy)_3^{2+}$ drops quickly. After 15 min, the channel is completely closed for $Fe(bpy)_3^{2+}$ due to the electrostatic repulsion from the surface as spiropyran is converted to positively charged merocyanine, as revealed by the ionic current changes in Figure 8.43.

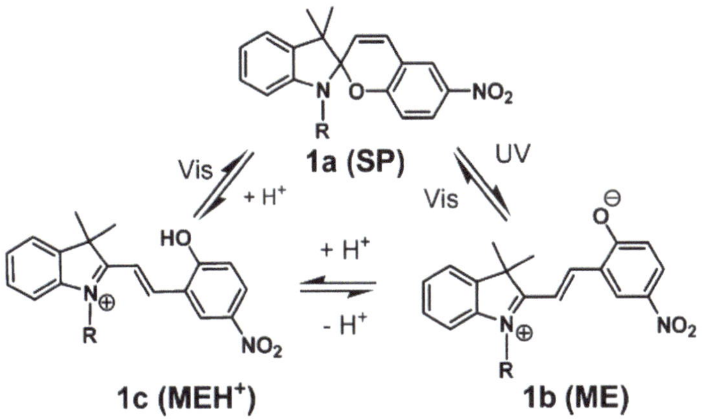

Figure 8.42 Transformation of spiropyran (1a, SP) to zwitterionic (1b, ME) or pronated merocyanine (1c, MEH+) under the irradiation of UV light. The conformational changes are reversed under visible light. (Reprinted with permission from Wang et al., 2006, Copyright (2006) American Chemical Society.)

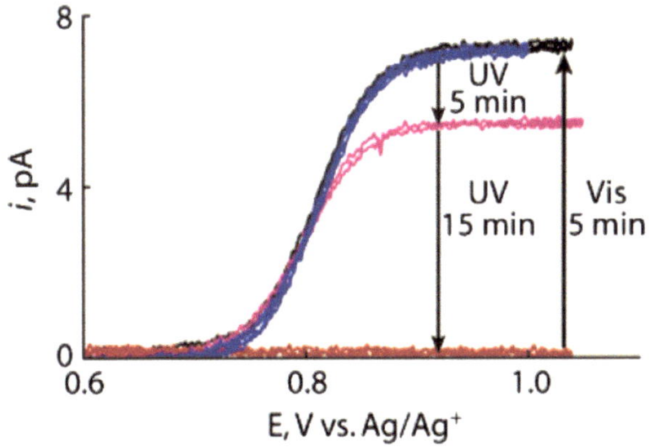

Figure 8.43 Ionic current of a 15-nm-radius, 2.8-mm-deep glass pore modified with spiropyran in a CH3CN solution with $Fe(bpy)_3^{2+}$ concentration of 5 mM. Blue, before the irradiation of UV light; magenta, 5 min after the irradiation of UV light; red, 15 min after the irradiation of UV light; black, 5 min after the irradiation of visible light. (Reprinted with permission from Wang et al., 2006, Copyright (2006) American Chemical Society.)

Another widely used photoswitch is azobenzene, which owns the properties of photoisomerization. The conformational change of azobenzene induced by light can also be used to react with other molecules to alter the charge polarity of a channel. A PET nanochannel modified with positively charged azobenzene molecules has been shown to rectify ion transport with the assist of negatively charged pillararene (P6A). Under darkness or visible light, azobenzene takes a configuration called *trans*-azobenzene, which interacts and forms a complex with P6A. This makes the channel surface negatively charged. Upon the irradiation of UV light, azobenzene is changed to the so-called *cis*-azobenzene and the azobenzene–P6A complex disassembles, resulting in a positively charged channel surface. Therefore, UV and visible light can control the surface charge of the channel to regulate ion transport. Figure 8.44a and b shows the ionic current changes when azobenzene–P6A complexes form and disassemble under the irradiation of UV and visible light. It can be seen that ionic current rectification is clear. At 2 V, the ratio of the current after UV irradiation to that before UV irradiation is about 5. The currents under alternating UV and visible light in Figure 8.44c show that this method has excellent reversibility.

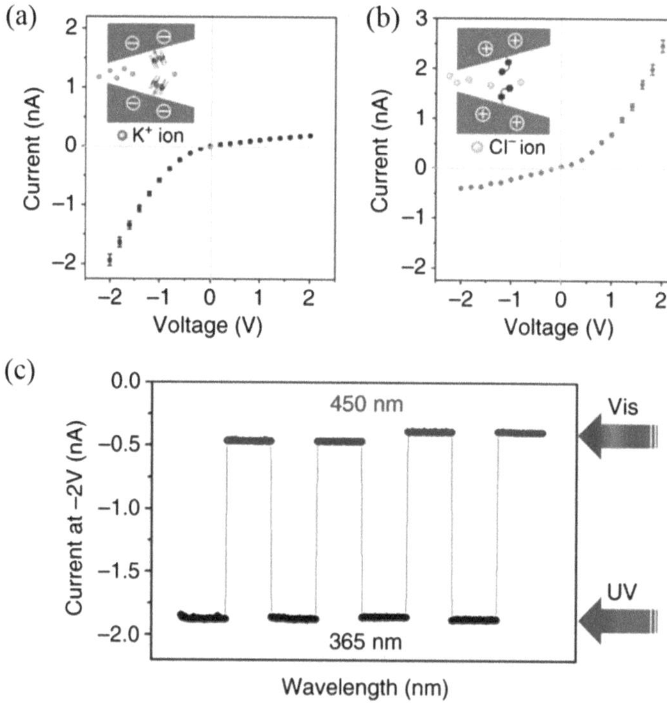

Figure 8.44 Light-gated ion selectivity using azobenzene-modified PET nanochannels. (a) *I–V* curve before UV light irradiation. (b) *I–V* curve upon UV light irradiation. (c) Currents under periodic irradiation of UV and visible light. (Reprinted with permission from Sun et al., 2017.)

Compared with biological ion channels, artificial nanochannels have better mechanical and chemical stability and are flexible in geometry and chemical modification. Moreover, the response time to light can be much shorter than biological ion channels, which are limited by the response time of proteins. Artificial ion channels can also bear higher light intensity than cells, offering more opportunities in ion transport control at the nanoscale. These advantages endow light-gated ion channels the potential for wide applications in many areas.

8.7.6 Voltage-gated nanochannels

Voltage-gated ion channels are quite common and play critical roles in physiological functions of certain biological systems, such as nerve, muscle, and neuroendocrine cells. Under a transmembrane potential, voltage-gated ion channels can undergo conformational changes or surface property variations to control the opening and closing of the channel. Voltage-gated ion channels are also ion selective. According to ion selectivity, they are classified into voltage-gated K^+ channels, Na^+ channels, Ca^{2+} channels, and Cl^- channels. Artificial nanochannels with similar voltage gating and ion selectivity properties possess promising potentials in many applications, such as functional nanodevices, drug delivery, and biodetection.

As an example, herein, a biomimetic voltage-gated Cl^- channel is introduced. It is a conical PI nanochannel with the channel surface functionalized by Cl^- responsive molecules, 5-aminoisophthalic acid (AIPA) and 4-butylbenzenamide (BBA) (Liu et al., 2016). The modified nanochannel shows a coupled dependence of ionic current on Cl^- activation and the applied voltage, as depicted in Figure 8.45, where the top panel illustrates the state changes of the channel when it experiences Cl^- activation under a high voltage. The bottom panel of Figure 8.45 shows the corresponding current–voltage characteristics.

Before the activation by Cl^-, the channel surface is covered by AIPA–BBA, which is nearly neutral and slightly hydrophilic (contact angle is about 80°). The channel exhibits a closed state with a very low ionic conductance. With the addition of Cl^- into the system, Cl^- binds with AIPA–BBA and causes the surface to be more hydrophilic (contact angle is about 63°) and negatively charged. This opens the channel and leads to a high ionic conductance. The effectiveness of surface modification and Cl^- activation is characterized by the current ratio I_{Cl^-}/I, where I_{Cl^-} and I are the currents measured with and without Cl^- activation, respectively. The current ratio I_{Cl^-}/I for the modified nanochannel is higher than that of unmodified nanochannel by one order of magnitude at 1 and –1 V.

In addition, the current ratio I_{Cl^-}/I for the modified nanochannel shows dependence on the voltage. It is about 9 at low voltages between –0.2 and –1.0 V, then drops to about 3 at medium negative voltages between –1 and –2.0 V, and levels off at high

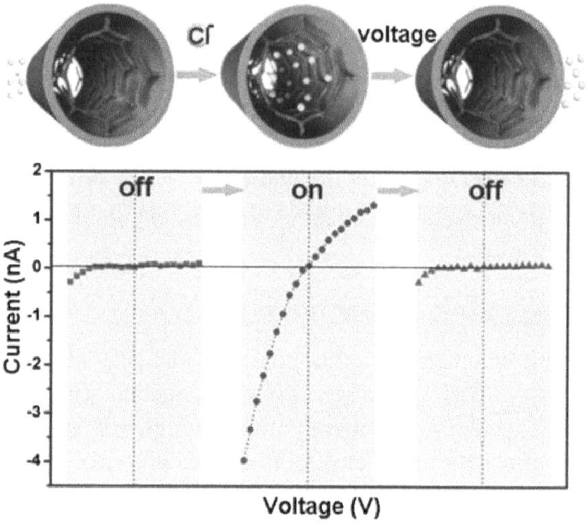

Figure 8.45 Voltage-gated Cl^- channel. Top: Schematic of the "open" and "close" states of the channel under Cl^- activation and a high voltage. Bottom: The corresponding current–voltage characteristics for the different states in the top panel. (Reprinted with permission from Liu et al., 2016.)

voltages, as shown in Figure 8.46. The decreasing current ratio, I_{Cl^-}/I, in Figure 8.46 is assumed to be caused by the electric field, which increases the surface wettability and enhances the current in the nanochannel without Cl^- activation. At a sufficiently high voltage, ~4 V, however, the binding between Cl^- and AIPA–BBA decomposes, which releases Cl^- from the channel into the solution and changes the channel into the closed state, as depicted in Figure 8.47 and the top panel of Figure 8.45. Therefore, the channel states can be switched by Cl^- activation and a high voltage, which makes the channel voltage-gated.

Other than the high-current ratios, I_{Cl^-}/I, at low and moderate voltages, the AIPA–BBA modified nanochannel is highly selective to Cl^- over other halogen ions, including F^-, Br^-, and I^-. The current ratio for Cl^- is more than five times higher than those for other halogen ions.

8.7.7 Duel-stimuli-responsive nanochannels

Most of the ionic gates introduced in previous sections are controlled by single stimulus. Some functional nanochannels can respond to multiple stimuli, typically dual stimuli, such as pH/temperature and pH/light. Duel-stimuli-responsive nanochannels offer more opportunities for the design of nanodevices with complex functions. Generally, there are two different approaches to design dual-stimuli

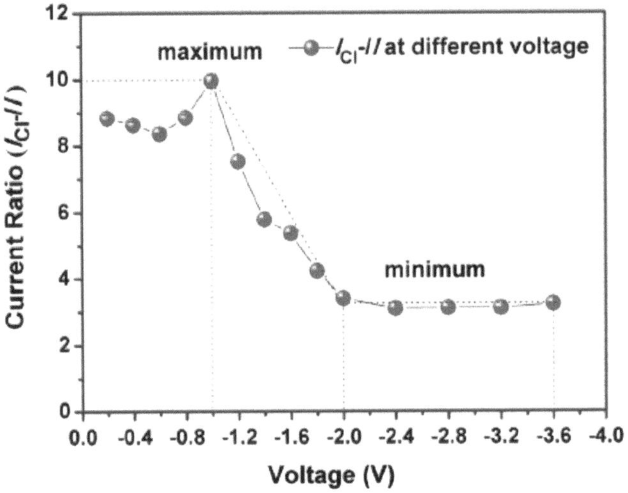

Figure 8.46 Current ratio, I_{Cl^-}/I, of an AIPA–BBA modified channel after Cl^- activation at different voltages. (Reprinted with permission from Liu et al., 2016.)

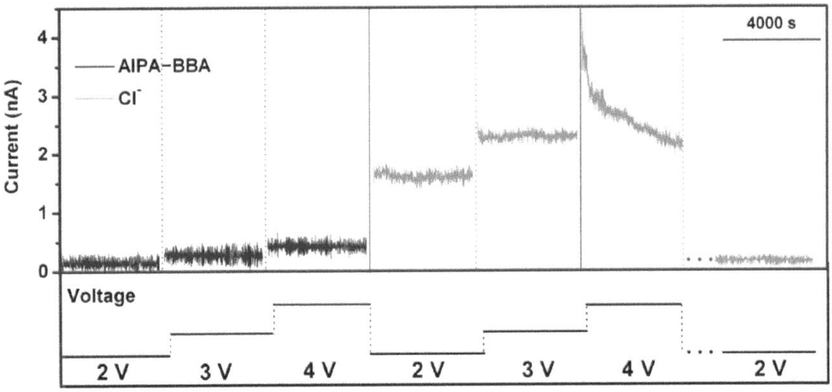

Figure 8.47 Ionic currents before (black) and after (gray) Cl^- activation at different voltages. (Reprinted with permission from Liu et al., 2016.)

responsive nanochannels. One is to incorporate dual responsive molecules into a nanochannel. The other is to modify a channel asymmetrically with different materials that respond to different stimuli.

pH/light dual-stimuli responsive nanochannels can be developed using the first strategy. Popular molecules that are sensitive to both pH and light include spiropyran 1′-(3-carboxypropyl)-3′,3′-dimethyl-6-nitro-spiro[2H-1]benzopyran-2,2′-indoline (SP–COOH) (Zhang et al., 2012) and malachite green derivative (MG–OH–NH_2) (Wen et al., 2012). Figure 8.48 illustrates three states of an SP–COOH modified PET conical nanochannel. Without the irradiation of UV light, the spiropyran is neutral and hydrophobic, and the channel is in the closed state (gray channels in Figure 8.48). Under the irradiation of UV light, the spiropyran becomes negatively charged at pH = 7 and the channel allows the passage of cations (blue channel in Figure 8.48). Due to the geometric shape of the channel, ionic current rectification is reached (see Section 8.6.1), as shown by the *I–V* curves in Figure 8.49a. At pH = 3, however, the spiropyran is turned to be positively charged and the channel selectively transport anions with ionic rectification (red channel in Figures 8.48 and 8.49b). Therefore, the cooperative gating of pH and UV light can efficiently control the on/off and the rectification direction of ion transport in the channel.

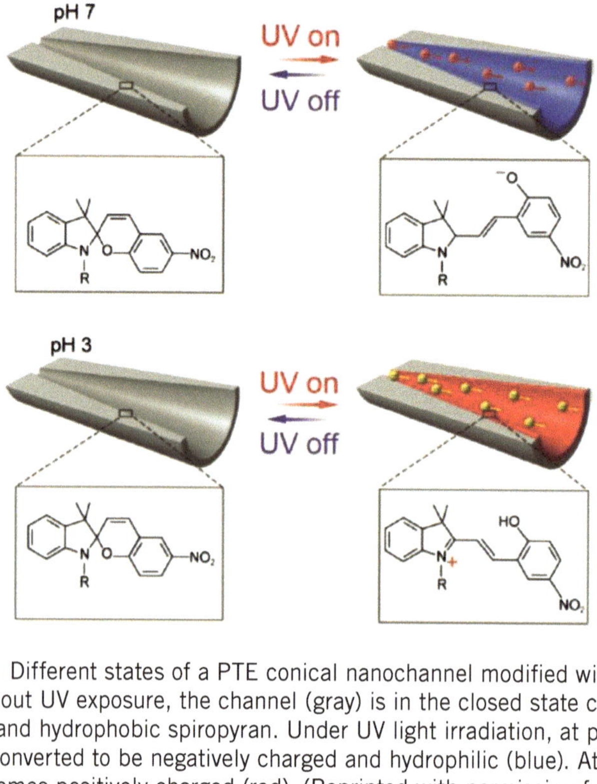

Figure 8.48 Different states of a PTE conical nanochannel modified with SP–COOH. Without UV exposure, the channel (gray) is in the closed state caused by the neutral and hydrophobic spiropyran. Under UV light irradiation, at pH 7, the channel is converted to be negatively charged and hydrophilic (blue). At pH 3, the surface becomes positively charged (red). (Reprinted with permission from Zhang et al., 2012.)

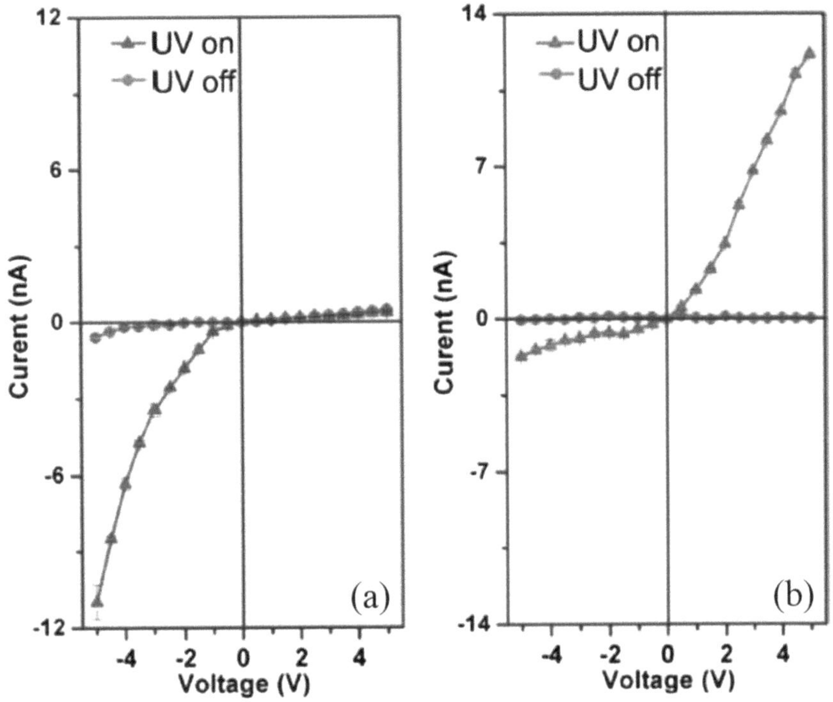

Figure 8.49 Current–voltage characteristics of the pH/light dual-stimuli responsive nanochannel in Figure 8.48. (a) pH 7, before and after the irradiation of UV light. (b) pH 3. (Reprinted with permission from Zhang et al., 2012.)

pH/temperature dual-stimuli responsive nanochannels can be designed by both approaches mentioned previously. For the first method, copolymer can be used as a pH/temperature-responsive material. For the second strategy, a nanochannel can be patterned with different materials, such as poly(*N*-isopropylacrylamide) (PNIPA) and poly(acrylic acid) (PAA), which are temperature and pH responsive, respectively.

Figure 8.50 shows a pH/temperature-responsive, PI nanopore modified with poly(*N*-isopropyl acrylamide-*co*-acrylic acid) [P(NIPAAm-*co*-AAc)] (Guo et al., 2010). The P(NIPAAm-*co*-AAc) brushes are composed of temperature-responsive NIPAAm monomers and pH-responsive AAc monomers. Temperature can induce structural changes of P(NIPAAm-*co*-AAc) between a swollen and a collapsed configurations, which correspond to the close and the open states. pH changes can lead to the protonation and deprotonation of P(NIPAAm-*co*-AAc), which alter the surface charge and change the rectification behavior of the nanopore. Figure 8.51 plots *I*–*V* curves at 25°C and 40°C for different pH values. Although

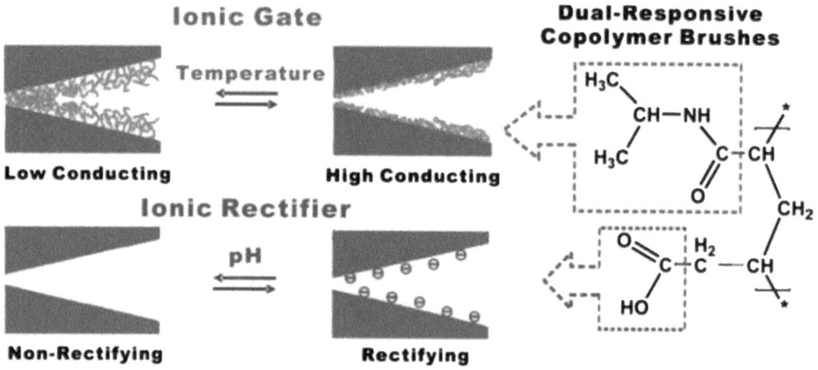

Figure 8.50 Working principle of a conical PI nanopore modified with P(NIPAAm-*co*-AAc). (Reprinted with permission from Guo et al., 2010.)

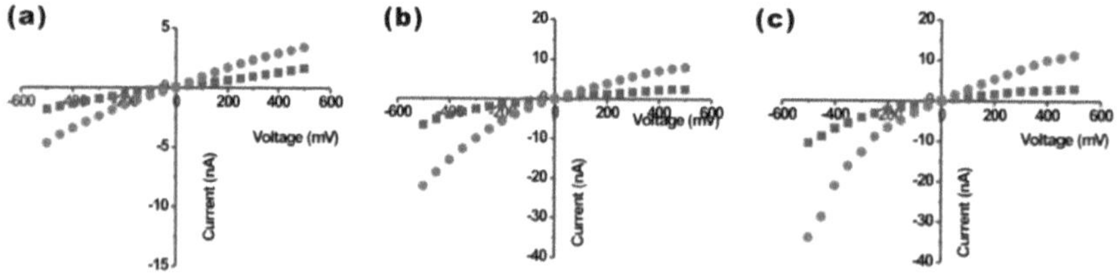

Figure 8.51 Current–voltage curves for the pH/temperature-responsive ionic gate in Figure 8.50 in a 0.1 M KCl solution at 25°C (squares) and 40°C (circles). (a) pH 3.6. (b) pH 7.2. (c) pH 9.4. (Reprinted with permission from Guo et al., 2010.)

the performance of the thermal- and pH gating is not exciting (the ionic current ratio between different states is ~3), the P(NIPAAm-*co*-AAc) modified nanopore exhibits joint features of temperature-gated ionic conductance and pH tunable ionic current rectification in one device.

pH/temperature-responsive nanochannels can also be fabricated by heterogeneously functionalizing a nanochannel. Figure 8.52a illustrates an hourglass-shaped PET nanopore modified by PNIPA and PAA at different sides (Hou et al., 2010). PNIPA is a temperature-responsive polymer, which assumes a swollen, hydrophilic state at 23°C and gradually changes to a collapsed, hydrophobic state when the temperature becomes higher than 32°C. PAA is a pH-responsive polymer, which is neutral, hydrophobic, and in a collapsed state at a pH lower than pKa 4.7. It is changed to a swollen, hydrophilic, and negatively charged configuration at a pH above pKa > 4.7. Figure 8.52b and c shows the ionic current ratio when a voltage bias is applied to the nanopore before (Figure 8.52b) and after (Figure 8.52c)

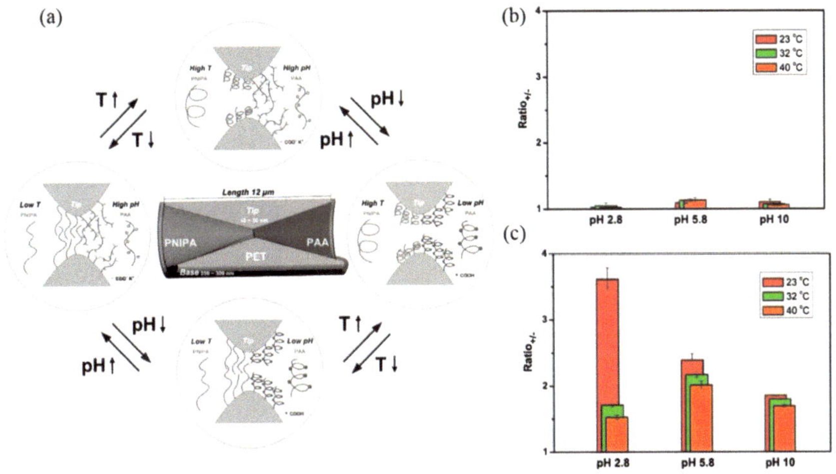

Figure 8.52 An hourglass-shaped PET nanopore asymmetrically modified with PNIPA and PAA as a pH/temperature-responsive ion channel. (a) Working principle. (b) Ionic current ratio of the nanopore without surface modification. (c) Ionic current ratio after surface modification. (Reprinted with permission from Hou et al., 2010, Copyright (2010) American Chemical Society.)

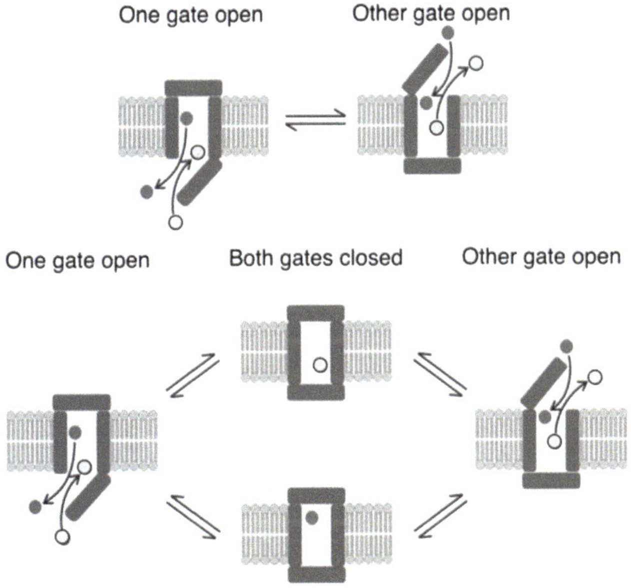

Figure 8.53 Modes of a biological ion pump. Top: Two gates open and close alternately. Bottom: Occlude state. (Reprinted by permission from Springer Nature, Gadsby, 2009, Copyright (2009).)

surface modification. It can be seen that the ionic rectification effect is greatly enhanced with the polymer functionalization, especially at low pH values.

8.7.8 Bioinspired ion pumps

Biological ion channels are capable of selectively transporting specific ions from a high concentration to a low concentration. In cells, to drive ions from a low concentration to a high concentration area, however, biological ion pumps are needed. Compared to an ion channel, which usually regulates ion transport through a single gate, an ion pump contains at least two cooperative gates located at the ends of a channel. At the normal working state, the two gates open and close alternatively. Target ions move into the pathway when one (the first) gate is open and leave the pathway at the other side of the pump when the second gate opens after the first gate closes, as illustrated in the top panel of Figure 8.53 (Gadsby, 2009). In some circumstances, both gates open simultaneously and the ion pump works as an ion channel. Besides, ion pumps also present a fail-safe mechanism that both gates close first and the bound ions are occluded in the pathway until the second gate opens (bottom panel of Figure 8.53).

A bioinspired, artificial PET nanochannel-based ion pump is schematically shown in Figure 8.54a (Zhang et al., 2013). This ion pump uses two cooperative pH-activated gates, which are fabricated by modifying the channel surface with pH-responsive polyelectrolytes, PVP (pKa = 5.2) and poly (acrylic acid) (PAA) (pKa = 4.7), at the ends of the channel (Figure 8.54a). As the environment pH is varied, both polymers can switch between two states, i.e., a swollen, positively charged, hydrophilic state and a collapsed, neutral, hydrophobic state, which correspond to the open and closed states of the channel. However, the conformational changes of these two polymers in response to pH variations are opposite. As the pH changes from an acid condition (pKa < 5.2) to a base condition (pKa > 5.2), PVP switches from the open state to the closed state. In contrast, PAA changes from the closed state to the open state as the pH varies from <4.7 to >4.7.

The distinct pH responses of the two gates result in a series of gating states of the nanochannel, i.e., alternative/simultaneous open and close states under symmetric/asymmetric pH stimuli, as demonstrated in Figure 8.54b. Under alternating and symmetric pH stimuli, the nanochannel works as an ion pump. Furthermore, the nanochannel shows the features of ion channels when both gates are open under asymmetric pH stimuli (Figure 8.54b). Moreover, the fail-safe mechanism can also be reached under symmetric and asymmetric pH stimuli, as shown in Figure 8.54b. A drawback of this ion pump is that it requires continuously manual pH changes, which are not desired in practical applications. If this is overcome, artificial ion pumps with similar functions to their biological counterparts possess promising potentials for precise control of ion transport, which can have wide applications in nanodevices, energy conversion, seawater desalination, and drug delivery.

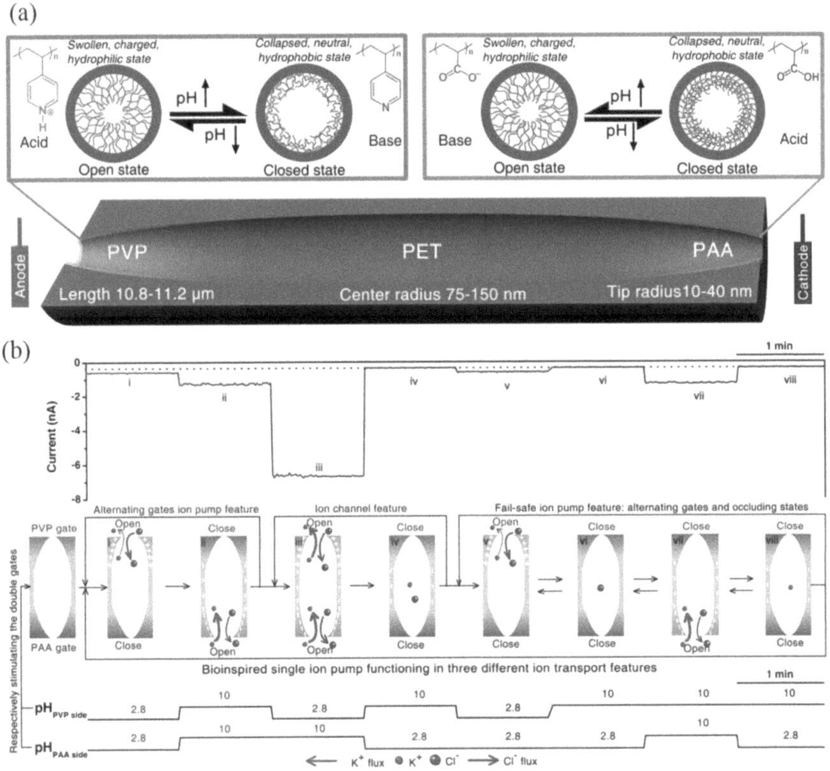

Figure 8.54 An artificial ion pump. (a) Schematic of the ion pump made of a PET nanochannel heterogeneously modified with PVP and PAA at the ends. (b) pH gating of three ion transport modes, i.e., ion pump, ion channel, and fail-safe ion pump. (Reprinted with permission from Zhang et al., 2013, Copyright (2013) American Chemical Society.)

8.8 Stochastic ionic current

In a macroscale system, ions or charged molecules move continuously under an external electric field. At the nanoscale, if the pore size is comparable to that of a migrating species, however, the current signal may fluctuate when the charged entity passes through the nanopore. Typically, a translocation event causes the current to drop, similar to the Coulter effect, which can be used to count the translocation events of charged particles or big molecules. In some cases, current fluctuations can also be observed for simple electrolytes (e.g., NaCl solutions). It has been reported that single cations, such as Li^+, Na^+, and K^+, can block a CNT nanofluidic channel and induce current drops. Such blockades of a nanopore are usually stochastic and the fluctuation of ionic current can be employed as an ion counter or for biodetection. In this section, some experimental studies on ionic current fluctuations in nanofluidic systems using simple electrolytes is introduced.

8.8.1 Electric field-induced wetting/dewetting

Wetting and dewetting are two common phenomena that can be frequently observed in nature. In biological systems, a transition between wetting and dewetting can play a critical role in an ion channel for gating the transport of ions. The mechanisms for wetting/dewetting have also been widely used in artificial nanopores for controlling ion migration. For a nanopore, usually a wet (or hydrophilic) surface promotes ion transport, while a hydrophobic surface hinders the passage of ions due to the large energy barrier at the entrance of the pore, as discussed in Chapter 5. If the surface of a nanopore can be controlled to switch between a hydrophilic state and a hydrophobic state, the nanopore can display current fluctuations, which can be used for various functional processes. One way to alter the wettability of a surface is to apply an external stimulus, such as an electric field, which can increase the surface wettability. A process that increases the wettability of a surface is termed electrowetting. In some cases, once a nanopore is changed from hydrophobic to hydrophilic by an external electric field, it can remain in the wet state even if the electric field is removed. This is undesired if continuous current fluctuation is needed.

Electric field-induced wetting and dewetting have been realized in PET nanopores (Powell et al., 2011). As PET nanopores are fabricated using an etching method, the nanopores are hydrophilic due to the carboxyl groups on the pore surface generated during the fabrication. When the pores are treated with (trimethylsilyl) diazomethane in methanol, the pores are changed to hydrophobic and the hydrophobicity of the pore surface can be tuned by controlling the treatment time and the concentration of (trimethylsilyl)diazomethane. With about 15-min treatment at dilute concentrations of (trimethylsilyl)diazomethane (0.1–12 mM), the PET nanopores exhibit voltage-gated ionic current fluctuations. As shown in Figure 8.55a, the nanopore is closed at a voltage of 1 V, as indicated by the zero ionic current. At higher voltages, 2 V and 3.5 V (or from –1 V to –2 V), current fluctuations take place intermittently, showing that the pore switches between an open state and the closed state occasionally. When the voltage is raised to 5 V (or –3 V), the nanopore mainly remains in an open state, but switches to the closed state stochastically. The open and the closed states are hypothesized to be caused by the wetting and dewetting of the pore surface due to the electric field. The current variations in Figure 8.55a indicate that the wetting and dewetting of the pore caused by the electric field are random, especially at relatively high voltages.

The electric field-induced wetting and dewetting of the pore are illustrated in Figure 8.55b and c. With a 24-hour treatment in 200 mM (trimethylsilyl)diazomethane, the PET nanopore surface is changed from hydrophilic (contact angle

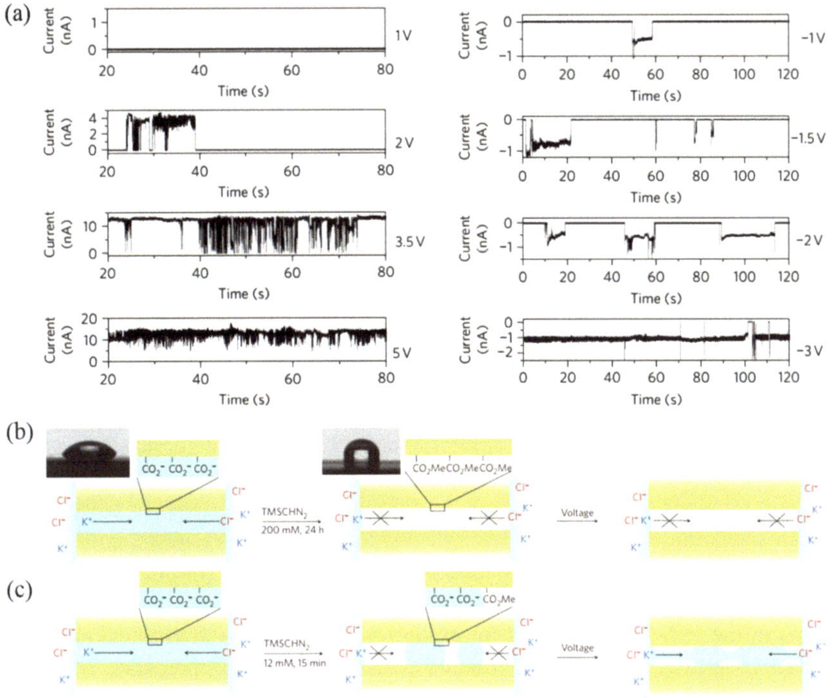

Figure 8.55 Stochastic current fluctuations in nanopores caused by electric field-induced wetting and dewetting. (a) Ionic currents recorded in a 16-nm-diameter conical PET nanopore modified in 12 mM (trimethylsilyl)diazomethane for 15 min. The electrolyte is 1 M KCl with pH 8. (b) Schematic of chemical treatment with a high concentration (trimethylsilyl)diazomethane for 24 hours. The contact angle for the surface is changed from ~60° to ~102°. The final hydrophobic surface blocks the transport of ions. (c) Short surface treatment in dilute solutions of (trimethylsilyl)diazomethane. The surface is hydrophilic with hydrophobic areas and can be wetted by external electric fields. (Reprinted by permission from Springer Nature, Powell et al., 2011, Copyright (2011)).

is ~60°) to hydrophobic (contact angle is ~102°), as shown in Figure 8.55b. Methyl groups functionalized on the pore surface and the exposed aromatic group of the polymer surface cause the increase in hydrophobicity. With a short treatment at a dilute (trimethylsilyl)diazomethane concentration (the case in Figure 8.55a), the nanopore surface is slightly hydrophilic (contact angle is ~75° in 12 nM (trimethylsilyl)diazomethane), which indicates that the chemical modification is incomplete and hydrophobic clusters may form on the pore surface, as illustrated in Figure 8.55c. These hydrophobic areas may lead to the formation of vapor pockets, which generate high energy barriers and hinder the transport of ions. When an electric field is applied, it facilitates the filling of water into the pore, which may further form continuous water threads as pathways for ion transport. The transition between the formation and breaking of the water threads possibly leads to the current fluctuations in Figure 8.55a. However, in a nanopore modified with 200-mM (trimethylsilyl)diazomethane for 2 h or longer time, ionic current is zero for voltages up to 10 V, indicating that ion transport is completely blocked. This is because the nanopore surface is hydrophobic (Figure 8.55b), which generates a high energy barrier such that the passage of ions requires a sufficiently high voltage.

Stochastic current variations are also observed in PDMS–glass nanopores. Figure 8.56 shows the currents for PDMS–glass pores with diameters ranging from 1.9 nm to 7.4 nm using various simple electrolytes, including KCl, NaCl, LiCl, CsCl, and $CaCl_2$. It can be seen that the currents fluctuate greatly. Further experiments show that the baseline current increases with increasing ion concentration, which suggests that ions are the major charge carriers. Therefore, the current changes (or pore blockages) are not caused by the ions. In addition, the dwell time (the duration of low conductance state) shows weak dependence on the applied voltage, especially at relatively low voltages. A physically reasonable explanation about the stochastic pore blocking is the wetting/dewetting transitions in the PDMS–glass nanopores, similar to that illustrated in Figure 8.55c. Under an electric field, the nanopores switch between an open state (pores are

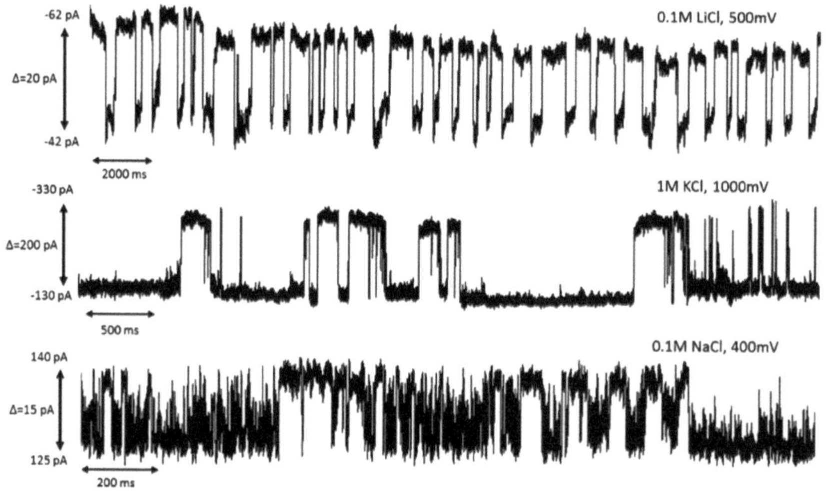

Figure 8.56 Representative current fluctuations recorded in PDMS–glass nanopores. (Reprinted with permission from Shimizu et al., 2013, Copyright (2013) American Chemical Society.)

filled with liquid) and the closed state (pores are occupied by vapor), leading to stochastic current fluctuations. This explanation is supported by theoretical analysis based on the variational field theory (Buyukdagli et al., 2010), which predicts that phase transitions can take place in weakly charged nanopores.

8.8.2 Current fluctuation in CNTs and graphene nanopores

Carbon nanotubes and graphene nanopores are special nanoconfinements, which have been widely employed in studying nanofluidics, including both simple fluid flows and ion transport. Section 6.5 (Chapter 6) discusses water flows in CNTs, where large slip lengths are paramount phenomena due to the smooth and hydrophobic surface properties. For ion transport, novel nanofluidic devices have been fabricated to investigate the motion of ions in CNTs and graphene nanopores. Stochastic current fluctuations are also a common phenomenon that has been observed in different CNT- and graphene nanopore-based flow systems. Unfortunately, the underlying mechanisms are not quite clear. In this subsection, experimental studies reporting stochastic current fluctuations in CNTs and graphene nanopores are discussed.

Stochastic ionic currents associated with ion transport in CNTs have been experimentally observed (Lee, 2010; Lee et al., 2010). In the fabrication of the nanofluidic device, ultralong, aligned SWCNTs (average diameter ~1.5 nm) are first grown on a Si/SiO_2 substrate through CVD. An epoxy frame with two compartments are then bonded onto the substrate. This is followed by oxygen plasma etching to open and keep a segment (~500 μm) of the CNTs, as shown in Figure 8.57. The two compartment of the epoxy frame serve as the reservoirs for electroosmotic flows. When an electrolyte containing simple cations, such as Li^+, Na^+, or K^+, is added to the reservoirs, the ionic current shows stochastic switching between two or three states above a threshold voltage, as shown in Figure 8.58a. Through control experiments, protons are recognized as the primary charge carriers and cations are identified as the dominant blocker. This claim is based on the following facts: (i) the increase of salt concentration reduces the baseline current; (ii) the addition of HCl enhances the baseline and blocking currents only when HCl is added to the anode; (iii) the pore blocking changes significantly when the cation type and concentration are varied; and (iv) the addition of solutions containing large cations, tetramethylammonium (TMA^+), does not show pore blocking. However, the conclusion of protons being the major charge carrier is inconsistent with other experiments, which report that ions are the major charge carriers (Liu et al., 2010; Pang et al., 2011; Liu et al., 2013; Geng et al., 2014; Amiri et al., 2017; Tunuguntla et al., 2017). Nevertheless, the cation mobilities calculated from the dwell time of the blocking events span a wide range for different cations. For Li^+ and K^+, the mobilities are more than two orders of magnitude higher than their bulk values, while the mobility of Na^+ is comparable to the bulk

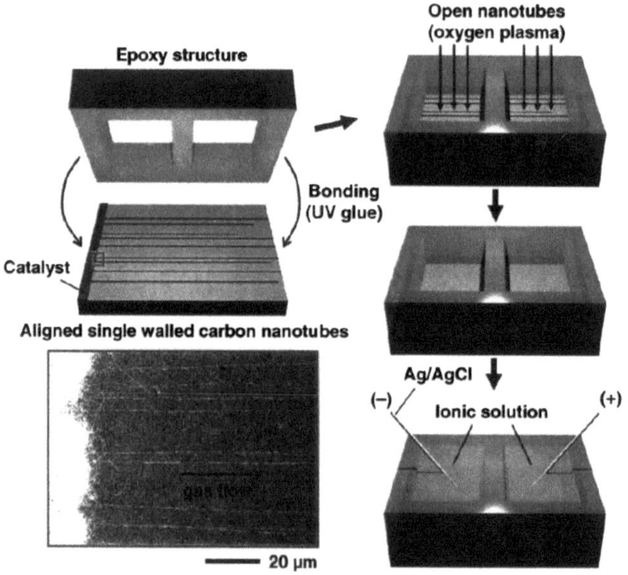

Figure 8.57 Fabrication of an SWCNT-based nanofluidic device for electroosmotic flow measurement. SWCNTs are grown on a Si/SiO_2 substrate using CVD method. An epoxy frame is then bonded with the substrate. The epoxy frame contains two compartments, which are used as reservoirs. Plasma etching is used to keep a segment of the CNTs for ion transport. (Reprinted from Lee, 2010, by courtesy of the author.)

mobility. The two-order-magnitude enhancement of ion mobility for Li^+ and K^+ is consistent with other experimental observation (Liu et al., 2010).

Under certain conditions, the pore blocking events show highly synchronized oscillations, as depicted in Figure 8.58b. Such resonant current oscillations are found to be the consequence of the interplay between the stochastic pore blocking and the proton-diffusion limitation at the CNT entrance. Under the unblocked state, the proton flux in the CNT is very high, which depletes the proton concentration and increases the blocking cation concentration at the CNT entrance. The accumulation of cations at the CNT entrance facilitates cation partitioning into the nanotube. As cations move in the CNT, the proton concentration at the entrance accumulates rapidly, which corresponds to the blocked state (low current). As the cations exit from the CNT, the high proton concentration at the entrance increases the proton flux again, leading to another cycle of current variation.

Stochastic current fluctuations in CNTs have also been observed in CNT–lipid bilayer systems, which are called CNT porins (Geng et al., 2014). The CNT–lipid bilayer systems are fabricated by inserting short CNTs (diameter ~1.5 nm, length

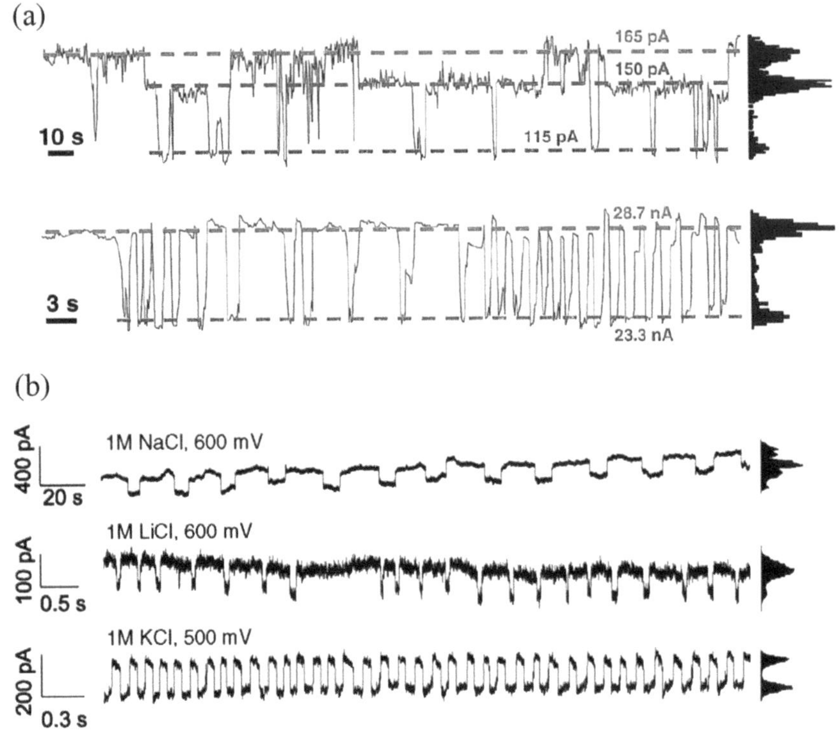

Figure 8.58 Stochastic current fluctuation in single CNTs. (a) Representative stochastic current fluctuation for NaCl, LiCl, and KCl solutions. (b) Resonant current oscillation. (Reprinted from Lee, 2010, by courtesy of the author.)

5–15 nm) into lipid bilayers, as shown in Figure 8.59a. For a part of the devices (less than 30%), the ionic current exhibits oscillation patterns between an open and a partially closed states, as depicted in Figure 8.59b. In contrast to the observation in Figure 8.58, the major charge carriers in the CNT–lipid systems are found to be cations instead of protons. Since the CNT porins only show low conductance instead of zero conductance at the partially closed state, the stochastic pore blocking is similar to the ionic gating of biological ion channels. The gating behavior of biological ion channels is usually induced by conformational changes. However, CNTs are rigid and the gating effect of the CNT porins might be caused by other mechanisms. A possible explanation for the current fluctuation is the transition of the system between ionic-penetration and ionic-exclusion states, as predicted for weakly charged or neutral pores. Nonetheless, further investigations are needed to elucidate the mechanisms of current variations in CNT–lipid systems.

Stochastic current fluctuations in sub-2-nm graphene nanopores have also been reported (Jain et al., 2015). Figure 8.60a illustrates a graphene nanopore. Figure 8.60b

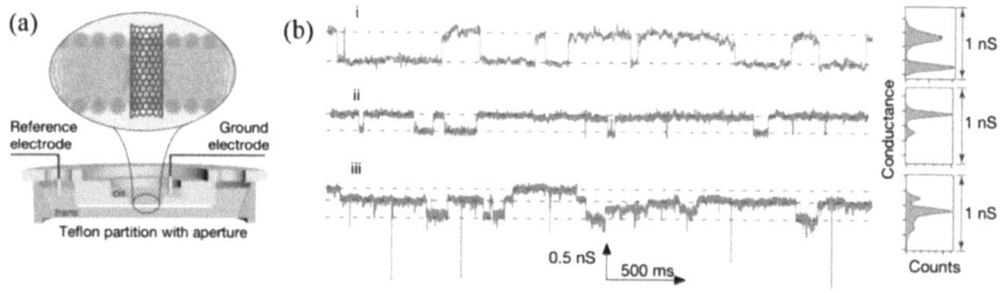

Figure 8.59 Stochastic current fluctuations in CNT–lipid bilayer systems. (a) Schematic of the CNT–lipid bilayer system. (b) Representative current fluctuations. (Reprinted by permission from Springer Nature, Geng et al., 2014, Copyright (2014).)

and c shows current variations within very short time intervals. It is seen that the current fluctuations are of very high frequencies, which are different from the current switching between constant levels, as found in CNTs and other nanopores. Such current fluctuations are possibly caused by the transition between protonation and deprotonation of the functional groups at the pore edge. This claim is supported by the power spectral analysis for the current fluctuations, which exhibits a characteristic timescale (Figure 8.60d) consistent with the characteristics of protonation and deprotonation in a protein ion channel.

All the stochastic current fluctuations reported in these work require two conditions, i.e., a sufficiently high ion concentration (≥1 M) and an electric field above a threshold, which suggest that there may be common mechanisms for the occurrence of stochastic current fluctuations, rather than the diverse mechanisms proposed in those studies. These common mechanisms require extensive investigations.

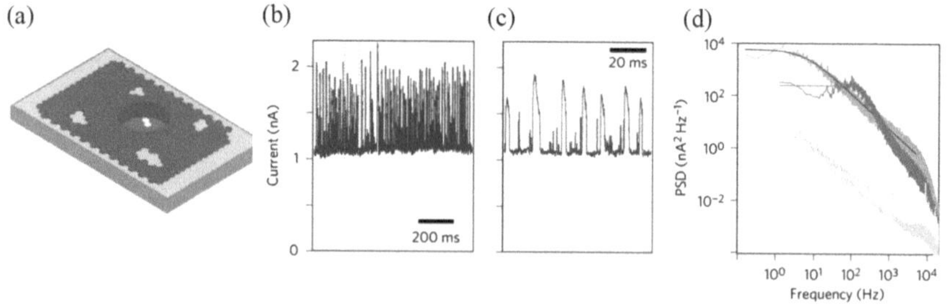

Figure 8.60 Stochastic current fluctuations in isolated graphene nanopores. (a) Schematic of an isolated graphene nanopore. (b)–(c) Currents measured in a graphene nanopore at −200 mV. (d) power spectral density for real-time current traces. The dark gray curve corresponds to the current traces in (b) and (c). (Reprinted by permission from Springer Nature, Jain et al., 2015, Copyright (2015).)

8.8.3 Chemisorption and nanoprecipitation

Similar to the adsorption of fluid molecules on a surface, ions can also be adsorbed by a nanochannel surface or functional groups immobilized on the channel surface. The adsorption of fluid molecules are typically physisorption caused by the van der Waals force. Although physisorption occurs for ions, the adsorption of ions by functional groups is mainly chemisorption because it usually results from covalent bindings. Once ion adsorption takes place, the accumulation of ions can greatly change the potential in a nanopore and affect the ion transport.

Ionic current fluctuations caused by ion adsorption are observed in PET conical nanopores, which are usually fabricated using an etching approach. The etching of PET using NaOH hydrolyzes ester bonds of PET and produces carboxylate groups on the pore surface. The carboxylate groups are negatively charged and the charge density is estimated to be around 1.5 e/nm^2 (Siwy et al., 2006b). For a KCl solution (0.1 M), the current generated by moving ions under a voltage is quite stable, as shown in Figure 8.61. However, if 0.3 mM $CaCl_2$ is introduced into the solution, the ionic current starts to fluctuate as the voltage is increased ($\Delta V < -300$ mV).

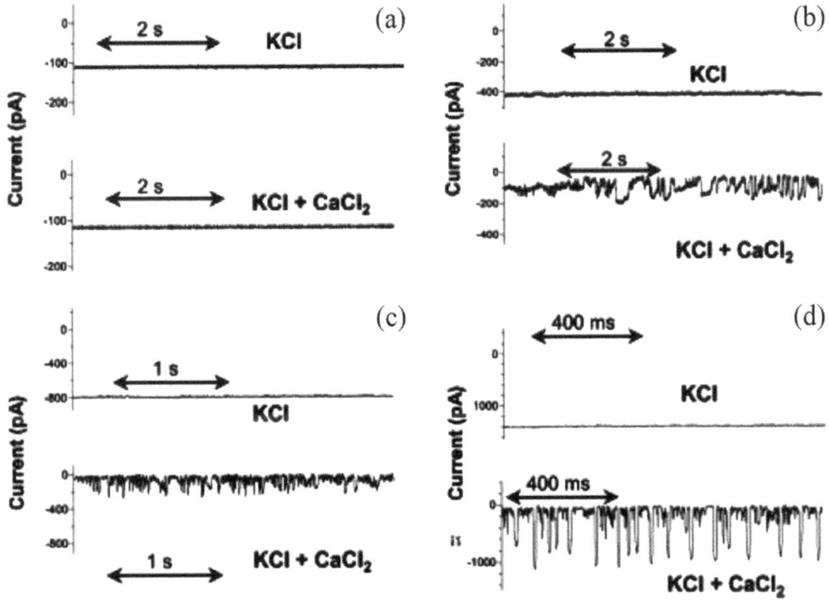

Figure 8.61 Time series of ionic current through a conical PET nanopore (the tip diameter is 4 nm) in a 0.1 M KCl solution with/without the addition of 0.3 mM $CaAl_2$. The voltage applied between the base and tip ends of the pore is −50 mV (a), −350 mV (b), −500 mV (c), and −900 mV (d) (Reprinted with permission from Siwy et al., 2006b, Copyright (2006) American Chemical Society.)

The current variation is attributed to the chemisorption of Ca^{2+} on the pore surface due to the binding between Ca^{2+} and the carboxylate groups on the surface. As the voltage increases, it affects the concentration of Ca^{2+} and the local electric potential, which alter the Ca^{2+}-carboxylate binding. As the electric potential in the pore experience stochastic variations, adsorption and desorption of Ca^{2+} take place, leading to current fluctuations. This explanation is supported by the fact that the introduction of Mn^{2+}, which binds with carboxylate groups much stronger than Ca^{2+}, results in current oscillations with a much lower frequency, such that the pore stays predominately in the closed state.

Somewhat similar to ion adsorption, nanoprecipitation presents another mechanism for ionic current variations. Nanoprecipitation is the precipitation of ions when the ion concentration exceeds the solubility limit. Nanopores fabricated through popular etching methods are usually negatively charged. This leads to the accumulation of counterions and exclusion of coions in the vicinity of the pore surface. Therefore, the ion concentration inside the nanopore differs significantly from that of the bulk solution. Moreover, an external electric field also greatly affects the ion concentration in the nanopore, which further amplifies the ion concentration difference between the bulk solution and the nanopore. Under certain conditions, functional groups on the pore surface may undergo transitions between protonation and deprotonation, resulting in a complex charge environment. When the ion concentration becomes sufficiently high, nanoprecipitation occurs.

The formation and redissolvation of nanoprecipitates are similar to the adsorption and desorption of ions and can cause ionic current fluctuations. In the absence of nanoprecipitation, a nanopore can conduct ions and remains in an open state. The occurrence of nanoprecipitation, however, prohibits the transport of ions, leading to the closure of the pore. If the nanoprecipitates are redissolved or released, the nanopore returns to an open state and ion transport recovers.

Ionic current variations caused by nanoprecipitation are observed in conical PET nanopores (Powell et al., 2008). For a 0.1 M KCl electrolyte containing sub-millimolar concentrations of divalent cations, such as Ca^{2+}, Co^{2+} and Mg^{2+}, the negatively charged pore surface and the external electric field cause ion accumulation in the pore, which facilitates the formation of nanoprecipitates at bulk concentrations lower than the saturation concentration. For a solution with sub-millimolar Ca^{2+} and 2 mM phosphate buffer (pH = 8), nanoprecipitates, calcium hydrogen phosphate ($CaHPO_4$), form in the nanopore above a threshold voltage, as illustrated in Figure 8.62a. Once the nanopore is blocked by the nanoprecipitates, the local ion concentration and distribution will change, which may redissolve the nanoprecipitates and recover the ionic current. As shown in Figure 8.62b, the ionic current decreases with increasing Ca^{2+} concentration, which indicates the occurrence of nanoprecipitation. Figure 8.62c–e shows the time series of ionic current oscillations for the three Ca^{2+} concentrations in Figure 8.62b. It can be seen that the oscillation frequency increases with increasing Ca^{2+} concentration. As nanoprecipitation is affected by the electric field, the voltage applied across the nanopore also causes the ionic current to fluctuate. As shown in Figure 8.62f, the ion current roughly remains constant at a low voltage of −100 mV. However, it becomes unstable when the voltage is increased to −200 mV (Figure 8.62g). At an even higher voltage (−1000 mV), the current fluctuates with a high frequency (Figure 8.62f). Current oscillations are reproduced with the precipitation of $CoHPO_4$, $Co(OH)_2$, or $Mg(OH)_2$, which further confirm the critical role of nanoprecipitation in current fluctuations, although the frequency and how the current oscillates are distinct for different nanoprecipitates.

To gain the molecular insights of nanoprecipitation-assisted current oscillations, microsecond MD simulations are performed for nanopores functionalized with carboxyl groups (Cruz-Chu and Schulten, 2010). It is found that the surface functional groups play a critical role in nanoprecipitation oscillations. Deprotonated carboxyl groups adsorb Ca^{2+} and serve as nucleation sites that trigger the formation of nanoprecipitates. The nanoprecipitates then tightly adhere to the functional groups, block the nanopore, and lead to a low ionic conductance. In contradiction to the argument of nanoprecipitate redissolvation, MD simulations suggest that the recovery of the ionic current is caused by the detachment of the

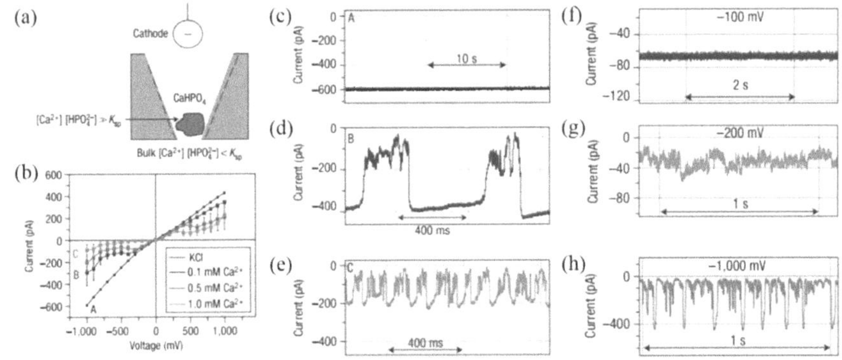

Figure 8.62 Nanoprecipitation-induced current fluctuations in nanopores. (a) Schematic of the formation of nanoprecipitates in a nanopore. (b) Current–voltage curves in 0.1 M KCl solutions with 2 mM PBS buffer alone and varying calcium concentrations. (c)–(e) Time series of the current. The three panels correspond to the cases, A, B, and C, in (b). (f)–(h) Current traces recorded in 0.1 M KCl, 0.2-mM $CaCl_2$, and 5-mM PBS at voltages of −100 mV, −200mV, and −1000mV. (Reprinted by permission from Springer Nature, Powell et al., 2008, Copyright (2008).)

nanoprecipitates induced by the reprotonation of the carboxyl groups. Though this deprotonation/reprotonation mechanism explains the current oscillation, other properties of nanoprecipitation-induced current oscillations, such as the oscillation shape, species-dependent frequency, and negative threshold voltages, require further explorations.

8.8.4 Summary

Stochastic current fluctuations are widely observed in nanotubes and nanopores, even in simple electrolytes. However, the current fluctuations may be caused by the coupling of different effects, rather than a single mechanism. In biological ion channels, stochastic pore blockages are typically caused by conformational changes. In nanotubes or nanopores, the reasons for stochastic pore blocking can be diverse, as discussed in previous sections. Although most of the explanations sound reasonable when interpreting the current variations in a specific system, few mechanisms are supported by solid evidence. To have a deep understanding of the current fluctuations in nanopores, further experimental and numerical studies are needed.

8.9 Ion transport in CNTs and BNNTs

In Section 6.5, water flows in CNTs are particularly discussed because of the special structure and surface properties of CNTs. Other than the cylindrical geometry and smooth surface, CNTs also possess extraordinary electrical and mechanical properties, which may interplay with ion transport and generate new ion transport phenomena. As discussed in Chapter 6, large slip lengths are a striking feature of simple fluid flows in CNTs. Flow slip can also enhance the migration of ions in CNTs. In addition, a CNT usually carries functional groups at its entrance, which can serve as a gate for selective transport of ions and charged molecules. The functional groups at the entrance can be easily changed through chemical modifications. These properties, together with the controllable CNT length and diameter, make CNTs ideal nanoconfinements for tunable transport of ions and molecules.

In the past decade, electrokinetic flows in CNTs have been extensively studied. Ion transport in CNTs has wide applications in biology, medicine, and engineering, including molecular detection and manipulation, drug delivery, flow sensing, energy conversion, and harvesting. Carbon nanotube-based nanofluidic platforms are quite diverse, from CNT membranes to novel devices containing single CNTs. In addition to stochastic current fluctuations introduced in the previous section, many other unique transport phenomena have been revealed, which will be discussed in this section.

BNNTs are another type of nanotubes, which own similar crystallographic structure to but dramatically different electronic properties from CNTs. The distinct properties of BNNTs result in fascinating transport characteristics for ions and fluid flows. Studies have shown that fluidic systems using BNNTs are promising for energy conversion, as will be discussed in the last part of this section.

8.9.1 CNT-based nanofluidic platforms

Generally, CNT-based nanofluidic platforms for ion transport can be categorized into two groups: CNT membranes and single CNT-based nanofluidic devices. CNT membranes are mainly composite membranes that contain highly ordered, aligned arrays of single-walled or multi-walled CNTs in a matrix film. CNT membranes can be prepared by several methods. On approach is to grow vertically aligned CNT arrays on a substrate first. The interstitial space between CNTs is then filled with a sealing material, such as polystyrene (PS) and silicon nitride (Si_3N_4), by spin coating or CVD. This is followed by removing the CNT composite membrane from the substrate and opening the ends of the CNTs through plasma treatment or ion milling. CNT membranes fabricated using this method have been employed to study water flows, which show large slip lengths, as mentioned in Chapter 6 (Majumder et al., 2005; Holt et al., 2006). Another method is to embed CNTs into a matrix, such as epoxy. This approach has been used to fabricate SWCNT-based membranes for studying ion mobilities and electroosmotic flows (Wu et al., 2012). In addition, CNT membranes can also be directly achieved through CVD growth of CNTs on a template with predefined nanopores, e.g., anodic aluminum oxide, as discussed in Chapter 4.

Single CNT-based nanofluidic platforms can be classified into embedded CNT platforms and on-substrate CNT platforms. A typical embedded CNT platform consists of a single CNT spanning in a matrix film, which can be epoxy, silicon nitride, or a lipid bilayer. A single CNT is first embedded in a matrix, which is then microtomed (perpendicularly to the CNT) into a number of membranes with a microscale thickness, as demonstrated in Figure 8.63 (Sun and Crooks, 2000). One of the CNT membranes is then mounted onto a Si_3N_4 or PDMS support with a μm-diameter hole to form a nanofluidic system (Figure 8.63). In Figure 8.63, the CNT is an MWCNT and the diameter of the hole is ~200 nm, which is drilled using focused ion beams in a silicon nitride membrane. The advantage of this fabrication method is that up to several hundreds of single CNT-based membranes can be produced with identical physical and chemical properties. At the early stage, limited by the CNT manipulation means, this fabrication method could only be applied to MWCNTs with diameters larger than 50 nm. With advanced nanomanipulation and fabrication techniques, embedded CNT platforms with CNT radius in the range of 3.5–35 nm and length in the range of 1–3 μm can be fabricated (Secchi et al., 2016b). The CNT insertion can be accomplished *in situ* by using a homemade nanomanipulator with piezoinertial step motors and a SEM. The interstitial space is then sealed with cracked naphthalene by local

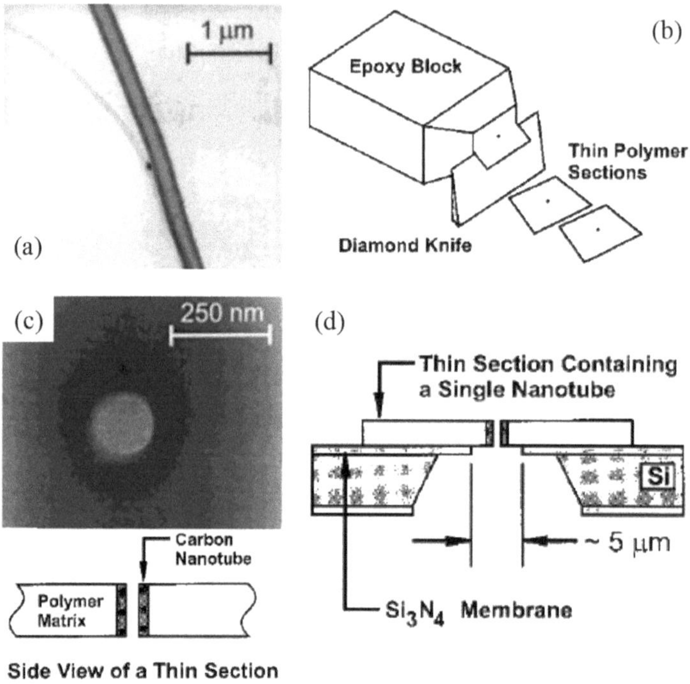

Figure 8.63 Fabrication of single CNT-based nanofluidic platform. (a) TEM image of an MWCNT. (b) After embedding the MWCNT into an epoxy block, single CNT-based membranes are produced through microtoming the epoxy (perpendicular to the CNT axis). (c) TEM image (top view) of a membrane (top panel) and the schematic of a side view (bottom panel). (d) Formation of a nanofluidic system by fixing a CNT membrane onto a Si_3N_4. (Reprinted with permission from Sun and Crooks, 2000, Copyright (2000) American Chemical Society.)

electron beams. Using such smaller diameter CNT platforms, the scaling behavior of ionic conductance at low salt concentrations has been investigated (Secchi et al., 2016b).

Since many unique transport phenomena only occur in sub-2-nm CNTs, embedded CNT platforms containing small CNTs offer great opportunities in nanofluidics. Such platforms have been realized by inserting a short SWCNT into a lipid bilayer (Liu et al., 2013). Ultrashort SWCNTs with diameter of 0.8–2.0 nm and length of 5–10 nm are obtained by cutting long SWCNTs. These short CNTs are then inserted into a lipid bilayer, 1,2-diphytanoyl-*sn*-glycero-3-phosphocholine, using a microinjection probe made of a glass capillary. The probe is put in close to the lipid bilayer and a solution containing CNTs is ejected at a certain flow rate until a current signal change is observed, indicating the accomplishment of CNT

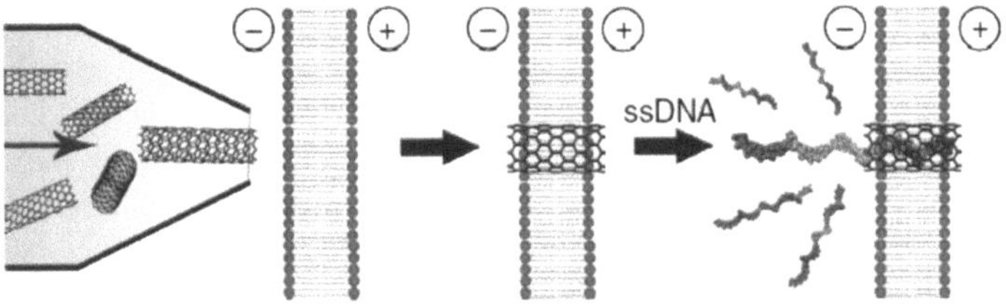

Figure 8.64 Illustration of the insertion of single CNT into a lipid bilayer. (Reprinted with permission from Liu et al., 2013.)

insertion, as illustrated in Figure 8.64. Two-thirds of the devices prepared using this method show stable current signals and have been employed to investigate ion transport and DNA translocation. Similar CNT–lipid bilayer systems have been developed by inserting CNTs (1.51 ± 0.21 nm in diameter, 5–15 nm in length) into 1,2-dioleoyl-*sn*-glycero-3-phosphocholine (Geng et al., 2014).

On-substrate CNT platforms are CNT-based fluidic devices that are directly fabricated on a substrate. In the fabrication, low-density horizontally aligned SWCNTs with diameters of 1–2 nm are grown on an oxidized silicon substrate. A layer of PMMA resist is then deposited on top of the CNTs with alignment markers created by standard electron beam lithography (EBL) for identifying a suitable CNT. After a CNT is selected, fluid reservoirs at proper locations are defined using an EBL procedure. This exposes parts of the selected CNT, which are removed through oxygen plasma treatment. Finally, the device is completed by sealing the substrate with a PDMS layer, on which embedded microchannels are etched to form flow pathways connecting the reservoirs and the external fluid source. Using these devices, ion and DNA transport driven by electric fields through single CNTs has been studied (Liu et al., 2010).

The nanofluidic device shown in Figure 8.57 represents another design of on-substrate CNT platform, where horizontally aligned, ultralong CNTs are grown on a Si/SiO_2 substrate and covered by an epoxy frame. This fabrication method results in multiple CNTs, with a maximum number of 45, in a single device. The ionic current variations can provide the transport information in single CNTs (Lee et al., 2010).

8.9.2 Charge carriers

For the study of ion transport in CNTs, an important issue is to clarify the major charge carriers responsible for the current observed. For a large diameter CNT, the major charge carriers are definitely ions because ions can easily enter into the tube and move with a relatively high speed. In CNTs with small diameters (e.g., sub-2-nm diameter), however, experimental results show that the major charge carrier could be either ions or protons. In many studies, the major charge carriers are determined as ions and most of them are cations, such as Li^+, Na^+, K^+, because the CNT surface and entrance are usually negatively charged. This conclusion is drawn from the fact that the ionic conductance of a CNT increases with increasing ion concentration and shows a sublinear or linear dependence on the ion concentration. Selective transport of cations in CNTs has been confirmed in experiments by changing the cations and anions. It has been shown that the addition of salts containing the same cations but different sized anions does not change the ionic conductance, while the introduction of salts containing the same anion but different sized cations leads to size-dependent conductance (Amiri et al., 2017). Moreover, lowering the pH to acidic conditions reduces the ionic conductance of a CNT with respect to that at the neutral pH (Geng et al., 2014). These observations verify that cations are the major charge carriers in CNTs. Nevertheless, some experiments suggest that CNTs with diameters <1 nm can selectively transport anions when the entrance functional groups are protonated, indicating that anions may also be important charge carriers under certain conditions (Tunuguntla et al., 2017).

In contrast, a study reveals that the major charge carriers in 500-μm-long CNTs are protons, instead of cations (Lee et al., 2010). In a series of control experiments, it is observed that an increase in the salt concentration reduces the baseline current. Besides, the addition of HCl increases the baseline current only when HCl is added to the anode, indicating that protons are the major charge carriers.

As different particles have been identified as the dominant charge carriers, an ionic current measured might be contributed by different charged species. If so, how the flow conditions affect the contribution of a charged particle to the ionic current becomes an important question. Surface charges, types of functional groups, and the protonation/deprotonation of functional groups may play a critical role. There might be other mechanisms that can alter the contributions of different charged particles to the ionic current. All these issues require intensive theoretical and experimental explorations.

8.9.3 Ionic conductance in CNTs

Ionic current and ionic conductance are important quantities for nanoscale ion transport. In CNTs, fast ion transport is expected due to flow slippage. Previous experiments show that ion transport in CNT membranes is different from that in isolated, long CNTs. Experiments on the ion transport in isolated, micrometer-long SWCNTs with diameters between 1 and 2 nm show that the ionic conductance is about two orders of magnitude higher than the bulk conductance (Liu et al., 2010), as shown in Figure 8.65. For a 1 M KCl electrolyte in a 2-μm-long CNT of 1.7 nm in diameter, the ionic conductance measured in experiments is about 9 nS, while the classic theory (Equation 8.64) predicts a value of 0.017 nS.

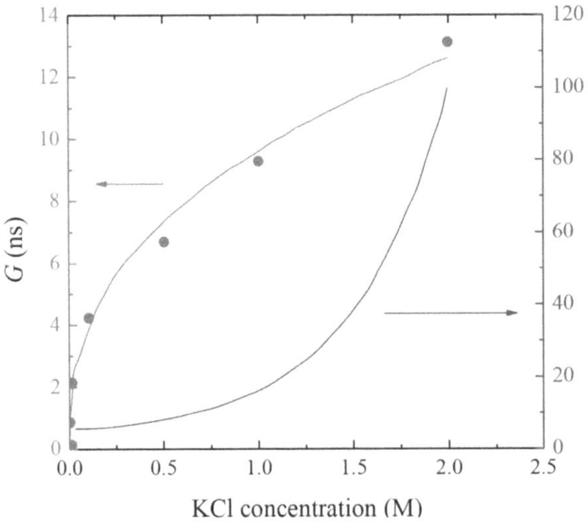

Figure 8.65 Ionic conductance as a function of KCl concentration in a 2-μm-long, 1.7-nm-diameter CNT. The upper line is a power-law fit to the data and the lower line is the conductance measured in a planar nanopore. (Data extracted from Liu et al., 2010.)

Such a high enhancement has not been observed in other nanoconfinements, including CNT membranes, solid-state nanopores, and lab-fabricated nanochannels. The significant enhancement of ion conductance is attributed to the fast electroosmotic flows in CNTs induced by surface charge and flow slip (Pang et al., 2011). Recently, molecular dynamics simulations show that the ultrafast ion transport might be caused by a new flow phenomenon, i.e., fluid breakup in CNTs, which will be discussed later.

Other than the ultrahigh ionic conductance, the dependence of ionic conductance G on the ion concentration in CNTs is also surprising. In a bulk solution, G increases linearly with increasing ion concentration n_0. In CNTs, however, a power-law dependence, $G \sim {n_0}^{\alpha}$, has been found in certain ion concentration range (Pang et al., 2011), as shown in Figure 8.66a. The exponent α lies in the range of 0.3–0.5. In a similar design of CNT-based fluidic system, a power-law dependence, $G \sim {n_0}^{0.5}$, is found in a 1.5-nm-diameter and 20-μm-long CNT in the salt concentration range of 0.01–2.0 M (Figure 8.66b) (Amiri et al., 2017). At high ion concentrations, $n_0 > 0.5$ M, the ionic conductance decreases as the ion concentration is raised, as shown in Figure 8.66c. This is because the transport of ions in small CNTs is dominated by single-file transport, i.e., ions form a chain in the CNT, for which the ion concentration in the CNT will saturate when n_0 becomes sufficiently high. In this situation, once an ion exits from the CNT and

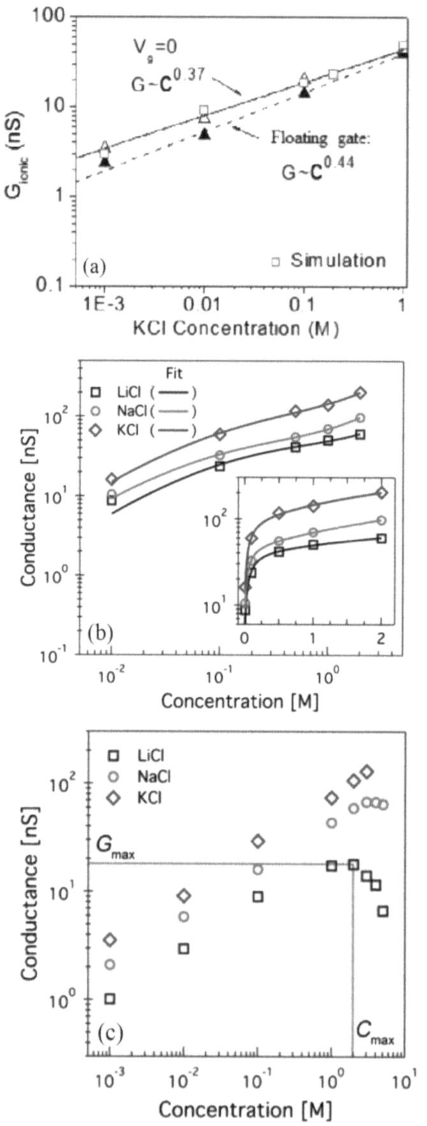

Figure 8.66 Power-law dependence of ionic conductance on ion concentration in CNTs. (a) Triangles, both open and filled, are experimental results for CNTs at different conditions. Squares are MD simulations with a proper surface charge. Lines are power-law fits (Reprinted with permission from Pang et al., ACS Nano 5, 7277–7283, 2011, Copyright (2011) American Chemical Society). (b) and (c) Ionic conductance for different electrolytes (Reprinted with permission from Amiri et al., 2017, Copyright (2017) American Chemical Society.)

enters into the downstream reservoir, another ion from the downstream reservoir may fill the position in the CNT, which inhibits the ion transport in the CNT (Amiri et al., 2017).

Ionic conductance is also measured in CNTs embedded in a lipid bilayer (Liu et al., 2013). In these CNTs, stochastic pore blocking found in epoxy wrapped SWCNTs (see Figure 8.58) is not observed. Instead, the ionic conductance varies significantly, over four orders of magnitude, and fall into three groups in terms of the CNT electronic properties and CNT size. In group 1, all the CNTs are pure metallic SWCNTs and the ionic conductance is high, 60–100 nS. In group 2, the CNTs are semiconductor CNTs with diameters between 1 nm and 2 nm. The ionic conductance in this group is intermediate, in the range of 3–24 nS. Group 3 contains semiconductor CNTs with diameters smaller than 1 nm and the ionic conductance is low, ranging from 0.1 nS to 2 nS. The ionic conductance in group 1 is about 1–2 orders of magnitude higher than the prediction of Equation 8.64, qualitatively consistent with the large enhancement observed in Figure 8.65. Experiments on the effect of pH on the ionic conductance show that the conductance increases with increasing pH, indicating that the functional groups at the CNT entrance or protons may play a critical role in ion transport.

In a similar CNT–lipid bilayer platform, ionic conductance in CNTs with average inner diameter of 1.51 ± 0.21 nm and lengths in the range of 5–15 nm is measured as 0.63 ± 0.12 nS (Geng et al., 2014), which is comparable to the bulk conductance. pH tests demonstrate that the ionic conductance at pH 7 is about two folds of that at pH 2, consistent with other experimental results. However, the power-law dependence of the ionic conductance on the ion concentration is not observed. Instead, linear relationships are found in the salt concentration range of 0.5–2.0 M, which is in line with the classic theory.

For smaller CNTs, 0.8 and 1.5 nm in diameter, the corresponding CNT–lipid bilayers are termed narrow CNT porins (nCNTPs) and wide CNT porins (wCNTPs), respectively (Tunuguntla et al., 2017). For the nCNTPs, the ionic conductance increases as the salt concentration is increased, as shown in Figure 8.67. The dependence of the ionic conductance on the ion concentration, however, shows different fashions at different pH values. At pH 7.5, the conductance increases quickly with increasing ion concentration when the ion concentration is <0.5 M and a power-law dependence, $G \sim n_0^{1/2}$, fits the data well. As the ion concentration becomes higher than 0.5 M, the conductance gradually levels off and approaches 70 pS. The saturation of the ionic conductance at high ion concentrations is similar to the behavior of biological ion channels, which is attributed to Michaelis–Menten kinetics induced by the interactions between ions and the functional groups at the CNT entrance. At pH 3, however, the ion conductance scales linearly with ion concentration.

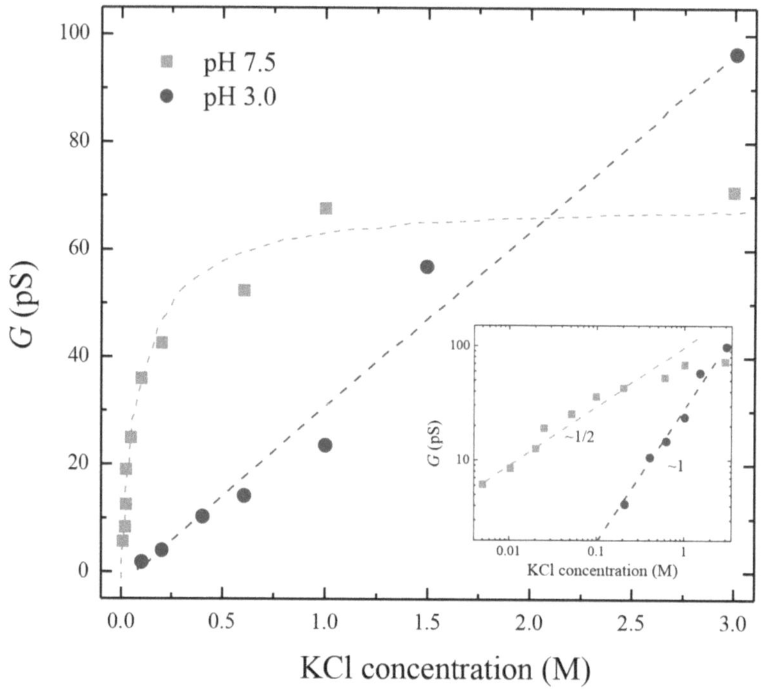

Figure 8.67 Ionic conductance of KCl electrolytes in nCNTPs at different pH values. Dashed lines are power-law (pH 7.5) and linear (pH 3.0) fits to the data. The inset shows the data on the logarithmic scale with dashed line being power-law fits to the data. (Data extracted from Tunuguntla et al., 2017.)

The scaling behavior of ionic conductance with the salt concentration in CNTs has also been investigated using an embedded CNT nanofluidic platform, where a CNT with a diameter of 3.5–35 nm is inserted into a silicon nitride membrane (Secchi et al., 2016b). It is found that at low salt concentrations, the ion conductance exhibits a power-law dependence on the salt concentration, $G \sim n_0^{1/3}$, with the magnitude depending on the pH, as shown in Figure 8.68. The power-law dependence of ionic conductance on salt concentration is the consequence of the surface-charge density, $|n_s|$, which, according to the Poisson–Boltzmann equation and the standard charge-regulation model, shows a power-law dependence on the salt concentration, $|n_s| \propto n_0^{\alpha}$.

8.9.4 CNT field-effect transistor

A field-effect transistor (FET) is an electronic device, where the electric behavior of the device is controlled by electric fields. With exceptional electronic

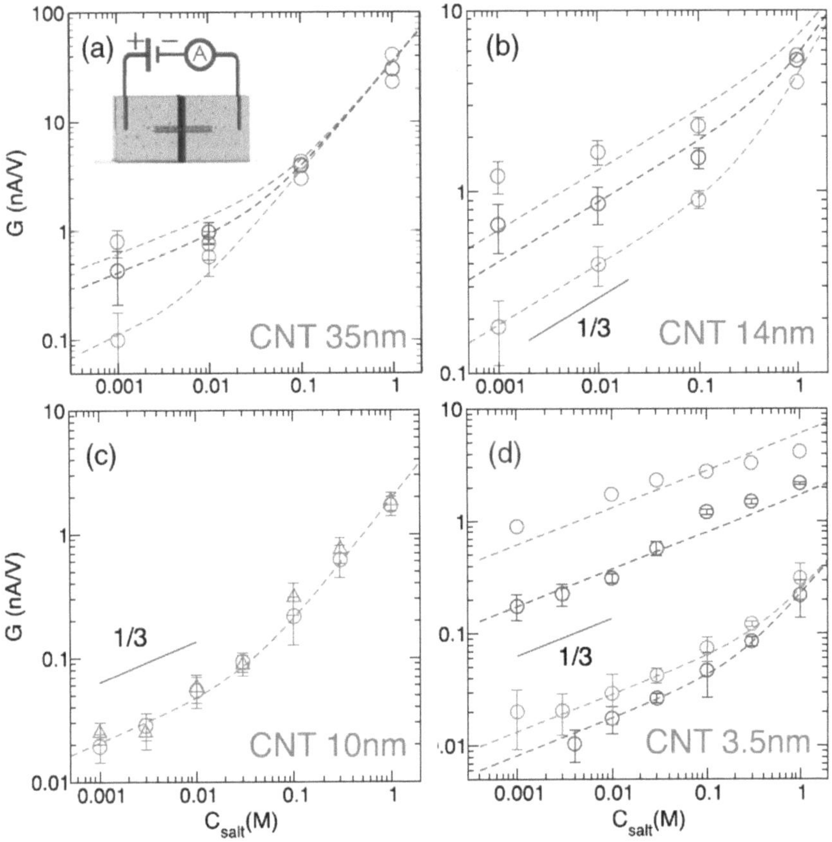

Figure 8.68 Ionic conductance of single CNTs with various radii, R, and lengths, L, at different pH values. (a) (R, L) = (35 nm, 1.5 μm). pH = 6, 9, 10 from bottom to top. (b) (R, L) = (14 nm, 2.0 μm). pH = 6, 9, 10 from bottom to top. (c) (R, L) = (10 nm, 2.5 μm). pH = 4. (d) (R, L) = (3.5 nm, 3.0 μm). pH = 4, 6, 8, 10 from bottom to top. (Reprinted with permission from Secchi et al., 2016b, Copyright (2016) by the American Physical Society.)

properties, CNT-based FETs have been demonstrated outperforming silicon-based metal-oxide-semiconductor FETs (Cao et al., 2017).

Figure 8.69a shows a single CNT-based nanofluidic system with a gate integrated on top of the device (Pang et al., 2011). Without a gate voltage, the device is just a regular nanofluidic device, which can generate electroosmotic flows and ionic currents. If a gate voltage is applied, the surface charge of the CNT is altered, leading to an enhanced ionic conductance. The CNT-based FET exhibits p-type FET characteristics such that negative gate voltages generate high ionic conductance, as shown in Figure 8.69b. To quantitatively characterize the gating performance, a gating efficiency $\Delta G/G_+$ is defined, where $\Delta G = G_- - G_+$ with G_- and G_+ being the ionic conductance at the highest negative and positive gate voltages. The value of $\Delta G/G_+$ ranges from 2 to 5. The variation might be caused by the electronic properties of the CNTs. It is assumed that metallic CNTs can effectively screen the gate field and show low gating efficiency. In contrast, semiconducting CNTs do not strongly screen the gate field and can accommodate a relatively high gating efficiency. Nevertheless, the gating efficiency of semiconducting CNTs spreads compared with that of metallic CNTs, indicating that the gating characteristics might be affected by external parameters. In addition, ion concentration and pH also affect the performance of the CNT-based FET.

The development of CNT nanofluidic FETs is still at its early stage. With further extensive explorations, precise manipulation of ions and molecules in CNT-based nanofluidic devices can be expected.

8.9.5 Ion selectivity

Molecular dynamics simulations about ion transport in CNTs have shown that ions can be rejected by or selectively allowed to pass through a CNT. In MD simulations, CNTs are typically treated as neutral, i.e., carbon atoms are not charged, and the van der Waals interactions are the only interactions between ions and carbon atoms, which are usually described by the LJ potential. Although the van der Waals force is weak, for CNTs with small diameters (e.g., sub-nanometer), the LJ potential may generate a large energy barrier at the entrance of a CNT. When the CNT diameter is comparable to the size of the hydration shell of ions, ions may be blocked due to the high dehydration energy. Molecular dynamics simulations on the transport of 250 mM NaCl solution through CNTs show that at an osmotic

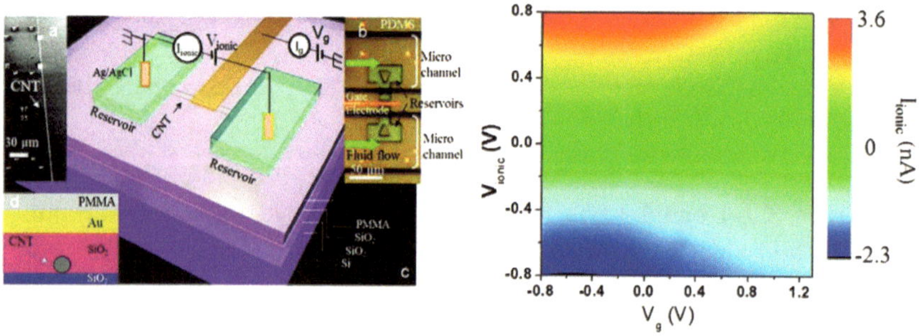

Figure 8.69 Single CNT-based nanofluidic system as an FET. Left: System layout. Right: Heat map of the ionic current as a function of the gate voltage V_g and drain-source voltage V_{ionic}, showing p-type FET characteristics. (Reprinted with permission from Pang et al., 2011, Copyright (2011) American Chemical Society).

pressure of 2.4 MPa, ions are completely rejected for (5,5) and (6,6) CNTs and the rejection rate drops to 95% and 58% for (7,7) and (8,8) CNTs, respectively, (Corry, 2008). In addition to the CNT size, ion rejection is also affected by the ion species, ion concentration, and the external electric field. Therefore, it is not surprising if certain ions are blocked in larger (e.g., (8,8)) CNTs (Beu, 2011).

The ion rejection shows a promising potential for water desalination. Using a membrane containing (7,7) CNTs with a density of 2.5×10^{11} CNTs/cm^2, it is estimated that the flow rate for water desalination can be improved by a factor of 6 compared with the existing technologies. Although high water fluxes in CNTs (see Chapter 6) have the potential to greatly reduce the cost of seawater desalination, CNTs with small diameters required for salt rejection can be easily clogged by impurities, which poses a challenge to the application of CNT membranes in desalination.

Aside from ion exclusion, it is also shown that CNTs with diameters in a certain range own intrinsic ion selectivity (Song and Corry, 2009). Due to the difference in dehydration energy and hydration number, different ions face distinct energy barriers to enter into a CNT. This can make a CNT an ion sieve. Under certain conditions, only particular ions can pass through a CNT, while other ions cannot. *Ab initio* MD simulations about ion transport in CNTs with radii ranging from 3 to 5 Å demonstrate that CNTs exhibit higher K^+ selectivity than Na^+ (Li et al., 2015). The ion selectivity originates from different hydration properties of Na^+ and K^+ and the interior environment of the CNTs. Na^+ has a smaller hydration shell but stronger hydration energy than K^+ and requires a high energy to enter into small CNTs. It is found that (8,8) CNTs are excellent artificial K^+ selective channels compared with nanotubes of other chiralities. The ion selectivity of K^+ over Na^+ in (8,8) CNTs is about 20 times higher than that of other synthetic nanopores. This observation is different from other MD simulations and the discrepancy might be caused by the molecular interaction potentials and the surface properties of CNTs. *Abi initio* MD simulations appear to be more accurate than regular MD simulations because they consider quantum many-body effects and the charge redistribution among ions induced by electric polarization, which could be important in nanoscale ion transport.

Actually, studies using density functional theory (DFT) challenge the validity of the simple molecular interactions between CNTs and ions commonly employed in MD simulations (Liu et al., 2015b; Pham et al., 2016). As the CNT surface is rich of aromatic π rings, cation–π interactions tend to cause the CNT surface to be electronegative and the interaction between a cation and a CNT is much stronger than the van der Waals interaction. The cation–π interaction can greatly reduce the entrance energy barrier of a CNT and promote the entry of cations into the CNT. Therefore, sophisticated numerical methods are necessary for precisely probing the ion transport in CNTs, though MD simulations are capable of providing qualitative descriptions.

8.9.6 Fluid breakup in CNTs

In most of previous MD simulations, the ultrafast ion transport in CNTs observed in experiments (Figure 8.65) is not found. The inconsistency between MD studies and experiments might be caused by different flow conditions. In CNTs, it is known that ion transport is greatly affected by the CNT size, electric field, ion concentration, and surface conditions. In MD simulations, the length of CNTs typically ranges from a few to tens of nanometers, while the length of CNTs in experiments is of micrometers. In addition, the CNT surface is usually negatively charged in experiments during the fabrication of nanofluidic devices. To reproduce experimental results, MD systems should consider using long CNTs with surface charges.

By duplicating experimental conditions, a recent MD study on the transport of a 1 M KCl solution in CNTs with the length ranging from 9.8 nm to about 1 μm shows that the CNT length and the external electric field significantly affect the dynamics of ions (Gao et al., 2017). In short CNTs or at low electric fields, K^+ and water molecules move continuously in the CNT as a liquid chain. In sufficiently long CNTs or under strong electric fields, however, the liquid chain breaks up into K^+-water clusters, which greatly enhance the ionic mobility and lead to ultrahigh ionic conductance.

Figure 8.70 illustrates the MD systems, where a CNT connects two reservoirs. Two different entrance structures, cylindrical and conical entrances, are considered, as shown in Figure 8.70a and b. The conical entrance is to imitate the

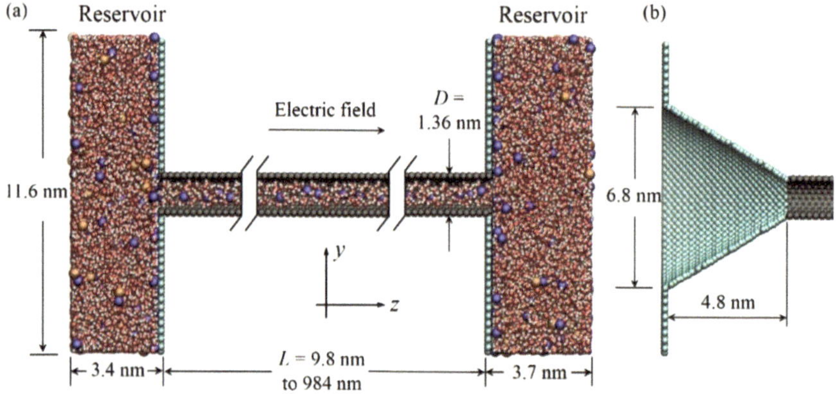

Figure 8.70 Molecular dynamics simulation systems for ion transport in CNTs (black, cyan, red, and white spheres represent carbon, reservoir wall, oxygen, and hydrogen atoms). Blue and orange spheres are K^+ and Cl^-. (a) Cylindrical entrance. (b) Conical entrance. (Reprinted from Gao, et al., 2017 with the permission of AIP Publishing.)

structure employed in previous experiments (Liu et al., 2010). To consider the negative surface charge of CNTs caused by aromatic π rings and surrounding materials in experiments, each carbon atom of the CNT is set to carry a partial charge of −0.005 e, which corresponds to a practical surface-charge density $n_s = -0.03\,C/m^2$. The walls of the reservoirs are also charged with a charge density of $-0.2\,C/m^2$, which is close to that of SiO_2 surfaces in contact with a high concentration electrolyte. Water molecules are modeled by the SPC/E potential and 1 M KCl solutions are employed as the electrolytes.

Figure 8.71a plots the ionic current I in CNTs of different lengths (cylindrical entrances) as the external electric field strength E is varied. In short CNTs, the I–E curve is linear, which agrees well with the classic theory. As the CNT length L is increased, however, the I–E relationship becomes nonlinear. In long CNTs, $L = 492$ nm, the dependence of I on E becomes linear when E is sufficiently strong, $E > 5$ mV/nm, where I increases much faster than in short CNTs. Similar fashions for the I–E characteristics are also found in CNTs with conical entrances, as shown in the inset of Figure 8.71a. The ionic conductance, $G = I/(EL)$, as a function of CNT length at $E = 20$ mV/nm is depicted in Figure 8.71b. It is seen that G changes nonlinearly as L is increased, which is different from the prediction of the classic theory, which states that G scales with $1/L$.

To understand the ultrafast ion transport, the flows in the CNTs are examined. Figure 8.72a shows the snapshots of flows in CNTs with different lengths at $E =$ 20 mV/nm. In short CNTs ($L < 50$ nm), K^+ and water molecules migrate as a liquid chain continuously with a relatively low velocity ($\overline{v}_{K^+} \approx 10$ m/s). In long CNTs, $L > 100$ nm, the liquid chain breaks into ion-water clusters, which move at a very high velocity ($\overline{v}_{K^+} > 100$ m/s). It is clear that the flow experiences a transition from the relatively slow passage of intact liquid chains to the fast transport of ion-water clusters. Further studies show that the electric field E plays a similar role to that of the CNT length in terms of the flow transition, as depicted in Figure 8.72b, where flows in a 196.8-nm-long CNT are shown as E is increased.

The breakup of the fluid dramatically affects the transport of ions. As ions and water molecules move continuously under a low electric field (or in a short CNT), the force due to the hydrogen bonds connecting water molecules acts as a drag force, which impedes the migration of ions. Once ion-water clusters form at a high electric field (or in a long CNT), hydrogen bonds between adjacent clusters break. This greatly reduces the resistance and leads to fast ion transport. Using the fluid breakup phenomenon, the ionic conductance in Figure 8.71b can be easily explained. In short CNTs, the flow is continuous and the flow resistance should be proportional to L, leading to $G \propto 1/L$. In CNTs with intermediate lengths ($50 < L < 300$ nm), the flow transition takes place and the occurrence of fluid breakup greatly enhances the mobility of ions. This is why the ionic conductance increases in this length range. In long CNTs, $L > 300$ nm, the ion-water clusters are fully

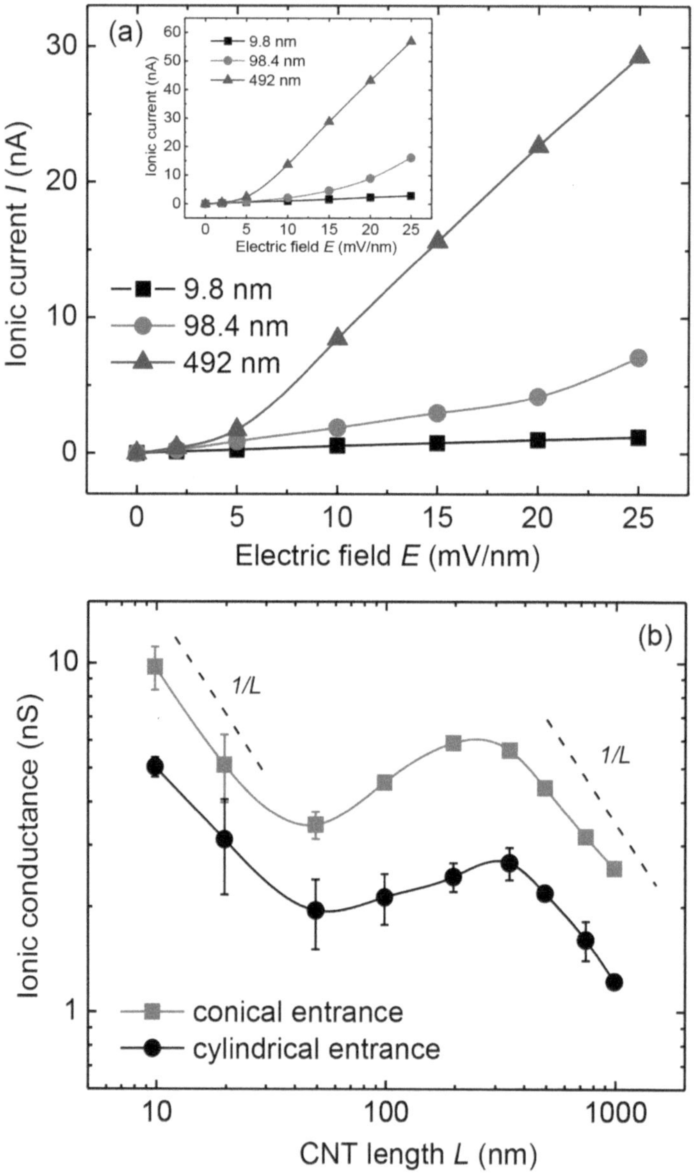

Figure 8.71 Ionic current I and conductance G in CNTs. (a) I versus electric field strength E in CNTs of different lengths with cylindrical entrances. Inset: I–E curves in CNTs with conical entrances. (b) G as a function of CNT length L at $E = 20$ mV/nm. (Reprinted from Gao, et al., 2017 with the permission of AIP Publishing.)

Figure 8.72 Flow snapshots in CNTs (long CNTs are partially shown). (a) $E = 20$ mV/nm. CNT length from top to bottom: 9.8, 19.7, 49.2, 98.4, and 196.8 nm. The mean velocity of K^+ from top to bottom: 10.4, 13.0, 24.7, 65.7, and 172.2 m/s. (b) $L = 196.8$ nm. E from top to bottom: 2, 5, 10 and 20 mV/nm. The mean velocity of K^+ from top to bottom: 4.9, 16.9, 47.9, 172.2 m/s. (Reprinted from Gao, et al., 2017 with the permission of AIP Publishing.)

developed and the flow is governed by the surface resistance, which is proportional to the CNT length resulting in $G \propto 1/L$.

The fluid breakup phenomenon is caused by the competition between the driving electrostatic force and the force due to the hydrogen bonds in water. The electrostatic force includes the forces caused by the external electric field and ion–ion interactions. To calculate the total electrostatic force on the ions in the CNT, the density distribution of the net charge $\rho_c(z) = e(n_+ - n_-)$ in the reservoirs is needed, which is obtained based on the model illustrated in Figure 8.73a. As the ion density in the CNT is determined by the CNT surface charge, it is assumed to be uniform in the CNT. If the thickness of the Stern layer close to the reservoir walls is δ and the ion density in the Stern layer is independent of the electric field, $\rho_c(z)$ can be obtained by solving the Poisson–Boltzmann equation (Gao et al., 2017):

$$\rho_c(z) = n_0 e\left[\left(\frac{1+\hat{\gamma}e^{z/\lambda_D}}{1-\hat{\gamma}e^{z/\lambda_D}}\right)^2 - \left(\frac{1-\hat{\gamma}e^{z/\lambda_D}}{1+\hat{\gamma}e^{z/\lambda_D}}\right)^2\right] \quad (8.77)$$

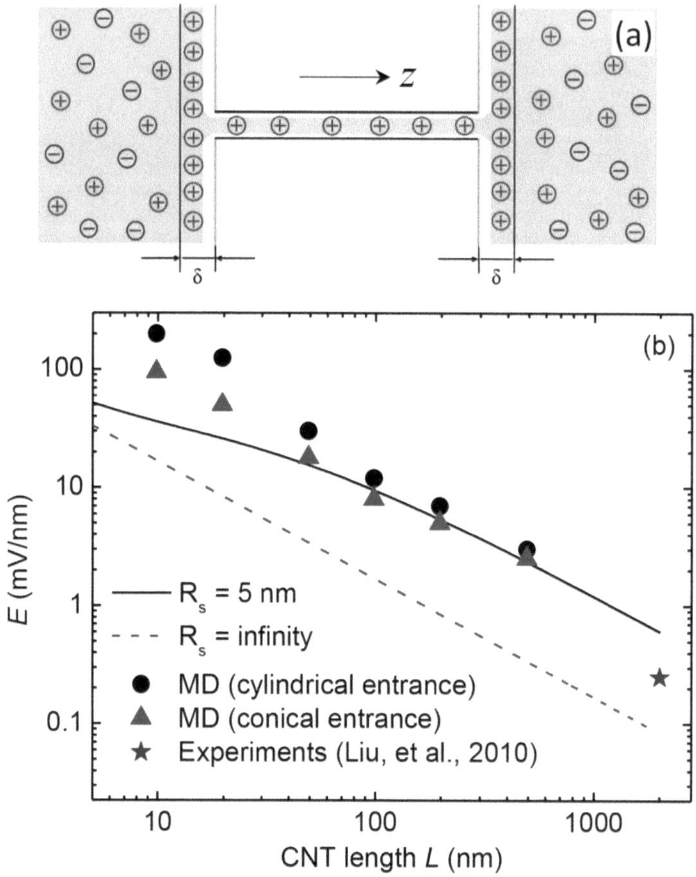

Figure 8.73 Theoretical analyses of fluid breakup in CNTs. (a) Theoretical model for ion transport in CNTs. (b) Critical electric field versus CNT length for the formation of ion-water clusters. (Reprinted from Gao, et al., 2017 with the permission of AIP Publishing.)

where $\hat{\gamma}$ is given by Equation 8.16. Assuming that the reservoirs are infinite, the electrostatic force acting on all the ions in the CNT by the ions in the upstream reservoir is then calculated as follows:

$$F_{\text{upstream}} = \int_{-\infty}^{-\delta} dz_1 \int_0^{+\infty} dr \int_0^L dz_2 k \frac{\rho_c(z_1) e 2\pi r \cdot 2\pi R_c |n_s|}{r^2 + (z_2 - z_1)^2} \frac{(z_2 - z_1)}{\sqrt{r^2 + (z_2 - z_1)^2}}$$
$$= 32\pi^2 R_c |n_s| Lkn_0 e\lambda_D \frac{\gamma e^{-\delta/\lambda_D}}{1 - \gamma^2 e^{-2\delta/\lambda_D}} \tag{8.78}$$

where k is Coulomb's constant and R_c is the CNT radius. The electrostatic force due to the ions in the downstream reservoir can be obtained in a similar fashion and the total electrostatic force on all the ions in the CNT is given by

$$F_d = 32\pi^2 R_c |n_s| L k n_0 e \lambda_D \left(\frac{\hat{\gamma}_1 e^{-\delta/\lambda_D}}{1-\hat{\gamma}_1^2 e^{-2\delta/\lambda_D}} - \frac{\hat{\gamma}_2 e^{-\delta/\lambda_D}}{1-\hat{\gamma}_2^2 e^{-2\delta/\lambda_D}} \right) + 2\pi R_c L |n_s| E \quad (8.79)$$

where $\hat{\gamma}_1$ and $\hat{\gamma}_2$ are for the upstream and downstream reservoirs, respectively. If the reservoirs are finite with radius R_s, the force is approximated as

$$F_d \approx 32\pi^2 R_c |n_s| k n_0 e \lambda_D \left(\frac{\hat{\gamma}_1 e^{-\delta/\lambda_D}}{1-\hat{\gamma}_1^2 e^{-2\delta/\lambda_D}} - \frac{\hat{\gamma}_2 e^{-\delta/\lambda_D}}{1-\hat{\gamma}_2^2 e^{-2\delta/\lambda_D}} \right) \left(L + R_s - \sqrt{R_s^2 + L^2} \right) + 2\pi R_c L |n_s| E \quad (8.80)$$

Fluid breakup will occur if this force is larger than the force due to the hydrogen bonds connecting water molecules in the CNT, which is estimated as follows:

$$F^* = 2\pi (R_c - \sigma) \gamma_{\text{water}} \quad (8.81)$$

where σ is the LJ collision diameter for water–CNT interactions and γ_{water} is the surface tension of water.

Therefore, the criterion for the formation of ion-water clusters is $F_d > F^*$, for which the critical electric field is plotted in Figure 8.73b as a function of the CNT length ($\delta = 4.0$ Å is used). It is seen that the theory (solid line) is in good agreement with MD results for long CNTs ($L > 50$ nm). In short CNTs, the theory underestimates the critical electric field. The discrepancy might be caused by the negligence of the repulsive forces from the downstream reservoir in the theory, which is important for short CNTs. It is also seen that previous experiments are in the breakup regime (above the dashed line for infinite reservoirs), which indicates that the ultrafast ion transport in the experiments might be caused by the fluid breakup phenomenon.

8.9.7 Proton transport in CNTs

Proton transport is a key issue in biological membranes and energy devices, such as fuel cells. In a bulk electrolyte, the solvation of excess protons is idealized as Eigen cations ($H_9O_4^+$) and Zundel ($H_5O_2^+$) cations. In an Eigen cation complex, there is a hydronium ion (H_3O^+) at the center, which is strongly hydrogen-bonded with three water molecules. A Zundel cation, however, has a symmetric structure with a proton shared by two water molecules. Protons usually transport via the well-known "Grotthuss mechanism," which suggests that excess protons shuttle through the hydrogen bond network of water molecules via the formation and breaking of covalent bonds with neighboring molecules. In a bulk solution, excess protons switch between Eigen cations and Zundel cations frequently, which lead to a popular mechanism for proton transport, i.e., the "Eigen – Zundel – Eigen" scenario.

At the nanoscale, a small confinement can affect the conformation of fluid molecules and generate ordered structures, such as the single-file chain and ring-like structures in CNTs introduced in Chapter 6. An ordered structure can promote unidirectional transport of protons and thus enhances the proton mobility.

In CNTs with diameters <1 nm, theoretical analysis shows that the proton transport mechanism differs from that in a bulk solution. Molecular dynamics simulations show that water molecules form an ordered structure, i.e., a one-dimensional chain, in (6,6) CNTs (diameter is 0.83 Å) (also see Chapter 6). In this case, protons transfer via a "Zundel – Zundel" mechanism, as shown in Figure 8.74, which is different from and more efficient than the "Eigen – Zundel – Eigen" mechanism in bulk water (Cao et al., 2010). The proton mobility along 1D water chains obtained in MD simulations exceeds that in bulk water by a factor of 40, while it decreases with the presence of orientational defects (Dellago et al., 2003).

Studies about the proton transport in CNT–lipid systems show that the CNT size greatly affects the mobility of protons. The diffusion coefficient of protons in

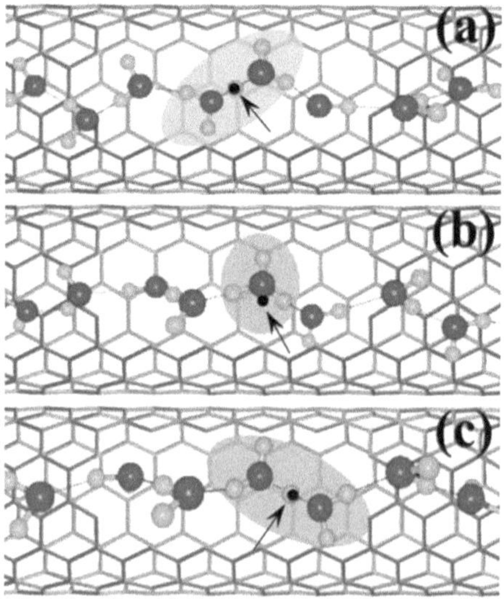

Figure 8.74 Snapshots of a proton transport event in a one-dimensional water chain formed in a (6,6) CNT. (a) Zundel state. (b) Intermediate $H_7O_3^+$. (c) Zundel state. The black balls indicated by arrows denote the center of excess charge (proton). (Reprinted with permission from Cao et al., 2010, Copyright (2010) American Chemical Society.)

0.8-nm-diameter CNTs is measured as 4.20 ± 0.91 Å^2/ps, which is one order of magnitude larger than that in bulk water (Tunuguntla et al., 2016). In 1.5-nm-diameter CNTs, however, the diffusion coefficient only doubles the bulk value. The large proton mobility in small CNTs is attributed to the fast transport of protons in one-dimensional hydrogen-bonded water chains. Furthermore, it is found that the proton conductance exceeds that in biological channels and can be modulated by the addition of other ions, such as Ca^{2+}, which make the CNT porins promising biomimetic proton conductors (Tunuguntla et al., 2016).

In very long CNTs (500 μm) with an average diameter of 1.5 nm, nanofluidic experiments have demonstrated that protons are the major charge carriers and the CNTs exhibit very high proton conductivity. A typical proton conductivity is as high as 5×10^2 S/cm, which is comparable to the maximum conductivity of highly efficient gramicidin proton channels (Lee et al., 2010).

In other nanochannels, the proton mobility enhancement is far less than that in CNTs. In 2-nm-high silica nanochannels, it is found that the proton mobility is enhanced by a factor of about 4 at low salt concentrations (Duan and Majumdar, 2010). In channels of 25 nm in height, the proton mobility becomes comparable to the bulk value and nanoconfinements paly a minor role.

8.9.8 Ion transport in BNNTs

BNNTs are structurally similar to CNTs but have distinct electronic properties. They have attracted growing attention for both fundamental investigations of ion transport and practical applications such as energy conversion and seawater desalination. A nanofluidic platform fabricated by inserting a BNNT (inner diameter ~15–40 nm, length ~1 μm) into a hole (100–200 nm in diameter) in a silicon nitride membrane (Siria et al., 2013), as shown in Figure 8.75a, shows that the osmotic current I_{osm} (~nA) generated by salinity gradients could be two orders of magnitude higher than the streaming current I_{stream} (~pA) produced by pressure drops. The high osmotic current is attributed to the extremely high surface-charge density of the BNNT in a high pH environment. Based on the measured ionic conductance, the surface-charge density n_s is expected to increase dramatically from a common value of 0.1 C/m^2 at pH 5 to a very high value of 1.0 C/m^2 at pH ~ 11, far exceeding the surface-charge density of regular surfaces. A relation for the osmotic current I_{osm} is derived as follows:

$$I_{osm} \approx \frac{2\pi R|n_s|}{L}\frac{k_B T}{\mu\lambda_B}\log\left(C_{s,h}/C_{s,l}\right) \tag{8.82}$$

where R and L are the tube radius and length, $\lambda_B = 0.7$ nm is the Bjerrum length, $C_{s,h}$ and $C_{s,l}$ are the high and low salt concentrations, respectively. The high osmotic current in BNNTs has the potential for energy conversion. Based on the osmotic

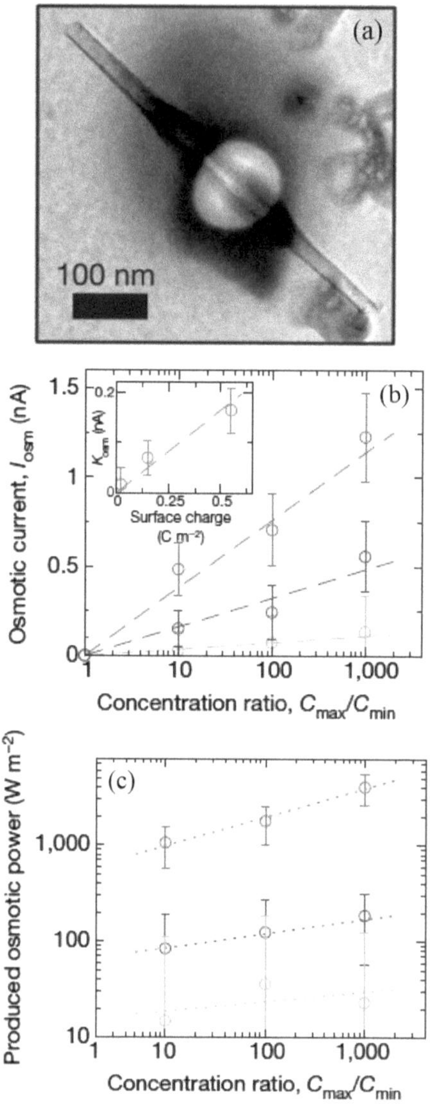

Figure 8.75 Ion transport in single transmembrane BNNTs under osmotic pressures. (a) TEM image of membranes embedded with a BNNT. (b) and (c) Osmotic current and power as a function of salt concentration ratio in a 40-nm-diameter, 1.25-μm-long BNNT at pH 5.5 (bottom circles), 9.5 (middle circles), and 11 (upper circles). Dashed lines are linear fits to the experimental data. The inset in (b) shows the osmotic mobility as a function of surface charge, which is determined from conductance measurement. (Reprinted by permission from Springer Nature, Siria et al., 2013, Copyright (2013).)

currents (Figure 8.75b) measured in the experiments, the maximum power density per unit tube area can reach 4 kW/m^2 for a single BNNT (Figure 8.75c). For a macroscopic membrane with a BNNT density of ~10^{10} cm^{-2}, the power density can be the order of kW/m^2, which exceeds that of commercial membranes by more than two orders of magnitude. If the experimental results are reliable, BNNTs are promising for harvesting osmotic power in nature using BNNT membranes.

8.10 Ion transport in two-dimensional nanopores

Ion transport through nanopores in 2D materials, such as graphene and MoS_2, has attracted great interest recently because of the small thickness and excellent physical and electrical properties of these materials. 2D nanopores have been shown promising in selective transport of ions, which can be used for seawater desalination, molecular separation, ion sieving, and energy harvesting.

8.10.1 Water desalination

For water desalination, a nanopore-containing 2D material should be capable of effectively screening ions, while owning good permeability for water molecules. The first demonstration of water desalination using 2D nanopores is performed through MD simulations using graphene nanopores (Cohen-Tanugi and Grossman, 2012). Compared with the commercial reverse osmosis (RO) membranes, graphene nanopores own several advantages. First, graphene is very thin (3.4 Å in thickness) such that the flow resistance across the graphene can be ignored. Second, graphene sheets have excellent mechanical strength that can stand a high pressure to ensure a high flux. Finally, the ion rejection ability of graphene nanopores can be managed by controlling the pore size and functionalization.

Molecular dynamics simulations have been performed to explore ion transport through hydrogenated and hydroxylated graphene nanopores with pore areas ranging from 1.5 to 62 $Å^2$. Hydrogenated pores are constructed by attaching each carbon atom at the pore edge with a hydrogen atom (Figure 8.76a). Hydroxylated pores are obtained by passivating the carbon atoms at the pore edge alternatively with H and OH groups, as depicted in Figure 8.76b. As the hydroxylated pores are hydrophilic, they exhibit high water permeability but relatively low ion rejection compared to the hydrophobic hydrogenated pores. Although the ion rejection rate decreases with increasing pore size and pressure, both types of nanopores demonstrate high ion rejection rate, ~100%, in the water permeability range of 10–100 $L/cm^2/day/MPa$, which surpasses conventional RO membranes by two to three orders of magnitude, as shown in Figure 8.77.

Experiments have also been carried out to validate the effectiveness of graphene nanopores for water desalination (Surwade et al., 2015). In the fabrication of

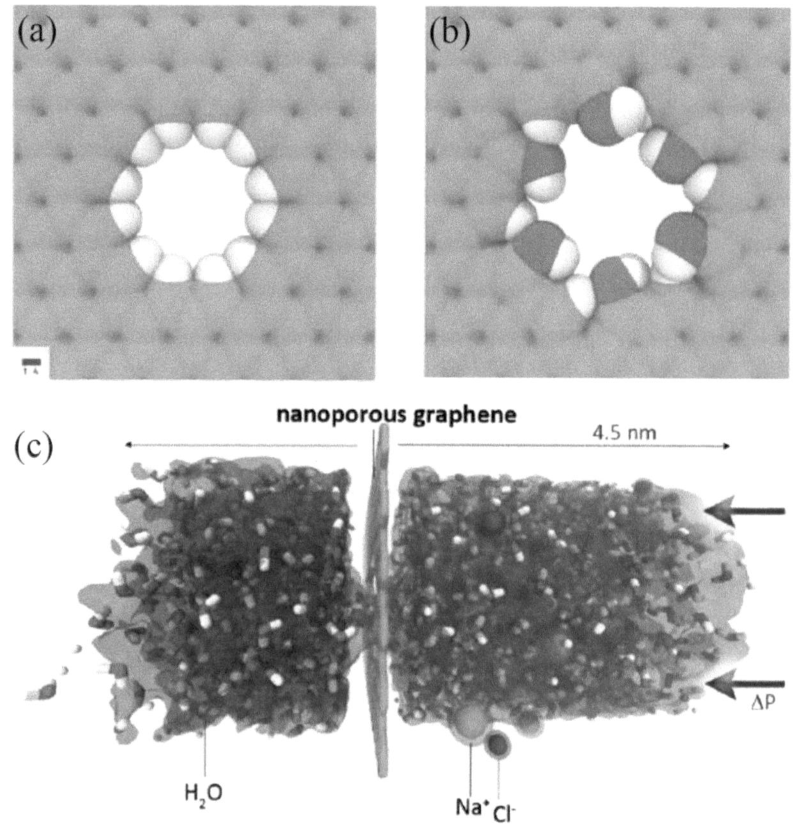

Figure 8.76 Graphene nanopores for water desalination. (a) Hydrogenated graphene nanopore. (b) Hydroxylated graphene nanopore. (c) Side view of the MD simulation system. (Reprinted with permission from Cohen-Tanugi and Grossman, 2012, Copyright (2012) American Chemical Society.)

graphene nanopore devices, single-layer graphene grown by CVD is transferred onto a silicon nitride film with a 5-μm-diameter hole. Nanopores with diameter between 0.5 and 1 nm are generated by oxygen plasma etching. Under an osmotic pressure, the water flux is measured as ~70 $gm^{-2}/s/atm$ based on an assumed pore density of $1 \times 10^{12} cm^{-2}$. This flux is about four times higher than that obtained in MD simulations (Cohen-Tanugi and Grossman, 2012). Using saturated water vapor pressure at 40°C as the driving force, however, the water flux reaches as high as 10^6 gm^{-2}/s, which is higher than the flux driven by osmotic pressures by several orders of magnitude. In addition to the high water permeability, graphene nanopores can almost completely reject ions (the rejection rate is about 100%).

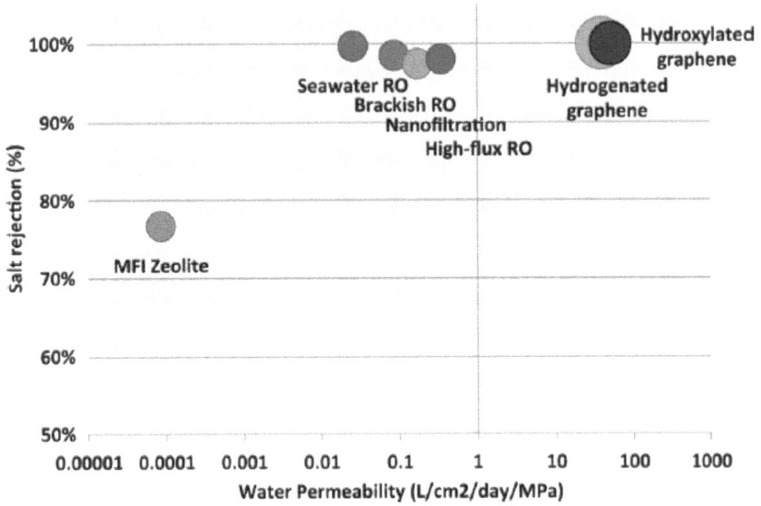

Figure 8.77 Salt rejection rate and water permeability for functionalized graphene nanopores and existing techniques. (Reprinted with permission from Cohen-Tanugi and Grossman, 2012, Copyright (2012) American Chemical Society.)

Although numerical and experimental results have demonstrated the conceptual effectiveness of graphene nanopores for water desalination, it is challenging to scale up nanoporous graphene with high-density pores of well-controlled size distribution and pore chemistry. If this cannot be achieved, applications of graphene nanopores in water desalination will be illusive.

A single-layer MoS_2 is another type of 2D nanomaterial, which has three atomic layers (two S layers and one Mo layer), as illustrated in Figure 8.78. The thickness of a single-layer MoS_2 is about 1 nm. MoS_2 nanosheets have very high mechanical strength with a Young's modulus of the order of 100 GPa, which is comparable to that of steel. Compared with graphene nanopores, MoS_2 nanopores are intrinsically charged due to the electron distributions of Mo and S atoms. As an MoS_2

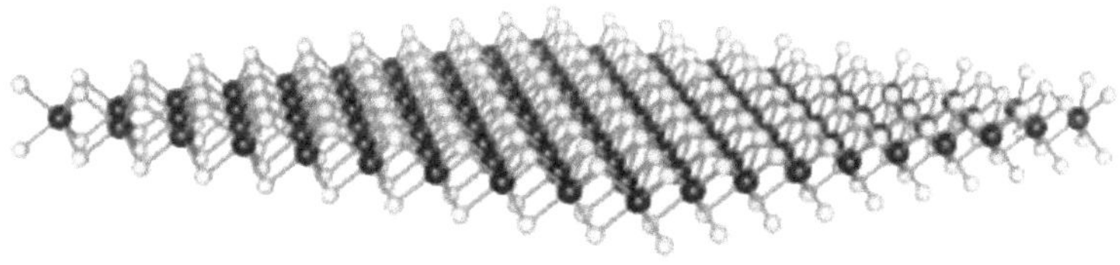

Figure 8.78 A single-layer MoS_2 sheet. White and black spheres are S and Mo atoms, respectively.

Figure 8.79 Different structures of MoS_2 nanopores. The innermost atoms are Mo (left panel), S (middle panel) and mixed Mo/S (right panel).

sheet contains three atomic layers, an MoS_2 nanopore can assume three different structures, depending on how Mo and S atoms are removed. Figure 8.79 shows three configurations of MoS_2 nanopores, where the innermost atoms are Mo, S, and mixed Mo/S in the left, middle, and right panels, respectively.

Molecular dynamics simulations show that the ion rejection rate of MoS_2 nanopores with pore area of about 20 Å^2 is almost 100% under an external pressure up to 350 MPa (Heiranian et al., 2015). Among the three types of pores, the pores with Mo being the innermost atoms exhibit the highest water permeability, as shown in Figure 8.80, which is about two to five orders of magnitude higher than

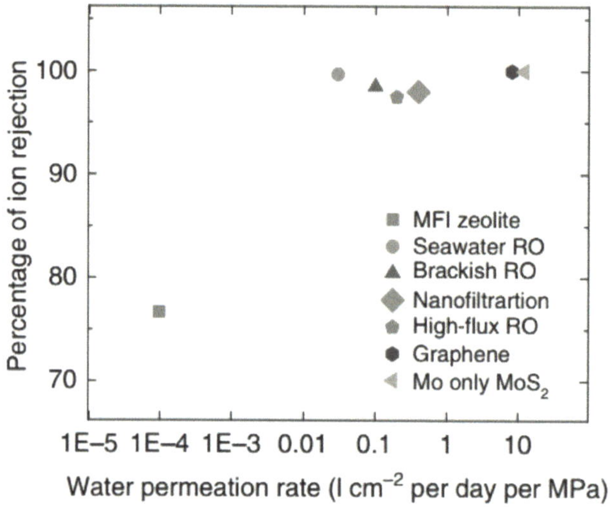

Figure 8.80 Ion rejection rate and water permeability for MoS_2 nanopres and other materials. The triangle denoted as "Mo only MoS_2" represent MoS_2 nanopores with Mo being the innermost atoms. (Reprinted with permission from Heiranian et al., 2015.)

common commercial RO membranes and 70% higher than that of graphene nanopores. The extraordinary water permeability is attributed to the coupling effects of the hydrophobic (S)–hydrophilic (Mo) arrangement and the hourglass-shaped pore geometry.

8.10.2 Ionic conductance and ion selectivity of graphene nanopores

Ion transport through isolated graphene nanopores has been experimental investigated (Jain et al., 2015). Nanopores are intrinsically generated as defects during the CVD growth of graphene and the diameters of the pores are less than 2 nm. The nanofluidic device is fabricated by transferring a pore-containing graphene sheet onto a silicon nitride membrane with a single 35–40 nm diameter hole. As the diameter of the hole is smaller than the mean distance among the graphene pores (70–100 nm), it is highly possible that there is only one graphene nanopore in the hole area, as demonstrated in Figure 8.60a. The current–voltage characteristics for the transport of 1 M KCl electrolytes through different pores are monitored. In addition to stochastic current fluctuations, as discussed in Section 8.8.2, the ionic conductance shows sensitive dependence on the pore size. It varies from about 0.01 nS to ~1 nS when the pore diameter is increased from 0.6 Å from 1 nm. Furthermore, the graphene nanopores display ion selectivity of K^+ over Li^+, Mg^{2+}, Ca^{2+}, and Ba^{2+}, as shown in Figure 8.81. This is not surprising as the dehydration energy of K^+ is much lower than the other ions.

Ion transport in large area single-layer graphene membranes with high pore densities has also been studied (O'Hern et al., 2014). The pores are created by ion bombardment and plasma etching. The pore size is controlled by the etching time. For relatively small pores, 0.2–0.35 Å in diameter, the graphene membranes show weak cation selectivity due to the negatively charged functional groups at pore edges. The cation selectivity diminishes in graphene membranes with large pore sizes (0.4 Å in diameter). The reason for the low ion selectivity might be caused by the existence of a small number of very large pores.

Another experimental work, where graphene nanopores are generated using an electrical pulse method, shows that K^+ is preferentially conducted over Cl^- with high selectivity ratios (over 100) (Rollings et al., 2016). The K^+/Cl^- selectivity ratio remains high for pores with diameters up to 20 nm. pH tests indicate that ion selectivity significantly depends on the deprotonation of the functional groups at the pore edge, similar to that in CNTs. Moreover, the nanopores exhibit valence selectivity for cations. They selectively conduct monovalent cations over divalent cations when the pore diameter is <10 nm and the selectivity ratio is up to 5 for 2-nm-pores. The good selectivity performance is the consequence of the negatively charged graphene surface at neutral pH. This work suggests that graphene

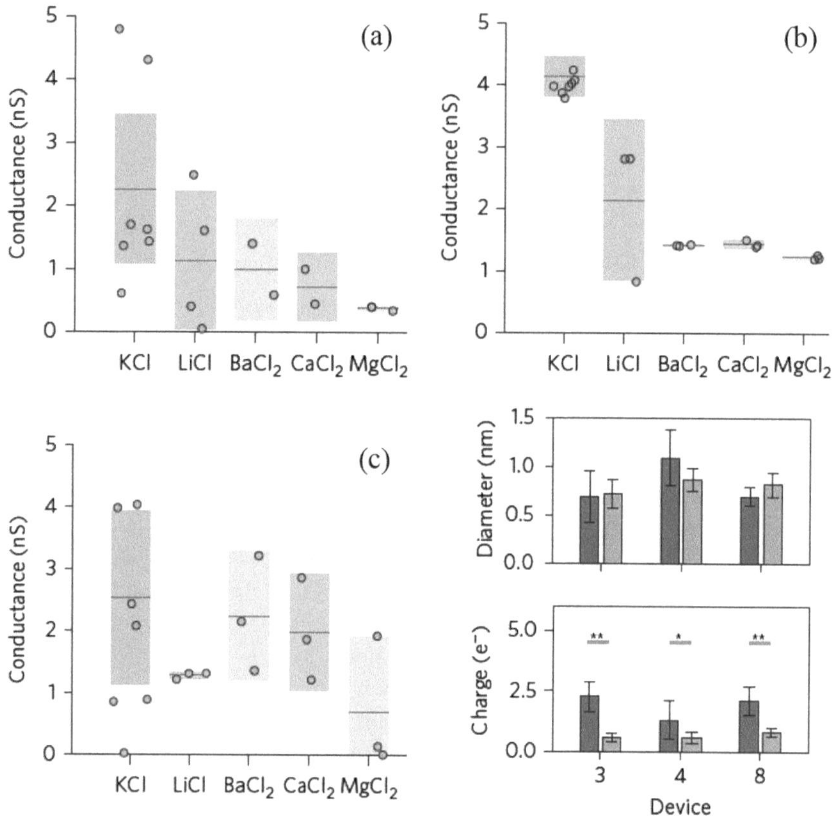

Figure 8.81 Ionic conductance of single graphene nanopores for different electrolytes at 100 mM chloride ion concentration. (a), (b), and (c) are for different devices (devices 3, 4, and 8, respectively, in the bottom right panel). Circles represent the conductance measured at a time point. Lines denote the mean conductance. Gray bars depict the standard deviation. The pore diameters and charges are given in the bottom right panel (dark and light gray bars are for monovalent and divalent ions, respectively). (Reprinted by permission from Springer Nature, Jain et al., 2015, Copyright (2015).)

nanopores may be applied for electrodialysis and the pore size is not necessary to be controlled at the sub-nanometer scale.

8.10.3 Ion transport in MoS_2 nanopores

Ion transport through MoS_2 nanopores is expected to be different from that through graphene nanopores due to different pore structures and electric properties. Figure 8.82a illustrates the transport of ions through an MoS_2 nanopore

(Feng et al., 2016a). In experiments, where MoS_2 nanopores are generated by electrochemical reactions and electron irradiation, the ionic current exhibits strongly nonlinear and voltage-activated behavior for a 0.6-nm-diameter pore in a wide range of ion concentrations (KCl solution), as shown in Figure 8.82b. A threshold voltage of ~0.4 V presents in the *I–V* curves (Figure 8.82b), below which the ionic current is greatly suppressed. The voltage-activated ion transport is observed for both monovalent ions (K^+, Na^+, and Li^+) and divalent ions (Mg^{2+} and Ca^{2+}). The voltage-activated transport behavior strongly depends on the pore size, as displayed in Figure 8.82c–e. For pores with diameters >1 nm, the *I–V* curve shows the common Ohmic, linear relationship.

The existence of a clear threshold voltage in Figure 8.82b indicates that the pore confinement imposes an energy barrier for ions to translocate, which could result from both the Coulomb interactions between ions and pore atoms and dehydration effects. The voltage-activated ion transport phenomenon is called ion Coulomb blockade as it is analogous to the Coulomb blockade of the electron conduction through quantum dots or single-molecule junctions. Nevertheless, further investigations are needed to elucidate the underlying mechanism of this phenomenon.

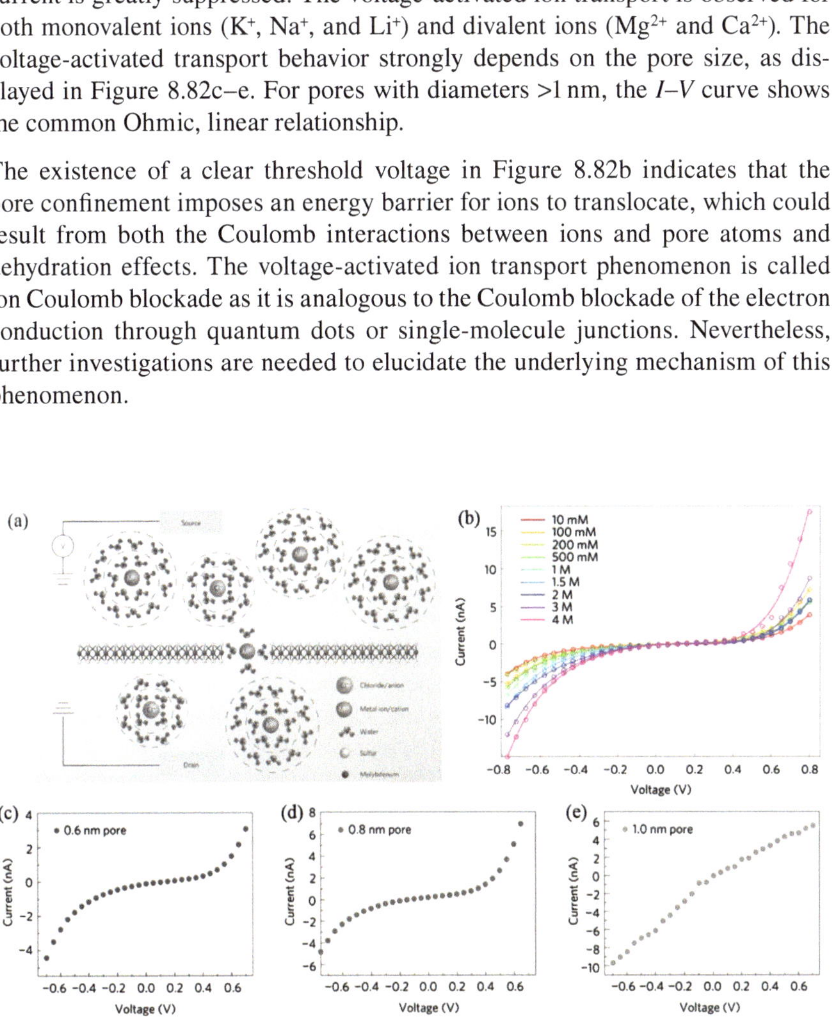

Figure 8.82 Ion transport through MoS_2 nanopores. (a) Schematic of ion transport through an MoS_2 nanopore. (b) Current–voltage curves for a 0.6-nm-diameter MoS_2 nanopore at various KCl concentrations. (c)–(e) Current–voltage curves for pores of different diameters. (Reprinted by permission from Springer Nature, Feng et al., 2016a, Copyright (2016).)

In a later experimental study conducted by the same group (Feng et al., 2016b), ion transport through MoS_2 nanopores driven by salinity gradient, i.e., osmotic pressure, is investigated. It is found that MoS_2 nanopores are negatively charged. The surface charge increases with increasing pH. It also depends on the pore size. The surface charges are −0.024, −0.053, and −0.088 C/m^2 for 2-, 6-, and 25-nm diameter pores, respectively. Therefore, the MoS_2 nanopores selectively conduct cations. The osmotic potential and osmotic current are found proportional to the salt concentration gradient and both increase with increasing pH. Notably, the osmotic current is about one order of magnitude higher than that observed in BNNTs. The high osmotic current is attributed to the coupling effect of the atomic scale thickness of the pore and the surface charge. The high currents show that MoS_2 nanopores are appealing for energy generation through osmosis. With a homogeneous pore size of 10 nm and a porosity of 30%, the power density of a single-layer MoS_2 membrane is estimated to be as high as 10^6 Wm^{-2} under a KCl salt gradient. This value surpasses the power density of BNNTs under similar conditions by two to three orders of magnitude (Siria et al., 2013).

It is also demonstrated that MoS_2 nanopores can act as a power source for a self-powered nanosystem, where an MoS_2 transistor can by powered by the energy supplied by the pores. If this is true, MoS_2 nanopores can be outstanding candidates for harvesting the abundant osmotic power in nature.

8.11 Ion sieving using graphene oxide membranes

GO membranes can be made by compressing layered GO laminates. The interlayer spacing between two adjacent laminates form 2D nanochannels, which allow the transport of sufficiently small molecules and ions while blocking the passage of species larger than the interlayer spacing. Compared with nanoporous graphene membranes discussed in Section 8.10, the preparation of GO membranes is much cost-effective. In addition, GO membranes are more scalable than nanopore-containing graphene sheets. GO membranes can be highly selective for fluid and ion transport. It has been shown that water can permeate through GO membranes, while other liquids and gases cannot (Nair et al., 2012). GO membranes can also be employed for sieving ions if the interlayer spacing is well controlled.

In a recent experimental study, microscale thick GO membranes are synthesized by vacuum filtration (Joshi et al., 2014). The GO membranes reject the transport of all species with a hydration radius larger than 4.5 Å and conduct those with a hydration radius smaller than 4.5 Å with a permeation rate much higher than simple diffusion. The permeation cutoff size, ~9 Å, is determined by the interlayer spacing (roughly 13.5 Å) of the swollen GO laminates in water. However, this permeation cutoff size is larger than the hydration diameter of most ions. Thus, to achieve ion sieving and rejection, the interlayer spacing needs to be reduced and stabilized when

GO membranes are immersed in water. A variety of methods have been proposed to reduce the swelling of GO in water by using reduced GO, reduced GO/titania hybrid membranes, crosslinks, physical confinements, and cationic control. Among these approaches, GO membranes using the physical confinement method can realize well-controlled interlayer spacing in the range of 6.4–9.8 Å with salt rejection rates up to 97% (Abraham et al., 2017). Such GO membranes are made by encapsulating stacked GO laminates into an epoxy structure, which stabilizes the interlayer spacing and prevents the swelling of the GO membrane, as shown in Figure 8.83a. The interlayer spacing can be tuned by changing the relative humidity. The reduction in interlayer spacing can lower the permeation cutoff size and the permeation rates for ions, as depicted in (Figure 8.83b). Specifically, by reducing the interlayer spacing from 9.8 to 7.4 Å, the permeation rates for K^+ and Na^+ exhibit an exponential decrease by about two orders of magnitude, while the permeation rate for water only shows a minor drop, a factor of ~2, as can be seen in Figure 8.83c. In addition, the permeation rate for water exceeds that predicted from the classic hydrodynamics with the stick boundary condition by three orders of magnitude, which is possibly caused by large slip lengths at the graphene surface.

Another method for controlling the interlayer spacing of GO membranes takes advantage of cation–π interactions by treating GO membranes in an electrolyte. In an experimental work, GO membranes prepared by the drop-casting method have an interlayer spacing of 12.8 ± 0.2 Å in pure water. When they are treated in 0.25 M solutions of KCl, NaCl, $CaCl_2$, LiCl, and $MgCl_2$ for an hour, the interlayer spacing is changed to 11.4 ± 0.1 Å, 12.1 ± 0.2 Å, 12.9 ± 0.2 Å, 13.5 ± 0.2 Å and

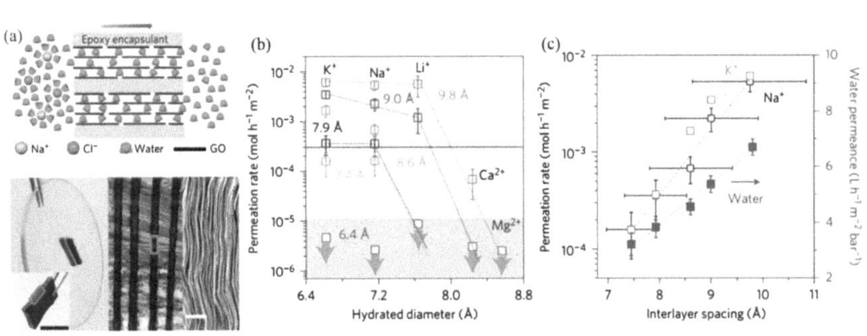

Figure 8.83 Ion sieving using GO membranes. (a) Schematic of ion transport in GO membranes (top) and macro-/microscopic images of physically confined GO membranes (bottom). (b) Permeation rates of ions, which have different hydration diameters, at different interlayer spacing (from 6.4 to 9.8 Å). (c) Exponential dependence of the permeation rates for Na^+ and K^+ versus linear dependence of water permeation rate on the interlayer spacing. (Reprinted by permission from Springer Nature, Abraham et al., 2017, Copyright (2017).)

13.6 ± 0.1 Å, respectively, as shown in Figure 8.44a (Chen et al., 2017). The spacing treated in KCl is quite stable as it remains almost unchanged upon the addition of other solutions, as shown in Figure 8.84b. The coupled effects of cation–π interactions and the interactions between cations and oxidized groups on the GO probably result in the fixed interlayer spacing. Compared with untreated GO membranes, ion permeation tests show that KCl treated GO membranes display more than two orders of magnitude higher ion rejection rates with the water flux being reduced only by a factor of 2 (0.36 lm^{-2}/h). It can be expected that other cations of different cation–π interaction strength can provide more choices for the precise control of the interlayer spacing of GO membranes.

8.12 Concluding remarks

For nanoscale electrokinetic phenomena, the classic theories, under certain conditions, may offer approximate predictions for some properties, such as the zeta potential and ion density distribution. However, they cannot precisely catch the physics underlying the new phenomena introduced in previous sections that are unique at the nanoscale. This is because the classic theories do not consider microscopic effects, which play important roles at the nanoscale.

Even though diverse ion transport scenarios have been observed, they may not present a complete picture of nanoscale electrokinetic phenomena. Due to the complex molecular interactions and the coupling between microscopic parameters and macroscopic properties, other new and interesting transport phenomena are expected.

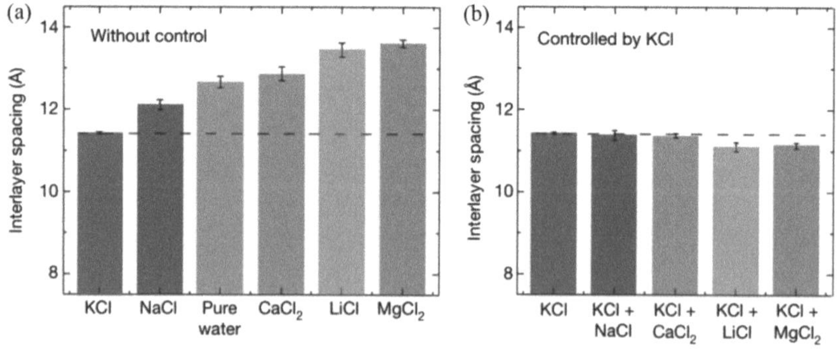

Figure 8.84 Cationic control of interlayer spacing in GO membranes. (a) Interlayer spacing for GO membranes in different solutions (0.25 M). (b) Interlayer spacing for GO membranes pretreated with KCl. (Reprinted by permission from Springer Nature, Chen et al., 2017, Copyright (2017).)

The intrinsic complexity of nanoscale fluid flows and ion transport makes it difficult to conduct theoretical analyses. This, however, should not be considered as a discouragement for understanding nanoscale ion transport. Instead, the rich ion transport phenomena accommodated by multiphysics should trigger extensive efforts in both science and engineering areas.

Finally, it is worth mentioning that certain transport behavior discussed in previous sections, observed either in experiments or simulations, may not be practically correct. Some experimental observations require further clarification and part of the numerical results could be artificial caused by the uncertainty in molecular interaction potentials.

Therefore, further systematic investigations are needed to provide a clear picture about nanoscale ion transport, which will find tremendous applications in a variety of areas.

References

Abraham, J., Vasu, K.S., Williams, C.D., Gopinadhan, K., Su, Y., Cherian, C.T., Dix, J., Prestat, E., Haigh, S.J., Grgorieva, I.V., Carbone, P., Geim, A.K., and Nair, R.R. 2017. Tunable sieving of ions using graphene oxide membranes. *Nat. Nanotech.* 12:546–550.

Amiri, H., Shepard, K.L., Nucholls, C., and Sánchez, R.H. 2017. Single-walled carbon nanotubes: Mimics of biological ion channels. *Nano Lett.* 17:1204–1211.

Beu, T.A. 2011. Molecular dynamics simulations of ion transport through carbon nanotubes. III. Influence of the nanotube radius, solute concentration, and applied electric fields on the transport properties. *J. Chem. Phys.* 135:044516.

Buchsbaum, S.F., Nguyen, G., Howorka, S., and Siwy, Z.S. 2014. DNA-modified polymer pores allow pH- and voltage-gated control of channel flux. *J. Am. Chem. Soc.* 136:9902–9905.

Burgreen, D., and Nakache, F.R. 1964. Electrokinetic flow in ultrafine capillary slitsl. *The J. Phys. Chem.* 68:1084–1091.

Buyukdagli, S., Manghi, M., Palmeri, J. 2010. Ionic capillary evaporation in weakly charged nanopores. *Phys. Rev. Lett.* 105:158103.

Cao, Z., Peng, Y., Yan, T., Li, S., Li, A., and Voth, G.A. 2010. Mechanism of fast proton transport along one-dimensional water chains confined in carbon nanotubes. *J. Am. Chem. Soc.* 132:11395–11397.

Cao, Q., Tersoff, J., Farmer, D.B., Zhu, Y., and Han, S.J. 2017. Carbon nanotube transistors scaled to a 40-nanometer footprint. *Science* 356:1369–1372.

Chen, L., Shi, G., Shen, J., Peng, B., Zhang, B., Wang, Y., Bian, F., Wang, J., Li, D., Qian, Z., Xu, G., Liu, G., Zeng, J., Zhang, L., Yang, Y., Zhou, G., Wu, M., Jin, W., Li, J., and Fang, H. 2017. Ion sieving in graphene oxide membranes via cationic control of interlayer spacing. *Nature* 550:380–383.

Cheng, L.J., and Guo, L.J. 2007. Rectified ion transport through concentration gradient in homogeneous silica nanochannels. *Nano Lett.* 7:3165–3171.

Cheng, L.J., and Guo, L.J. 2009. Ionic current rectification, breakdown, and switching in heterogeneous oxide nanofluidic devices. *ACS Nano* 3:575–584.

Cohen-Tanugi, D., and Grossman, J.C. 2012. Water desalination across nanoporous graphene. *Nano Lett.* 12:3602–3608.

Corry, B. 2008. Designing carbon nanotube membranes for efficient water desalination. *J. Phys. Chem. B* 112: 1427–1434.
Cruz-Chu, E.R., and Schulten, K. 2010. Computational microscopy of the role of protonable surface residues in nanoprecipitation oscillations. *ACS Nano* 4:4463–4474.
Daiguji, H., Oka, Y., and Shirono, K. 2005. Nanofluidic diode and bipolar transistor. *Nano Lett.* 5:2274–2280.
Dekker, C. 2007. Solid-state nanopores. *Nat. Nanotech.* 2:209–215.
Dellago, C., Naor, M.M., and Hummer, G. 2003. Proton transport through water-filled carbon nanotubes. *Phys. Rev. Lett.* 90:105902.
Duan, C., and Majumdar, A. 2010. Anomalous ion transport in 2-nm hydrophilic nanochannels. *Nat. Nanotech.* 5:848–852.
Feng, J., Liu, K., Graf, M., Dumcenco, D., Kis, A., Di Ventra, M., and Radenovic, A. 2016a. Observation of ionic Coulomb blockade in nanopores. *Nat. Mater.* 15:850–855.
Feng, J., Graf, M., Liu, K., Ovchinnikov, D., Dumcenco, D., Heiranian, M., Nandigana, V., Aluru, N.R., Kis, A., and Radenovic, A. 2016b. Single-layer MoS_2 nanopores as nanopower generators. *Nature* 536:197–200.
Gadsby, D.C. 2009. Ion channels versus ion pumps: The principal difference, in principle. *Nat. Rev. Mol. Cell Biol.* 10:344–352.
Gao, X., Zhao, T., and Li, Z. 2017. Fluid breakup in carbon nanotubes: An explanation of ultrafast ion transport. *Phys. Fluids* 29:092003.
Geng, J., Kim, K., Zhang, J., Escalada, A., Tunuguntla, R., Comolli, L.R., Allen, F.I., Shnyrova, A.V., Cho, K.R., Munoz, D., Wang, Y.M., Grigoropoulos, C.P., Ajo-Franklin, C.M., Frolov, V.A., and Noy, A. 2014. Stochastic transport through carbon nanotubes in lipid bilayers and live cell membranes. *Nature* 514:612–615.
Gong, X., Li, J., Xu, K., Wang, J., and Yang, H. 2010. A controllable molecular sieve for Na^+ and K^+ ions. *J. Am. Chem. Soc.* 132:1873–1877.
Guan, W., Fan, R., and Reed, M.A. 2011. Field-effect reconfigurable nanofluidic ionic diodes. *Nat. Commun.* 2:506.
Guo, W., Xia, H., Cao, L., Xia, F., Wang, S., Zhang, G., Song, Y., Wang, Y., Jiang, L., and Zhu, D. 2010. Integrating ionic gate and rectifier within one solid-state nanopore via modification with dual-responsive copolymer brushes. *Adv. Funct. Mater.* 20:3561–3567.
He, Z., Zhou, J., Lu, X., and Corry, B. 2013. Bioinspired graphene nanopores with voltage-tunable ion selectivity for Na^+ and K^+. *ACS Nano* 11:10148–10157.
Heiranian, M., Farimani, A.B., and Aluru, N.R. 2015. Water desalination with a single-layer MoS_2 nanopore. *Nature Commun.* 6:8616.
Holt, J.K., Park, H.G., Wang, Y., Stadermann, M., Artyukhin, A.B., Grigoropoulos, C.P., Noy, A., and Bakajin, O. 2006. Fast mass transport through sub-2-nanometer carbon nanotubes. *Science* 312:1034–1037.
Hou, X., Yang, F., Li, L., Song, Y., Jiang, L., and Zhu, D. 2010. A biomimetic asymmetric responsive single nanochannel. *J. Am. Chem. Soc.* 132:11736–11742.
Hunter, R.J. 1981. *Zeta potential in Colloid Science: Principles and Applications*. Academic Press.
Jain, T., Rasera, B.C., Guerrero, R.J.S., Boutilier, M.S.H., O'Hern, S.C., Idrobo, J., and Karnik, R. 2015. Heterogeneous sub-continuum ionic transport in statistically isolated graphene nanopores. *Nat. Nanotech.* 10:1053–1057.
Jang, D., Idrobo, J.C., Laoui, T., and Karnik, R. 2017. Water and solute transport governed by tunable pore size distributions in nanoporous graphene membranes. *ACS Nano* 11:10042–10052.
Joshi, R.K., Carbone, P., Wang, F.C., Kravets, V.G., Su, Y., Grigorieva, I.V., Wu, H.A., Geim, A.K., and Nair, R.R. 2014. Precise and ultrafast molecular sieving through graphene oxide membranes. *Science* 343:752–754.

Kang, Y., Zhang, Z., Shi, H., Zhang, J., Liang, L., Wang, Q., Ågren, H., and Tu, Y. 2014. Na^+ and K^+ ion selectivity by size-controlled biomimetic graphene nanopores. *Nanoscale* 6:10666–10672.

Karnik, R., Fan, R., Yue, M., Li, D., Yang, P., and Majumdar, A. 2005. Electrostatic control of ions and molecules in nanofluidic transistors. *Nano Lett.* 5:943–948.

Karnik, R., Duan, C., Castelino, K., Daiguji, H., and Majumdar, A. 2007. Rectification of ionic current in a nanofluidic diode. *Nano Lett.* 7:547–551.

Lee, C.Y. 2010. Electronic detection of molecules on the exterior and molecular transport through the interior of single walled carbon nanotubes. Ph.D. thesis, MIT.

Lee, C.Y., Choi, W., Han, J., and Strano, M.S. 2010. Coherence resonance in a single-walled carbon nanotube ion channel. *Science* 329:1320–1324.

Levine, S., Marriott, J.R., and Robinson, K. 1975. Theory of electrokinetic flow in a narrow parallel-plate channel. *J. Chem. Soc., Faraday Trans.* 2(71):1–11.

Li, H., Francisco, J.S., and Zeng, X.C. 2015. Unraveling the mechanism of selective ion transport in hydrophobic subnanometer channels. *Proc. Natl. Acad. Sci. U.S.A.* 112:10851–10856.

Li, L. 2016. Experimental investigation of fluid flow in nanochannels. Ph.D. thesis, The Hong Kong University of Science and Technology.

Liu, H., He, J., Tang, J., Liu, H., Pang, P., Cao, D., Krstic, P., Joseph, S., Lindsay, S., and Nuckolls, C. 2010. Translocation of single-stranded DNA through single-walled carbon nanotubes. *Science* 327:64–67.

Liu, L., Yang, C., Zhao, K., Li, J., and Wu, H.C. 2013. Ultrashort single-walled carbon nanotubes in a lipid bilayer as a new nanopore sensor. *Nat. Commun.* 4:2989.

Liu, Q., Xiao, K., Wen, L., Lu, H., Liu, Y., Kong, X., Xie, G., Zhang, Z., Bo, Z., and Jiang, L. 2015a. Engineered ionic gates for ion conduction based on sodium and potassium activated nanochannels. *J. Am. Chem. Soc.* 137:11976–11983.

Liu, J., Shi, G., Guo, P., Yang, J., and Fang, H. 2015b. Blockage of water flow in carbon nanotubes by ions due to interactions between cations and aromatic rings. *Phys. Rev. Lett.* 115:164502.

Liu, Q., Wen, L., Xiao, K., Lu, H., Zhang, Z., Xie, G., Kong, X., Bo, Z., and Jiang, L. 2016. A biomimetic voltage-gated chloride nanochannel. *Adv. Mater.* 28:3181–3186.

Majumder, M., Chopra, N., Andrews, R., and Hinds, B.J. 2005. Nanoscale hydrodynamics: Enhanced flow in carbon nanotubes. *Nature* 438:44.

Nair, R.R., Wu, H.A., Jayaram, P.N., Grigorieva, I.V., and Geim, A.K. 2012. Unimpeded permeation of water through helium-leak–tight graphene-based membranes. *Science* 335: 442–444.

O'Hern, S.C., Boutilier, M.S.H., Idrobo, J.-C., Song, Y., Kong, J., Laoui, T., Atieh, M., and Karnik, R. 2014. Selective ionic transport through tunable subnanometer pores in single-layer graphene membranes. *Nano Lett.* 14:1234–1241.

Pang, P., He, J., Park, J.H., Krstic, P.S., and Lindsay, S. 2011. Origin of giant ionic currents in carbon nanotube channels. *ACS Nano* 5:7277–7283.

Pham, T. A., Mortuza, S. G., Wood, B. C., Lau, E. Y., Ogitsu, T., Buchsbaum, S. F., Siwy, Z.S., Fornasiero, F., and Schwegler, E. 2016. Salt solutions in carbon nanotubes: The role of cation-π interactions. *J. Phys. Chem. C* 120:7332–7338.

Powell, M.R., Sullivan, M., Vlassiouk, I., Constantin, D., Sudre, O., Martens, C.C., Eisenberg, R.S., and Siwy, Z.S. 2008. Nanoprecipitation-assisted ion current oscillations. *Nat. Nanotech.* 3:51–57.

Powell, M.R., Cleary, L., Davenport, M., Shea, K.J., and Siwy, Z.S. 2011. Electric-field-induced wetting and dewetting in single hydrophobic nanopores. *Nat. Nanotech.* 6:798–802.

Rice, C.L., and Whitehead, R. 1965. Electrokinetic flow in a narrow cylindrical capillary. *J. Phys. Chem.* 69:4017–4024.

Rollings, R.C., Kuan, A.T., and Golovchenko, J.A. 2016. Ion selectivity of graphene nanopores. *Nat. Commun.* 7:11408.

Ruan, Y., Zhu, Y., Zhang, Y., Gao, Q., Lu, X., and Lu, L. 2016. Molecular dynamics study of Mg^{2+}/Li^{+} separation via biomimetic graphene-based nanopores: The role of dehydration in second shell. *Langmuir* 32:13778–13786.

Schneider, G.F., Xu, Q., Hage, S., Luik, S., Spoor, J.N.H., Malladi, S., Zandbergen, H., and Dekker, C. 2013. Tailoring the hydrophobicity of graphene for its use as nanopores for DNA translocation. *Nat. Commun.* 4:2619.

Schoch, R.B., and Renaud, P. 2005. Ion transport through nanoslits dominated by the effective surface charge. *Appl. Phys. Lett.* 86:253111.

Schoch, R.B., Han, J., Renaud, P. 2008. Transport phenomena in nanofluidics. *Rev. Mod. Phys.* 80:839–863.

Secchi, E., Marbach, S., Niguès, A., Stein, D., Siria, A., and Bocquet, L. 2016a. Massive radius-dependent flow slippage in carbon nanotubes. *Nature* 537:210–213.

Secchi, E., Niguès, A., Jubin, L., Siria, A., and Bocquet, L. 2016b. Scaling behavior for ionic transport and its fluctuations in individual carbon nanotube. *Phys. Rev. Lett.* 116:154501.

Shimizu, S., Ellison, M., Aziz, K., Wang, Q.H., Ulissi, Z., Gunther, Z., Bellisario, D., and Strano, M.S. 2013. Stochastic pore blocking and gating in PDMS–glass nanopores from vapor-liquid phase transitions. *J. Phys. Chem. C* 117:9641–9651.

Siria, A., Poncharal, P., Biance, A.L., Fulcrand, R., Blase, X., Purcell, S.T., and Bocquet, L. 2013. Giant osmotic energy conversion measured in a single transmembrane boron nitride nanotube. *Nature* 494:455–458.

Siwy, Z.S., Heins, E., Harrell, C.C., Kohli, P., and Martin, C.R. 2004. Conical-nanotube ion-current rectifiers: The role of surface charge. *J. Am. Chem. Soc.* 126:10850–10851.

Siwy, Z.S. 2006a. Ion-current rectification in nanopores and nanotubes with broken symmetry. *Adv. Funct. Mater.* 16:735–746.

Siwy, Z.S., Powell, M.R., Petrov, A., Kalman, E., Trautmann, C., and Eisenberg, R.S. 2006b. Calcium-induced voltage gating in a single conical nanopore. *Nano Lett.* 6:1729–1734.

Siwy, Z.S. and Davenport, M. 2010. Nanopores: Graphene opens up to DNA. *Nat. Nanotech.* 5:697–698.

Song, C. and Corry, B. 2009. Intrinsic ion selectivity of narrow hydrophobic pores. *J. Phys. Chem. B* 113:7642–7649.

Stein, D., Kruithof, M., and Dekker, C. 2004. Surface-charge-governed ion transport in nanofluidic channels. *Phys. Rev. Lett.* 93:035901.

Sun, L., and Crooks, R.M. 2000. Single carbon nanotube membranes: A well-defined model for studying mass transport through nanoporous materials. *J. Am. Chem. Soc.* 122:12340–12345.

Sun, Y., Ma, J., Zhang, F., Zhu, F., Mei, Y., Liu, L., Tian, D., and Li, H. 2017. A light-regulated host-guest-based nanochannel system inspired by channel rhodopsins protein. *Nat. Commun.* 8:260.

Surwade, S.P., Smirnov, S.N., Vlassiouk, I.V., Unocic, R.R., Veith, G.M., Dai, S., and Mahurin, S.M. 2015. Water desalination using nanoporous single-layer graphene. *Nat. Nanotech.* 10:459–464.

Thomas, M., and Corry, B. 2016. A computational assessment of the permeability and salt rejection of carbon nanotube membranes and their application to water desalination. *Phil. Trans. R. Soc. A* 374:20150020.

Tunuguntla, R.H., Allen, F.I., Kim, K., Belliveau, A., and Noy, A. 2016. Ultrafast proton transport in sub-1-nm diameter carbon nanotube porins. *Nat. Nanotech.* 11:639–644.

Tunuguntla, R.H., Henley, R.Y., Yao, Y., Pham, T.A., Wanunu, M., and Noy, A. 2017. Enhanced water permeability and tunable ion selectivity in subnanometer carbon nanotube porins. *Science* 357:792–796.

Verwey, E.J.W. and Overbeek, J.T.G. 1948. *Theory of the Stability of Lyophobic Colloids.* Amsterdam: Elsevier.
Vlassiouk, I., and Siwy, Z.S. 2007. Nanofluidic diode. *Nano Lett.* 7:552–556.
Wang, L., Boutilier, M.S.H., Kidambi, P.R., Jang, D., Hadjiconstantinio, N. G., and Karnik, R. 2017. Fundamental transport mechanisms, fabrication and potential applications of nanoporous atomically thin membranes. *Nat. Nanotech.* 12: 509–522.
Wang, G., Bohaty, A.K., Zharov, L., and White, H.S. 2006. Photon gated transport at the glass nanopore electrode. *J. Am. Chem. Soc.* 128:13553–13558.
Wei, C., Bard, A.J., and Feldberg, S.W. 1997. Current rectification at quartz nanopipet electrodes. *Anal. Chem.* 69:4627–4633.
Wen, L., Liu, Q., Ma, J., Tian, Y., Li, C., Bo, Z., Jiang, L. 2012. Malachite green derivative-functionalized single nanochannel: Light-and-pH dual-driven ionic gating. *Adv. Mater.* 24:6193–6198.
Wu, J., Gerstandt, K., Zhang, H., Liu, J., and Hinds, B.J. 2012. Electrophoretically induced aqueous flow through single-walled carbon nanotube membranes. *Nat. Nanotech.* 7:133–139.
Xiao, K., Xie, G., Zhang, Z., Kong, X.Y., Liu, Q., Li, P., Wen, L., and Jiang, L. 2016. Enhanced stability and controllability of an ionic diode based on funnel-shaped nanochannels with an extended critical region. *Adv. Mater.* 28:3345–3350.
Yameen, B., Ali, M., Neumann, R., Ensinger, W., Knoll, W., and Azzaroni, O. 2009a. Synthetic proton-gated ion channels via single solid-state nanochannels modified with responsive polymer brushes. *Nano Lett.* 9:2788–2793.
Yameen, B., Ali, M., Neumann, R., Ensinger, W., Knoll, W., and Azzaroli, O. 2009b. Ionic transport through single solid-state nanopores controlled with thermally nanoactuated macromolecular gates. *Small* 5:1287–1291.
Zhai, Q., Zhang, S., Jiang, H., Wei, Q., Wang, E., and Wang, J. 2014. Biomimetic nanopore for sensitive and selective detection of Hg(II) in conjunction with single-walled carbon nanotubes. *J. Mater. Chem. B* 2:6371–6377.
Zhang, M., Hou, X., Wang, J., Tian, Y., Fan, X., Zhai, J., and Jiang, L. 2012. Light and pH cooperative nanofluidic diode using a spiropyran-functionalized single nanochannel. *Adv. Mater.* 24:2424–2428.
Zhang, H., Hou, X., Zeng, L., Yang, F., Li, L., Yan, D., Tian, Y., and Jiang, L. 2013. Bioinspired artificial single ion pump. *J. Am. Chem. Soc.* 135:16102–16110.
Zhu, Y., Ruan, Y., Zhang, Y., Chen, Y., Lu, X., and Lu, L. 2017. Mg^{2+}-channel-inspired nanopores for Mg^{2+}/Li^{+} separation: The effect of coordination on the ionic hydration microstructures. *Langmuir* 33:9201–9210.

Appendix A

Table A.1 Reduced Collision Integral $\Omega_s^{(1,1)*}$ for Gas–Nanoparticle Interactions

	σ'														
T^*	**0.01**	**0.03**	**0.05**	**0.07**	**0.09**	**0.11**	**0.13**	**0.15**	**0.17**	**0.19**	**0.2**	**0.3**	**0.4**	**0.5**	**0.6**
0.10	1.05	1.13	1.23	1.33	1.40	1.51	1.60	1.69	1.76	1.85	1.90	2.41	2.85	3.35	3.82
0.15	1.04	1.12	1.20	1.29	1.36	1.44	1.53	1.60	1.67	1.74	1.79	2.23	2.63	3.07	3.48
0.20	1.04	1.11	1.18	1.26	1.32	1.40	1.48	1.55	1.61	1.68	1.73	2.12	2.49	2.89	3.25
0.25	1.04	1.10	1.17	1.24	1.30	1.37	1.45	1.51	1.57	1.63	1.68	2.04	2.39	2.75	3.09
0.30	1.04	1.10	1.16	1.23	1.28	1.35	1.42	1.48	1.54	1.60	1.64	1.98	2.31	2.65	2.95
0.35	1.03	1.10	1.15	1.22	1.27	1.33	1.40	1.45	1.51	1.57	1.61	1.93	2.25	2.56	2.84
0.40	1.03	1.09	1.14	1.21	1.26	1.32	1.39	1.43	1.50	1.55	1.58	1.89	2.19	2.48	2.75
0.45	1.03	1.09	1.14	1.20	1.25	1.31	1.37	1.42	1.48	1.53	1.56	1.86	2.14	2.42	2.67
0.50	1.03	1.09	1.13	1.19	1.24	1.30	1.36	1.41	1.47	1.52	1.55	1.83	2.10	2.36	2.60
0.55	1.03	1.08	1.13	1.19	1.24	1.29	1.35	1.40	1.45	1.50	1.53	1.81	2.07	2.31	2.54
0.60	1.03	1.08	1.13	1.18	1.23	1.28	1.34	1.39	1.44	1.49	1.52	1.78	2.03	2.27	2.49
0.65	1.03	1.08	1.12	1.18	1.22	1.28	1.33	1.38	1.43	1.48	1.51	1.76	2.00	2.23	2.44
0.70	1.03	1.08	1.12	1.18	1.22	1.27	1.33	1.37	1.42	1.47	1.49	1.75	1.98	2.19	2.40
0.75	1.03	1.07	1.12	1.17	1.21	1.26	1.32	1.36	1.42	1.46	1.48	1.73	1.95	2.16	2.36
0.80	1.02	1.07	1.12	1.17	1.21	1.26	1.31	1.36	1.41	1.45	1.48	1.71	1.93	2.13	2.33
0.85	1.02	1.07	1.12	1.17	1.21	1.26	1.31	1.35	1.40	1.45	1.47	1.70	1.91	2.11	2.30
0.90	1.02	1.07	1.11	1.16	1.20	1.25	1.30	1.34	1.40	1.44	1.46	1.68	1.89	2.09	2.28
0.95	1.02	1.07	1.11	1.16	1.20	1.25	1.30	1.34	1.39	1.43	1.45	1.67	1.87	2.07	2.25
1.00	1.02	1.07	1.11	1.16	1.20	1.24	1.29	1.33	1.38	1.43	1.44	1.66	1.86	2.05	2.23
1.10	1.02	1.06	1.11	1.15	1.19	1.24	1.29	1.33	1.37	1.41	1.43	1.64	1.83	2.01	2.19
1.20	1.02	1.06	1.11	1.15	1.19	1.23	1.28	1.32	1.36	1.40	1.42	1.62	1.80	1.98	2.16
1.30	1.02	1.06	1.11	1.14	1.19	1.23	1.27	1.31	1.35	1.39	1.41	1.60	1.78	1.96	2.14
1.40	1.02	1.06	1.10	1.14	1.18	1.22	1.27	1.30	1.35	1.38	1.40	1.59	1.76	1.94	2.11
1.50	1.02	1.06	1.10	1.14	1.18	1.22	1.26	1.30	1.34	1.38	1.39	1.58	1.75	1.92	2.09

(Continued)

Table A.1 (*Continued*) Reduced Collision Integral $\Omega_s^{(1,1)*}$ for Gas–Nanoparticle Interactions

	σ'														
T^*	**0.01**	**0.03**	**0.05**	**0.07**	**0.09**	**0.11**	**0.13**	**0.15**	**0.17**	**0.19**	**0.2**	**0.3**	**0.4**	**0.5**	**0.6**
1.60	1.02	1.06	1.10	1.14	1.18	1.21	1.26	1.29	1.33	1.37	1.39	1.56	1.73	1.90	2.07
1.70	1.02	1.06	1.10	1.13	1.18	1.21	1.25	1.29	1.33	1.36	1.38	1.55	1.72	1.89	2.06
1.80	1.02	1.06	1.10	1.13	1.17	1.21	1.25	1.28	1.32	1.36	1.37	1.54	1.71	1.87	2.04
1.90	1.02	1.05	1.10	1.13	1.17	1.20	1.24	1.28	1.32	1.35	1.37	1.53	1.70	1.86	2.03
2.00	1.02	1.05	1.10	1.13	1.17	1.20	1.24	1.27	1.31	1.34	1.36	1.53	1.69	1.85	2.02
3.00	1.02	1.05	1.08	1.12	1.15	1.18	1.21	1.24	1.28	1.31	1.32	1.47	1.62	1.77	1.93
4.00	1.02	1.05	1.08	1.11	1.14	1.17	1.20	1.23	1.26	1.28	1.30	1.44	1.58	1.73	1.89
5.00	1.01	1.05	1.07	1.11	1.13	1.16	1.19	1.21	1.24	1.27	1.28	1.42	1.56	1.71	1.86
6.00	1.01	1.05	1.07	1.10	1.13	1.15	1.18	1.21	1.23	1.26	1.27	1.41	1.54	1.69	1.83
7.00	1.01	1.04	1.07	1.10	1.12	1.15	1.17	1.20	1.23	1.25	1.26	1.40	1.53	1.67	1.81
8.00	1.01	1.04	1.06	1.09	1.12	1.14	1.17	1.19	1.22	1.24	1.26	1.39	1.52	1.66	1.80
9.00	1.01	1.04	1.06	1.09	1.12	1.14	1.17	1.19	1.21	1.24	1.25	1.38	1.51	1.65	1.79
10.00	1.01	1.04	1.06	1.09	1.11	1.14	1.16	1.19	1.21	1.24	1.25	1.37	1.50	1.64	1.77
20.00	1.01	1.03	1.06	1.08	1.10	1.12	1.14	1.17	1.19	1.21	1.22	1.34	1.46	1.58	1.71
30.00	1.01	1.03	1.05	1.07	1.09	1.12	1.14	1.16	1.18	1.20	1.21	1.32	1.43	1.55	1.67
40.00	1.01	1.03	1.05	1.07	1.09	1.11	1.13	1.15	1.17	1.19	1.20	1.31	1.42	1.53	1.65
50.00	1.01	1.03	1.05	1.07	1.09	1.11	1.13	1.15	1.17	1.19	1.20	1.30	1.41	1.52	1.63
60.00	1.01	1.03	1.05	1.07	1.09	1.11	1.12	1.14	1.16	1.18	1.19	1.29	1.40	1.51	1.62
70.00	1.01	1.03	1.05	1.06	1.08	1.10	1.12	1.14	1.16	1.18	1.19	1.29	1.39	1.50	1.61
80.00	1.01	1.03	1.05	1.06	1.08	1.10	1.12	1.14	1.16	1.18	1.19	1.28	1.39	1.49	1.60
90.00	1.01	1.03	1.05	1.06	1.08	1.10	1.12	1.14	1.16	1.17	1.18	1.28	1.38	1.48	1.59
100.00	1.01	1.03	1.04	1.06	1.08	1.10	1.12	1.14	1.15	1.17	1.18	1.28	1.38	1.48	1.58

Table A.2 Reduced Collision Integral $\Omega_d^{(1,1)*}$ for Gas–Nanoparticle Interactions

	σ′														
T^*	**0.01**	**0.03**	**0.05**	**0.07**	**0.09**	**0.11**	**0.13**	**0.15**	**0.17**	**0.19**	**0.2**	**0.3**	**0.4**	**0.5**	**0.6**
0.10	1.50	1.69	1.85	2.02	2.15	2.33	2.48	2.61	2.74	2.88	2.96	3.71	4.36	5.08	5.82
0.15	1.48	1.65	1.80	1.95	2.07	2.21	2.35	2.47	2.59	2.71	2.78	3.44	4.03	4.66	5.29
0.20	1.47	1.63	1.77	1.91	2.02	2.14	2.28	2.38	2.49	2.60	2.67	3.27	3.82	4.39	4.96
0.25	1.47	1.62	1.76	1.87	1.98	2.09	2.22	2.31	2.43	2.53	2.59	3.15	3.66	4.19	4.71
0.30	1.47	1.62	1.74	1.85	1.95	2.05	2.18	2.26	2.37	2.47	2.52	3.05	3.54	4.04	4.52
0.35	1.46	1.61	1.72	1.83	1.93	2.02	2.15	2.22	2.33	2.42	2.47	2.98	3.44	3.91	4.36
0.40	1.46	1.61	1.71	1.81	1.91	2.00	2.12	2.19	2.30	2.39	2.43	2.91	3.36	3.80	4.23
0.45	1.46	1.60	1.70	1.80	1.89	1.98	2.09	2.17	2.27	2.35	2.40	2.86	3.29	3.71	4.12
0.50	1.46	1.60	1.69	1.79	1.87	1.97	2.07	2.14	2.24	2.33	2.37	2.82	3.23	3.63	4.02
0.55	1.46	1.60	1.68	1.78	1.86	1.95	2.05	2.12	2.22	2.30	2.34	2.78	3.17	3.56	3.94
0.60	1.46	1.60	1.67	1.77	1.85	1.94	2.04	2.11	2.20	2.28	2.32	2.74	3.12	3.50	3.87
0.65	1.46	1.59	1.66	1.77	1.84	1.93	2.02	2.09	2.19	2.26	2.30	2.71	3.08	3.45	3.81
0.70	1.46	1.59	1.65	1.76	1.83	1.92	2.01	2.08	2.17	2.24	2.28	2.68	3.04	3.40	3.75
0.75	1.46	1.59	1.65	1.75	1.82	1.91	2.00	2.06	2.15	2.23	2.26	2.65	3.00	3.36	3.70
0.80	1.46	1.59	1.64	1.75	1.81	1.90	1.99	2.05	2.14	2.21	2.24	2.63	2.97	3.32	3.66
0.85	1.46	1.58	1.64	1.74	1.81	1.89	1.98	2.04	2.13	2.20	2.23	2.60	2.94	3.28	3.62
0.90	1.46	1.58	1.64	1.74	1.80	1.88	1.97	2.03	2.12	2.19	2.22	2.58	2.92	3.25	3.58
0.95	1.46	1.58	1.63	1.73	1.80	1.87	1.96	2.02	2.10	2.18	2.21	2.56	2.89	3.22	3.55
1.00	1.46	1.57	1.63	1.73	1.79	1.87	1.95	2.02	2.09	2.17	2.20	2.55	2.87	3.20	3.52
1.10	1.46	1.57	1.62	1.71	1.79	1.85	1.94	2.00	2.08	2.15	2.17	2.51	2.83	3.15	3.47
1.20	1.46	1.56	1.62	1.70	1.78	1.84	1.92	1.99	2.06	2.13	2.16	2.48	2.80	3.11	3.43
1.30	1.45	1.56	1.62	1.70	1.77	1.83	1.91	1.98	2.04	2.11	2.14	2.46	2.77	3.07	3.39
1.40	1.45	1.55	1.62	1.69	1.77	1.82	1.90	1.97	2.03	2.09	2.13	2.44	2.74	3.04	3.36

(Continued)

Table A.2 (*Continued*) Reduced Collision Integral $\Omega_d^{(1,1)*}$ for Gas–Nanoparticle Interactions

	σ′														
T^*	**0.01**	**0.03**	**0.05**	**0.07**	**0.09**	**0.11**	**0.13**	**0.15**	**0.17**	**0.19**	**0.2**	**0.3**	**0.4**	**0.5**	**0.6**
1.50	1.45	1.55	1.61	1.68	1.76	1.81	1.89	1.96	2.02	2.08	2.11	2.42	2.72	3.02	3.34
1.60	1.45	1.54	1.61	1.67	1.76	1.81	1.88	1.95	2.01	2.07	2.10	2.40	2.70	2.99	3.31
1.70	1.45	1.54	1.61	1.67	1.75	1.80	1.87	1.94	2.00	2.06	2.09	2.38	2.68	2.97	3.29
1.80	1.45	1.53	1.61	1.66	1.75	1.80	1.86	1.93	1.99	2.04	2.08	2.37	2.66	2.95	3.27
1.90	1.45	1.53	1.61	1.66	1.74	1.79	1.86	1.92	1.98	2.03	2.06	2.36	2.65	2.94	3.26
2.00	1.45	1.52	1.61	1.65	1.74	1.79	1.85	1.92	1.97	2.02	2.06	2.34	2.63	2.92	3.24
3.00	1.45	1.50	1.59	1.63	1.69	1.76	1.80	1.86	1.93	1.96	1.99	2.27	2.53	2.83	3.14
4.00	1.45	1.48	1.59	1.62	1.66	1.74	1.78	1.82	1.90	1.93	1.95	2.23	2.48	2.78	3.08
5.00	1.45	1.48	1.58	1.61	1.65	1.73	1.76	1.80	1.87	1.91	1.93	2.21	2.45	2.76	3.04
6.00	1.45	1.47	1.57	1.60	1.63	1.71	1.75	1.78	1.85	1.90	1.92	2.19	2.42	2.74	3.01
7.00	1.45	1.47	1.56	1.60	1.62	1.70	1.75	1.77	1.83	1.90	1.91	2.17	2.41	2.73	2.98
8.00	1.45	1.47	1.55	1.60	1.62	1.69	1.74	1.76	1.82	1.89	1.90	2.15	2.40	2.72	2.97
9.00	1.45	1.47	1.54	1.59	1.61	1.67	1.74	1.76	1.81	1.89	1.89	2.14	2.39	2.72	2.95
10.00	1.45	1.47	1.53	1.59	1.61	1.66	1.74	1.75	1.80	1.88	1.89	2.12	2.38	2.71	2.94
20.00	1.45	1.46	1.49	1.58	1.60	1.62	1.72	1.74	1.75	1.87	1.88	2.06	2.36	2.67	2.89
30.00	1.45	1.46	1.48	1.58	1.59	1.61	1.72	1.73	1.74	1.86	1.87	2.05	2.35	2.63	2.88
40.00	1.45	1.46	1.47	1.58	1.59	1.60	1.72	1.73	1.74	1.86	1.87	2.04	2.34	2.60	2.86
50.00	1.45	1.46	1.47	1.58	1.59	1.60	1.72	1.73	1.74	1.85	1.86	2.03	2.33	2.57	2.86
60.00	1.45	1.46	1.47	1.58	1.59	1.60	1.71	1.72	1.73	1.85	1.86	2.03	2.33	2.55	2.85
70.00	1.45	1.46	1.46	1.58	1.59	1.60	1.71	1.72	1.73	1.84	1.86	2.03	2.33	2.54	2.85
80.00	1.45	1.46	1.46	1.58	1.59	1.59	1.71	1.72	1.73	1.83	1.86	2.02	2.32	2.53	2.84
90.00	1.45	1.46	1.46	1.58	1.59	1.59	1.71	1.72	1.73	1.82	1.86	2.02	2.32	2.52	2.84
100.00	1.45	1.46	1.46	1.58	1.59	1.59	1.71	1.72	1.73	1.81	1.86	2.02	2.32	2.51	2.83

Appendix B

Table B.1 Reduced Collision Integral $\Omega_s^{(1,2)*}$ for Gas–Nanoparticle Interactions

	σ'													
T^*	**0.01**	**0.03**	**0.05**	**0.07**	**0.09**	**0.11**	**0.13**	**0.15**	**0.17**	**0.19**	**0.2**	**0.3**	**0.4**	**0.5**	**0.6**
0.10	1.04	1.12	1.20	1.29	1.36	1.45	1.53	1.61	1.67	1.75	1.81	2.25	2.66	3.10	3.52
0.15	1.04	1.11	1.17	1.26	1.32	1.39	1.47	1.53	1.60	1.66	1.71	2.09	2.47	2.85	3.21
0.20	1.04	1.10	1.16	1.23	1.29	1.35	1.43	1.48	1.55	1.61	1.65	2.00	2.34	2.68	3.00
0.25	1.03	1.10	1.15	1.22	1.27	1.33	1.40	1.45	1.51	1.57	1.61	1.93	2.24	2.56	2.84
0.30	1.03	1.09	1.14	1.20	1.25	1.31	1.38	1.42	1.49	1.54	1.57	1.88	2.17	2.45	2.71
0.35	1.03	1.09	1.13	1.19	1.24	1.30	1.36	1.41	1.46	1.52	1.55	1.83	2.11	2.37	2.60
0.40	1.03	1.08	1.13	1.19	1.23	1.28	1.35	1.39	1.45	1.50	1.52	1.80	2.06	2.30	2.52
0.45	1.03	1.08	1.12	1.18	1.22	1.27	1.34	1.38	1.43	1.48	1.51	1.77	2.01	2.24	2.44
0.50	1.03	1.08	1.12	1.17	1.22	1.27	1.32	1.37	1.42	1.47	1.49	1.74	1.97	2.18	2.39
0.55	1.02	1.07	1.12	1.17	1.21	1.26	1.31	1.36	1.41	1.46	1.48	1.72	1.94	2.14	2.33
0.60	1.02	1.07	1.11	1.17	1.21	1.26	1.31	1.35	1.40	1.45	1.47	1.70	1.91	2.10	2.29
0.65	1.02	1.07	1.11	1.16	1.20	1.25	1.30	1.34	1.39	1.44	1.46	1.68	1.88	2.07	2.26
0.70	1.02	1.07	1.11	1.16	1.20	1.24	1.29	1.34	1.38	1.43	1.45	1.66	1.86	2.04	2.22
0.75	1.02	1.06	1.11	1.15	1.19	1.24	1.29	1.33	1.38	1.42	1.44	1.65	1.83	2.02	2.20
0.80	1.02	1.06	1.11	1.15	1.19	1.24	1.28	1.32	1.37	1.41	1.43	1.63	1.82	1.99	2.17
0.85	1.02	1.06	1.11	1.15	1.19	1.23	1.28	1.32	1.36	1.40	1.42	1.62	1.80	1.97	2.15
0.90	1.02	1.06	1.11	1.15	1.19	1.23	1.27	1.31	1.36	1.40	1.41	1.61	1.78	1.96	2.13
0.95	1.02	1.06	1.11	1.14	1.18	1.22	1.27	1.31	1.35	1.39	1.41	1.59	1.77	1.94	2.11
1.00	1.02	1.06	1.10	1.14	1.18	1.22	1.26	1.30	1.35	1.38	1.40	1.58	1.76	1.92	2.10
1.10	1.02	1.06	1.10	1.14	1.18	1.22	1.26	1.29	1.34	1.37	1.39	1.57	1.73	1.90	2.07
1.20	1.02	1.06	1.10	1.13	1.18	1.21	1.25	1.29	1.33	1.36	1.38	1.55	1.71	1.88	2.04
1.30	1.02	1.05	1.10	1.13	1.17	1.21	1.25	1.28	1.32	1.35	1.37	1.54	1.70	1.86	2.03
1.40	1.02	1.05	1.10	1.13	1.17	1.20	1.24	1.27	1.31	1.35	1.36	1.52	1.68	1.84	2.01

(Continued)

Table B.1 (*Continued*) Reduced Collision Integral $\Omega_s^{(1,2)*}$ for Gas–Nanoparticle Interactions

	σ'														
T*	**0.01**	**0.03**	**0.05**	**0.07**	**0.09**	**0.11**	**0.13**	**0.15**	**0.17**	**0.19**	**0.2**	**0.3**	**0.4**	**0.5**	**0.6**
1.50	1.02	1.05	1.10	1.12	1.17	1.20	1.24	1.27	1.31	1.34	1.35	1.51	1.67	1.83	1.99
1.60	1.02	1.05	1.09	1.12	1.16	1.20	1.23	1.26	1.30	1.33	1.35	1.50	1.66	1.82	1.98
1.70	1.02	1.05	1.09	1.12	1.16	1.19	1.23	1.26	1.29	1.33	1.34	1.49	1.65	1.81	1.97
1.80	1.02	1.05	1.09	1.12	1.16	1.19	1.22	1.26	1.29	1.32	1.33	1.49	1.64	1.80	1.96
1.90	1.02	1.05	1.09	1.12	1.16	1.19	1.22	1.25	1.28	1.31	1.33	1.48	1.63	1.79	1.95
2.00	1.02	1.05	1.09	1.12	1.15	1.18	1.22	1.25	1.28	1.31	1.32	1.47	1.62	1.78	1.94
3.00	1.01	1.05	1.08	1.11	1.14	1.17	1.19	1.22	1.25	1.28	1.29	1.43	1.57	1.72	1.87
4.00	1.01	1.05	1.07	1.10	1.13	1.15	1.18	1.21	1.23	1.26	1.27	1.41	1.55	1.69	1.84
5.00	1.01	1.04	1.07	1.10	1.12	1.15	1.17	1.20	1.22	1.25	1.26	1.39	1.53	1.67	1.81
6.00	1.01	1.04	1.06	1.09	1.12	1.14	1.17	1.19	1.22	1.24	1.25	1.38	1.51	1.65	1.79
7.00	1.01	1.04	1.06	1.09	1.11	1.14	1.16	1.19	1.21	1.23	1.25	1.37	1.50	1.64	1.78
8.00	1.01	1.04	1.06	1.09	1.11	1.13	1.16	1.18	1.21	1.23	1.24	1.37	1.49	1.63	1.76
9.00	1.01	1.04	1.06	1.08	1.11	1.13	1.15	1.18	1.20	1.23	1.24	1.36	1.49	1.62	1.75
10.00	1.01	1.04	1.06	1.08	1.11	1.13	1.15	1.18	1.20	1.22	1.23	1.35	1.48	1.61	1.74
20.00	1.01	1.03	1.05	1.07	1.10	1.12	1.14	1.16	1.18	1.20	1.21	1.32	1.44	1.56	1.68
30.00	1.01	1.03	1.05	1.07	1.09	1.11	1.13	1.15	1.17	1.19	1.20	1.31	1.42	1.53	1.65
40.00	1.01	1.03	1.05	1.07	1.09	1.11	1.13	1.14	1.16	1.18	1.19	1.30	1.40	1.51	1.62
50.00	1.01	1.03	1.05	1.06	1.08	1.10	1.12	1.14	1.16	1.18	1.19	1.29	1.39	1.50	1.61
60.00	1.01	1.03	1.05	1.06	1.08	1.10	1.12	1.14	1.16	1.18	1.19	1.28	1.38	1.49	1.59
70.00	1.01	1.03	1.04	1.06	1.08	1.10	1.12	1.14	1.15	1.17	1.18	1.28	1.38	1.48	1.58
80.00	1.01	1.03	1.04	1.06	1.08	1.10	1.12	1.13	1.15	1.17	1.18	1.27	1.37	1.47	1.57
90.00	1.01	1.03	1.04	1.06	1.08	1.10	1.11	1.13	1.15	1.17	1.18	1.27	1.36	1.46	1.56
100.00	1.01	1.03	1.04	1.06	1.08	1.09	1.11	1.13	1.15	1.17	1.17	1.27	1.36	1.46	1.56

Table B.2 Reduced Collision Integral $\Omega_d^{(1,2)*}$ for Gas–Nanoparticle Interactions

	σ′														
T*	**0.01**	**0.03**	**0.05**	**0.07**	**0.09**	**0.11**	**0.13**	**0.15**	**0.17**	**0.19**	**0.2**	**0.3**	**0.4**	**0.5**	**0.6**
0.10	1.41	1.58	1.72	1.88	1.99	2.13	2.27	2.38	2.49	2.61	2.68	3.33	3.91	4.53	5.15
0.15	1.40	1.55	1.69	1.82	1.92	2.03	2.16	2.25	2.37	2.47	2.53	3.10	3.63	4.17	4.70
0.20	1.40	1.54	1.67	1.77	1.87	1.97	2.10	2.17	2.29	2.38	2.43	2.95	3.44	3.93	4.40
0.25	1.40	1.54	1.65	1.74	1.84	1.93	2.05	2.12	2.23	2.32	2.36	2.85	3.30	3.75	4.17
0.30	1.39	1.53	1.63	1.72	1.81	1.90	2.01	2.08	2.18	2.27	2.31	2.76	3.19	3.60	4.00
0.35	1.39	1.53	1.61	1.71	1.79	1.88	1.98	2.05	2.15	2.23	2.26	2.70	3.10	3.48	3.86
0.40	1.39	1.53	1.60	1.70	1.77	1.86	1.96	2.02	2.12	2.19	2.23	2.64	3.02	3.39	3.74
0.45	1.39	1.52	1.58	1.69	1.76	1.84	1.93	2.00	2.09	2.16	2.20	2.60	2.96	3.31	3.65
0.50	1.39	1.52	1.57	1.68	1.74	1.83	1.92	1.98	2.07	2.14	2.17	2.56	2.90	3.24	3.57
0.55	1.39	1.52	1.57	1.67	1.73	1.82	1.90	1.97	2.05	2.12	2.15	2.52	2.86	3.18	3.50
0.60	1.39	1.51	1.56	1.67	1.72	1.81	1.89	1.95	2.04	2.10	2.13	2.49	2.81	3.13	3.45
0.65	1.39	1.51	1.55	1.66	1.71	1.79	1.88	1.94	2.02	2.09	2.11	2.46	2.78	3.09	3.40
0.70	1.39	1.51	1.55	1.65	1.71	1.78	1.87	1.93	2.00	2.07	2.10	2.44	2.74	3.05	3.36
0.75	1.39	1.50	1.55	1.65	1.70	1.77	1.86	1.91	1.99	2.06	2.08	2.41	2.71	3.02	3.32
0.80	1.39	1.50	1.55	1.64	1.70	1.77	1.85	1.91	1.98	2.05	2.07	2.39	2.69	2.99	3.29
0.85	1.39	1.50	1.54	1.63	1.69	1.76	1.84	1.90	1.97	2.03	2.06	2.37	2.67	2.96	3.26
0.90	1.39	1.49	1.54	1.62	1.69	1.75	1.83	1.89	1.96	2.02	2.05	2.35	2.64	2.94	3.23
0.95	1.39	1.49	1.54	1.62	1.69	1.74	1.82	1.88	1.95	2.01	2.04	2.34	2.63	2.91	3.21
1.00	1.39	1.49	1.54	1.61	1.69	1.74	1.82	1.87	1.94	2.00	2.03	2.32	2.61	2.89	3.19
1.10	1.39	1.48	1.54	1.60	1.68	1.73	1.80	1.86	1.92	1.98	2.01	2.29	2.58	2.86	3.16
1.20	1.39	1.47	1.53	1.59	1.67	1.72	1.79	1.85	1.91	1.96	2.00	2.27	2.55	2.83	3.13
1.30	1.39	1.46	1.53	1.58	1.67	1.71	1.78	1.84	1.89	1.95	1.98	2.25	2.53	2.81	3.11
1.40	1.39	1.46	1.53	1.58	1.66	1.70	1.77	1.83	1.88	1.93	1.97	2.23	2.51	2.78	3.09

(Continued)

Table B.2 (*Continued*) Reduced Collision Integral $\Omega_d^{(1,2)*}$ for Gas–Nanoparticle Interactions

	σ′														
T*	**0.01**	**0.03**	**0.05**	**0.07**	**0.09**	**0.11**	**0.13**	**0.15**	**0.17**	**0.19**	**0.2**	**0.3**	**0.4**	**0.5**	**0.6**
1.50	1.39	1.45	1.53	1.57	1.66	1.70	1.76	1.82	1.87	1.92	1.95	2.22	2.50	2.76	3.07
1.60	1.39	1.44	1.53	1.56	1.65	1.69	1.75	1.82	1.86	1.91	1.94	2.21	2.48	2.75	3.06
1.70	1.39	1.44	1.53	1.56	1.65	1.69	1.74	1.81	1.86	1.90	1.93	2.20	2.47	2.73	3.04
1.80	1.39	1.44	1.53	1.56	1.64	1.69	1.73	1.80	1.85	1.89	1.92	2.19	2.45	2.72	3.03
1.90	1.39	1.43	1.52	1.55	1.63	1.68	1.73	1.79	1.85	1.88	1.91	2.18	2.44	2.71	3.02
2.00	1.39	1.43	1.52	1.55	1.63	1.68	1.72	1.78	1.84	1.88	1.90	2.17	2.43	2.70	3.01
3.00	1.38	1.41	1.51	1.54	1.58	1.66	1.69	1.73	1.81	1.83	1.85	2.13	2.35	2.65	2.93
4.00	1.38	1.40	1.51	1.53	1.56	1.64	1.67	1.70	1.78	1.82	1.83	2.10	2.32	2.62	2.88
5.00	1.38	1.40	1.49	1.53	1.54	1.62	1.66	1.68	1.75	1.81	1.81	2.07	2.30	2.61	2.85
6.00	1.38	1.40	1.48	1.52	1.54	1.61	1.66	1.68	1.73	1.80	1.81	2.05	2.28	2.60	2.83
7.00	1.38	1.40	1.47	1.52	1.53	1.59	1.66	1.67	1.71	1.80	1.80	2.03	2.28	2.60	2.81
8.00	1.38	1.40	1.46	1.52	1.53	1.58	1.65	1.67	1.70	1.80	1.80	2.02	2.27	2.59	2.80
9.00	1.38	1.39	1.45	1.52	1.53	1.57	1.65	1.66	1.69	1.79	1.80	2.00	2.27	2.59	2.79
10.00	1.38	1.39	1.44	1.52	1.53	1.56	1.65	1.66	1.69	1.79	1.80	1.99	2.26	2.58	2.79
20.00	1.38	1.39	1.40	1.51	1.52	1.53	1.64	1.65	1.66	1.78	1.79	1.95	2.25	2.54	2.76
30.00	1.38	1.39	1.40	1.51	1.52	1.52	1.64	1.65	1.66	1.78	1.78	1.94	2.24	2.48	2.75
40.00	1.38	1.39	1.40	1.51	1.52	1.52	1.64	1.65	1.65	1.77	1.78	1.94	2.23	2.45	2.74
50.00	1.38	1.39	1.39	1.51	1.51	1.52	1.64	1.65	1.65	1.76	1.78	1.94	2.23	2.43	2.73
60.00	1.38	1.39	1.39	1.51	1.51	1.52	1.64	1.64	1.65	1.75	1.78	1.93	2.23	2.42	2.73
70.00	1.38	1.39	1.39	1.51	1.51	1.52	1.64	1.64	1.65	1.74	1.78	1.93	2.22	2.41	2.72
80.00	1.38	1.39	1.39	1.50	1.51	1.52	1.64	1.64	1.65	1.72	1.78	1.93	2.22	2.40	2.72
90.00	1.38	1.39	1.39	1.50	1.51	1.52	1.63	1.64	1.65	1.71	1.77	1.93	2.22	2.40	2.72
100.00	1.38	1.39	1.39	1.50	1.51	1.52	1.63	1.64	1.65	1.70	1.77	1.93	2.22	2.39	2.71

Appendix C

Table C.1 Reduced Collision Integral $\Omega^{(l,q)*}$ Based on the Lennard–Jones Potential[a]

T^*	$\Omega^{(1,1)*}$	$\Omega^{(1,2)*}$	$\Omega^{(1,3)*}$	$\Omega^{(2,2)*}$	T^*	$\Omega^{(1,1)*}$	$\Omega^{(1,2)*}$	$\Omega^{(1,3)*}$	$\Omega^{(2,2)*}$
0.30	2.662	2.256	1.962	2.785	2.70	0.9770	0.8833	0.8347	1.069
0.35	2.476	2.078	1.795	2.628	2.80	0.9672	0.8767	0.8290	1.058
0.40	2.318	1.931	1.663	2.492	2.90	0.9576	0.8700	0.8237	1.048
0.45	2.184	1.808	1.556	2.368	3.00	0.9490	0.8640	0.8187	1.039
0.50	2.066	1.705	1.468	2.257	3.10	0.9406	0.8580	0.8138	1.030
0.55	1.966	1.618	1.396	2.156	3.20	0.9328	0.8520	0.8093	1.022
0.60	1.877	1.543	1.336	2.065	3.30	0.9256	0.8473	0.8048	1.014
0.65	1.798	1.479	1.285	1.982	3.40	0.9186	0.8420	0.8007	1.007
0.70	1.729	1.423	1.242	1.908	3.50	0.9120	0.8373	0.7967	0.9999
0.75	1.667	1.375	1.205	1.841	3.60	0.9058	0.8327	0.7928	0.9932
0.80	1.612	1.332	1.172	1.780	3.70	0.8998	0.8287	0.7892	0.9870
0.85	1.562	1.295	1.144	1.725	3.80	0.8942	0.8240	0.7857	0.9811
0.90	1.517	1.261	1.119	1.675	3.90	0.8888	0.8200	0.7822	0.9755
0.95	1.476	1.231	1.096	1.629	4.00	0.8836	0.8167	0.7790	0.9700
1.00	1.439	1.204	1.076	1.587	4.10	0.8788	0.8127	0.7758	0.9649
1.05	1.406	1.179	1.058	1.549	4.20	0.8740	0.8093	0.7727	0.9600
1.10	1.375	1.157	1.041	1.514	4.30	0.8694	0.8060	0.7697	0.9553
1.15	1.346	1.137	1.027	1.482	4.40	0.8652	0.8027	0.7668	0.9507
1.20	1.320	1.119	1.013	1.452	4.50	0.8610	0.7993	0.7640	0.9464
1.25	1.296	1.102	1.000	1.424	4.60	0.8568	0.7960	0.7613	0.9422
1.30	1.273	1.086	0.9887	1.399	4.70	0.8530	0.7933	0.7585	0.9382
1.35	1.253	1.072	0.9780	1.375	4.80	0.8492	0.7907	0.7560	0.9343

(Continued)

Table C.1 (*Continued*) Reduced Collision Integral $\Omega^{(l,q)*}$ Based on the Lennard–Jones Potential[a]

T^*	$\Omega^{(1,1)*}$	$\Omega^{(1,2)*}$	$\Omega^{(1,3)*}$	$\Omega^{(2,2)*}$	T^*	$\Omega^{(1,1)*}$	$\Omega^{(1,2)*}$	$\Omega^{(1,3)*}$	$\Omega^{(2,2)*}$
1.40	1.233	1.059	0.9680	1.353	4.90	0.8456	0.7873	0.7535	0.9305
1.45	1.215	1.046	0.9588	1.333	5.00	0.8422	0.7847	0.7510	0.9269
1.50	1.198	1.034	0.9502	1.314	6.00	0.8124	0.7607	0.7295	0.8963
1.55	1.182	1.023	0.9420	1.296	7.00	0.7896	0.7420	0.7210	0.8727
1.60	1.167	1.013	0.9345	1.279	8.00	0.7712	0.7260	0.6973	0.8538
1.65	1.153	1.004	0.9272	1.264	9.00	0.7556	0.7127	0.6848	0.8379
1.70	1.140	0.9947	0.9205	1.248	10.00	0.7424	0.7013	0.6735	0.8242
1.75	1.128	0.9860	0.9142	1.234	20.00	0.6640	0.6293	0.6048	0.7432
1.80	1.116	0.9780	0.9082	1.221	30.00	0.6232	0.5909	0.5680	0.7005
1.85	1.105	0.9707	0.9023	1.209	40.00	0.5960	0.5651	0.5432	0.6718
1.90	1.094	0.9633	0.8968	1.197	50.00	0.5756	0.5459	0.5248	0.6504
1.95	1.084	0.9567	0.8917	1.186	60.00	0.5596	0.5307	0.5100	0.6335
2.00	1.075	0.9500	0.8867	1.175	70.00	0.5464	0.5181	0.4980	0.6194
2.10	1.057	0.9380	0.8775	1.156	80.00	0.5352	0.5075	0.4878	0.6076
2.20	1.041	0.9267	0.8688	1.138	90.00	0.5256	0.4984	0.4790	0.5973
2.30	1.026	0.9167	0.8612	1.122	100.00	0.5130	0.4903	0.4713	0.5882
2.40	1.012	0.9073	0.8538	1.107	200.00	0.4644	0.4403	0.4233	0.5320
2.50	0.9996	0.8987	0.8470	1.093	300.00	0.4360	0.4135	0.3975	0.5016
2.60	0.9878	0.8907	0.8407	1.081	400.00	0.4170	0.3955	0.3802	0.4811

[a] Hirschfelder, J.O., Curtiss, C.F., and Bird, R.B. 1954. *Molecular Theory of Gases and Liquids*. New York: Wiley.

Index

D

I

K

L

M

R

S

T

Index

U

V

W

Y

Z